RICE SCIENCE

Dr. B. CHANDRASEKARAN
Dr. K. ANNADURAI & Dr. R. KAVIMANI
Tamil Nadu Agricultural University
Coimbatore-641 003 (Tamil Nadu)

Published by:

Scientific Publishers (India)
5 A, New Pali Road, P.O. Box 91
Jodhpur 342 001 (India)

Branch Office
Scientific Publishers (India)
4806/24, Ansari Road, Daryaganj
New Delhi - 110 002 (India)

E-mail: info@scientificpub.com
Website: www.scientificpub.com

Print: 2013

ISBN: 978-81-7233-504-5

Typeset by Rajesh Ojha
Printed in India

FOREWORD

"There is no greatest happiness than knowledge. Knowledge stretched – is book"

Agronomy is a science that helps the advancement of agricultural technologies to feed the world. We can call the Agronomy as backbone of all agricultural sciences, because, the management of soil and water, with a view to achieve the production potential of high yielding varieties in green revolution, is exclusively under an agronomic domain. It may not appear as glamorous as nuclear science working miracles but like Ayurvedic medicines, it has the capacity to reach the poorer section of the society to bring out the desired results. Agronomists can synthesize the production practices from several fields of specialization.

Fortunately the anticipated famine during these years, due to manifolding demographic pressure on inelastic, ever shrinking crop production resource base, has not been experienced only, because of higher food productivity and production in the country as a result of advancement made in Agricultural Science and Technology leading to green revolution. But still, we are unable to be enchanted / complacent because of multi faceted prevalent predicaments in the agricultural production system as a whole. Food and protein shortage, mal and under nutrition have become incessant problems and perpetual phenomena across various states in the country and in globe, apart from escalation of human population in a geometrical proportion.

Around 198.9 million people are below poverty line in the semi arid tropics alone. About 17 million children under the age of 5 die every year because of mal nourishment. About 180 million children, mostly in developing countries are seriously underweight. Nearly 800 million people, majority of them are women and children, go hungry to bed every day, paradoxically, the largest pockets of hunger existing in rural areas where food is grown. Though food production in the country crossed 209 million tones from the rock bottom level of 52 million tones immediately after independence, still there are problems of 3As. *viz.,* Affordability, Acceptability and Availability of food especially to the rural people who live below poverty line. Hungry people listen not to reason, not care for justice, nor are bent by any prayers said a Roman Philanthropist at Seneca Roman 2000 years before. Under the predicament situations, the year 1966 was promulgated as International year of Rice (IYR) by Food and Agricultural Organization, Rome, mainly to relegate famine. Again the year 2004 was also declared as IYR to mainly accomplish sustainability in rice production by United Nations Organizations and FAO.

However, the essential factors in food production such as cultivated land and fresh water are decreasing continuously. Current trends on world agriculture shows that it is imperative to find a scientific and rational way to develop it, a way that cannot only steadily increase the output but also ensure long term sustainable use of resource in the process of promoting agricultural development.

I am happy that the authors have made efforts to compile the available information on rice science. It covers a wide range of topics. In this connection, publication of **"Rice Science"** by **Dr. B. Chandrasekaran, Dr. K. Annadurai** and **Dr. R. Kavimani** of TNAU, Coimbatore is quite appropriate and timely. In particular, I strongly believe that the book would help youngsters to channel their efforts so as to make inroads into some frontier areas such as evolving new plant types, precision farming, Integrated Farming systems, Organic farming, evolving new varieties like nutrient fortified and aromatic rice suitable for export market, system of rice intensification (SRI), integrated crop production technologies (ICM) etc.

The authors deserve commendation for their strenuous efforts and my congratulations to them. I am sure that the publication will prove to be a useful volume for students and researchers.

Place : Coimbatore

Prof. C. Ramasamy
Vice-Chancellor
Tamil Nadu Agricultural University

PREFACE

"Through understanding of what, pray does all this world become understood, sir?

Mundaka Uppanisad, 1-1-3

Agronomy deals with the principles and practices of crop production and soil management. In its broader sense, it includes crop ecology, crop production, crop nutrition, soil fertility, water management, weed control, seed technology etc. To be a good agronomist, one needs to have a sound knowledge of all these agronomic aspects and also some related aspects from other sciences.

Rice continues to be the most important food source in the world, with about 20% of the global population depending on it as a staple food. A compilation of available information on rice has been a felt need of students, teachers, research workers and administrators in Agronomy. This book is aimed at providing a comprehensive text on rice cultivation/production with major emphasis on rice based integrated farming system models, organic farming aspects, alternate cropping, new techniques like SRI, role of biotechnology etc., in an easily understandable manner.

We believe that this book will help to enrich the knowledge of young researchers in various fields of agriculture and in particular, agronomy, as well as to the teachers and researchers of the Agricultural Universities / Research Organizations. We believe that bringing available information together in one volume will contribute to the advancement of knowledge on Rice Science.

It is difficult to name and acknowledge each and every reference used in this compilation. We sincerely thank the authors known and unknown for whatever help we have received from the works.

We place on record our sincere gratitude to TNAU, Coimbatore for granting permission to bring out this book and to **Prof. Dr. C. Ramasamy, Ph.D., Vice Chancellor, TNAU** for providing the Foreword.

We are most grateful to our teachers and researchers. We thank **Dr. K. Alagusundaram**, National Fellow (ICAR) for his encouragement; **Dr. R. Marimuthu** Professor (Plant Breeding and Genetics) for his critical comments and suggestions for hybrid rice and breeding approaches; **Dr. A. Tajuddin** Professor (Farm Machinery and Power) for his valuable comments on rice mechanization, **all Scientists of Tamil Nadu Rice Research Institute, Aduthurai** in particular, **Prof. Dr. A. Arockiaraj, Prof. K. Natarajan and Prof. Dr. R. Rajendran,** as well as fellow scientists of Department of Agronomy, TNAU for their comments and critical suggestions. We thank **Dr. S. Ramasamy**, Professor (Agronomy) and **Dr. E. Somasundaram,** Associate Professor (Agronomy) for correcting the proof and their suggestions. In spite of the best efforts, it is possible that some errors may have crept into the compilation. The readers are requested to kindly let us know the mistakes so that these could be taken care of in the further edition. Finally we thank our publishers for bringing out this book so efficiently and promptly.

Authors

CONTENTS

CHAPTER 1

INTRODUCTION

Rice is a member of the grass family. Archeological evidence indicates a sophisticated rice cultivation system existed in China over 7,000 years ago. There are 42 rice producing countries in the world, ranging from mountainous Himalayan to lowland delta areas. Rice is the staple food in Asia, Latin America, parts of Africa and the Middle East. Rice grains were initially gathered and consumed by prehistoric people in the humid tropics and subtropics from self-propagating wild stands. Cultivation began when men or, more likely, women, deliberately dropped rice grains on the soil in low-lying spots near their homesteads, kept out the weeds and animals, and manipulated the water supply. The association between rice and human community was clearly indicated in the exciting excavations at He-mu-du, Luo-jia-jiao, and Pen-tou-shan in China where rice was a principal food plant in the developing human settlements there more than 7,000 years ago.

It has been cultivated in China and then, it was introduced into India before the time of the Greeks. Now, rice cultivation has been carried into all regions with necessary warmth and abundant moisture favorable to its growth, mainly sub-tropical rather than hot or cold. It has been estimated that half the world's population subsists wholly or partially on rice. After wheat, rice is the second most important crop of the world and more than 90% is currently grown in Asia, because in Asia, it is the main item of the diet.

Rice first entered the diet as a supplement to other food plants as well as to game fish, and shellfish. As rice cultivation expanded and became more efficient, it replaced other cereals (millets, sorghum and even wheat), root crops and forage plants. The continuous expansion of rice cultivation owed much to its unique features as a self-supporting semi-aquatic plant. These features include the ability of seed to germinate under both aerobic and anaerobic conditions and the series of air-conducting aerenchymatous tissues in the leaf sheaths, stems and roots that supply air to roots under continuous flooding. Also important are soil microbes in the root zone that fix nitrogen to feed rice for growth and the wide adaptability of rice to both wetland and dry land soil-water regimes. It is for these reasons that rice is the only subsistence crop whose soil is poorly drained. And these factors, in turn, account for the broad rice-growing belt from the Sino-Russian border along the Amur River (53°N latitude) to central Argentina (35°S). At world level, the rice cultivation area stretches from 39^0S latitude to 50^0N latitude (i.e. from Australia to China). In India, the rice cultivation areas extends from 8 to 34^0S latitude i.e. extending almost through out the country. Rice cultivation can be taken up in areas lying below sea level (Kerala) and up to an altitude of 2000m MSL (Kashmir) (Somasundaram *et.al.*, 2000). To meet out the rice requirement of the growing population of India, the growth rate should be maintained at a minimum of 2.35%. Though we reached self-sufficiency in food production, we have to produce more i.e. at least 60% m over the coming years to meet out the demands of the growing population.

During the last 45 years (1967-2004), while the rice area has increased only by one half times from 115.50 million ha to about 153.26 million ha, but the rice production had increased more than three times from 215.66 million tones to 608.50 million tones and the productivity has increased two and half times from 1.87 t/ha to 3.97 t/ha. At the world level, Asia occupied the highest area of rice cultivation (134.54 million ha) with the production of 549.46 million tones. Almost 90% of the world's rice is produced in Asia. China is the world's leading producer, growing nearly two-fifths of Asia's total on 32 million ha. India possesses the largest rice area (44 million ha) producing nearly a quarter of Asia (Moya *et al.,* 2004). In terms of productivity, North and Central America ranks the top most level of 6.27 t/ha (Table 1). Among the various Asian rice producing countries even though the area of rice cultivation is less in Europe (0.59 million ha), it recorded higher rice productivity (5.69 t/ha). Due to the lack of high yielding varieties and advanced technologies, African country could able to attain less productivity of 1.88 t/ha. Forty percent of world population uses rice as a major source of calories. The average consumption of rice varies from one country to another. Vietnamese are on the top with consumption of 240 Kg of rice per person followed by Thais with 204 Kg of rice per person. In Pakistan, it is only 20.78 Kg in the year 1999-2000.

In India, it is an important stable food for more than 2/3rd of population. As this crop plays a vital role in our National food security and is a means of livelihood for millions of rural households, the slogan" Rice for life" is most appropriate. In India, a number of festivals such as Bihu in Assam, Pongal in Tamil Nadu and Onam in Kerala are associated with rice harvest. While celebrating these festivals, the people use the newly harvested rice and there is rejoice and feasting. During the period from 1961 to 2005, the rice area has increased from 35 m. ha to 44 m. ha, production from 54 m. t to 124 m. t and productivity from 1.54 tones to 2.93 tones/ha. Among the Asian countries, the highest yield level/productivity of 6.35 t/ha was recorded in China (Table 2). This is mainly because of cultivation of hybrid rice varieties in major areas with advanced rice production technologies and mechanization. Next to China, Indonesia recorded the rice productivity level of 4.52 t/ha.

It is estimated that the rice demand in 2025 will be 140 million tons in India. The demand for food grains is expected to rise not only as a function of population growth but also as more and more people cross the poverty line (30% of more than one billion population). Morcovcr, thc natural calamities such as drought, the recently faced Tsunami in coastal areas of Tamil Nadu, Andhra Pradesh, Kerela and Andaman Nicobar islands and the recently faced floods in Maharastra, can dwindle the production trend. In the post-WTO era, adequate rice is to be produced not only for the self-sufficiency but also for export purposes. The exportable surplus of good quality rice is to be produced at the competitive price. This increase in production has necessarily to come from increased productivity under depleting and diminishing resources, and scarce and costly lobour decreasing total factor productivity and has to meet the demands of sustainability and preservation of environmental quality. Rice research has to be geared up to surmount the technological challenges in breaking the genetic yield barriers, improving input use efficiency and developing environmentally acceptable strategies for alleviating losses due to pest and diseases. To face challenges of globalization, we have to integrate our action plans to enhance unit area productivity; developing new plant type (NPT) by changing its architectures; crop diversification; precision farming; processing and value added addition of rice and rice products; developing transgenic in rice-golden rice, aromatic/scented rice and basmati rice; organic farming and enhancing exports through quality improvement; adoption of INM and ICM technologies; farming systems models for different agro climatic regions, etc.,. So enhancing the productivity and improving quality without any detrimental effect to environment and health hazards are the two main priorities to reorient our task plans for rice crop.

Table 1

Table 1. Area, Production and Productivity of Rice (1961 - 2004) in the World- Continentwise

Year	World			Europe			South America			North Central America			Africa			Asia		
	Area	Production	Yield	Area	Production	Yield	Area	Production	Yield	Area	Production	Yield	Area	Production	Yield	Area	Production	Yield
	m.ha	m.t	t/ha	m.ha	m.t	t/ha	m.ha	m.t	t/ha	m.ha	m.t	t/ha	m.ha	m.t	t/ha	m.ha	m.t	t/ha
1961	115.50	215.66	1.87	0.33	1.62	4.90	3.91	7.14	1.82	1.24	3.44	2.77	2.91	4.32	1.48	106.96	198.78	1.86
1962	119.58	226.56	1.89	0.31	1.55	4.94	4.18	7.52	1.80	1.32	3.97	3.01	3.23	5.68	1.76	110.39	207.41	1.88
1963	120.28	247.14	2.05	0.32	1.50	4.72	4.48	7.53	1.68	1.30	4.20	3.23	3.17	5.83	1.84	110.83	227.56	2.05
1964	125.22	263.02	2.10	0.34	1.59	4.73	5.09	8.47	1.66	1.27	4.26	3.37	3.35	5.93	1.77	114.95	242.17	2.11
1965	124.99	254.08	2.03	0.33	1.33	4.03	5.59	9.87	1.76	1.26	4.50	3.56	3.29	5.50	1.67	114.26	232.17	2.03
1966	125.84	251.19	1.08	0.33	1.50	4.53	4.99	8.10	1.62	1.35	4.93	3.65	3.39	5.57	1.64	115.49	240.23	2.08
1967	127.71	277.46	2.17	0.35	1.67	4.83	5.23	9.21	1.76	1.40	5.23	3.74	3.72	6.70	1.80	116.69	253.59	2.17
1968	129.45	288.71	2.23	0.36	1.52	4.17	5.37	8.98	1.67	1.57	5.94	3.78	3.85	7.06	1.83	117.94	263.98	2.24
1969	134.37	295.65	2.25	0.39	1.86	4.75	5.54	8.98	1.62	1.51	5.48	3.64	3.88	7.21	1.86	119.68	270.81	2.26
1970	133.10	316.38	2.38	0.40	1.79	4.50	5.96	10.44	1.75	1.40	5.26	3.75	4.02	7.33	1.82	120.92	290.10	2.40
1971	134.74	317.76	2.36	0.40	1.81	4.57	5.69	9.37	1.65	1.38	5.34	3.89	4.13	7.41	1.79	122.70	292.17	2.38
1972	132.41	307.33	2.32	0.40	1.59	3.97	5.41	9.56	1.77	1.38	5.30	3.85	4.06	7.34	1.81	120.69	281.75	2.33
1973	136.82	334.99	2.45	0.40	1.96	4.88	5.74	10.26	1.79	1.60	5.79	3.62	4.23	7.41	1.75	124.35	307.60	2.47
1974	137.14	332.02	2.42	0.40	1.89	4.75	5.75	10.48	1.82	1.78	6.89	3.87	4.31	7.74	1.80	124.33	302.82	2.44
1975	141.97	357.00	2.51	0.37	1.90	5.08	6.49	11.93	1.84	2.01	7.94	3.95	4.51	8.14	1.80	128.00	324.81	2.54
1976	143.10	347.77	2.45	0.37	1.64	4.35	7.76	13.51	1.74	1.77	7.08	4.00	4.54	7.86	1.73	127.05	315.39	2.48
1977	143.89	369.48	2.57	0.40	1.48	3.74	7.21	13.23	1.84	1.66	6.48	3.90	4.71	7.98	1.69	129.27	337.70	2.61
1978	143.65	385.13	2.68	0.40	1.81	4.55	6.87	11.69	1.70	1.88	7.94	4.22	4.68	8.12	1.74	129.14	353.10	2.73
1979	141.24	375.14	2.66	0.38	2.02	5.27	6.76	12.27	1.81	1.89	8.06	4.26	4.88	8.38	1.72	126.59	341.47	2.70
1980	144.66	396.87	2.74	0.37	1.85	5.02	7.53	14.33	1.90	2.02	8.75	4.33	4.96	8.61	1.74	129.00	360.11	2.79

1981	145.29	410.06	2.82	0.35	1.77	5.11	7.50	13.40	1.79	2.29	10.70	4.67	5.05	8.56	1.70	129.35	372.52	2.88
1982	141.81	421.91	2.98	0.36	1.89	5.21	7.51	15.31	2.04	2.05	9.19	4.49	5.18	8.92	1.72	125.93	383.41	3.04
1983	143.07	447.86	3.13	0.34	1.74	5.10	6.40	12.50	1.95	1.58	6.86	4.34	5.08	8.95	1.76	128.93	414.84	3.22
1984	144.47	465.16	3.22	0.38	1.97	5.14	6.82	14.57	2.14	1.86	8.68	4.68	5.10	8.95	1.76	129.50	427.80	3.30
1985	143.90	467.96	3.25	0.39	2.18	5.54	6.19	14.32	2.32	1.81	8.30	4.86	5.12	9.41	1.84	129.58	429.95	3.32
1986	144.55	468.40	3.24	0.41	2.28	5.52	6.94	15.29	2.20	1.67	8.38	5.03	5.25	9.83	1.87	129.54	429.49	3.32
1987	141.33	461.11	3.26	0.42	2.15	5.13	7.49	16.26	2.17	1.65	8.20	4.97	5.21	10.15	1.95	125.79	421.33	3.35
1988	146.32	487.06	3.33	0.44	2.21	5.07	7.48	17.73	2.37	1.83	9.38	5.12	5.52	10.33	1.87	130.26	444.02	3.41
1989	148.85	514.05	3.45	0.42	2.12	5.10	6.88	17.55	2.55	1.79	9.31	5.19	6.42	12.72	1.98	132.58	469.27	3.54
1990	146.98	518.22	3.53	0.45	2.40	5.35	5.52	13.42	2.43	1.80	9.24	4.12	6.04	12.35	2.05	132.44	477.70	3.61
1991	146.64	518.40	3.54	0.42	2.30	5.52	5.66	15.34	2.71	1.76	9.31	5.28	6.51	13.45	2.07	131.59	475.20	3.61
1992	147.28	528.26	3.59	0.69	3.12	4.55	6.36	16.57	2.60	1.95	10.35	5.30	6.60	13.71	2.08	131.55	483.37	3.67
1993	145.84	529.58	3.63	0.65	2.80	4.32	6.14	16.91	2.75	1.69	8.87	5.24	6.53	14.23	2.18	130.70	485.79	3.72
1994	147.16	538.66	3.66	0.60	2.86	4.73	6.26	18.15	2.90	1.99	11.00	5.52	6.78	13.87	2.04	131.38	491.68	3.74
1995	149.50	547.22	3.66	0.55	2.70	4.95	6.30	19.23	3.05	1.88	9.97	5.29	6.90	14.71	2.13	133.74	499.46	3.73
1996	150.19	569.04	3.79	0.63	3.19	5.09	5.20	16.94	3.26	1.87	10.15	5.42	6.97	16.09	2.31	135.37	521.69	3.85
1997	151.03	577.41	3.82	0.60	3.16	5.24	4.97	17.03	3.43	2.00	10.86	5.42	7.37	16.88	2.29	135.91	528.06	3.89
1998	151.70	579.50	3.82	0.58	3.19	5.54	4.97	16.00	3.22	2.05	10.64	5.20	7.54	16.13	2.14	136.42	532.20	3.90
1999	156.96	611.32	3.89	0.59	3.24	5.47	6.02	22.06	3.66	2.10	11.80	5.62	7.67	17.73	2.31	140.43	555.09	3.95
2000	154.12	598.98	3.89	0.61	3.19	5.23	5.65	20.48	3.62	1.97	11.16	5.68	7.62	17.65	2.32	138.14	545.38	3.95
2001	151.65	598.03	3.94	0.57	3.15	5.54	5.12	19.78	3.86	2.02	12.26	6.06	7.58	16.62	2.19	136.17	544.57	4.00
2002	147.58	571.08	3.87	0.56	3.21	5.27	5.03	19.61	3.90	2.00	12.15	6.08	8.38	17.37	2.07	131.45	517.53	3.94
2003	152.24	586.25	3.85	0.57	3.22	5.62	5.18	20.29	3.92	1.87	11.47	6.13	10.24	19.46	1.90	134.33	531.34	3.96
2004	153.26	608.50	3.97	0.59	3.38	5.69	5.80	23.17	4.00	2.03	12.71	6.27	10.22	19.22	1.88	134.54	549.46	4.08

Table 2

Table 2. Area, production and productivity of rice (1961 - 2004) in the important Asian growing Countries

Year	Asia			India			China			Indonesia			Myanmar			Pakistan			Thailand		
	Area	Production	Yield	Area	Production	Yield	Area	Production	Yield	Area	Production	Yield	Area	Production	Yield	Area	Production	Yield	Area	Production	Yield
	M.ha	m.t	t/ha	M.ha	m.t	t/ha	M.ha	m.t	t/ha	M.ha	m.t	t/ha	M.ha	m.t	t/ha	M.ha	m.t	t/ha	M.ha	m.t	t/ha
1961	106.96	198.78	1.86	34.69	53.49	1.54	27.05	56.22	2.08	6.86	12.08	1.76	4.25	6.83	1.61	1.21	1.69	1.39	6.12	10.15	1.66
1962	110.39	207.41	1.88	35.70	49.83	1.40	27.71	65.68	2.37	7.28	13.00	1.79	4.65	7.67	1.65	1.19	1.64	1.39	6.54	11.25	1.72
1963	110.83	227.56	2.05	35.81	55.50	1.55	28.49	76.44	2.68	6.73	11.60	1.72	4.88	7.78	1.60	1.29	1.79	1.39	6.50	12.17	1.87
1964	114.95	242.17	2.11	36.46	58.96	1.62	30.35	85.85	2.83	6.98	12.31	1.76	4.98	8.51	1.71	1.36	2.03	1.49	6.31	11.60	1.84
1965	114.26	232.17	2.03	35.47	45.88	1.29	30.58	90.71	2.97	7.33	12.98	1.77	4.85	8.06	1.66	1.39	1.98	1.42	6.27	11.16	1.78
1966	115.49	240.23	2.08	35.25	45.66	1.30	31.29	98.40	3.14	7.69	13.65	1.77	4.52	6.64	1.47	1.41	2.05	1.45	7.35	13.50	1.84
1967	116.69	253.59	2.17	36.44	56.42	1.55	31.20	96.74	3.10	7.56	13.22	1.76	4.71	7.77	1.65	1.42	2.25	1.58	6.40	11.20	1.75
1968	117.94	263.98	2.24	36.97	59.64	1.61	30.66	97.72	3.19	8.02	17.16	2.14	4.76	8.02	1.68	1.56	3.05	1.96	6.94	12.41	1.79
1969	119.68	270.81	2.26	37.68	60.65	1.61	31.20	98.00	3.14	8.01	18.02	2.25	4.67	7.99	1.71	1.62	3.60	2.22	7.26	13.41	1.85
1970	120.92	290.10	2.40	37.59	63.34	1.68	33.11	113.10	3.42	8.14	19.33	2.38	4.81	8.16	1.70	1.50	3.30	2.19	6.85	13.85	2.02
1971	122.70	292.17	2.38	37.76	64.60	1.71	35.64	118.13	3.31	8.32	20.19	2.43	4.76	8.18	1.72	1.46	3.39	2.33	7.10	13.74	1.94
1972	120.69	281.75	2.33	36.69	58.87	1.60	35.85	116.43	3.25	7.90	19.39	2.46	4.53	7.36	1.62	1.48	3.50	2.36	6.78	12.41	1.83
1973	124.35	307.60	2.47	38.29	66.08	1.73	35.78	124.58	3.48	8.40	21.49	2.56	4.88	8.60	1.76	1.51	3.68	2.44	7.74	14.90	1.92
1974	124.33	302.82	2.44	37.89	59.65	1.57	36.26	127.01	3.50	8.51	22.47	2.64	4.88	8.58	1.76	1.60	3.47	2.16	7.33	13.39	1.83
1975	128.00	324.81	2.54	39.48	73.35	1.86	36.48	128.73	3.53	8.50	22.34	2.63	5.03	9.21	1.83	1.71	3.93	2.30	8.36	15.30	1.83
1976	127.05	315.39	2.48	38.51	63.05	1.64	36.97	129.23	3.50	8.37	23.30	2.78	4.91	9.32	1.90	1.75	4.11	2.35	8.17	15.07	1.84
1977	129.27	337.70	2.61	40.28	79.01	1.96	36.27	131.92	3.64	8.36	23.35	2.79	4.86	9.46	1.95	1.90	4.42	2.33	8.75	13.92	1.59
1978	129.14	353.10	2.73	40.48	80.61	1.99	35.17	140.02	3.98	8.93	25.77	2.89	5.01	10.53	2.10	2.03	4.91	2.42	8.94	17.47	1.96
1979	126.59	341.47	2.70	39.41	63.48	1.61	34.56	146.85	4.25	8.80	26.28	2.99	4.44	10.45	2.35	2.04	4.82	2.37	8.65	15.76	1.82
1980	129.00	360.11	2.79	40.15	80.31	2.00	34.48	142.88	4.14	9.01	29.65	3.29	4.80	13.32	2.77	1.93	4.69	2.42	9.20	17.37	1.89
1981	129.35	372.52	2.88	40.71	79.88	1.96	33.93	146.96	4.33	9.38	32.77	3.49	4.81	14.15	2.94	1.98	5.15	2.60	9.11	17.77	1.95

1982	125.93	383.41	3.04	38.26	70.77	1.85	33.97	164.74	4.89	8.99	33.58	3.74	4.56	14.37	3.15	1.98	5.17	2.61	8.94	16.88	1.89
1983	128.93	414.84	3.22	41.24	90.05	2.18	33.78	172.01	5.09	9.16	35.30	3.85	4.66	14.29	3.07	2.00	5.01	2.51	9.61	19.55	2.04
1984	129.50	427.80	3.30	41.16	87.55	2.13	33.77	181.10	5.36	9.76	38.14	3.91	4.60	14.26	3.10	2.00	4.97	2.49	9.63	19.91	2.07
1985	129.58	429.95	3.32	41.14	95.82	2.33	32.63	171.32	5.25	9.90	39.03	3.94	4.66	14.32	3.07	1.86	4.38	2.35	9.83	20.26	2.06
1986	129.54	429.49	3.32	41.17	90.78	2.21	32.80	174.72	5.33	9.99	39.73	3.98	4.67	14.13	3.03	2.07	5.23	2.53	9.99	18.87	2.05
1987	125.79	421.33	3.35	38.81	85.34	2.20	32.69	176.66	5.40	9.92	40.08	4.04	4.48	13.64	3.04	1.96	4.86	2.48	9.15	18.43	2.01
1988	130.26	444.02	3.41	41.74	106.37	2.55	32.46	171.44	5.28	10.14	41.68	4.11	4.53	13.17	2.91	2.04	4.80	2.35	9.91	21.26	2.15
1989	132.58	469.27	3.54	42.17	110.31	2.62	33.18	182.49	5.50	10.53	44.73	4.25	4.73	13.81	2.92	2.11	4.83	2.29	9.88	20.60	2.09
1990	132.44	477.70	3.61	42.69	111.52	2.61	33.52	191.62	5.72	10.50	45.18	4.30	4.76	13.97	2.94	2.11	4.89	2.32	8.79	17.19	1.96
1991	131.59	475.20	3.61	42.65	112.04	2.63	33.02	185.69	5.62	10.28	44.69	4.35	4.58	13.20	2.89	2.10	4.87	2.32	9.05	20.40	2.25
1992	131.55	483.37	3.67	41.78	109.00	2.61	32.49	188.29	5.80	11.10	48.24	4.34	5.06	14.84	2.94	1.97	4.67	2.37	9.16	19.92	2.17
1993	130.70	485.79	3.72	42.54	120.40	2.83	30.75	179.75	5.85	11.01	48.18	4.38	5.49	16.76	3.06	2.19	5.99	2.74	8.48	18.45	2.17
1994	131.38	491.68	3.74	42.81	122.64	2.86	30.54	177.99	5.83	10.73	46.64	4.35	5.74	18.20	3.17	2.13	5.17	2.43	8.98	21.11	2.35
1995	133.74	499.46	3.73	42.80	115.44	2.70	31.11	187.30	6.02	11.44	49.74	4.35	6.03	17.96	2.98	2.16	5.95	2.75	9.11	22.02	2.42
1996	135.37	521.69	3.85	43.40	122.50	2.02	31.75	197.00	6.20	11.57	51.10	4.42	5.77	17.68	3.06	2.25	6.46	2.87	9.27	22.33	2.41
1997	135.91	528.06	3.89	43.47	123.70	2.85	32.13	202.77	6.31	11.14	49.38	4.43	5.41	16.65	3.08	2.32	6.50	2.80	9.91	23.58	2.38
1998	136.42	532.20	3.90	44.80	129.06	2.88	31.57	200.57	6.35	11.73	49.24	4.20	5.46	17.08	3.13	2.42	7.01	2.89	9.51	23.45	2.47
1999	140.43	555.09	3.95	45.16	134.50	2.98	31.64	200.40	6.33	11.96	50.87	4.25	6.21	20.13	3.24	2.52	7.73	3.07	9.97	24.17	2.42
2000	138.14	545.38	3.95	44.71	127.40	2.85	30.30	189.81	6.26	11.79	51.90	4.40	6.30	21.32	3.38	2.38	7.20	3.03	9.89	25.84	2.61
2001	136.17	544.57	4.00	44.62	139.90	3.14	29.14	179.31	6.15	11.50	50.46	4.39	6.41	21.91	3.42	2.11	5.82	2.75	####	26.52	2.62
2002	131.45	517.53	3.94	40.41	108.90	2.69	28.51	176.34	6.19	11.52	51.49	4.47	6.20	22.78	3.67	2.23	6.72	3.02	9.99	26.06	2.61
2003	134.33	531.34	3.96	44.00	130.40	2.69	26.78	162.30	6.06	11.48	52.08	4.54	6.65	24.64	3.71	2.46	7.27	2.95	####	27.24	2.67
2004	134.54	549.46	4.08	42.50	124.40	2.93	29.42	186.73	6.35	11.75	53.10	4.52	6.00	23.00	3.83	2.21	7.57	3.43	9.80	25.20	2.57

1.1. THE ECOLOGICAL ASPECTS OF RICE

Rice is a hydrophytic plant and often cultivated as a semi-aquatic crop. However, there are varieties and strains, which are grown under, dry or rain fed and semi-dry conditions. For the efficient physiological functions of the plant, a saturated condition of the soil is sufficient. It is not correct to assume that rice crop requires standing water. The practice of maintaining a standing sheet of water is for checking the growth of weds. Rice is essentially a crop of sub-tropics and on higher elevations up to 6000 feet above sea level, it is also cultivated. Besides the abundant supply of fresh water needed for irrigation, the rice growing areas are characterized by high temperatures during the growing season and high atmospheric humidity. High altitudes and low temperatures delay flowering and maturity. The conducive temperature range is from 21°C to 31°C, although extremes over these are also tolerated by the crop. Rice can be grown successfully on a variety of soils. The most important requirement of the soil is its ability to hold moisture for a considerable period. Rice alluvial soils with impervious sub-soils are ideal for the crop. Rice thrives over a wide range of soil reaction ranging from 4.5 to 8.5 pH. The most favourable range is 5.5 to 7 pH. Only few rice varieties possess tolerance for saline and alkaline conditions. The details are discussed in the subsequent chapters.

1.2 HISTORY AND ORIGIN OF RICE

The cultivation of plants is one of the man's oldest occupations and probably began when he discovered plants that are useful to him. There is no precise record about when man started domestication of plants. The earliest evidence of genuinely cultivated forms so far discovered, dates from about 7000 BC, with fresh evidence, plant domestication is known to have originated independently and simultaneously 6000 to 9000 years ago in America, Africa, South Western Asia and South Eastern Asia. The first successful domestication of plants has recently been suggested to have occurred in Thailand. Rice is a member of the grass family. Archeological evidence indicates a sophisticated rice cultivation system existed in China over 7,000 years ago.

1.2.1 History of cultivated plants

Although man domesticated plants at an early date, he did not study them seriously for a long time but for the occasional sketches made in the caves. Greek and Roman Physician Botanists laid the real foundation of the scientific study of plants during the fourth century. Theophrastus was considered as "Father of Botany". It was Dioscorides' materia medica, which gave detailed account about plants. One of the oldest belief regarding the origin of cultivated plants was that they came to man as gifts from God. It was Alphonse de Candole (during 1883) a Swiss botanist who first attempted to solve the mystery about the ancestral form, region of domestication and history of cultivated plants. According to him rice is one of the plants cultivated at least 4000 years and possibly longer period ago. It was Vavilov an outstanding Russian Scientist who proposed the theory of centres of origin of cultivated plants. According to his theory, crop plants evolved from wild species in the area exhibiting great diversity and termed them as primary centre of origin. Vavilov divided the entire world into twelve centres of origin and according to this the primary centre of origin for rice is Hindustan centre of origin, which includes Burma, Assam, Malaya, Java, Borno, Sumatra and Philippines.

1.2.2. Rice

There is a general consensus that rice was domesticated some where in India or Indochina, probably in Southern India, where there are marshy areas which are ideally suited to cultivation of rice. The great varietal diversity of cultivated rice, the presence of the nearest known wild relatives and many dominant genes in Indian rice varieties lend support to this view. Exploration in Jeypore tract, Orissa have shown that the rice cultivated by the hill tribes at elevations between 500 and 1000 m included a large varietal diversity containing the whole range of variations from the wild perennial rice *Oryza perennis, Oryza longistaminata, Oryza spontanea* to cultivated indica and japonica races. There has been no large-scale introduction of rice biotypes into these areas.

Similarly in the Western Ghats, various biotypes of hill rice are being cultivated. The identification of *Oryza malampuzaensis* and also rice with various grain colours indicate the diversity of forms available to conclude about centre of origin. Rice spread eastward from India to china and then to Japan and Westward into Iran, Iraq, Turkestan and Egypt. Alexander the Great introduced it to Europe from India by AD 300. Rice was introduced into the American colonies in 1647 and rice cultivation in Australia began towards the end of 19th century. Other details are discussed under the head-botany of rice.

1.3 NUTRITIVE VALUE

Rice ranks second position in terms of area harvested or cultivated. In terms of importance as a food crop or in terms of calorific value rice crop occupies first place than any other cereal crops. For example, one ha of rice crop can able to sustain 5.7persons/ yr where as a ha of maize crop can sustain 5.3 persons / yr, and a ha of wheat crop can sustain 4.1 persons/yr.

100g of grains of rice contain,

Carbohydrate - 74.8g

Protein - 8.4g

Fat - 2.6g

The remaining are other minerals, amino acids and fibre content. Thus 100g of rice can supply about 477 kcal of energy. The nutritive value of rice is given in Table 3.

Table 3. Nutritive value of rice (per 100 g)

Food	Energy K.cal.	Protein g.	Fat g.	Carbohy-drate g.	Calcium mg.	Iron mg.	Carotene mg.	Thiamine mg.	Riboflavin mg.	Niacin mg.
Rice, parboiled, handpounded	349	8.5	0.6	77.4	10	2.8	9	0.27	0.12	4.0
Rice, parbo-iled, milled	346	6.4	0.4	79.0	9	1.0	-	0.21	0.05	3.8
Rice raw, handpounded	346	7.5	1.0	76.7	10	3.2	2	0.21	0.16	3.9
Rice, raw, milled	345	6.8	0.5	78.2	10	0.7	0	0.06	0.06	1.9

Source: Gopalan *et al.,* 1991

Composition

It is influenced by genetic and environmental factors. The germ, the pericarp and aleurone layers that are richer than endosperm in nutrients like protein, minerals and vitamins are separated from the grain during milling along with the husk.

Carbohydrate: The major carbohydrate of rice is starch, which is 72-75%. The amylose content of starch varies according to the grain type. The longer grain and superior types containing upto 17.5% amylose while some coarse types are completely devoid of it. Glutinous rice consists almost entirely of amylopectin. Rice also contains some free sugars like glucose, sucrose, dextrin, fructose and raffinose. The fibre of rice is the hemi cellulose made up of pentoses, arabinose and xylose.

Protein: The protein content of rice is 7-8.5%. It is much lower than that of the wheat. Glutelin, which is also known as oryzenin, is the principal protein of rice. Rice also contains small quantities of albumin, globulin and prolamines. The proteins of polished rice have a lower biological value but a higher digestibility than those of rice bran and rice proteins. Parboiling has no effect on the biological value or digestibility of the proteins. The nutritive value of rice protein is of high order being superior to that of the wheat and other cereal products. The rice proteins are richer in arginine compared to other cereal proteins. Rice is deficient in lysine and threonine. The biological value of rice protein is 80 whereas wheat protein has 66 and maize protein has 50.

Minerals : In its mineral content, rice resembles other cereals. Most of the minerals present in the rice are located in the pericarp and germ and poor in calcium and iron. Coloured types of rice contain more iron than the white rice. The phosphorus content is high, about 4% of which is present as phytic acid. Rice also contains some trace elements.

Enzymes: Rice contains amylases, lipases, oxidases, peroxidases and phenolases. On storage the oxidase activity of rice remains constant but the amylase, lipase and peroxidase activities decrease. The activity of the alpha amylase in fresh rice is probably responsible for its sticky consistency after cooking. This enzyme gets inactivated during storage and cooked grains get separated easily.

Pigments : Coloured rice contains anthocyanins and carotenoids.

1.4 THE GEOGRAPHY OF RICE

1.4.1 Origin and Diffusion of Cultivated Rice

Two species of domesticated cultivated rice evolved from a pool of wild Oryza species (Chang, 1976).

1. *Oryza glaberrima* (Steud) is found in West Africa's middle Niger valley and in a few secondary centres. That species mostly grows wild as a deep-water rice in seasonally flooded portions of West African river valleys. The same pattern is seen in Brazil's Amazon basin. Only locally *O. glaberrima* managed as a dry land crop in the high rainfall zone of West Africa. More frequently, it is considered as a weed in recently developed fields of *O. sativa*.

2. *Oryza sativa* was domesticated in tropical and subtropical Asia but the centre of domestication is matter of contention. One school assumes its simultaneous domestication in various centres extending from the plains

below the eastern foothills of the Himalayas, through upper Burma, northern Thailand, Laos, and northern Vietnam to southwest and west China (Chang, 1976). Another school postulates a more limited centre of origin, notably the inland valleys in northern Thailand, the Shan states of Burma, and the adjacent area of Laos. The dependable monsoon rains of that area, its warm and humid environment, and the specific physiography of the valleys with their multitude of wet and seasonally shallowly flooded grassland depressions offered an ideal environment for *O. sativa's* domestication. From its initial point or points of cultivation, *O. sativa* has spread over a considerable part of the earth and has become the major food crop for a large part of the world's population.

Varietal adaptation including the development of the main subspecies of *indica* and *japonica* and the improvement of land and water management practices, which changed and optimizes rice's edaphic and climatic environment, have pushed rice growing well beyond the conditions of its point of origin. Rice now grows as far north as latitude 53^0 N in the Amur River valley on the China-Russia boundary, requiring varieties with a far different photoperiodicity and temperature tolerance than those that grow close to the equator. Rice's requirements for water have, in the process of diffusion, been met by selection of areas with sufficient rainfall, by extension of the culture in wet lowlands, by water conservation, and ultimately by irrigation. Thus, the limits of rice's diffusion throughout the world are imposed mainly by the temperature regime, while within that zone, the availability of an ample water supply has always been a determinant of the distribution in the landscape of rice as a food crop.

1.4.2 Rice in the Landscape

Rice, through varietal adaptation, can be grown as a dry land crop in the more humid areas, but it is by origin and by preference of most cultivators – mainly a wetland crop. Agricultural systems based on rice as the food staple are thus clearly lowland system. The major areas are the extensive lowlands in southern and eastern Asia areas almost exclusively river basins and deltas with their contiguous coastal plains. The distribution of rice in the landscape is best understood through knowledge of the semi aquatic or amphibian nature of the rice plant. Rice is not an aquatic plant in the botanical sense, as proven by its root system, but it thrives in waterlogged soils where no other grain crop survives. Only certain aroids and sago palms (*Metroxylon sp.*) can compete with rice as a food crop adapted to wetland conditions, but they are of minor importance.

The semi aquatic character of rice was the key to the development of wet lowlands in Asia at an early stage in the history of agriculture. Rice grew in those lowlands without extensive drainage and flood protection works, such as those required to develop wetland areas for nonaquatic upland crops. There are, of course, areas where neither rice as a colonizing crop, nor the necessary techniques for flood protection and drainage to grow upland crops were available. Such wet lowlands, mainly in tropical Africa and South America, remain largely underutilized. A typical example is the coastal delta of the Niger river and its adjacent lowlands, which is virtually unused for agriculture. Nonutilized lowlands in the basins of major South American rivers form an even greater reserve of potential wet rice lands.

In the presumed heartland of rice growing, the highlands of central South- East Asia, the general relief is rugged and determined by series of more or less parallel ridges, separated by flat-bottomed valleys. Agriculture is mainly concentrated in those valleys, a considerable portion of which have wetland or hydromorphic soils. In sharp contrast, the steep slopes of the ridges bordering the valleys remained under forest. Only recently, the valley farmers, due to population pressure encroached on the steep land, but no sedentary agriculture has developed. In many places, the encroachment on the valley slopes has accelerated erosion and let to rapid land deterioration. On the crests of the ridges, and on the small high plateaus, the staple food crop is mainly rice in a shifting cultivation system on well-drained soils. The relatively low importance of the rice grown in shifting cultivation's is illustrated in northern Thailand, where the lowland population exceeds the hill-tribe population by a factor of 50, but where the total surface of the lowlands is less than 15% of total land surface. Clearly, an agriculture based mainly on the use of well-defined and confined lowlands, where rice grows in wet soil, is natural adaptation of cropping patterns to land conditions in an area.

Below the hilly heartland of *O. sativa*, rice growing extended to the major river valleys, deltas, and coastal plains. Physiographically, the land area to the south and east of the Himalayas- Tibet highlands and its eastern continuation is characterized by broad lowland plains. These include the plains of Yangtze, Mekong, and Brahmaputra-Ganges river systems, the sources of which are in the Himalayas- Tibet highlands. There are also the shorter rice systems of the Red River, the Chao Phraya, and the Irrawaddy, which start from the hill areas southeast of the Himalayas. All these ricers have extensive middle and lower valleys as well as wide deltas. It is here that, over the centuries, the major extension of rice cultivation took place mostly on lands that because of their hydromorphic nature were exclusively suited for rice as the major food crop.

From a global point of view, there is no other area in the rice-growing zone – as it is limited by temperature restraints – with major river valleys and deltas comparable with those east and south of the Himalayas. The deltas and adjacent river plains of the major African rivers are much smaller. The two major equatorial rivers, the Congo and the Amazon, have small deltas, although large zones of wet lowlands exist in their middle course. The Indus River valley in Asia is an exception, in as much as, its total surface devoted to rice growing is small compared to the valleys and deltas of the east Asian rivers. The restriction of the Indus is mainly climatic. Most of the Indus system is in arid zones. While most of the large rice-growing areas in Asia are thus found in association with major river, rice cultivation has spread also into smaller river valleys. Almost all depression landforms in South and Southeast Asia are used for rice growing, to the point that paddies are found in even the narrowest valleys. Small-scale land –use maps make this abundantly clear. Water conditions within the lowlands are, of course, not uniformly optimal for rice. Of fundamental importance for the spread of rice growing in the lowlands was the artificial creation of aquatic (flooded) conditions by leveling and bunding of fields to conserve water from flooding, groundwater, or rain. The combined measures of leveling- bunding, which developed early in the history of rice growing in Asia, tremendously increased the hectarage of rice land with potential for sustained crop production. The leveling – bunding technique was mostly adapted to wetlands and to lower concave slopes, where many of the soils have slow permeability and where excess water from streams and from adjacent higher lands collects in the higher bunded rice lands is distributed by overflow to the lower rice lands.

The leveling and bunding of rice-growing lands, which is second nature to the rice farmers of Asia, was adopted as a basic land management technique in newer rice–growing areas such as those around the Mediterranean and in parts of the United states. The practice is not, however, widespread in Africa and South America. The next logical measure for extending the possibility of rice growing in the landscape is, of course, irrigation from an outside source. Irrigation of rice land was possibly established as early as was the bunding and leveling technique, and again resulted by and large in extension and improvement of the aquatic environment optimal for rice. With irrigation, rice will grow in many arid zones where the natural water supply is insufficient. Moreover, with irrigation, rice can be grown on existing rice lands in the dry season. Hence, irrigation leads to an expansion of the rice hectarage, both in space and in time. Wetland rice growing also expanded from its natural lowland position to higher lands, and even to steep slopes, provided that a source of irrigation water was harnessed at a yet higher elevation. Thus, the use of sloping uplands for terraced paddies is clearly related to the presence of such water resources in the higher aspects of the landscape. To a certain extent, the extension of wetland rice up the slopes can be found anywhere in Asia where irrigation water can be made available. But the major impact of irrigation has been in the low lands of the rice-growing zones of the world, both in the small valleys and on the major plains and deltas. While most rice is aquatic, an estimated one-sixth of the total rice land is reported as upland (De Datta, 1975), i.e. land where rice grows in fields not bunded or leveled and that depends exclusively on rainfall for moisture. This estimate may be exaggerated, because in the newer rice-growing areas such as West Africa, most of the so-called upland rice-growing areas rice in reality lowlands with temporary high groundwater (hydromorphic soils). The main difference between African and Asian rice lands is that Asian techniques for water retention by leveling and bunding are not applied in Africa. If only the rainfed rice grown on well-drained land is included under the term upland, such land may not be more than 10% of the total rice land.

The occurrence of rice as a crop in freely drained uplands is restricted to zones with a sufficient water supply from rain. Thus, semiarid or drier zones are excluded from this type for rice cultivation. Within the climatic restrictions, however, rice is grown on a wide range of landscapes and topographic positions. A much larger proportion is found on moderately to gently sloping plateaus and peneplains as, for instance, the volcanic plateaus in Vietnam and the Philippines (Batangas area of Luzon), and the extensive savanna (cerrado) regions of central Brazil. Intensified cultivation of rice in freely drained uplands, without or with only short periods of fallow, is restricted to much flatter terrains and to lands that are not too severely subjected to erosion. Therefore, while dry land rice, land can be found in most aspects of the landscape, it is restricted mostly to lands that are gently sloping.

1.4.3 Major Landforms of Lowland Rice – Growing Areas

The geomorphology of the major and minor lowland rice-growing areas is diverse and their surface relief varies from place to place. More or less detailed geomorphological studies, frequently made in conjunction with soils and hydrologic studies, are now available. Examples are the studies by Brammer (1971 a,b) of the plains in the Ganges – Brahmaputra river system in Bangladesh; by Van der Kevie and Yenmanas (1972) and Takaya (1971) in the Chao Phraya basin of Thailand; by Moormann (1961) covering several parts of the lower Mekong basin; by Brinkman and Pons (1968) for the coastal plans of the Guiana's in S. America, and by ILACO-NEDECO (1966) for the lower Niger River delta in Nigeria.

1.4.3.1 Inland valleys

A considerable proportion of wetland rice is grown in valleys of modest to very small size. Inland valleys are the lowlands of the upper reaches of major and minor watersheds, where the sediments – either alluvial or colluvial – are derived from the adjacent uplands. Soils, therefore, frequently reflect one or more characteristics of the dry land soils, even though certain properties such as base saturation and clay mineralogy may have altered considerably under the specific hydromorphic regime of the valleys. In many areas, inland valleys form the major resource – if not the only one – for wetland rice growing. A typical example is seen in Sri Lanka, where the larger part of wetland rice growing is confined to inland valleys. In all major and minor river basins, inland valleys are preferred sites for rice growing, especially in most of Asia. Inland valleys vary widely in a number of characteristics influence among others, by inequalities of slope, rock resistance, structure, and geologic geomorphologic history of a region. Inland valleys conform to the drainage pattern of the physiographic region in which they occur.

- **Dendritic**: A nonsystematic or tree like branching pattern of valleys extending in many directions. This pattern is formed on rocks of uniform resistance, often in peneplains of horizontal sedimentary units and complex metamorphics without a dominant slope or structural control
- **Trellis**: A branching pattern controlled by the presence of folded and faulted strata. The major primary and secondary valleys are parallel to the rock strata and are connected by short and often steep and narrow cross-valleys.
- **Rectangular**: A pattern where the branching is controlled by the presence of approximately right – angle faults or joints. This pattern is often found in areas of old metamorphized crystalline rocks.
- **Radial:** A pattern of drainage valleys radiating outward from a common higher centre or, alternatively, converging in a common circular basin.

The general relief of the terrain in which rice-growing inland valleys occur, influences slope, cross section, and general hydrology of the valleys. Moreover, the general relief is an important factor regarding accessibility to the field and general cropping patterns of the area. For descriptive and practical purposes, the general relief of the land systems with inland valleys can be described as mountainous, hilly, rolling, or undulating. Inland valleys vary widely in their longitudinal profile. Generally, gradients are steepest in the mountainous and hilly areas, especially those close to the valley head. Terrace forms, size of rice fields, and the hydrology of the rice lands in the valley bottom are strongly influenced by the longitudinal gradient. A typical example of this can be found in the valleys of structured and joined metamorphic rocks in rolling to hilly terrains, as in the mid country of Sri Lanka and in rolling basement – complex areas of West Africa.

1.4.3.2 Alluvial fans and piedmont plains

The alluvial fans and piedmont plains are found between hilly areas and lowlands where sediment –charged water from the rivers and creeks abruptly loses much of its velocity as it enters the flatter lowland. A well-formed alluvial fan shows a definite gradient in texture and overall soil drainage. The fan's upper part is generally coarser textured, often with gravel beds at the surface or at shallow depth, and is often freely to excessively drained. Soils become more clayey and more poorly drained toward the fan's lower, wider part. There are, of course, variations on the general alluvial fan pattern, depending on the speed of discharge of the water from the hills and the nature of the sediment load as determined by the soils and the surface formations in the watershed area. In the more arid climates, where discharge from the highland is in the form of flash floods, the alluvial fans – and often the piedmont areas – are coarse textured and excessively drained throughout. In such areas little or no rice is grown. Where sediments are more clayey and flooding is gentler, the whole of an alluvial fan may have a medium texture or finer. That is the case for many alluvial fans of volcanic areas, such as in Java, where aquatic rice cultivation has extended from the lower part of the fan to its apex. In general, wetland rice in alluvial fans is found mainly in the lower, more clayey aspects of the landform. Terracing and irrigation are common on this land in Asia.

1.4.3.3. Meander floodplains

Meander floodplains are found mostly in the middle course of rivers. Land forms composing this type of landscape are riverbeds with stream channels, river levees and river basin. Sedimentation on the levees results from floods, which have a relatively high velocity close to the channels. Hence, such river levees commonly have a fine sandy or medium texture. Away from the river, where the velocity of the floodwater is less or nil, the finer particles remaining in the floodwater settle, and the soils formed are considerably more clayey. Because of slow sedimentation, the basins are lower and more hydromorphic in nature than the levees. There is thus a distinct gradient away from the river, starting with medium-textured, well-drained soils, going to very fine, clayey, poorly-drained soils, and eventually to transitions to peaty soils in those parts of the basins that never dry.

1.4.3.4. Lacustrine floodplains

Lacustrine floodplains occur in several landscape positions and may form part of major river valleys. An example of a well-developed lacustrine floodplain used for wetland rice growing is the plain surrounding the Top Sap in Cambodia. The landscape of such plains of flat and the sediments are usually clayey but may be fine-layered at some depth (carved sediments), representing recurrent peaks of flooding and sedimentation in the lacustrine environment. The sediments in lacustrine floodplains vary, however, in accordance with the lithology of the surrounding area. Lacustrine formations in volcanic areas (e.g. around Laguna de Bay, Luzon, Philippines) are dominantly composed of fine montmorillonitic clay. In contrast, kaolinitic clays dominate in such lacustrine areas as are found in the basin of the Moon River valley in northeastern Thailand, which received their sediments from highly weathered formations and from acidic sedimentary rocks in the adjacent highlands.

1.4.3.5. Marine floodplains

In the coastal lowlands of the rice-growing area, two broad landscape can be distinguished. The first type is that of the **rapidly aggrading coastal plains** found in the lower deltas of many minor and major rivers. These are broad clayey plains, mostly separated by consecutive sandy beach ridges, and without a dendritic tidal creek system. An important aspect of the rapid growth of this type of coastal plains is that the accumulation of pyrite (FeS_2) under mangrove vegetation is restricted, and such plains have few acid sulphate soil areas. Rapidly aggrading coastal plains are found mainly in the lower delta areas of river systems with a considerable sediment load and with shallow coastal waters (continental shelves). A typical example of a rapidly aggrading lower delta is that of the Mekong River system in Vietnam.

The second type of landscape in the coastal lowlands is the **slowly aggrading or stationary coastal plain**. Such coastal plains are characterized mainly by broad mangrove zones and relatively little development for rice land. The tidal land is cut up by numerous winding tidal rivers and creeks, and sandy beach ridges are either absent or rare. The estuary of the Saigon River is a typical example of a stationary coastal plain. These plains are found mainly in the deltaic areas of rivers with a low silt load. Many of the equatorial coastal plains are of this type, e.g. the coastal plains of Sumatra and Kalimantan in Indonesia. Another type of relatively stable coastal plain in the Sunderbans is India (Bengal) and Bangladesh, where the coastal landscape undergoes a continuous subsidence and mass movement because of the presence of a deep trough (so called swath of no ground) in the sea bottom close to the coastline. The coast line has been reclaimed for rice growing, as in the Sunderbans, the rice land usually has more or less severe hydrological and soil problems-results in a high accumulation of organic matter and of pyrite and leads to acidification of the soil (acid sulfate soils) after drainage and reclamation.

1.4.3.6. Alluvial terraces

Alluvial terraces are former river, lake, or marine floodplains that, because of a relative lowering of the base level of drainage, are above the level of regular flooding. In some cases, alluvial terraces are formed by orogenetic lifting of blocks of the landscape (uplifted peneplains). But most alluvial terraces are related to changes in the base level of erosion, corresponding to cyclic climatic changes in recent geological times. All major and most minor river floodplains are accompanied by river terraces, with an almost level or slightly undulating topography. The river growing area on alluvial terraces is considerable in the major river basins of Southeast Asia. For the Cho Phraya basin in Thailand, Fukui (1971) estimated that 40% of the paddy lands are found on river terraces and their associated alluvial –fan complexes. Some flatter parts of older Pleistocene terraces, the so-called middle terrace, are also in use for rice growing under a rain fed regime in bunded and leveled paddy fields. Rice is grown extensively on the Mekong terraces, especially in northeast Thailand, and also on the Vientiane plain in Laos, the area surrounding Ton Let Sap in Cambodia as well as parts of the low terrace areas east of the Mekong in southeastern Cambodia, and in Victnam. Low river terraces, with conditions more or less favorable for rice growing belt. Typically, however, their extension relative to the higher land is less favourable than in the major rice-growing areas in Asia. Africa's Niger River and its tributaries, for example, have a narrow floodplain and their accompanying level terraces are not extensive. Nevertheless, rice growing has expanded considerably on the Niger terraces in the last 10 years.

A characteristics type of valley incision is observed on terraces composed mainly on unconsolidated, permeable sediments as, for instance, the fan terrace complexes of the foot slopes of volcanoes of the Philippines and Java. There, the flat summit areas used for rain fed or irrigated rice are maintained intact while a deep gorge-like erosion takes place at the site of rivers and lesser drainage ways. In other formations, the erosion pattern is of a more lateral nature, with wide shallow valleys giving a more undulating aspect to the landscape. That is the case in the wide low terrace areas of northeastern Thailand. Various forms of dissection of the terrace areas in Bangladesh have been described by Brammer (1971b). Dissection of the Madhupur and Barind terrace tracts has resulted in four major forms:

- level, poorly drained areas, with meandering streams about 9 km part;
- flat summit areas, with deeply cut valleys and dissected gullied land bordering the valleys
- areas of deeply weathered terrace formations crossed by broad and wide shallow valleys, most of which are stream less, although considerable areas remains wet as a result of damming of valleys; and
- areas of shallowly weathered terrace formations, dissected by a close network of small, branching, stream less valleys.

These examples indicate the complexity of landforms that are found on alluvial terraces.

CHAPTER 2

RICE BOTANY

There are two divergent views regarding the origin of these two cultivated rices. One is polyphyletic origin i.e. origination from several species and monophyletic i.e. evolving from a single species. According to polyphyletic theory, *O. sativa* and *O. glaberrima* have evolved independently in their respective regions viz. Asia and Africa. *O. rufipogan* of Asia has given rise to most varieties of *O. sativa* but some forms are derived from *O.minuta* while *O. barthii* of Africa is the progenitor of *O. glaberrima*. Most modern taxonomists, however, consider the cultivated rice to be monophyletic, having originated from a common ancestral stock. The close similarity between *O. sativa* and *O. glaberrima* (the only differences are in glumes pubescence, ligule size and colour of pericarp which is red in *O. glaberrima*) and the existence of intermediate forms between the two provide forcible reason for this view (Palanisamy, 1998).

(i) Polyphyletic : Originated from several species. According to theory, the two forms of cultivated rice viz., Asian rice

Common ancestor

South & South East Africa ***Tropical Africa***

Perennial *O. rufipogan* *O. longistaminata*

O. sativa and African rice; *O. glaberrima* have evolved independently in their respective regions from several species.

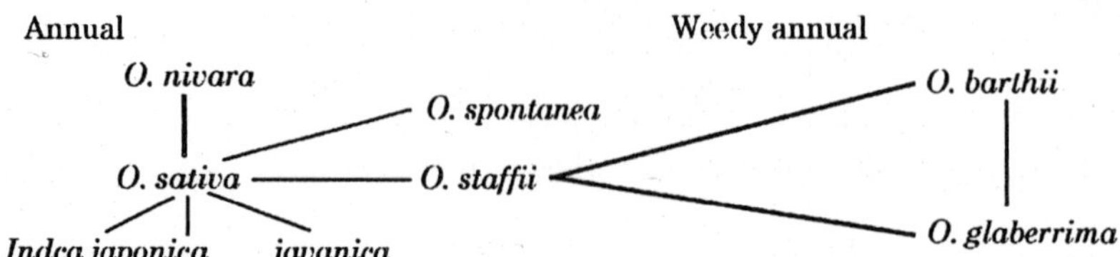

(ii) Monophyletic: According to this theory, both Asian rice and African rice arose from a common parent (*O.perenis*). This view is the most accepted one, because both Asian rice and African rice are similar except in glume pubescence, ligule size and colour of pericarp, which is red in African rice.

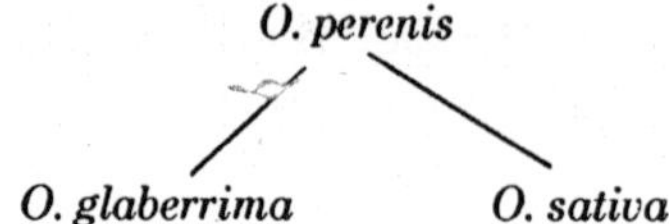

a. Species in the genus Oryza

The genus Oryza has 24 species of which only two are cultivated diploids viz., *Oryza sativa,* Asian rice and *Oryza glaberrima*, African rice, and the rest are wild species, which include both diploid and tetraploid forms.

Botanical name	Chromosome No.	Genome	Origin
O. sativa	24	AA	Asia
O. nivara	24	AA	Asia
O. meridionalis	24	—	Australia
O. longistaminata	24	AA	Africa
O. rufipogan	24	AA	Asia
O. glumaepatula	24	—	America
O. grandiglumis	48	CCDD	America
O. glaberrima	24	AA	Africa
O. barthii	24	AA	Africa
O. australiensis	24	EE	Australia
O. latifolia	48	CCDD	America
O. alta	48	CCDD	America
O. eichingeri	24	CC	Africa
O. minuta	48	BBCC	
O. punctata	48	BBCC	Asia
O. officinalis	48	BBCC	Asia
O. granulata	24	CC	Asia
O. meyeriane	24	—	Asia
O. ridleyi	24	—	Asia
O. longiglumis	48	—	Asian
O. brachantha	48	—	New Guinea
O. schlechter	24	FF	Africa
O. malampuzaensis	—	—	New Guinea

b. Subspecies in cultivated Oryza

Rice has been in cultivation for long period and adapted well under diverse climatic conditions and soils. This has resulted in the evolution of three geographical races which has been given subspecies status. The three subspecies are:

1. *O. sativa sub sp. indica* : Tall, spreading, moderate tillering, awnless
2. *O.sativa sub sp. japonica* : Short, erect, more tillering, awnless
3. *O.sativa sub sp javanica* : Tallest, erect, poor tillering, awned.

Marked sterility barriers occur between the subspecies. It ranges up to 80 percent in case of indica x japonica where as it is less in case of indica x javanica.

2.1. CLASSIFICATION

The genus Oryza belongs to the sub-tribe Oryzineae and tribe Oryzeas. This tribe Oryzeae is an isolated group in Gramineae and is characterized by an aquatic mode of life. Altogether a total number of 27 species have been enumerated up to date under Oryza. Of these, only two species, namely, *O. sativa L.* and *O. glaberrima* are cultivated species and these two come under the section sativa.

O. sativa is divided into three sub-species, though they are considered as geographical races. Hybrids between them show varying degrees of sterility (Table 4).

1. The javanica or bulu sub-species includes a small number of varieties with a sparse tillering habit, stiff, straw, long panicles, awned spikelets long duration and low sensitivity to differences in natural day length. These varieties are cultivated mainly in Indonesia.
2. The japonica sub-species are characterized by compact panicles, pubescent spikelets and oval or round grains. Theses varieties are found mainly in Japan and are adapted for cultivation in the sub-tropical and warm temperate regions.
3. The indica sub-species include all the other varieties cultivated all over tropical Asia, including India. In this group, the range of variation in both morphological characters is very large. As rice is a predominantly self- fertilized crop, the varieties are homozygous and the differences between them are only due to one or more genes.

Table 4. Differences between the three sub-species of *O. sativa*

	indica	japonica	javanica
Leaves	Broad, light Green leaves	Narrow, dark green leaves	Broad, stiff, light green leaves
Tillering	Profuse tillering	Medium tillering	Low tillering
Height	Tall plant stature	Short plant stature	Tall plant stature
Hairs on Lemma and Palea	Thin and short hairs on lemma	Dense and long hairs on lemma and palea	Long hairs on lemma and palea
Awns	Mostly awnless	Awnless to long Awned	Awnless or long awned
Grains	Slender, somewhat Flat grains	Short, roundish grains	Broad, thick grains
Grains	Fast shattering	Low shattering	Low shattering
Tissues	Soft plant tissues	Hard plant tissues	Hard plant tissue
Photosensitivity	Varying sensitivity to photoperiod	Varying sensitivity to photoperiod	Low sensitivity to photoperiod

2.2. MORPHOLOGY

Cultivated rice is generally considered as a semi-aquatic annual grass, although in the tropics it can survive as a perennial, producing new tillers from nodes after harvest (ratooning) (Fig. 1). At maturity, the rice plant has a main stem and a number of tillers. Each productive tiller bears a terminal flowering head or panicle. Plant height varies by variety and environmental conditions, ranging from approximately 0.4 m to over 5 m in some floating rices. The morphology of rice is divided into the vegetative phases (including germination, seedling, and tillering stages) and the reproductive phases (including panicle initiation and heading stages).

a. Seeds

The rice grain, commonly called a seed, consists of the true fruit or brown rice (caryopsis) and the hull, which encloses the brown rice (Fig. 2). Brown rice consists mainly of the embryo and endosperm. The surface contains several thin layers of differentiated tissues that enclose the embryo and endosperm. The palea, lemmas and rachilla constitute the hull of indica rices. In japonica rices, however, the hull usually includes rudimentary glumes and perhaps a portion of the pedicel. A single grain weighs about 10-45 mg at 0% moisture content. Grain length, width, and thickness vary widely among varieties. Hull weight averages about 20% of total grain weight. Further details are given in the subsequent sections of this chapter.

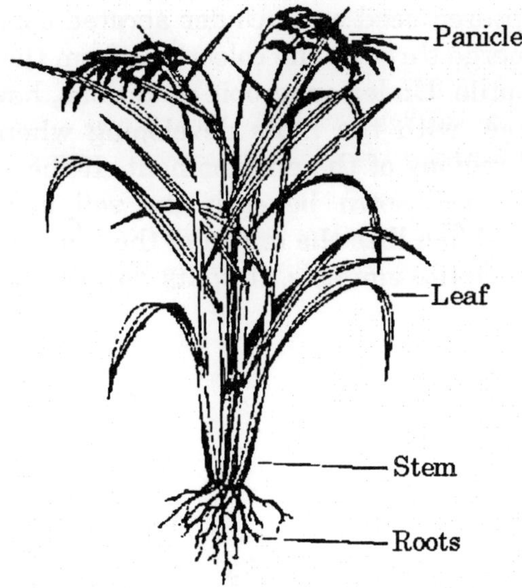

Figure 1. Rice plant (Source : Arunachalam, 1984)

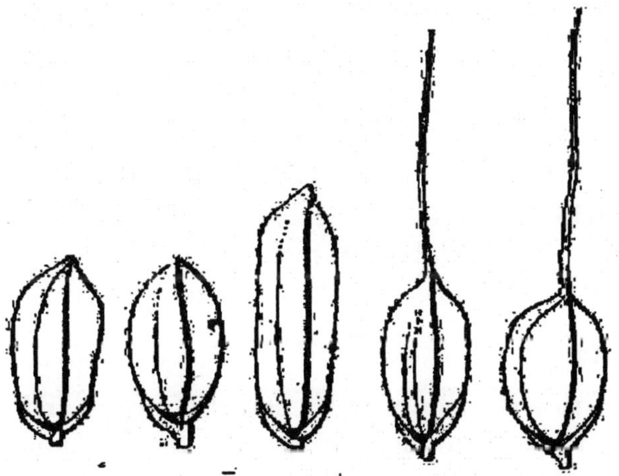

Paddy seeds vary in their size, shape and color

Figure 2. Seed/grain (Source : Arunachalam, 1984)

b. Seedlings

Germination and seedling development start when seed dormancy has been broken and the seed absorbs adequate water and is exposed to a temperature ranging from about 10 to 40^0 C. The physiological definition of germination is the time when the radicle or coleoptile (embryonic shoot) emerge from the ruptured seed coat. Under aerated conditions, the seminal root is the first to emerge through the coleorhiza from the embryo, and this is followed by the coleoptile. Under anaerobic conditions, however, the coleoptile is the first to emerge, with the roots developing when the coleoptile has reached the aerated regions of the environment. If the seed develops in the dark as when seeds are sown beneath the soil surface, a short stem (mesocotyl) develops, which lifts the crown of the plant to just below the soil surface. After the coleoptile emerges, it splits and the primary leaf develops (Fig. 3).

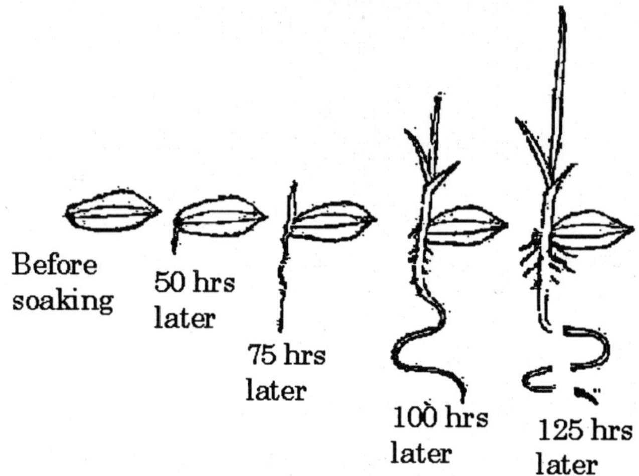

Figure 3. Stages of seed germination (Source : Arunachalam, 1984)

c. Tillering plants

Each stem of rice is made up of a series of nodes and internodes. The internodes vary in length depending on variety and environmental conditions, but generally increase from the lower to upper part of the stem. Each upper node bears a leaf and a bud, which can grow into a tiller. The number of nodes varies from 13 to 16 with only the upper 4 or 5 separated by long internodes. Under rapid increase in water level, some deepwater rice varieties can also increase the lower internode lengths by over 30 cm each. The leaf blade is attached at the node by the leaf sheath, which encircles the stem, where the leaf blade and the leaf sheath meet is a pair of claw like appendages, called the auricle, which encircle the stem. Coarse hairs cover the surface of the auricle. Immediately above the auricle is a thin, upright membrane called the ligule. The tillering stage starts as soon as the seedling is self-supporting and generally finishes at panicle initiation. Tillering usually begins with the emergence of the first tiller when seedlings have five leaves. This first tiller develops between the main stem and the second leaf from the base of the plant. Subsequently when the 6th leaf emerges, the second tiller develops between the main stem and the 3rd leaf from the base. Tillers growing from the main stem are called primary tillers. These may generate secondary tillers, which may in turn generate tertiary tillers. These are produced in a synchronous manner. Although the tillers remain attached to the plant, at later stages, they are independent because they produce their own roots. Varieties and races of rice differ in tillering ability. Numerous environmental factors also affect tillering including spacing, light, nutrient supply and cultural practices.

The rice root system consists of two major types: crown roots (including mat roots) and nodal roots. In fact, both these roots develop from nodes, but crown roots develop from nodes below the soil surface. Roots that develop from nodes above the soil surface usually are referred to as nodal roots. Nodal roots are often found in rice cultivars growing at water depths above 80 cm. Most rice varieties reach a maximum depth of 1 m or deeper in soft upland soils. In flooded soils, however, rice roots seldom exceed a depth of 40 cm. That is largely a consequence of limited O_2 diffusion through the gas spaces of roots (aerenchyma) to supply the growing root tips.

d. Panicle and spikelets

The major structures of the panicle are the base, axis, primary and secondary branches, pedicel, rudimentary glumes, and the spikelets. The panicle axis extends from the panicle base to the apex; it has 8-10 nodes at 2 to 4cm intervals from which primary branches develop. Secondary branches develop from the primary branches. Pedicels develop from the nodes of the primary and secondary branches; the spikelets are positioned above them. Since rice has only one fully developed floret (flower) per spikelet, these terms are often used interchangeably. The flower is enclosed in the lemma and palea, which may be either awned or awnless. The flower consists of the pistil and stamens, and the components of the pistil are the stigmas, styles, and ovary.

2.3. BOTANICAL DESCRIPTION

Annual, tillering freely and forming clumps; stems erect, leaves long, prominently 1 – nerved, striated, scaberulous, sheath enclosing the culm; ligule prominent, long and 2- partite, auriculate, auricles hairy and clasping the stem tightly. Panicle erect or drooping, open or compact. Spikelets jointed on a slender rachilla arising at the ends of the branches of the panicle; single flowered; laterally compressed, oblong; rachilla thickened and annular at the top. Glumes two, small unequal. Lemma chartaceous, obtuse or acute, muticous or mucronate or awned; strongly three to five – nerved, scaberulous. Palea as long as the lemma, thick in texture, keeled. Lodicules two, each two lobed. Stamens six, anther linear. Styles two, free. Stigmas plumose laterally exerted. Grain elongated or round enclosed by the lemma and palea, endospermous, endosperm mealy or flinty; embryo lateral.

A. Roots

As the rice grain germinates, the radicle pierces its way out through the coleorhiza and the enveloping glume and forms the seminal root. Lateral roots are formed from the seminal root. As the seedlings begin to grow, adventitious roots arise from the basal nodes. When seedling are pulled out, most of the roots get torn off but very soon fresh roots develop. The root system in rice varies much between the different varieties. Short duration varieties have less extensive and less dense root development than the long duration varieties. In the same variety, variations in root growth occur due to changes in soil types, manuring, irrigation and drainage facilities, cultural operations and spacing between plants. As a general rule, taller varieties have a greater penetration and lateral spread in the root system than the shorter varieties. Similarly varieties with good tillering have a well-developed root system. Root development is promoted by organic manures and phosphatic fertilizers. Clayey soil seems to be ideal for the normal development of the roots. Pruning of all the roots of the seedlings before planting does not in any way affect the growth of the crop. The broadcast crop develops a deeper but poorly developed root system, whereas the transplanted crop has a shallow but dense root system. Nodal roots occurs in the deep water or floating rice varieties. These nodal roots develop on the higher nodes and help the plant to absorb the food material from the surrounding water.

B. Stem

The stem of rice, popularly known as the haulm or the culm, is generally erect, cylindrical, hollow at the internodes and solid at the nodes. The number and length of the internodes vary with the different varieties. There may be ten to twenty internodes. The lower internodes are generally short, the length increasing with the upper ones, and the longest being the terminal node carrying the inflorescence. In deep water rice varieties and in floating types, the culms have generally very long internodes. Each node bears a leaf and a bud, which may grow in to a shoot or tiller (Fig. 4) and productive tillers grow out of the main culm in alternate order, and the secondary tillers from the primary tillers also follows the same pattern. The productive tillers may be calculated as shown in the Fig. 5.

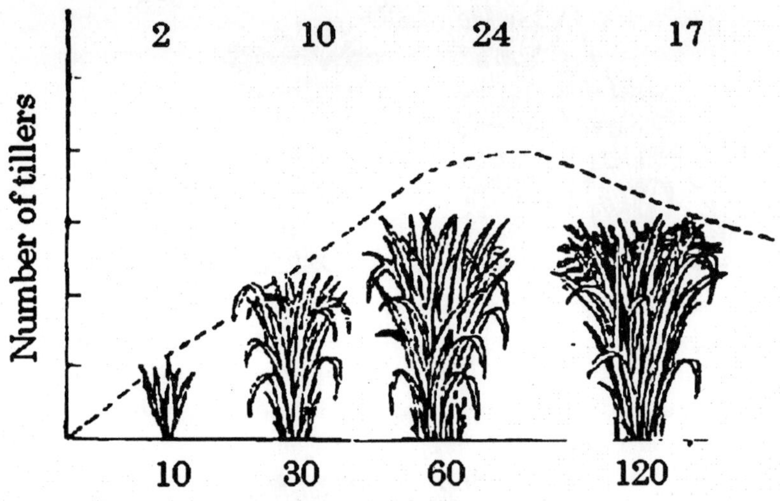

Figure 4. Formation of tillers (Source : Arunachalam, 1984)

C. Tillers

When the seed germinates and the plumule grows, the hypocotyl region becomes a little thickened forming an erect cylindrical rootstock from which the leaves continuously arise at closely set nodes. At the same time, buds are formed in the axils of these leaves on the rootstock. These buds begin to grow into tillers. Though each of the nodes on the culm carries a bud, it is mostly the lowermost buds on the crowded nodes just at the ground level or below that develop into tillers. The tillering habit is dependent on varieties, spacing, manuring and cultural conditions. It has been found that rice plants begin to develop tillers around two weeks after transplanting and continue developing more tillers for three to five weeks thereafter. The maximum production of tillers is generally attained thirty to forty days after transplanting. In rice, it is not the total production of tillers that is important, but the number of functional tillers. Greater the number of productive tillers, the greater is the yield as shown in Fig. 5.

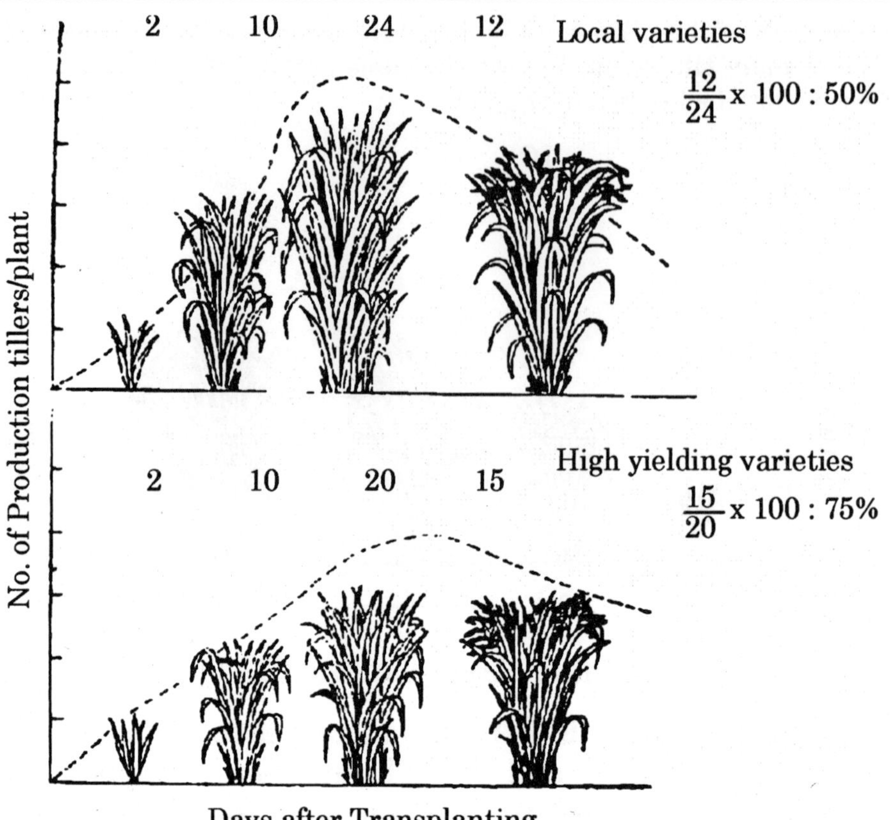

$$\textit{Productive tillers (\%)} = \frac{\text{Productive tillers}}{\text{No. of tillers}} \times 100$$

Figure 5. Productive tillers percentage (Source: Arunachalam, 1984)

D. Leaf

The leaf generally linear though variations are met within the length and breath of leaves. A certain amount of correlation has been found to exist between the width of the grain and the width of the leaf. Narrower grain types tend to have narrow leaves. Arising on either side of the attachment of the leaf blade to the sheath, are two linear ciliate structures known as the auricles. The auricles clasp the stem. The sheath as it joins the lamina has well developed thin membranous ligule, tending to split into two as it develops (Fig. 6).

E. Development of the ear / Panicle

The panicle initiation of the ear formation commences only, after the tillering phase is completed. There is an interval between the completion of tillering and the commencement of panicle initiation. This interval depends upon the varieties for eg. in short duration varieties, tillering and panicle initiation is overlapping.

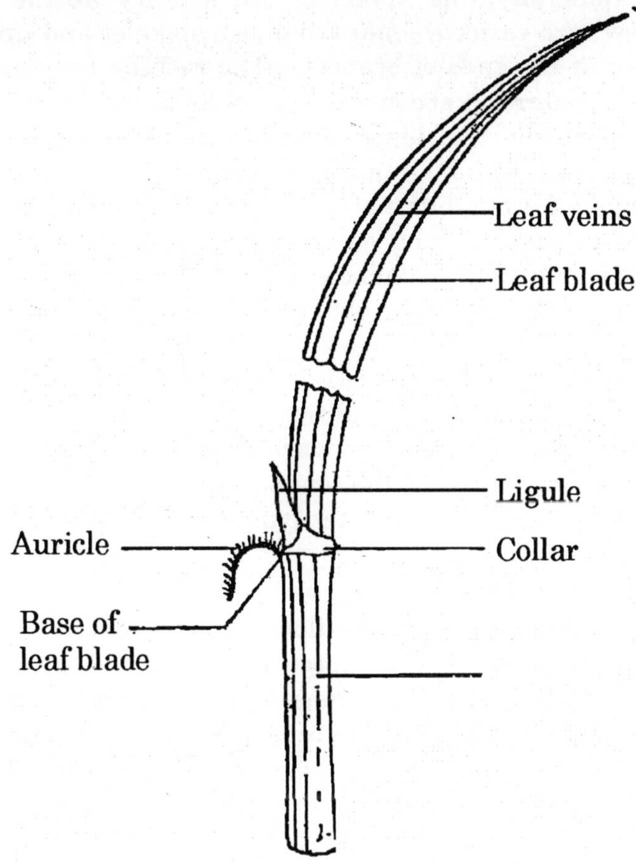

Figure 6. Leaf (Source: Arunachalam, 1984)

F. Panicle

The panicle is borne at the end of the vegetative shoot. It is branched with a number of spikelets attached to small rachilla. The portion of the peduncle from the junction of the flag leaf to the lowermost node of the rachis is called exsertion. Most of the cultivated varieties posses loose and open ear heads, while a few have dense compact panicles. The density of a panicle is measured by the number of grain per unit length. The number of branches on the rachis, the arrangements of the spikelets on the branches and sub branches, the length of the rachillae of each spikelet and the size of the grain are factors, which go to make up the density of the panicle.

G. Spikelets

The spikelets are carried on small rachillae at the ends of the branches of the panicle. Generally, the spikelets are solitary at the ends of the branches. In a few rare varieties, more than one spikelet and up to seven are clustered together at the ends of branches. The rachilla bearing the spikelet has a little bent or contortion above which is an enlarged annular portion. On this enlarged region, the two lower glumes are placed. In the cultivated varieties, the lower sterile glumes of the spikelets are small and inconspicuous, about 1/4 to 1/5 the size of the lemma. In a few varieties and species, these glumes are as long as the lemma and palea and sometimes even longer. Such long glumes spikelets are called winged spikelets. The lemma and palea are chartaceous, obtuse or acute, and scaberulous on the nerves. The lemma is strongly three to five – nerved whereas the palea is deeply keeled, three – nerved and with the margins overlapped by those of the lemma. The tip of the lemma is muticous, mucronate or awned. The awn character helps in the grouping of varieties. The awn is actually the extension of the central nerve of the lemma. Most of the wild rice growing under adverse conditions and deep water types has fully developed awns.

H. Flower

The spikelet in the rice is single flowered enclosed by the lemma and palea (Fig. 7). The lodicules, which represent the aborted perianth, are two, broad and fleshy. There are six stamens, in two whorls of three each, the filaments being very slender and delicate and having sudden exertion of growth at the time of anthesis. The anthers are linear. The pistil has different shades of purple. The ovary is tricarpellary, single celled, single ovuled and with basal placentation. In some of the varieties, a third style may be present as a very much reduced non stigmatic process. Though the single ovary within a spikelet is a common feature, in certain varieties up to nine ovaries have been recorded. A single grain producing more than one embryo resulting in poly embreyony has also been recorded in a number of cases.

I. Anthesis

The opening of the spikelet or anthesis commences shortly after the panicle emerges from the enclosing leaf sheath. Generally the spikelets at the top of the panicle open first. Anthesis proceeds in an irregular succession towards the base of the panicles. All the spikelets of the panicle complete their anthesis in six or seven days. The maximum anthesis takes place on the second, third and fourth days. Though anthesis is dependent upon the atmospheric conditions such as temperature, humidity, light and rainfall, the spikelets open between 10 AM and 12 Noon in the cultivated varieties.

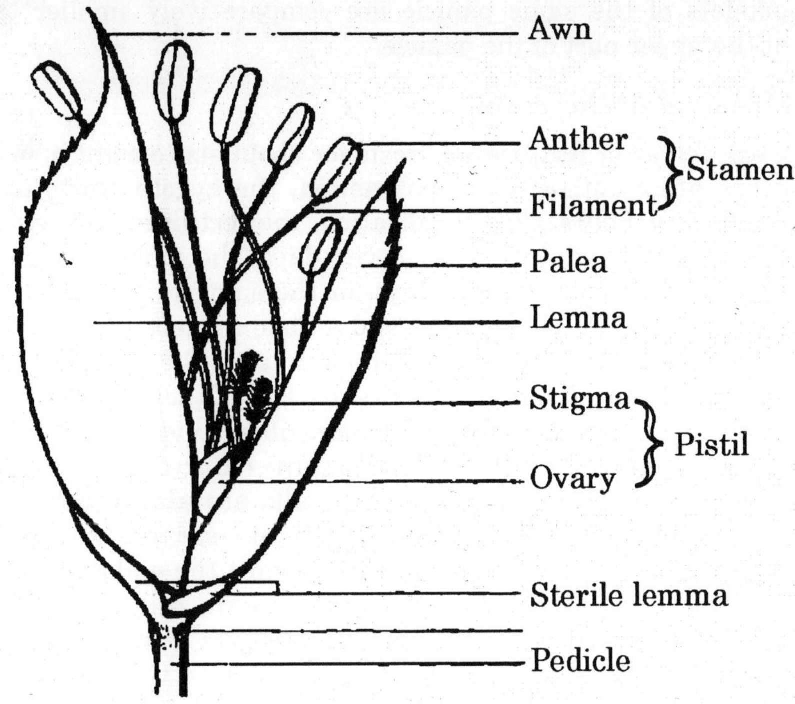

Figure 7. Every grain in the panicle is a flower in the beginning

J. Pollination and fertilization

Along with the lengthening of the filaments, the stigmas also protrude. The stigmas remain receptive from the time of anthesis up to three days. In most cases the plumose stigmas are dusted with pollen when they emerge out as the dehiscence of the anthers takes place in many varieties just at the time of opening. Though self pollination is the rule, nature cross pollination also occurs, varying from 0.1 to 4.0 per cent. In some varieties, cross pollination goes even up to 10 per cent. Pollen grains on the stigma germinate in about three minutes after pollination. About an hour after pollination, the pollen tube enters the embryo sac and fertilization is effected. The ovary with the single ovule inside develops into the fruit or the caryopsis. The ovary wall or pericarp gets fused with the developing seed inside. The endosperm constitutes the bulk of the grain with the embryo laterally placed at its base. The size and shape of the unhusked grain of rice is often an index of the size and shape of the grain inside. Besides the genetical factors, nitrogenous fertilizers such as ammonium sulphate and heavy and rich soils tend to increase the size and, coarseness of rice. The grains formed on the lower spikelets of the same panicle are comparatively smaller than those formed at the upper part of the panicle.

K. Development of the grain

The egg cell after fertilization, rests for about six to eight hours. In eight to ten hours after anthesis and pollination, the zygote starts dividing. In about twenty four hours after fertilization, the fertilized egg cell is divided into two to four cells and in about seven days, the embryo is more or less formed but it takes about twenty days for the embryo to be fully developed. The triple fusion of two polar nuclei and a male nucleus occurs about two to three hours after anthesis and pollination and the primary endosperm nucleus start its division by mitosis. The division of the endosperm nucleus is rather active and in about twenty four hours about seventy endosperm nuclei have been noted. In about forty eight hours, the embryo–sac is lined with one layer of endosperm nuclei and their cell walls are also formed. These new endosperm cells form a layer around the embryo – sac and by active division, new cells of the endosperm are multiplied toward the embryo. In about four days, the embryo–sac is filled with the endosperm cell tissue. It has been reported that the starch grain formation also begins at this stage. After this, the starch grains accumulate in the endosperm cells, which multiply inside, and in about ten days starch grains become compact within the grain.

The first formed layer of the endosperm tissue, which lies outermost, gradually changes its form and becomes the aleurone layer in about seven to nine days after flowering. In the early stages of the ovary development, plenty of starch is present in the cells of the pericarp layers. As the embryo and the endosperm of the fertilized ovule develop into the seed, the starch from the pericarp gets exhausted and is formed in the endosperm. By the time the grain is fully developed, there are only traces of starch in the pericarp, scutellum and embryo. The grain coat develops coloration by the seventh day after fertilization. The pericarp has small quantities of protein up to the sixth day after fertilization after which period, the proteins are translocated to the aleurone layer and the starch cells below. One or two layers of cells adjacent to the aleurone layer are also rich in proteins. In the purple grained varieties there is a maximum development of the aleurone layer while in the white grained types there is a very poor development. Similarly, there appears to be correlation between the coarseness of the grain and the thickness of the aleurone; the coarser the grain, the thicker is the aleurone layer. The richness of the soil and good manuring increase the thickness of the aleurone layer whereas in poor soils the layer is thin. In the young developing grain, the endosperm completely envelops the embryo. But when the grain becomes mature, the embryo is displaced to one side. By this time, reserve food material in the form of carbohydrates, usually starch or dextrins are formed inside the endospermic cells. Glutinous types have more of dextrin than in the ordinary types. The outermost layer of the endosperm is the aleurone layer in which the proteins are formed. In the early stages, there is much inorganic phosphate in the endosperm cells. This decreases gradually until only traces could be detected in the mature grain. A large quantity of organic phosphate accumulates in the cells of the aleurone layer.

L. Dormancy of rice grains

Short duration rice grains generally show no dormancy whereas medium and long duration rice grains require varying dormant or rest periods. The grains of most of the short term rice varieties germinate immediately after harvest and if there is heavy rainfall or loading in the moist field, the grains begin to germinate even on the panicle. In the wild species and types, dormancy is well pronounced. Seed dormancy also called as 'after ripening process' appears to be a genetically controlled factor. However, there is the possibility that this is physiological phenomenon brought about by the enclosing lemma and palea; dehusking in some dormant types helps towards good germination.

M. Germination in rice

The first sign of germination is indicated by the appearance of the plumule and not the radicle. Once the radicle pushes its way, its growth becomes quicker than that of the plumule. The radicle elongates piercing the lemma and coleoptile. The coleoptile also comes out piercing the lemma. After growing a little, the coleoptile is split through by the growing plumule. The radicle grows and forms the seminal root from which other lateral roots arise. If the rice seeds are sown a little deep in the soil, the plumule and coleoptile are raised above by the elongation of the mesocotyl. The secondary roots begin to develop from the lower nodes. The essential food nutrients for the germination of the grain are obtained from the endosperm. The epithelial layer, which is found lining the scutellum towards the endosperm, helps in the absorption of the digested food material from the endosperm to the growing embryo. For this process of digestion of starch of the endosperm into simple nutrients for the growing seedling, the enzymes produced in the embryo play an important part.

N. Structure of the rice grain

The ovule after fertilization develops into the seed with its coats completely fused together with the developing ovary wall or pericarp. Under such circumstances, the fruit coat or the pericarp is fused with the seed coat. The rice grain has the following structure as shown in Fig. 8.

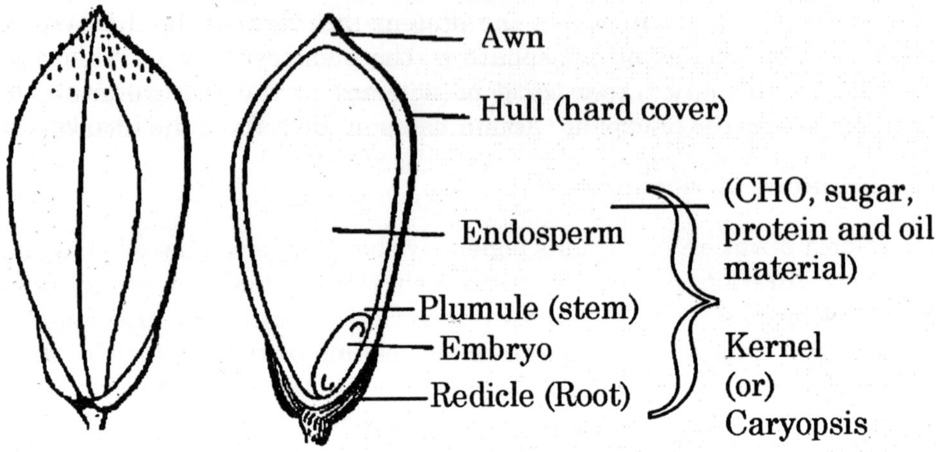

Figure 8. Seed and its parts (Source: Arunachalam, 1984)

1. **The pericarp of fruit coat:** The pericarp is made up of distinct layers of quadrangular cells, which form the epicarp. These cells have slight thickening and are followed by cells, which are much compressed and form the mesocarp consisting of two to three layers. The endocarp is a single layer of tube cells. The colour in the rice grain is found in the pericarp layer in the mature stage.
2. **The seed coats** : Due to the pressure brought out by the developing seed on the pericarp, the testa and tegmen become much pressed down and out of shape. A few layers of such cells below the pericarp can be diagnosed as the integuments of seed coats.
3. **The aleurone layer**: A prominent layer of rectangular cells, which contain protein lies next to the seed coat. This layer is known as the aleurone layer. This layer in rice is not coloured unlike in the case of maize. It has been observed that in coloured varieties of rice, the aleurone layer is thicker than in the white rice varieties. The coarse rice generally has a larger aleurone layer than the finer rice. It has also been found that in poor soils, the aleurone layer is thin and improves the thickness with the fertility of the soil and manure.

4. **The endosperm:** The entire mass of tissue below the aleurone layer is made up of cells, which contain plenty of starch grains and these forms the endosperm.
5. **The embryo** : The scutellum has an upper free part which has a fresh projection known as the ventral scale. Below this upper ventral scale and almost at the middle of the free part there is another outgrowth, which can be called as the 'inner ventral scale', and this inner ventral scale is peculiar to rice embryos only. On the surface of the embryo, this outgrowth along with the epiblast forms a continuous covering around the plumule. The structure between the scutellum and the plumule is the mesocotyl.

CHAPTER 3

RICE GROWTH AND PHYSIOLOGY

The growth duration of the rice plant is 3-6 months, depending on the variety and the environment under which it is grown. During this time, rice completes two distinct growth phases: vegetative and reproductive. The vegetative phase is subdivided into germination, early seedling growth and tillering; the reproductive phase is subdivided into the time before and after heading, i.e., panicle exsertion. The time after heading is better known as the ripening period.

Potential grain yield is primarily determined before heading. Ultimate yield, which is based on the amount of starch that fills the spikelets, is largely determined after heading. Hence, agronomically, it is convenient to regard the life history of rice in terms of three growth phases: vegetative, reproductive and ripening. A 120-days variety, when planted in a tropical environment, spends about 60 days in the vegetative phase, 30-days in the reproductive phase and 30-days in the ripening phase (Fig. 9).

A. Vegetative phase

The vegetative phase is characterized by active tillering, gradual increase in plant height, and leaf emergence at regular intervals. Tillers that do not bear panicles are called ineffective tillers. The number of ineffective tillers is a closely examined trait in plant breeding, since it is undesirable in irrigated varieties, but sometimes an advantage in rainfed lowland varieties where ineffective tillers may be lost due to unfavourable conditions.

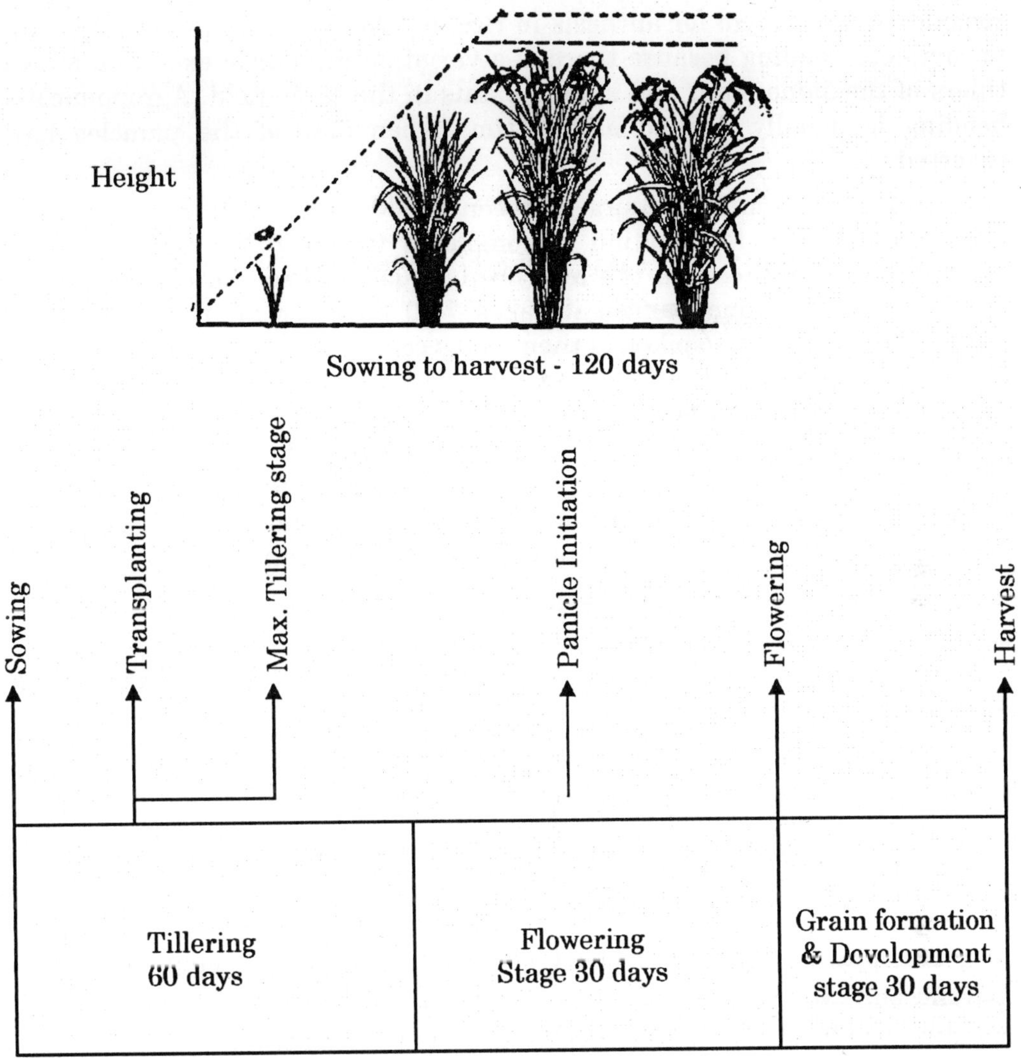

Figure 9. Rice growing stages. E.g. 120 days crop. (Source: Arunachalam, 1984)

B. Reproductive phase

The reproductive growth phase is characterized by culm elongation (which increases plant height), decline in tiller number, emergence of the flag leaf (the last leaf), booting, heading and flowering of the spikelets. Panicle initiation is the stage about 25 days before heading when the panicle has grown to about 1 mm long and can be recognized visually or under magnification following stem dissection. Spikelet anthesis (or flowering) begins with panicle exsertion (heading), or on the following day. Consequently, heading is considered a synonym for anthesis in rice. It takes 10-14 days for a rice crop to complete heading because there is a variation in panicle exsertion among tillers of the same plant and among plants in the same field. Agronomically, heading is usually defined as the time when 50% of the panicles have exserted.

Anthesis normally occurs between 10.00 and 13.00 hr in tropical environments and fertilization is completed within 6 hr. Only very few spikelets have anthesis in the afternoon, usually when the temperature is low. Within the same panicle it takes 7-10 days for all the spikelets to complete anthesis; the spikelets themselves complete anthesis within 5 days. Ripening follows fertilization, and may be subdivided into milky, dough, yellow-ripe and maturity stages. These terms are primarily based on the texture and colour of the growing grains. The length of ripening varies among varieties from about 15 to 40 days. Ripening is also affected by temperature with ranges from about 30 days in the tropics to 65 days in cool, temperate regions such as Hokkaido, Japan, and Yanco, NSW, Australia.

3.1 PHYSIOLOGICAL PHASES

The entire life cycle of the rice plant may be divided into three main phases as indicated above.

1. Vegetative phase (from seed germination to panicle initiation)
2. Reproductive phase (from panicle initiation to flowering)
3. Ripening phase (from flowering to full maturity)

I. Vegetative phase

The phase begins with seed germination, which is signified by the emergence of the radicle or coleoptile in the germinating embryo. During the vegetative phase the plant undergoes the following stages.

a. Seedling Stage

This stage follows seed germination at the seedling stage, the seedling develops seminal and lateral roots. During this stage, the seedling absorbs most of the endosperm.

b. Transplanting stage

Transplanted seedlings undergo this stage. Transplanting is done by many ways viz., manual planting, machine planting-transplanter etc., This stage covers the period from uprooting of seedling to full recovery following transplanting.

c. Tillering stage

The tillering stage follows the seedling stage. It starts with the appearance of the first tiller from axillary bud in one of the lowermost nodes. The increase in number of tiller continues until the maximum tiller number (maximum tillering stage) is reached. Then some tiller die and the number of tiller declines and then levels off. The plants stop producing tillers after the tertiary tillers have been produced.

There are two vegetative stages namely Basic Vegetative Phase (BVP) or active vegetative phase and Photoperiod Sensitive Phase (PSP) or lag vegetative phase. The basic vegetative phase is the minimum period of vegetative growth required by the plant before it will initiate a panicle primordium. The photoperiod sensitive phase is that portion of the vegetative phase in which the flowering date is determined by number of hours of day light per day to which the plant is exposed. Early maturing and non-seasonal varieties have very short photoperiod sensitive phase.

II. Reproductive phase

a. Panicle initiation stage

The reproductive phase begin before the maximum tiller number is reached, at about the time of the highest by the initiation of the panicle primordium of microscope dimension at the growing shoot. This stage reached 60-70 days after seeding but it varies among varieties.

b. Internode elongation and booting stage

The time of occurrence of internode elongation stage differs among varieties. With late maturing varieties the accelerated elongation of the upper internode may begin considerably earlier than the reproductive phase. With early maturing varieties, elongation may begin after panicle initiation. As the young panicle develop, it becomes visible to the naked eye in a few days. The tiny panicle is a transparent structures measuring 1 to 2 mm long with a fuzzy or spongy tip. This occurs about 55 days before the expected date of maturity of a variety and makes the beginning of the booting stage. The increase in the size of panicle and its upward extension inside the flag leaf sheath causes the sheath to bulge.

c. Heading stage

The booting stage is followed by the emergence of panicle tip (heading) out of the panicle flag leaf sheath. Emergence continues until 90% percent of the panicles are out of the sheaths.

d. Flowering stage

Anthesis (blooming or flowering) begins with protrusion of the first dehiscing anthers in the terminal spikelets on the panicle branches. Flowering occurs about 25 days after booting regardless of variety. Flowering continues successively until all spikelets in the panicle's bloom.

III. Ripening phase

The ripening phase occupies a period of 25 to 35 days regardless of variety. The rice grain develops after pollination and fertilization. Grain development is a continuous process and the grain undergoes distinct changes before it fully matures.

a. Milky stage : The contents of the caryopsis (the starchy portion of the grain) are first watery but later turn milky in consistency.

b. Dough stage : The milky caryopsis turns into soft dough and later turns into hard dough.

c. Maturation stage : The individual grain is matured when the caryopsis is fully developed in size and is hard, clear and free from greenish tint. The maturation stage is completed, when more than 90 per cent of grains in the panicles are fully ripened. Mature grains may have a greenish tint if harvested with moisture content of 26 per cent or more. As the grains ripen, the leaves become senescent and turn yellowish in ascending order. The non-functioning leaves and culm tissues are termed dead straw. In some varieties like IR 8, the culm and upper leaves may remain green even when the grains have ripened. The straw dies and the over ripened grains shatter from the panicle. Cracking of the caryopsis within the lemma and palea may occur while the kernels are still attached to the panicle with a resulting loss unmilling recovery and market grade. The short, medium and long duration varieties growth stages are given below.

Duration of Growth stages of rice

	Short duration	Medium duration	Long duration
Growth stage	ADT.36 (105 days)	CO.43 (135 days)	CR.1009 (150 days)
Vegetative			
Germination	3-5	3-5	3-5
Seedling	20-22	25-30	30-35
Active tillering	30	40-45	55-60
Maximum tillering	40	55-60	70-75
Total	40	70	85

Reproductive :			
Panicle initiation	45	75	90
Booting	60	90	105
Heading	75	105	120
Total	30	30	30
Ripening :			
Soft dough	80	110	125
Hard dough	90	120	135
Mature grain	105	135	150
Total	30	30	30

3.2 PHYSIOLOGICAL PROCESSES

A high grain yield of any crop can be achieved only when a proper combination of variety, environment and agronomic practices are obtained. Understanding the physiological processes involved in grain production such as vegetative growth, formation of storage organs and grain filling helps to determine the best combination of the above three factors and also suggest what improvement can be made to achieve a further increase in grain yield.

3.2.1. Dry matter production and grain yield

The economic yield (Y) is the function of Harvest Index (HI) and total dry matter production (TDMP) i.e. biological yield and normally expressed as Y=HI x TDMP. This implies that the yield can be increased either by improving dry mater production or harvest index. Leaf area (source size) and photosynthetic efficiency (source strength) are the two parameters, which determine dry matter production. Light interception by a canopy of leaves is strongly influenced by size, shape and angle of leaves. In terms of dry matter produced in rice, leaf area index (LAI), photosynthetic rate and leaf angle is the major determinants of crop growth. Of these three, LAI is the most variable and can be manipulated through agronomic practices. Increasing LAI raises dry matter production but this relationship does not hold indefinitely because of mutual shading of leaves thereby decreasing mean photosynthetic rate per unit area. Respiration also increases linearly with increasing LAI. Since TDMP is the balance of photosynthesis and respiration, there exists a possibility of arriving a critical or optimum LAI, above which any increase of LAI reduces the photosynthesis or increases respiration thereby affecting dry matter production. Further, the leaf angle also finds a place in determining the photosynthesis and in rice, erect leaves were found to increase the leaf surface area, thereby permitting more even distribution of incident light.

Total dry matter production is the integral part of crop growth rate over the entire growth period, and it is related to grain yield by the harvest index and this is difficult to change for a given variety under most conditions. Harvest index tends to be lower as total dry matter production increases.

3.2.2. Yield capacity

In rice the storage organs are formed after a period of vegetative growth and before panicle emergence. The yield capacity can be formulated as

Yield capacity = No. of grains per unit area X Potential size of grain

The number of grains per unit area is derived from the number of panicles and number of spikelets per panicle. Beyond a certain plant density, there always exists a negative correlation between components of yield capacity and thus it tends to become constant under a given set of conditions. The potential size of rice grain is physically restricted by the size of the hull and the 1000-grain weight is a quite stable varietal character. However heavy shading and CO_2 enrichment are found to affect grain weight to some extent.

3.2.3 Panicle development

The development of panicle is well studied in rice. The differentiation of panicle neck node starts about 32 days before flowering. The differentiation of spikelets proceeds for the period from 13-15 days before flowering, which determines the maximum number of spikelets. Degeneration of formed spikelets occurs 13 days before flowering. The hull size is determined by 1 week before flowering. These developmental stages are more variable as they are influenced by environmental factors. Degeneration of spikelets is greatly affected by nitrogen status at differentiation stage of spikelets and by solar radiation at the reduction division stage. The number of grains is found to be positively correlated with the nitrogen absorption during the end of spikelet initiation stage or by flowering. A close correlation exists between LAI at flowering and number of spikelets as LAI determines the formation of yield capacity or production of carbohydrate for grain filling or both. The number of spikelets is affected by cultural practices (spacing and nitrogen application), growth characters (tillering) and climatic conditions (solar radiation, temperature).

3.2.4. Duration of panicle growth

The length of growth period affects the growth of panicle. The variation in growth duration is largely due to differences in vegetative growth period. There is however a positive correlation between growth duration and the length of period from panicle initiation to heading. Thus an early maturing variety has a relatively short period for panicle growth.

3.2.5. Grain filling

Carbohydrates such as sugar, starch and other polysaccharides reach a maximum concentration in plant vegetative parts around heading. The concentration begins to decrease as ripening proceeds and may rise again lightly near maturity. The accumulated carbohydrate has three possible functions in grain production.

- To supply a portion of the grain carbohydrate.
- To support sustained growth under variable weather conditions.
- To stabilize grain yield under favourable weather conditions during ripening. The contribution of accumulated carbohydrates to grains can be arrived by

$$\text{Contribution (\%)} = \frac{\text{CHO present in veg. parts at flowering} - \text{CHO present in veg. parts at maturity}}{\text{CHO in grain}} \times 100$$

This does not take into account the respiration losses during grain filling, which also uses carbohydrates and tends to give an over estimated contribution. Under normal field conditions, around 60-70% of stored carbohydrate is translocated to grains, 20% is respired and rest stays in vegetative parts. The accumulated carbohydrate translocated into the grain equals to only 20% of the grain carbohydrate. The possible contribution of stored carbohydrate to grain is large only at low nitrogen levels or at low light intensities as they affect the current photosynthesis. Since large amounts of nitrogen are applied to achieve higher yields, the grain carbohydrate of high yielding rice varieties is most derived from photosynthesis after heading.

3.2.6 Current photosynthesis

The contribution of photosynthesis during ripening by different plant parts to the grain are based on the potential photosynthetic activity, longevity of the tissue during ripening and light environment in a crop canopy. The photosynthetic rate of different plant parts vary and in rice, the net photosynthesis of ear and leaf sheath is very low when compared with leaf blades and can be neglected. The potential net photosynthesis of leaves is up to 90% of the total. All leaves from the flag leaf down to the third leaf from the top export assimilates to the panicle and lower leaves send their assimilates to roots. The longevity of the green tissue must be considered when assessing the photosynthetic contribution of different plant parts during ripening. The panicle becomes yellow relatively early in ripening and leaves remain green much longer. The flag leaf remains green and active until maturity. At later stages of maturity however photosynthesis may not make much of a net contribution to grain production because rapid grain growth occurs earlier and slows down towards maturity. The light environment of different plant parts in a crop canopy is extremely important for determining real photosynthesis. In canopy with flat leaves the whole area of the top is more exposed to sunlight than lower leaves. But in a canopy with erect leaves, the tips of lower leaves may receive more sun light than basal part of the flag leaf, thus contributing to total photosynthesis of crop canopy.

3.2.7 Plant characters in relation to yielding ability

The close association between certain morphological characters of rice varieties and yielding ability led to the development of new plant – type.

A. Plant height

It is the predominant factor determining the nitrogen response of rice plant. It determines the lodging behaviour thus deciding yield. Lodging disturbs leaf display resulting in shading thus increasing the percentage of unfilled grains. Shorter culm reduces the respiratory loss. On the other hand, tall stature facilitates light penetration. Thus an optimum plant height must exist and in rice, presently short stature is preferred because of greater yield loss due to lodging.

B. Leaf characters

The desirable characters are a thick, short and small and erect leaves as they are found to be associated with the net photosynthesis. Thick leaves have higher photosynthetic rate per unit area. Short and small leaves favour even distribution of leaves and erect leaves increases the surface area exposed to solar incidence.

C. Tillering habit

Upright tillers are favoured as they permit greater penetration of light. High tillering is essential for transplanted rice, as it permits faster leaf area development.

D. Panicles

As the panicles determine the ultimate yield of the crop, the panicles should be having high ripening percentage and high grain to straw ratio (Harvest Index). Based on the present knowledge of crop photosynthesis, mineral nutrition and yield components, high rice yield needs the following requisites.

1. The variety must have a short, stiff culm.
2. Erect upper leaves and drooping gradually towards the lower region (canopy)
3. An optimum LAI of 5- to achieve max photosynthesis during ripening and maintenance of greenness until completion of grain growth.
4. Exposure to high solar radiation during later phases of growth.
5. Yield components to meet the targeted yield
6. Application of essential nutrients (especially nitrogen) and other micronutrients like Zn & Cu.

3.2.8. Factors limiting grain yield

Rice yield is determined by the interactions between photosynthesis and yield capacity. But their interactions with climatic factors make it difficult to determine which usually limits grain yield. But under favourable climate and with an abundant supply of nutrients, it is more likely that the yield capacity limits the grain yield. Evidences indicate that change in photosynthesis alter yield capacity by

a. Degenerating spikelets and reducing grain size due to low light intensity
b. Shading during panicle initiation to heading.

This implies that reduced photosynthesis due to low intensity decreases the spikelet number per plant. Hormonal effects also seem to be involved in determination of yield. The partitioning of assimilates between developing panicles and leaves are probably under hormonal control. Greater distribution may produce larger panicles and understanding the mechanism of partitioning and finding the means to control it may change the yield ultimately. Thus attempts to increase yield capacity and to increase photosynthetic capacity appears to be equally important for increased rice yield.

CHAPTER 4

RICE- ECOSYSTEMS

4.1 EVOLUTIONARY PATHWAY

Rice was initially grown as a rainfed crop in low-lying areas where rainwater could be retained. Such areas were located in marshy, but flood-free sites around river beds, as found in Honan and Shenxi Provinces of north China and at larger sites between small rivers, as represented by the He-mu-du site in east China. Early community efforts led to irrigation or drainage projects. The earliest of such activities in the historical record were flood-control efforts in the Yellow River area under Emperor Yu at about 2000 BC. Irrigation works, including dams, canals, conduits, sluices, and ponds, were in operation during the Yin period (1400 BC). A system of irrigation and drainage projects of various sizes was set up during the Chou Dynasty. Large-scale irrigation works were built during the Warring States period (770-21 BC). By 400 BC, "rice [Tao] men" were appointed to supervise the planting and water management operations. The famous Tu-Cheng-Yen Dam was constructed near Chendu in Sichuan (Szechuan) Province about 250 BC, which made western Sichuan the new rice granary of China.

Further developments during the Tang and Sung dynasties led to extensive construction of ponds as water reservoirs and of dams in a serial order to impound fresh water in rivers during high tides. Dykes were also built around lakeshores to make use of the rich alluvial soil and the importance of water quality was recognized. Broadcasting rice grains into a low-lying site was the earliest method of planting and can still be seen in the Jaipur tract of India and many parts of Africa. In dry soils, the next development was to break through the soil with implements, mainly the plough, whereas in wetland culture, it was to build levees (short dikes or bunds) around a field in order to impound the water. In the latter case, such an operation also facilitated land leveling and soil preparation by puddling the wet soil in repeated rounds. The next giant step came in the transplanting (insertion) of young rice seedlings into a well-puddled and leveled wet field. Transplanting requires the raising of seedlings in nursery beds then pulling them out from those beds, bundling and transporting to the field where the seedlings are thrust by hand into the softened wet soil. A well-performed transplanting operation also requires seed selection, the soaking of seeds prior to their initial sowing, careful management of the nursery beds, and proper control of water in the nursery and in the field. The transplanting practice began in the late Han period (AD 23-270) and subsequently spread to neighbouring countries in Southeast Asia as a package comprised of the water buffalo, plough, and the spike-tooth harrow. Transplanting is a labourer-consuming operation. Depending on the circumstances, between 12 and close to 50 days of an individual's labourer is required to transplant one hectare of rice land. On the other hand, transplanting provides definite advantages in terms of a fuller use of the available water, especially during dry years, better weed control, more uniform maturation of the plants, higher grain yield under intensive management, and more efficient use of the land for rice and other crops in cropping sequence.

Despite these advantages, however, in South Asia the transplanting method remains second in popularity to direct seeding (broadcasting or drilling) due to operational difficulties having to do with farm implements, water control, and labour supply (Chang, 1976). Variations of the one-step transplanting method were (1) to interplant an early maturing variety and a late one in alternating rows in two steps (once practiced in central China) and (2) to pull two-week-old seedlings as clumps and set them in a second nursery until they were about one meter tall. At this point, they were divided into smaller bunches and once more transplanted into the main field. This method, called double transplanting, is still practiced in Indochina in anticipation of quickly rising floodwaters and a long rain season.

Weeds in rice fields have undoubtedly been a serious production constraint since ancient times. The importance of removing weeds and wild rice plants was emphasized as early as the Han Dynasty. Widely practiced methods of controlling unwanted plants in the southern regions involved burning weeds prior to ploughing and pulling them afterward, complemented by maintaining proper water depth in the field. Fallowing was mentioned as another means of weed control, and midseason drainage and tillage has been practiced since Eastern Chou as an effective means of weed control and of the suppression of late tiller formation by the rice plant. Different tools, mainly of the hoe and harrow types, were developed for tillage and weed destruction. Otherwise, manual scratching of the soil surface and removal of weeds by hand was practiced by weeders who crawled forward among rows of growing rice plants. Short bamboo tubes tipped with iron claws were placed on the fingertips to help in the tedious operation. More efficient tools, one of which was a wooden frame shape with a long handle and rows of spikes beneath it, appeared later. This was surpassed only by the rotary weeder of the twentieth century.

Insect pests were mentioned in Chinese documents before plant diseases were recognized. The Odes (Sixth century BC) mentioned stem borers and the granary moth. During the Sung Dynasty, giant bamboo combs were used to remove leaf rollers that infest the upper portions of rice leaves. A mixture of lime and tung oil was used as an insect spray. Kernel smut, blast disease and cold injury during flowering were recognized at the time of the Ming Dynasty. Seedling rot was mentioned in the Agricultural Manual of Chen Fu during South Sung (Chinese Academy of Agricultural Sciences, 1986). The relationship between manuring and increased rice yield was observed and recorded more than two thousand years ago. The use of compost and plant ash was advocated in writings of the first and third centuries. Boiling of animal bones in water as a means to extract phosphorus was practiced in Eastern Han. Growing a green manuring crop in winter was advised in the third century. The sixth century agricultural encyclopedia Chi-Min-Yao-Shu (Ku undated) distinguished between basal and top dressings of manure, preached the use of human and animal excreta on poor soils, and provided crop rotation schemes.

Irrigation practices received much attention in China because of the poor or erratic water supply in many rice areas. Therefore, the labourer inputs on water management in Wushih County of Jiangsu Province in the 1920s surpassed those of weeding or transplanting by a factor of two, whereas in monsonal Java, the inputs in water management were insignificant. Because of the cooler weather in north China, irrigation practices were attuned to suitable weather conditions as early as the Western Han: Water inlets and outlets were positioned directly opposite across the field so as to warm the flowing water by sunlight during the early stages of rice growth. Elsewhere, the inlets and outlets were repositioned at different intervals in order to cool the water during hot summer months. The encyclopedia Ch'i-Min-Yao-Shu devoted much space to irrigation practices:

Watering should be attuned to the weather; the fields should be drained after tillage so as to firm the roots and drained again before harvesting.

In order to supplement the unreliable rainfall, many implements were developed to irrigate individual fields. The developments began with the use of urns to carry water from creeks or wells. The urn or bucket was later fastened to the end of a long pole and counterbalanced on the other end by a large chunk of stone. The pole was rested on a stand and could be swung around to facilitate the filling or pouring. A winch was later used to haul a bucket from a well. The square-pallet chain pump came into use during the Eastern Han; it was either manually driven by foot pedaling or driven by a draft animal turning a large wheel and a geared transmission device. The chain pump was extensively used in China until it was replaced by engine-driven water pumps after the 1930s. The device also spread to Vietnam. During hot and dry summers, the pumping operation required days and nights of continuous input. Other implements, such as the water wheel in various forms, were also used.

Although deepwater rice culture in China never approached the scale found in tropical Asia, Chinese farmers used floating rafts made of wooden frames and tied to the shore so as to grow rice in swamps. Such a practice appeared in Late Han, and the rafts were called feng (for frames) fields. Many rice cultivars are capable of producing new tillers and panicles from the stubble after a harvest. Such regrowth from the cut stalks is called a ratoon crop. Ratooning was practiced in China as early as the Eastern Tsin period (AD 317-417), and it is now an important practice in the southern United States. Ratooning gives better returns in the temperate zone than in the tropics because the insects and diseases that persist from crop to crop pose more serious problems in the tropics.

Seed selection has served as a powerful force in cultivar formation and domestication. Continued selection by rice farmers in the field was even more powerful in fixing new forms; they used the desirable gene-combinations showing up in the plantings to suit their farmer's different needs and fancies. The earliest mention of human-directed selection in Chinese records during the first century BC was focused on selecting panicles with more grains and fully developed kernels. Soon, varietal differences in awn color and length, maturity, grain size and shape, stickiness of cooked rice, aroma of milled rice, and adaptiveness to dry land farming were recognized. The trend in selection was largely toward an earlier maturity, which reduced cold damage and made multiple cropping more practical in many areas. The encyclopedia Ch'i-Min-Yao-Shu advised farmers to grow seeds in a separate plot, rotate the seed plot site in order to eliminate weedy rice, and select pure and uniformly colored panicles. The seeds were to be stored above ground in aerated baskets, not under the ground. Seed selection by winnowing and floatation in water was advised. Dry land or hill rice was mentioned in writings of the third century BC. During Eastern Tsin, thirteen varieties were mentioned; their names indicated differences in pigmentation of awn, stem and hull, maturity, grain length, and stickiness of cooked rice. Varieties with outstanding grain quality frequently appeared in later records. Indeed, a total of 3,000 varieties was tallied, and the names were a further indication of the differences in plant stature and morphology, panicle morphology, response to manuring, resistance to pests, tolerance to stress factors (drought, salinity, alkalinity, cool temperatures, and deep water), and ratooning ability. The broad genetic spectrum present in the rice varieties of China was clearly indicated.

Harvesting and processing rice is another labourious process. The cutting instruments evolved from knives to sickles or scythe. Community efforts were common in irrigated areas, and such neighbourly cooperation can still be seen in China, Indonesia, the Philippines, Thailand, and other countries. Threshing of grains from the panicles had been done in a variety of ways: beating the bundle of cut stalks against a wooden bench or block; trampling by human feet or animal hoofs; beating with a flail; and, more recently, driving the panicles through a spiked drum that is a prototype of the modern grain combine. Other important post harvest operations are the drying of the grain (mainly by sun drying), winnowing (by natural breeze or a hand-cranked fan inside a drum winnower), dehusking (dehulling), and milling (by pestle and mortar, stone mills, or modern dehulling and milling machines). Grains and milled rice are stored in sacks or in bulk inside bins. In Indonesia and other hilly areas, the long-panicle Javanica rice are tied into bundles prior to storage.

To sum up the evolutionary pathway in wetland rice cultivation on a worldwide scale, cultivation began with broadcasting in rain-fed and unbunded fields under shifting cultivation. As the growers settled down, the cultivation sites became permanent fields. Then, bunds were built to impound the rainwater, and the transplanting method followed. As population pressure on the land continued to increase, irrigation and transplanting became more imperative. The entire range of practices can still be seen in the Jaipur Tract and the neighbouring areas. The same process was retraced in Bang Chan (near Bangkok) within a span of one hundred years. In the twentieth century, further advances in agricultural engineering and technology have to do with several variations in seeding practices that have been adopted to replace transplanting. Rice growers in the southern

United States drill seed into a dry soil. The field is briefly flushed with water and then drained. The seeds are allowed to germinate, and water is reintroduced when the seedlings are established. In northern California, pregerminated seeds are dropped from airplanes into cool water several inches deep. The locally selected varieties are able to emerge from the harsh environment.

Recently, many Japanese farmers have turned to drill-plant pregerminated seed on wet mud. Anti-oxidant is applied to the seed before sowing so as to obtain a uniform stand of plants. For the transplanted crop, transplanting machines have been developed not only to facilitate this process but also to make commercial raising of rice seedlings inside seed boxes a profitable venture. As labourer costs continue to rise worldwide, direct seeding coupled with chemical weed control will be the main procedures in the future. For deepwater rice culture, rice seeds are broadcasted on dry soil. The seeds germinate after the monsoon rains arrive. The crop is harvested after the rains stop and the flooding water has receded. For dry land rice, seeds are either broadcasted, drilled, or dropped (dibbled) into shallow holes dug in the ground. Dibbling is also common in West Africa. Dry land (hill or upland) rice continues to diminish in area because of low and unstable yield. It has receded largely into hilly areas in Asia where tribal minorities and people practicing shifting cultivation grow small patches for subsistence.

4.2 SYSTEMS OF RICE CULTIVATION IN INDIA

Rice is grown on a wide variety of climate, soil and hydrology regimes, which are ground into various '**rice ecosystem**' 'or 'rice agroecological zones'. In India, rice is grown in all the states and union territories; from Srinagar in the north to Kanyakumari in the south and from Arunachal Pradesh in the east to Rajasthan in the west. The rainfall varies from about 700 mm to 2000 mm per annum and above, received mostly in a few months (June – September) by southwest monsoon. The eastern coastal area of the peninsular India receives rains also during October – February by northeast monsoon, rather more in this part of the year than in June – September as in the rest of the country. Rice soils also vary considerably, from loamy sands in Punjab to heavy clays in several parts of the country (Rajendra Prasad, 1999). A very broad based classification of rice environment or ecosystems used in India includes five rice ecosystems, namely, irrigated wet season (kharif), irrigated dry season (rabi), rainfed lowland, rainfed upland and deep water (Table 5) and these rice ecosystems are subdivided in to several sub eco systems based on hydrological/ soil factors (Table 6) (IRRI, 1985).

Table 5. Broad based rice-ecosystems in India and their important features

Ecosystem	Region	Yield t ha^{-1}	Cultural Practices	Constraints to Production
Irrigated wet season (kharif)	Punjab, Haryana Western UP & Southern India	2-5	bunding, pudding, transplanting	1. Occasional drought if tube-well water not available. 2. Micronutrients deficiencies 3. Weeds 4. Moderate diseases/pest problem
Irrigated dry season (rabi)	Eastern India (Less than 1 million ha) & Southern India	3-5	-do-	1. Occasional drought if irrigation water not available 2. Micronutrient deficiencies 3. Weeds
Rainfed lowland (max. water depth tillering to flowering 30cm – 1 m)	East India (40-50% of rice area in this region) Southern India (5% of total area in this region)	less than 2	-do-	1. Flood/drought 2. Heavy incidence of diseases and pests 3. Weeds-too much 4. Problem in fertilizer application
Rainfed upland (Max. water depth tillering to flowering (0-30 cm)	Eastern India (22-26% of rice area in this region and 90% of India's area under rainfed conditions)	1	Often unbunded broadcast	1. Drought/floods 2. Incidence of diseases & pests 3. Weeds-too-much 4. Poor soil fertility
Deep water (max water depth from tillering to flowering 1m – 6m)	Eastern India (6-9% of rice area in India) 12% of rice area in eastern U.P.	1	Broadcast in dry unbunded fields before onset of monsoon	1. Diseases & pests 2. Unavailability of high yileding varieties 3. Poor soil fertility 4. Fertilizer application problematic

Source : Widawsky and O'Toole (1996); Ramasamy *et al.,* (1996).

*Despite the upper limit of 30 cm, rainfed uplands have no standing water for most of the rice growth period for more than 48 hours.

Table 6. Terminology for rice-growing environments

Environment	Sub-ecosystem
Irrigated	Irrigated, with favourable temperature
	Irrigated, low temperature, tropical zone
	Irrigated, low temperature, temperature zone
Rainfed lowland	Rainfed shallow, favourable
	Rainfed shallow, drought-prone
	Rainfed shallow, drought and submergence-prone
	Rainfed shallow, subermergence-prone
	Rainfed medium deep, waterlogged
Upland	Favourable upland with long growing season
	Favourable upland with short growing season
	Unfavourable upland with long growing season
	Unfavourable upland with short growing season
Deepwater	Deepwater
	Very deepwater
Tidal wetlands	Tidal wetlands with perennially fresh water
	Tidal wetlands with seasonally or perennially saline water
	Tidal wetlands with acid sulphate soils
	Tidal wetlands with peat soils

Source : IRRI (1984)

Table 7. Geographical deliniation ofagroecologicalzones (AEZ's) in Asia.

Agroecological zone	Geographical boundaries
Tropics*	
Warm sermi-arid tropics (AEZ 1)	South-western India (A.P., Karnataka, Tamil Nadu, Maharashtra, Gujarat).
Warm sub humid tropics (AEZ 2)	Thailand, Myanamar, Eastern India (MP, Orissa, Bihar).
Warm humid tropics (AEZ 3)	Indonesia, Malaysia, Philippines, Vietnam, Cambodia, Laos, Srilanka, Banladesh, parts of India (Assam, NE States, W.B., Kerala).
Sub-tropics	
Warm sermi-arid subtropics (AEZ 4)	Pakistan, Parts of India (Rajasthan, Haryana), part of China (Helong, Laioning, Jilin, Tianjin, Sandong, Hebei).

Warm subhumid subtropics (AEZ 5)	Northwestern India (UP, Punjab), Nepal, Parts of China (Jiangshu, Anhui, Hubei, Sichuan, Henan, Ghizhou, Yunan), North Korea, South Korea.
Warm humid-subtropics (AEZ 6)	Parts of China (Shanghai, Zehjiang, Fujian, Jianxi, Hunan, Guandong, Guanxi), Taiwan.
Cool subtropics temperate (AEZ 7)	Parts of India (HP, J & K), Parts of zone, China (Beijing, Shani, Gansu, Ningia, Xinjiang, Quinghai).

Note : Exclude AEZ 4 regions covering Middle East and Central Asia

*Arid – Length of growing period (LGP) <75 days, Semiarid – LGP75-180 days, Sub-humid-LGP 180-270 days, Humid >270 days; Tropics- mean monthly temperature (MMT) > 18°C for all months, Subtropics-MMT 5-18°C for one or more months, Temperate MMT <5°C for one or more months; Warm – Daily mean temp. during the growing period of >20°C, Cool – Daily temp. during the growing period of 5-20°C.

Source : TAC (1990) & Garrity *et al.,* (1996).

The Technical Advisory Committee (TAC) of the Consultative Group of International Agricultural Research (CGIAR) has suggested seven agroecological rice zones in Asia and Southeast Asia (Garrity *et al.*, 1996). The major criteria are -

(i) The moisture regime based on the length of growing period (LGP) determine from the data on the seasonal distribution of rainfall and PET at different times of the year (arid LGP<75-180days; sub humid LGP 180 – 270 days; humid LGP > 270days) and

(ii) The thermal regime based on mean monthly temperature (MMT) (tropics > 18^{0}C for all months; subtropics 5-18^{0}C for 1 or more months; temperate < 5^{0}C for 1 or more months; warm daily mean temperature (DMT) during the growing period (DMT GP>20^{0} C and cool DMT GP 5-20^{0}C). The details are given in the Table 8.

The general classifications of rice environments can be useful only for mega or at the best for macro level analysis. For developing location specific agronomy package, one needs to prioritize the research and for that purpose micro level analysis is a must. Garrity *et al.*, (1996) have further divided each rice – growing environments into sub ecosystems based on duration of rice varieties grown (long, medium, short), water balance (favourable, drought prone) in the region and soil fertility class (fertile, adverse).

4.2.1. Factors

The systems of rice cultivation in a region largely depends on factors such as situation of land, soil type, irrigation sources, availability of labourers and intensity and distribution of rainfall etc.

i. Situation of land

If the land is situated at higher altitude where water stagnation will not be there, upland system of cultivation is followed, and where the land is at lower level and water stagnation is common, low land system of cultivation is adopted.

ii. Soil type

In low land situation, the soil type varies from loamy or alluvial to heavy clay soils and in upland condition, the soil available will be sandy loam.

iii. Irrigation sources

If the irrigation source is canal, low land rice cultivation is adopted and if the irrigation source is rainfall or well water or tank, the system of cultivation is upland.

iv. Labour availability

Whenever the labour is available in plenty at low cost, wetland cultivation is adopted and if labour availability is scarce, upland cultivation is followed.

v. Rain fall

Areas receiving high amount of rainfall with more intensity will have upland cultivation.

4.2.2. Rice ecosystems in Asia

Unlike many other crops, rice is grown in diverse land and water management systems. Therefore, tillage practices for land preparation in rice vary considerably starting with the cultivation practices. The cultivation practices for rice differ with the rice ecosystems and are broadly categorized as follows:

4.2.2.1. Irrigated rice ecosystems

Irrigated rice ecosystems are those areas that have bunded, puddled fields with assured irrigation for one or more crops per year. This ecosystem, which is concentrated mostly in the humid and sub humid subtropics, accounts for about 80 million ha (55%) of the harvested rice area and contributes 76% of global rice production, with average yields varying from 3 to 9 t ha^{-1}. There are three rice environments (Table 6) in this ecosystem, namely,

i. Irrigated with favourable temperature,

ii. Irrigated with unfavourable temperature-tropical zone and

iii. Irrigated with unfavourable temperature-temperate zone.

Improved rice cultivars that have been developed for irrigated rice lands are of short duration and are responsive to nitrogen fertilizer, and incorporate resistance to several biological stresses and some tolerance to adverse soils. It is characterized by high cropping intensity and intensive use of agrochemicals. Production constraints in low (less than 3 t ha^{-1}) and medium yield (3-5 t ha^{-1}) areas are delayed plantings, shortage of labour, use of low to medium potential varieties, low input rates, poor input management, poor weed management, inadequate water supply, unavailability of irrigation at one or more critical stage, salinity/alkalinity problem; associated poor drainage and environmental stresses. Problems associated with high production (more than 5 t ha^{-1}) areas are yield instability due to pests, multiple cropping systems – delaying transplanting and environmental degradation

4.2.2.2. Rainfed lowland rice ecosystems

Rain fed lowland rice grows in budded fields that are flooded for at least part of the cropping season to water depths that do not exceed 50 cm for more than ten consecutive days. Rain fed lowlands are characterized by lack of water control, with floods and drought being potential problems. About 40 million ha (25%) of the global rice area is rain fed, contributing 18% of the total production. Rainfall is often erratic, so that conditions are diverse and unpredictable. Typically, traditional photoperiod-sensitive cultivars, which do not respond to fertilizer, are grown. No water harvesting/conservation facilities are available on the farm and there is no irrigation water available. Five major environments have been identified within the rainfed lowland ecosystem. These are

i. Rainfed shallow favourable environments: No drought or waterlogging

ii. Rainfed shallow drought prone environments: These areas have frequent and severe water-deficit problems and are drought prone. The rainy season may be short (90-100 days) as in central and eastern India or long with erratic distribution as in Bangladesh and Thailand.

iii. Rainfed shallow drought and submergence prone environments: Areas subjected to both drought and waterlogging.

iv. Rainfed shallow submergence prone environments: Frequent short-term flooding that may damage or destroy the crop.

v. Rainfed medium deep rice environment: Water 25-30 cm deep may stand for a substantial period of growing season.

4.2.2.3. Upland rice ecosystems

Of the 148 m. ha total rice area, this ecosystem covers about 18 million hectares of which 64 per cent is located in Asia, 25 per cent in Latin America and 11 per cent in Africa. However, the area of the upland rice ecosystem is much larger than the area under rice, because rice is grown in rotation with other crops. Yields are typically I t ha^{-1}. In Asia, most upland rice is grown on rolling and mountainous land with slopes varying from 0% to more than 30%. Dry soil preparation and direct seeding in fields that are generally unbunded are characteristic of the upland rice ecosystem. Surface water does not accumulate for any significant time, although in some countries, such as in eastern India and Bangladesh, upland rice fields are frequently bunded to save scarce water. Shifting cultivation in common in Indonesia, Laos, the Philippines, northern Thailand and Vietnam. These are rainfed rice areas with bunded or unbunded fields. Soils are well drained and therefore no surface water accumulates. Soils are low in fertility and weeds are a major constraint; in many regions soils tend to be acid. In the hilly regions, mono cropping of rice is practiced and it is detrimental to soil and environment. Four major rice environments identified are:

i. Favourable uplands with long growing season,

ii. Favourable environments with short growing season,

iii. Unfavourable environment with long growing season and

iv. Unfavourable environment with short growing season.

Major constraints are drought, weeds, low rates of inputs, poor input management, poor soil fertility and environmental stresses.

Since water is the major constraint in upland rice ecosystem, a major part of area under rainfed upland ecosystem is direct seeded, generally at the onset of rain. Most of the crop establishment operations are done when temperatures are high (generally above 30°C) and days are long and despite low crop canopy PET is quite high. This is called as ***Moist I period*** (rainfall is less than PET but more than ½ PET). After this period, heavy y rains come and rainfall is more than PET; this period is called as ***Humid period.***

Humid period is again followed by ***Moist II period*** (rainfall is less than PET but more than ½ PET). During this period, temperature is lower and days are shorter than during Moist I and humid periods.

4.2.2.4. Deep water and tidal wet land rice ecosystems

About 13 million hectares are under deep water and tidal wetland rice, mostly in the back swamps of the flood plains and deltas and on the slopes of natural levees in South Asia. These are the regions where rice grows under rainfed dry land or shallow flooding conditions for 1-3 months and is then subjected to flooding with water depth of more than 50 cm for a month or longer. These areas are found in the river basins of Ganges and Brahmaputra in India and Bangladesh, the Irrawaddy in Myanmar, the Mekong Delta of Vietnam and Cambodia and the Chao Pharaya in Thailand. When water levels are over 100 cm, the crop is known as floating rice, while Tidal wetlands occur where water levels are fluctuating and are influenced by tide in coastal belts. Floating rice yields range from 0.5 to about 2.5 t h^{-1} , while deep water rice may produce up to 4 t ha^{-1}. Despite these low yields, these flood-prone areas support over 100 million people. The major problems are rice is the only crop that is possible; traditional low yield potential varieties; deep flooding and droughts both are possible; variable flooding patterns; tidal wetlands are subjects to salinity and other soil stresses (acid, acid sulphate and peat soils); susceptibility to stem borer; no resistance available and nematode problem (ufra disease).

4.2.3 Rice ecosystems

Rice ecosystems can be classified in to various types based on the climatic conditions, resource availability, method of planting and cropping systems. The following are the some of the most important classifications followed.

I. Based on the resource availability (especially water),

The ecosystem can be classified into

a. Irrigated rice ecosystem

Lowland

Upland (semidry)

b. Rainfed

Shallow (0-30 cm water depth)

Intermediate (30-100 cm water depth)

c. Deep water

II. Based on the method of planting

(a) Transplanted rice

(b) Direct sown rice

III. Based on cropping system

(a) Single crop rice ecosystem

(b) Double crop rice and multi crop intensive rice ecosystem

The area and production of rice in India under different rice ecosystems is given below.

Ecosystem	Area (% of total area under rice)	Production (% to total)
1. Single crop irrigated	13	19
2. Double crop irrigated	19	31
3. Upland	10	5
4. Shallow rainfed	35	33
5. Medium deep rainfed and deep water	15	8
6. Floating rice	8	4

The above Table indicated that, shallow rain fed ecosystem accounts for 35% of total area under rice cultivation in India, but is able to provided only 33% of total rice production. Intensive rice cropping either under single crop rice system or double crop rice system could able to provide more rice output than their per cent area occupation. Hence there is a huge potential exists in shallow rainfed, medium deep rainfed, deep water and floating rice ecosystems for yield improvement.

4.2.4 Agro-climatic zones of rice cultivation

a. Regions of rainy season of six months or more

This regions consists of Kerala and North east India and this part of the country receives more quantity of rainfall during south west monsoon season. This occupies about 3.62 million ha of area under rice, accounts for 8.4% of total area and yields 5.16 m.t (7.1%). The average yield levels are lesser than National average (1433 kg/ha).

b. Regions of rainy season of four to five months

i. West coast of India– 1 million ha and average yield is 1,400 kg/ha

ii. North eastern India-parts of Assam, sub-Himalayan, West Bengal, eastern parts of Bihar. Most of the area in this region are under rain fed condition, and the average yield is lesser than the national average.

c. Region with rainy season of three months

i. Southern India – Rice area is concentrated in intensive irrigated rice ecosystems, and the yield level is > 2000 kg/ha

ii. Central India – Single season rice is dominated with low input use and hence the average yield is about 1100 kg/ha.

iii. The Gangetic plains consists of West Bengal and Bihar. Rain fed rice cultivation is also practiced and average yield of Bihar is only 814 kg/ha.

iv. The hilly area- the average yield is lower than the national average (1080 kg/ha) because of improper soil and crop management and climatic constraints.

d. Region with rainy season of two months

The area under this region consists of Southern India, Central India, the Gangetic plains, and the Indus plains of Punjab. Though the rainfall is only for two months, the other climatic parameters rice cultivation are more appropriate and the watering is done through canal irrigation by canals and wells.

4.2.5 Environmental issues of rice ecosystems

The environmental cost of human activities are often hidden and not sufficiently considered in assessing productivity and ecological sustainability of the intensive rice system. As a side-effect of the Green Revolution, for example, assured water supply and control, prolonged period of submergence, intensive land preparation and fertilization, and resultant improved growth have favoured methane production and emission, CO_2 and N_20 emission, nitrate pollution, and nitrogen fixation with phenols. Global methane emission from flooded rice fields has been estimated at 20-100 Tg/year or approximately 10% of total global methane emission. Nitrous oxide emissions from the fields occur as a result of nitrification-denitrification during periods of alternate wetting and drying. Due to enhanced decomposition of pesticide residues in tropical flooded rice soil, small amounts of pesticides applied at recommended rates and intervals do not persist beyond the crop-growing period and rarely have a detrimental effect on microbial populations or their activities. Water contamination by pesticide is probably higher in multiple cropping areas compared with that in more traditional one crop per year areas, particularly in light-textured soils. Pesticide leaching causes ground water contamination. The potential also exists for nitrate pollution of ground water, especially if fertilizer N is inefficiently managed. Nitrate leaching and contamination of water is a well-known problem in rain fed rice areas or rice - non rice cropping systems on course textured soils. Salinity and Hydrogen sulphide emissions are other environmental concerns in these ecosystems.

Table 8. Important morphological characteristics of rice ideotypes for major rice ecosystems

S.N.	Plant part	Directed seeded	Rainfed lowland	Annual upland	Perennial upland	Deep water	Tidal wetland
1	Root system	Vigorous	Extensive	Deep thick	Deep thick with adventitious rhizomes	Earty develo-pment	Wide shallow
2	Stem	Very study	Very study	Very study	Study	Upper nodal root-ing tilleri-ng ability	Very study
3	Tallness	90 cms	130 cms	130 cms	130-150 cm	Internode elogation for 1 m flood	130 cm
4	Leaves	Dark green, erect thick	Dark green, erect or moderately droopy	Upper erect, lower droopy	Upper erect, lower droopy	Dark green, long erect	Erect to moderately droopy
5	Tillers	No unpro-ductive	No unpro-ductive	—	—	—	—
6	Panicles plant^{-1}	3-4	6-10	5-8	5-7	5-7	5-7
7	Grains panicle^{-1}	200-250	150-200	150-200	100-150	150-200	150-200
8	Duration (days)	75-110	120-150	100	3-5 yr	—	—
9	Yield potential (t ha^{-1})	1.3-1.5	5-7	3-4	3-4	1.5-2	1.5-2

10	Other qualities	a. Strong grain dormancy b. Strong sub-mergence tolerance	—	—	1. N fixing ability 2. Aromatic grain	1. Photo period sensitivity 2. grain dormancy	1. Photo period sensivity 2. Sub-mergence tolerance
11	Multiple disease and insect resistant	Required	Required	Required	Required	Required	Required

Source: Adopted from IRRI (1989)

4.3. IDEOTYPES

It is generally accepted that there is an ideotype (ideal plant type) in each species grown. The scientists at the International Rice Research Institute (IRRI, 1989) have given out some morphological characteristics of rice ideotypes for different major rice ecosystem. These are given in Table 8.

CHAPTER 5

RICE – AGRONOMY

Since the rice crop is a semi-aquatic plant, it needs certain requirements for its growth and development. The basic requirements for rice cultivation are listed below.

a. Optimum temperature, optimum relative humidity, rainfall, bright sunshine hours (mainly during ripening period).

(i) Short duration varieties	:	$30.4^0 - 32.6^0C$
(ii) Medium duration varieties	:	$29.7^0 - 30.4^0C$
(iii) Long duration varieties	:	$29.7^0 - 30.4^0C$

b. Cultivation should be free from weather or Natural Calamities, i.e. occurrence of frost, fog, cyclone, flood, heavy wind, severe cold and extreme drought.

- The best-suited soil for rice cultivation is loamy or alluvial soil with a pH ranging from 4.8 – 8.0, with out salinity and alkalinity problems.
- There should be assured availability of water to meet out the requirement of 5cm depth of irrigation from transplanting to harvest, mainly during critical stages of crop growth.
- Necessary facilities for proper drainage
- There should be availability of all fertilizers, farm manures etc., which will be able to supply all major and micronutrients, required for plant growth.

- There should be more availability of implements for preparatory, primary cultivation, secondary cultivation, harvest and post harvest techniques.
- All inputs, insecticides, pesticides and other biofertilizer should be available.
- Availability of labourers during the cultivation and marketing facilities should be available.

Considering the above requirements, the following are discussed in detail for successful rice cultivation.

5.1 CLIMATIC REQUIREMENTS

Rice can be grown in different locations under a variety of climates. As discussed earlier, it is grown below sea level in Kerala and elevations above 2000 m in Kashmir. The *indica types* are widely grown in tropical regions; *Japonica* types, which are adapted to cooler areas, are largely grown in temperature countries. Both *indica* and *japonica* rice are grown in subtropical regions. There are various weather parameters that influence rice production as given below.

(i) Temperature

(ii) Solar radiation

(iii) Rainfall

(iv) Wind velocity.

(v) Light intensity and

(vi) Relative Humidity.

The climatic parameters directly influence the yield of rice crop by influencing the various physiological processes like respiration, photo-synthesis, transpiration etc., involved in rice production. Climate indirectly influences the yield of the rice crop by the occurrence of pest and disease. For example, if IR50 is cultivated in cooler climatic condition it is susceptible to blast disease. Occurrence of leaf roller will also be high.

5.1.1. Temperature

Extreme temperatures are always destructive to crop growth. Temperature influences the crop by influencing various phenomenons like tillering, spikelet formation, ripening, etc. Temperature has a linear effect on rice crop. The low, high and optimum temperature required for rice crop is given in Table 9.

Table 9. Critical temperature for rice crop

Growth stage	Temperatures °C		
	Low	High	Optimum
Germination	10	45	20-35
Seedling and emergence	12-13	35	25-30
Rooting	16	35	25-28
Leaf elongation	7-12	35	31
Tillering	9-16	33	25-31
Panicle initiation	15	-	-
Anthesis	22	35	30-33
Ripening	12-18	30	20-25

Temperature influences the growth rate just after germination and increases almost linearly with increasing temperature within a range of 22-31°C. At later stage temperature only slightly affecting tillering rate and the relative growth rate. During reproductive stage, the spikelet number per plant increases as the temperature drops. The number of spikelets formed will be more under low temperature condition and will be less under higher temperature. The temperature during ripening appears to affects the weight per grain. Indica varieties are better adapted to high temperature, while japonica varieties require low temperature for better ripening. Rice is most sensitive to low temperature and the common cold injuries are failure to germinate, delayed seedling emergence, stunting, leaf discoloration, panicle tip degeneration, incomplete panicle exertion, delayed flowering, high spikelet sterility and irregular maturity. High temperature causes high percentage of spikelet sterility in dry season crops.

5.1.2. Solar Radiation

Solar radiation influences the crop growth by occurrence of bright sunshine hours. The solar radiation requirements of a rice crops differ from one growth stage to another. Shading during vegetative stage only slightly affects yield and yield components. Shading during reproductive stage however has pronounced effect on spikelet number. During ripening, it reduces grain yield considerably because of a decrease in the percentage of filled spikelets. Solar radiation at the reproductive stage has the greatest effect on grain yield. In tropics, solar radiation is higher in the dry season than in wet season. Consequently the dry season yield is low usually than wet season crop.

5.1.3. Rainfall

Rice crop is highly sensitive to water stress condition as well as water stagnation condition. If the rice crop is grown under both conditions will affects the rice yield. If the crop is grown under water stress condition it affects the rice crop. The stress condition during critical stages of the plant like tillering, panicle, initiation, flowering, maturity etc., will affect the yield. Under rainfed rice culture, rainfall is the most limiting factor in rice cultivation. The rice plant is most sensitive to water deficit from the reduction division stage to heading. Water stress before heading reduces yield significantly by causing a high percentage of sterility. Once sterility occurs, there is no way for the plant to compensate. On the other hand water stress during vegetative stage may not affect the yield. If the rice crop is exposed for longer period of stagnation, it affects the plant leaf area, rooting etc. When rice crop is submerged to different degrees at different growth stages, the grain yield decreases in different proportions. The yield decreasing effects of partial submergence could be attributed to impaired tillering and decreased area of photosynthetic leaf surface. To overcome this, varieties with intermediate plant height are replacing semi dwarf varieties, intermediate in shallow water areas that are subject to occasional floods. In temperate regions, one or two rice crops can be grown annually. In the tropics, three to four crops a year are possible provided irrigation is available. In terms of annual rice production, the tropics have a distinctly greater potential than the temperate regions.

5.1.4. Wind Velocity

Rice crop requires a gentle wind breeze or low wind speed for a high yield. The rate of photosynthesis is directly correlated with wind speed. If the wind speed increases, the rate of photosynthesis will also increase. Over a certain limit of speed, it does not affect photosynthesis. The optimum speed of wind is 0.9m/sec. However high speed wind causes injuries to leaf. i.e. the tip of leaf will show dryness.

5.1.5. Light Intensity

The rice crop requires high light intensity for normal growth and high yield. Low light intensity will affect the yield of rice crop especially the rice crop grown during monsoon period season.

5.1.6 Relative Humidity

Relative humidity affects the yield by influencing the flowering of rice crop. For normal flowering, the optimum relative humidity required is 70 –

80%. Whenever relative humidity is less than 40%, it affects the flowering of rice crop.

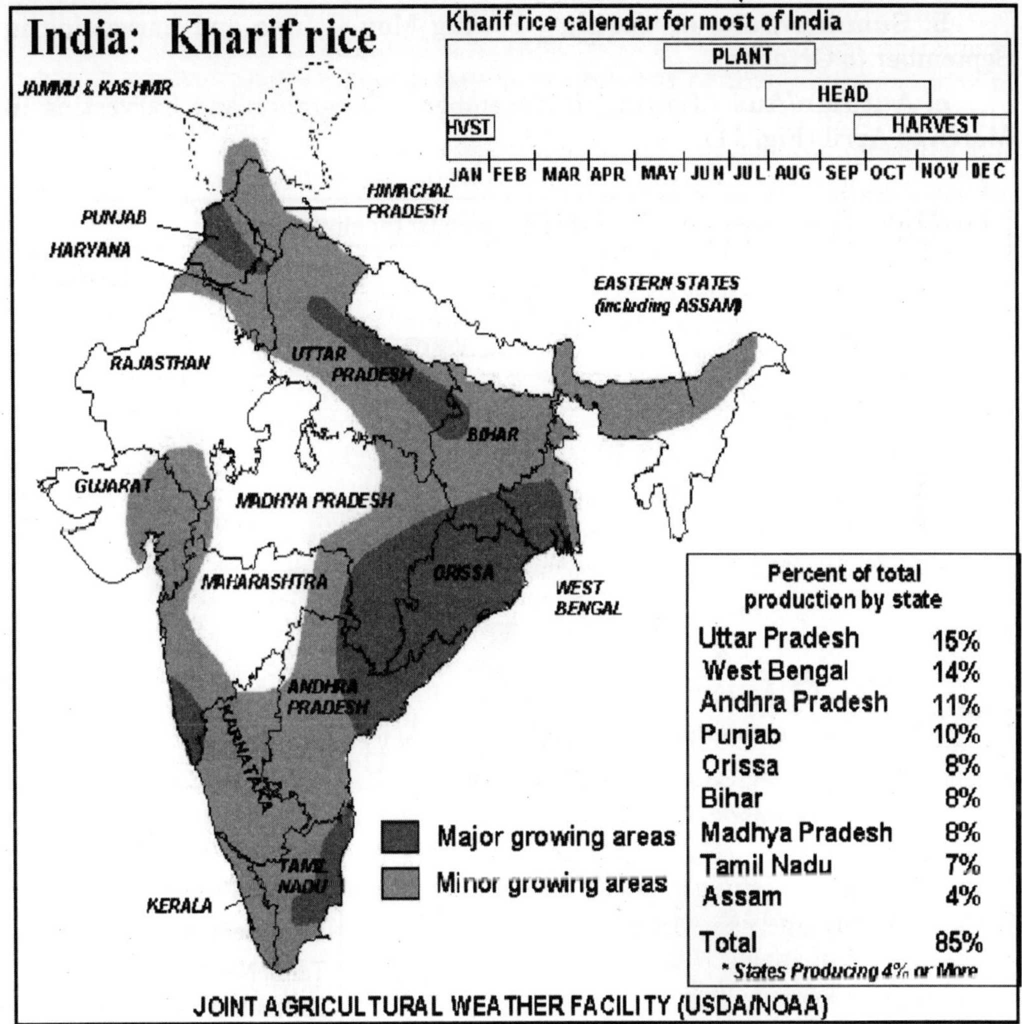

Figure 10. Kharif rice

5.2. SEASONS AND VARIETIES

5.2.1. Seasons of Rice

Rice growing seasons vary indifferent parts of India depending on temperature, rainfall and other local climatic conditions. In general, there are three seasons of growing rice in India, which is as follows.

a. Kharif - Major season–sowing in June –July and harvesting in November to December (Fig. 10)

b. Summer/boro or spring - Sowing May – June and harvesting in September to October

c. Autumn/Aus - Sowing in November – December and harvesting in March to April (Fig. 11)

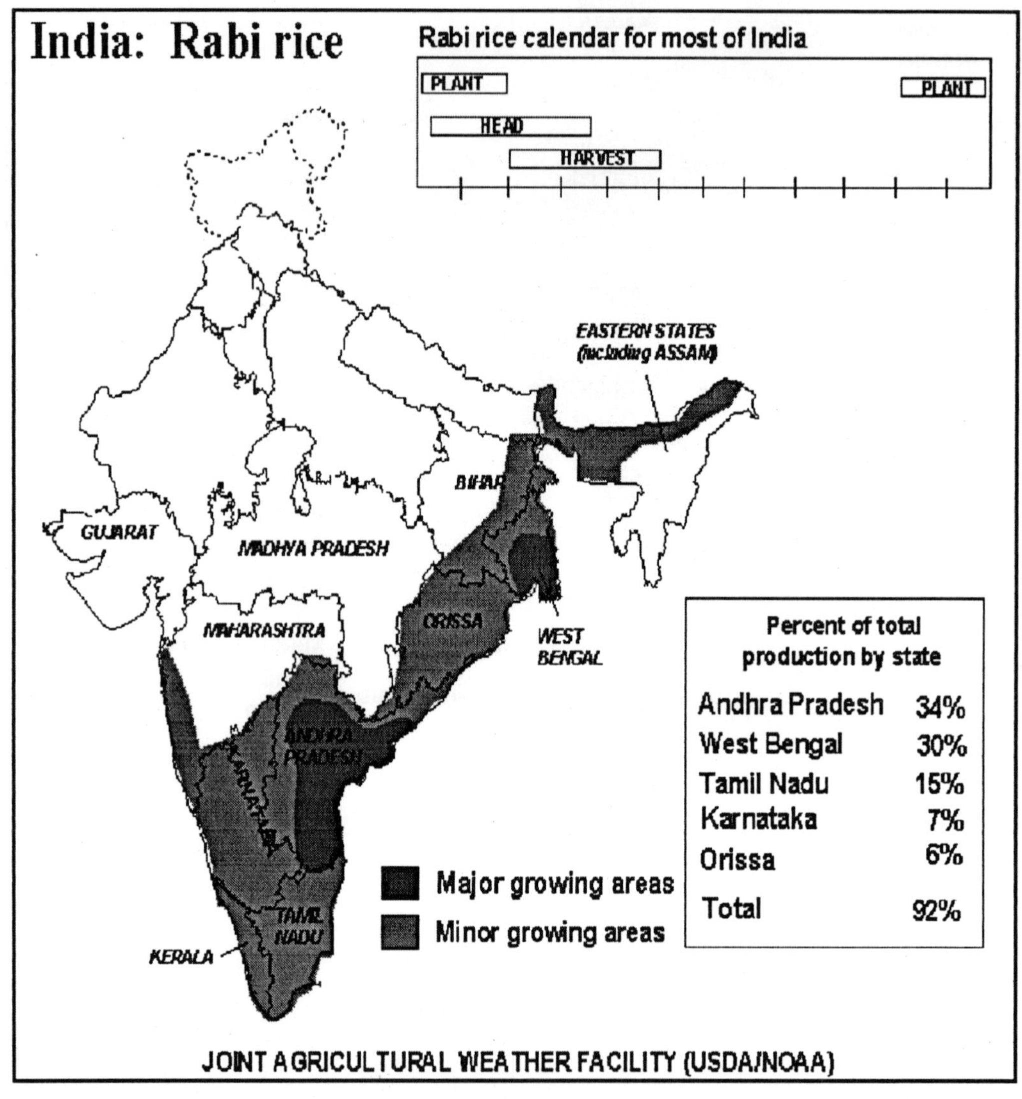

Figure 11. India rabi rice

Rice cropping seasons in different states are given in Table 10.

Table 10. Rice cropping seasons in different states of India

State	Local name of season	Period
Andhra Pradesh	Sarwa, Abi	May-June to Oct.-Nov.
	Dalwa	Nov.-Dec. to April-May
	Tabi	Dec.-Jan. to April-May
Assam	Ahu	April-May to Aug.-Sept.
	Sali	June-July to Nov.-Dec.
	Boro	Nov.-Dec. to April
Bihar, Eastern U.P.	Kharif	June-July to Nov.-Dec.
	Rabi	March to June
Jammu & Kashmir	*Kharif*	March-April to Sept.-Dec.
Karnataka	Haine	June to Nov.-Dec.
Kerala	Viruppa	April to Sept.
	Mundakan	Sept.-Oct. to Jan.-Feb.
	Poonja	Dec.-Jan. to April
Orissa	Beali	May-June to Sept.-Oct.
	Sarad	June-July to Nov. – Dec.
	Dalua	Dec.-Jan. to April-May
Tamil Nadu	Swarnavari,	May to Sept.
	Kar, Kuruvai	June to Sept.
	Samba,	July to Jan-Feb.
	Thaladi, Pishanan,	Oct. to March
	Karthigai, Navarai, Manavari	Jan. to May
West Bengal	Aus	April-May to Aug.-Sept.
	Aman	June-July to Nov. – Dec.
	Boro	Dec.-Jan. to April-May
Gujarat, Haryana, Himachal Pradesh, Madhya Pradesh, Punjab, Rajasthan, Western UP	Kharif	June-July to Oct.-Nov.

Source: Adapted from Rao (1985)

5.2.2. Varieties

The popular rice varieties grown under different rice ecosystems are given in Table 11.

Table 11. Popular rice varieties for different ecosystems and rice-growing situations in India.

Ecosystem or rice-growing situaion	Variety
Rainfed upland	Tulasi, Aditya, Prasanna, Rasi, Satya, Rudrama, PNR 381, Birsadhan 101, Birsadhan 104, Birsadhan 105, Birsadhan 201, Birsadhan 202, Heera, Kanchan, Kalinga III, Annada, GR 2, GR 5, GR 6, Himdhan, Nagardhan, VLK Dhan, VI. Dhan 221, RP 2421, Himalaya 741, Himalaya 2216, Amrut, Mukti, IET 7564, Suvarna, Modan, Onam, JR 75, R 281-31-

	1, Tulijapur 4, Ratnagiri 73-1-41, Teena, Kranti, Ngoba, Parijat, Pathara, Kalyani II, Sattari, Neela, Rudra, Vanaprabha, Khandagiri, Nilagiri, Ghanteswari, Sneha, Patnai 23, Aswani, Renu, Saket 4, Narendraghan 18, Narendradhan 80, Narendradhan 97, Narendradhan 118, Khitish, Kiron, Bhupen.
Rainfed shallow lowland	Manasarovar, Swarndhan, Phalguna, Madhya Vijaya, Nandi, Swarna, Sambamahsuri, Pinakini, Krishnaveni, Thikkana, Chaitanya, Pothans, Orugallu, Sri Nanga, Sagarasamba, Simhapuri, Laksmi, Salivahana, Bahadur, Kushal, Ranjit, Manoharsali, Mahsuri, Moniram Katekijoha, Rangili, Bhogali, Pankaj, Savitri, Sita, Jayashree, Janaki, Radha, CR 1002, Rajashree, Kanak, Vaidehi, Hemavathi, Nathravathi, IET 7191, Abhilash, Intan, KPH-2, Neerja, Rashmi, Kayamkulam, Shyamala, Sotril 7, Kranti, Surekha, Ratnagiri 3, R 374-11, Jagannath, NEH Megha Rice 1, Darna, Megha Rice 2, Rajeswari Seema, Parijat, CR 1014, Jajati, Urbashi, Samalei, Pratp, Saladhi, Gauri, Daya, Mahalakshmi, Lakhmi, Savitri, Madhukar, Jayalakshmi, Suresh, Dinesh, Biraj, Bipsa, IR 42.
Semi-deep and deep water situation	Amulya, Nalini, Jogen, Sabita, Biraj, Jaladhi 1, Suha, Jaladhi 2, Vaidehi, Jalmagna, Rambha, Utkal, Prabha, Manika, Mahalaxmi, Kanchan, Panidhan, FR 13A, Jallahri, Jalnidhi, Jalapriya, Jitendra, Madhukar, Chakia 59, Natina, Mandira, Matangiri, Purendu.
Irrigated	Rasi, Vikas, Tella Hamsa, Prabhat, Abhaya, Rajendra, Divya, Sasyashree, Bhaddrakali, Suraksha, IR 64, Ajaya, Vikramarya, Vibhava, Satua, Saleem, Surekha, Kavya, Erramalley, Pusa 2-21, Pusa 44, Gautam, CR 1002, Archana, Rajendradhan 201, Rajendradhan 202, Jaya, Vikram, Karjat 2, Karjat 3, Ratnagiri 3, Sugandha, Smbika, Jaya, Narmada, Ratna, IR 20, IR 36, HR 4, Gaur 10, GR 102, GR 103, HKR 102, Haryana Basmati, Taroari Basmati Basmati 370, Pusa 44, Pusa Basmati 1, Kasturi, PR 103, HKR 126, RP 732, Himalaya 732, Ranbir Basmati, Jhelum, K78-13, SKAU 23, SKAU 27, Akash, Avinash, Karna, Mahaveer, Prakash, Vikram, Vidhava, Sonasali, IR 30864, Red Annapurna, Mandya Vani, Triveni, Jyothi, Pavizham, Athira, Kartika, Makom, Sabari, Jayathi, Kanchana, Swarnaprabha, Aishwarya, Aruna, Remya, Konakam, Mata Triveni, Kairali, Ruchi, Mahamaya, Patel 85, Madhuri, Pawana, Sugandha, Punshi, Maniphouibi 1, Maniphouibi 2, IR 8, PR 106, PR 109, PR 108, Basmati 385, PR 111, Bharatidasan, Jawahar, Aravindar, Punithavathi, Savitri, Puduvai Ponni, BK 79, BK 190, Mahisugandha, IR 50, TKM 9, PMK 1, ADT 36, ADT 37, IR 20, Co 41, Co 43, Co 44, TPS 1, MDU 2, MDU 3, ADT 39, ADT 42, JJ 92, ASD 17, White Ponni, Ponni, Paiyur 1, ADT 40, Co 42, Co 45, TRC Boro Dhan, Saket 4, Jarjoo 52, Narendradhan 359, Govind, Madhukar, Jayalakshmi, Suresh, Dinesh, Biraj, Bipsa, IR 42, Givind, Pantdhan 4, Pandhan 6, Pandhan 4, Pandhan 6, Pandhan 10, Pandhan 12, Manhar, Narendradhan 2, Kunti, Lakhmi, Munal.
Saline areas	CST 7-1, Lunishree, CSR 10, Panvel 1, Panvel 2, Vytilla, Vytilla 3, Vytilla 4, CSR 5
Alkaline areas	Heera, Luit, Kalinga III
Post-flood situation Hilly regions	Pantdhan 1, Majhera 3, VL Dhan 39, VL Dhan 163, VL Dhan 206, VL Dhan 221

Source : Krishnaiah (1998)

5.2.2.1. Seasons, varieties, area, production and yield of rice in Tamil Nadu

A. *Rice seasons of Tamil Nadu*

Table 11a. Rice seasons of Tamil Nadu

Month of sowing	Season	Districts	Duration of the varieites recommended
Dec.-Jan.	Navarai	Chengai, Vellore, Cuddalore, Tiruchy, Tanjore, Madurai, Salem, Dingugal, Dharmapuri, Coimbatore, Erode & Pudukottai	Short duration below 120 days
April-May	Sornavari	Chengai, Vellore, Cuddalore and Vilupuram	Short duration below 120 days
April-May May-June	Early Kar Kar	Tirunelveli, Kanyakumari, Tuticorin, Erode, Coimbatore, Madurai, Dindugal, Salem, Dharmapuri and Nammakkal	Short duration below 120 days
June-July	Kuruvai	Tiruchy, Tanjore, Nagapattinam, Thiruvarur, Karur, Pudukottai and Erode	Short duration below 120 days
July-Aug.	Early Samba	Chengai, Vellore, Salem, Cuddalore, Madurai, Theni, Ramnad, Dharmapuri, Erode, Coimbatore, Nammakkal, Thiruvannamalai, Pudukottai & Nilgiris	Medium duration below 120 days
August	Samba	All districts	Medium duration 130-135 days and long duration 150 days
Sept.-Oct.	Late samba/ Thaladi/ Pishnam	Chengai, Madurai, Theni, Coimbatore and Erode	Medium duration 130-135 days
Nov.-Dec.	Late thaladi	Tanjore, Nagapattinum, Thiruvarur, Tiruchy, Madurai, Dindugal and Kanyakumari	Short duration 115-120 days Medium duration 130-135 days
	Late Sornavari	Tirunelveli, Madurai, Theni, Kanyakumari and Tuticorin.	Short duration 115-120 days Medium duration 130-135 days

Note: TKM9 – It is susceptible to blast when sown in Navarai (when exposed to cold weather at any stage of the crop), IR-50 is recommended for kar, Sornavari and kuruvai seasons. It should not be grown during cold weather period.

B. Varieties for Tamil Nadu

At present, Tamil Nadu Government has fixed a target of 7 t/ha paddy to increase the rice production in the State. The lands under rice cultivation in Tamil Nadu are not located in places with favourable conditions throughout the year. Their location, situation, climatic conditions and environmentals factors are entirely different. Achieving 7 t/ha from all these lands needs proper planning and systematic approach in recommending varieties & package of practices to suit the needs. Rice cultivation in Tamil Nadu is not having a smooth take off as it is cultivated in varied climatic, soil & stress conditions under eight different seasons with rice varieties of five different groups.

1. Very early duration (100 days)
2. Short duration (110-115 days)
3. Medium duration (120-140 days)
4. Long duration (140-160 days)
5. Very long duration – above 160 days.

I. Short duration varieties

Short duration varieties are generally recommended for Kar/Kuruvai (April-July), Sornavari, Navarai and late thaladi seasons. Some of the important varieties are ADT 36, ADT 37, ADT 42, ADT 43, ASD 16, ASD 18, ASD 20, IR 50, TKM 9, IR 36, MDU 5, ADTRH 1, and CORH 1 (Hybrids).

- ADT 36 (Tiruveni x IR 20) 110 days, 4.5-5 t/ha, medium slender white rice resistant to BPH and blast.
- ADT 37 (BG – 280-1-2/PTB 33) 110 days, 6 t/ha, short bold white rice, resistant to many pests and diseases, dormancy for 60 days.
- ADT 42 (AD 9246 / ADT 29) 115 days, 5.9 t/ha, long slender white rice resistant to GLH.
- ADT 43 (IR 50/Improved white ponni) 110 days, 5.9 t/ha, medium slender white rice, high tillering and fine rice, resistant to GLH.
- ASD 16 (ADT 31/CO 39) 115 days, 5.5 t/ha, short bold white rice, medium tillering, more number of grains per panicle, resistant to blast.
- ASD 18 (ADT 31/IR 50) 110 days, 6-7 t/ha, medium slender white, rice, resistant to blast.
- ASD 20 (IR 18348-38/IR 25863-61-3-2/IR 58) 115, t.t t/ha, long slender, white rice, resistant to many pests and diseases.
- IR 50 (IR 2153-14-1-6-2/IR 28/IR36) 110 days, 5.5 t/ha, long slender white rice, susceptible to BPH and blast.

- TKM 9 (TKM 7/IR8) 110 days, 5-6 t/ha, short bold red kernel resistant to BLB drought tolerant.
- IR 36 (IR 15 61-228/IR 24-4 *Oyrza nivara*/CR 94-3) 115 days, 5.5 t/ha, long slender white rice, resistant to many pests and diseases.
- MDU 5 (*O. glaberimma* / Pokkali), 100 days, 5 t/ha, medium slender, white rice, resistant to many pests and diseases.
- ADTRH 1 (Hybrid) (IR 58025A/IT 66R) 115 days, 7.9 t/ha, long slender white rice, mild aroma, resistant to stem borer and leaf folder.
- CORH 1 (Hybrid) (IR 62829 A/IR 10198-66 2R) 115 days, 6 t/ha medium duration, white rice, resistant to BPH and GLH.

II. Medium duration varieties

Medium duration varieties are generally recommended for samba (early and late), thaladi and Pishnam (early and late) seasons. Important varieties are CO43, CO 45, CO46, MDU 2, MDU4, ASD 19, IR20, ADT 38, and CORH 2 (Hybrid). CO 43, CO 45, ADT 38 and CORH 2 (Hybrid) should be sown during September, IR 20 and ASD 19 are to be sown during first fortnight to October.

- ADT 38 (IR 1529-680-3-2/IR 443252-6-4/IR 7963-30-2), 135 days, 6.5 t/ha, long slender, white rice, resistant to BPH, GLH leaf folder, susceptible to BLB. There is a problem of reduction of viability after 6 months storage. For this, the seed should be brought down to 12% moisture content. After soaking it should be incubated for 2-3 days before sowing to allow the seed to sprout.
- CO 43 (Dasal/IR 20) 135 days, 5.5 t/ha, medium slender white rie, tolerant to saline and alkaline soils.
- MDU 2 (CO 25/IR 8) 135 days, 5 t/ha, medium slender, white rice cold tolerant.
- MDU 4 (AC 2836/Jaganath) 125 days, 6 t/ha, long slender, white rice cold tolerant.
- CO 46 (T7/IR 20) 130 days, 6 t/ha, long slender, white rice, resistant to BPH suitable for BPH endemic areas of Coimbatore and Erode districts only.
- IR 20 (IR 262 / TKM 6) 130 days, 5.5 t/ha, medium slender white rice, resistant to stem borer.
- ASD 19 (Lalnakanda / IR 30) 125 days, short slender, 5.8 t/ha, white rice, drought tolerant.

- CORH 2 (Hybrid) (IR 58025 A/C2OR) 130 days, 6.1 t/ha, medium slender white rice suited to saline and alkaline soils also.

III. Long duration varieties

The important long duration varieties are CR 1009 (Augsut sowing), ADT 40 (September sowing), Paiyur-1 (150 days), PY-4-jawahar (150 days)

- CR 1009 (Pankaj/Jaganath) 160 days, 6 t/ha, short bold, white rice, resistant to RTV.
- ADT 40 (Swarnadhan/R6-3/sona) 150 days, 5 t/ha, short bold white rice suited to water logged situation.

Choice of varieties according to the location/need will be highly useful to accomplish more than 7 t/ha-1. Some varieties are highly potential and able to produce >10 t/ha. To quote, the following are the examples of high yield potentials of the currently ruling rice varieties.

- CR 1009-27 t/ha (Gangai kondan, Thirunelveli district)
- ASD 16-20 t/ha (Alwarthirunagai, Tuticorin district)
- ADT 36-17 t/ha (Madurai district)
- IR 64-18 t/ha (Theni district)
- MDU 2-10 t/ha (Theni district)
- IR 20-10 t/ha (Salem district).

5.3 METHODS OF RICE GROWING

There are two methods of growing rice.

1. Direct seeding – Seeds are sown directly in an upland or lowland field
2. Transplanting – seedling are first grown in seedbeds before they are planted in the fields, as given for different situations.

Criterion	Methods		
	Wet bed	Dry bed	Dapog
Seed rate kg/ha	35-45	35-45	
Seed area m²/ha	800	500	25-30
Soaking time (hours)	24	24	24
Incubation time (hours)	24-48	24	24
Sowing to transplanting (days)	20-25	20-30	9-14
Transplanting depth (cm)	1.5-3	1.5-3	1.5
Seedling /hill	2-3	2-3	5-12

5.3.1. Direct seeding

In Tamil Nadu, the southwest monsoon has failed and the storage position in the Mettur Reservoir was meager for the past three years (2001-2004) and kuruvai crop was called off in the delta region. There was uncertainty in the release of water from the dam even for samba rice crop. As a measure of crisis management, the farmers of the delta region were advised to undertake direct seeding of rice (Plate 1). Summer ploughing should be mainly- to economize water and to hasten the preparatory cultivation of mainland besides deriving other benefits.

Plate 1. A view of direct sown rice using line marker developed at AEC & RI, Kumulur, Trichirappalli, Tamil Nadu

5.3.1.1 Direct seeding technologies (Wet seeding)

a. Management strategy

1. Use of hardened seeds to overcome water stress
2. Sowing of hardened dry seeds in dry soil
3. Application of enriched farm yard manure (5t/ha) as basal, application of P and split application N and K after water is ponded

4. Use of herbicides to suppress weed growth

b. Selection of varieties

In Tamil Nadu, the following varieties are selected depending upon the area.

CO.43, CR 1009, ADT 38, IR 20, IR 60, ADT 36, ADT 37, ADT 39, Ponni, White Ponni.

c. Seed hardening

Seed hardening with 1% KCl solution for 18-24 hours (and then shade dried) is suggested for the following reasons.

- It accelerates rapid germination (and growth rate of seedlings)
- Plants from hardened seeds recover much more quickly
- Hardened seeds are able to withstand high temperature/drought
- Early emergence of seedlings keeps to compete more effectively with seeds.

d. Seed rate

The seed rate is 80-100 kg /ha. If the seeds are sown in moist soil, the covering will be difficult. Sprouted seeds are not recommended as the seeds will desiccate, if rains failed. Seed treatment with thiram at 4 g/kg of seed can be done 24 hours prior to sowing.

e. Methods of sowing

- Pre monsoon sowing
- Broadcasting and covering with country plough
- Line sowing – goru / seed drill

Depth of sowing will be at 2.5 – 3.0 cm and wooden plank is used to compact the soil. It is very necessary to protect the seeds from ants, rats etc.

f. Weed management

Application of Thiobencarb / Butachlor at 2.5 lit/ha as pre-emergence is being followed one day after wetting/soaking (with in 10 days of sowing). This has to be followed by hand weeding on 30 DAS.

g. Gap filling

After the receipt of canal water / coinciding with rain, thinning and gap filling operations are to be done, before N-fertilizer application.

h. Nutrient management

Application of enriched FYM @ 750 kg with 50 kg/ha of P as basal or application of 12.5 t/ ha of FYM or compost is recommended. The organic manures will help in forming soil aggregates and increase water retention capacity. The manures and fertilizers are best applied in the plough sole before seeding. The first split application of N is suggested immediately after impounding water in the field. The hand weeding and thinning and gap filling should be done before N fertilizer application. The same recommended dose of N for lowland rice (100-125 kg/ha) is suggested. Top dressing of K along with N is recommended.

5.3.2 Transplanting

5.3.2.1 Types of Nursery

Nursery: Nursery is a place where seedlings are raised for transplanting

Selection of Site for Nursery

- The nursery area should be selected nearer to an irrigation source to facilitate free irrigation
- Nursery should be located in a fairly elevated area.
- The area should be inherently fertile
- The soil should be well drained in order to avoid water stagnation even after heavy rain.
- It should be far away from the trees to avoid the roots and shade effect.
- It should be away from clcctric poot to avoid the incidence of pest like Green Leaf Hopper

5.3.2.1.1 Wet Nursery

This is normally practiced in an area where water is abundant. Before letting in water, FYM or compost is to be applied at the rate of 1 t /20 cents. The organic manure is to be applied uniformly on the field followed by ploughing until to get a puddled condition. Pre- germinated seeds are to be sown in the nursery field. Instead of compost or FYM, application of green leaf manure 10 days prior to sowing operation may be followed.

Advantages

- Less seed is required per unit area
- Growth is quick, uniform and strong
- Tolerate salinity
- It may be grown in any type of soil

Disadvantages

- Copious water is essential
- Seeds are easily carried away by rain
- Preparation laborious

5.3.2.1.2 Dry Nursery

Nursery preparation is normally followed under rain fed condition also in areas where water is not sufficient to raise the wet nursery. The land is to be prepared repeatedly until to get a fine tilth condition. One ton of FYM or compost is to be applied during the course of land preparation. Dry seeds are sown where there is sufficient moisture in the soil or in the anticipation of rainfall. The seeds are covered using light wooden plank or with the help of country plough. Small beds are formed to wet the nursery area during pulling out the seedlings.

Advantages

- Timely planting is possible
- Preparation is easier
- Pulling out is easier
- Establishment is easy in the main field
- Seedling hardy and can tolerate adverse conditions

5.3.2.1.3. Semidry Nursery

At the beginning, the nursery is maintained as dry and later it is maintained as wet nursery. This is mostly followed in southern districts of Tamil Nadu. Before sowing, beds and channels are formed for irrigation & weeding.

5.3.2.1.4. Dapog Nursery (Or) Mat Nursery

This method of raising nursery has been introduced in India from Philippines. Some farmers in Andhra Pradesh are adopting. It saves almost half of this practice. The main advantage of this method is that less area is required to raise seedlings. 25-30 m^2 of area is enough to grow seedlings for transplanting one ha. The seedlings grow at faster rate in this method. In this method, the beds could be prepared on an even but slightly raised 4-5 cm surface in an open field or an even on cement floor. A width of 1.5 m has been found easily manageable. About 1 m² of seedbed should be required for every 3 kg of seed (Plate 2). Cover the soil surface of the seedbed evenly or completely with banana leaves with the protruding mid ribs removed or preferably with plastic sheets or coconut sheath. Carefully lay strips of banana bracts along the edge of seedbed. Push bamboo pegs through the bracts in the soil to keep bracts up right and firm. The sprouted seeds are broadcasted uniformly @ 1.5 kg/m² area. The beds are kept moist constantly and pressed slightly 2-3 times a day with a smooth wooden plank for the first 3 days, so that root remain in contact with water. When the seedlings attain 2-3 cm height, a constant film of water should be maintained. In about 12 days, roots are well developed and are entangled with another so that nursery can be cut into strips and rolled like a mat and then transported to the main field easily. Seedlings of 1 square metre area can be transplanted in 200-m² area. Use of Dapog nursery seedlings for SRI method is discussed in the subsequent chapters.

Plate 2. Mat nursery seedlings ready for transplanting

Advantages

- Requires half time period compared to other type of nursery.
- Area required is very minimum.

- Area to be raised to about 25 to 30 m²/ha.
- Seedlings are easily rolled like a mat and freely transported to the long distance fields.
- Seedlings will grow very fast.

Disadvantages

- Seedlings are very thin, slender and short in nature.
- Seedlings are very dedicate and cannot survive for a longer period in the Dapog nursery. For more than 2 weeks, seedlings cannot be placed in the dapog nursery.
- For dapog nursery preparation, special skill is needed

5.3.2.2 Nursery Preparation

The following steps are involved for selection of good quality seeds before preparing nursery.

5.3.2.2.1 Seed character and seed rate

Seeds selected should be true to type, healthy, viable with good germination percentage. Using certified seeds from authorized agents, all the characters are taken care of. Bold seed varieties require high seed rate than fine seed varieties.

The Seed rate will be recommended as follows

SD-short duration	: 60kg/ha.
MD-medium duration	: 40 kg/ha
LD-long duration	: 30kg/ha

5.3.2.2.2 Seed Testing

Seed quality evaluation is done in the Seed Testing Laboratories and they play a pivotal role in seed certification. There are 90 such laboratories including a Central Seed Testing Laboratory, out of which is the notified under Section 4(2) of the Seeds Act (1966). Private seed companies have also set up 17 seed testing laboratories. Thus, rapid progress has been made since 1967 when 17 Seed Testing Laboratories existed. The main aim of 'Seed Testing' is to obtain accurate and reproducible results about the planting value of the seed. The planting value of the seed is dependent on genetic identity and purity, physical purity, germination, moisture and seed health status of seed lot. In the testing laboratories, seed samples are tested for the determination of seed quality in accordance with the standardized seed testing procedures provided by the International Seed Testing Association. The seed testing aspects are the physical purity analysis, germination, moisture content and seed health testing are being done for the Breeder Seed samples produced in the University. Department and Cooperative Institutes are being analyzed in the Department of Seed Technology. In addition grow-out test is also conducted under National Seeds Project to test the genetic purity.

5.3.2.2.2.1. Seed Testing Procedures for rice

a. Seed standards

Factors	Foundation	Certified seed
Pure seed (Minimum)	98.0%	98.0%
Inert matter (Maximum)	2.0%	2.0%
Huskless seed (Maximum)	2.0%	2.0%
Other crop seeds (Maximum)	10%	20%
Other Distinguishable varieties (Maximum)	10%	20%
Total weed seeds (Maximum)	10%	20%
Objectionable weed seeds (Maximum)	2%	5%
Seeds infected (Maximum)	0.10% (by number)	0.5% (by number)
Germination (Minimum)	80%	80%
Moisture (Maximum)	13.0%	13.0%
For Vapour –Proof Containers (Maximum)	8.0%	8.0%

- Objectionable weed : Wild rice *Oryza sativa var. fatua*

b. Minimum weight of sample required for testing

For submitted/deposited sample = 400 g
For working sample = 40 g
For count of other species = 400 g

c. Germination conditions

a. Medium : Sand, Paper
b. Temperature : 20-30°C
c. Relative humidity : 90-95%
d. Days required for counting : First count on 5th day, Final count on 14th day.
e. Treatment to break dormancy : overnight soaking in 0.2% KNO_3

d. Moisture estimation

a. Weight of the submitted sample: 100 grams (in polythene bag of 700 gauge)

b. Methods

I. Air oven method : 130°C for 13 hours

II. Moisture meters : Use of moisture meters is a very convenient and rapid way for determining seed moisture content. In moisture meters the seed moisture is determined by their conductivity because moisture content is directly proportional to resistance and the dielectric constant of the sample.

e. Biochemical test for seed viability

The tetrazolium test commonly known as the T2 test is used for quick estimates of seeds germinability. The test is very useful in processing, handling, storing and marketing large quantities of seeds in a short time, testing dormant, seed lots and vigour rating of seeds.

I. Reagent : The chemical used for this test is a cream or light yellow coloured water soluble powder called 2,3,5-triphenyl Tetrazolium chloride.

II. Principle : All living tissues, which respire, are capable or reducing the colourless chemical, 2,3,5 – triphenyl tetrazolium chloride into a red coloured compound (formazan) formed by H Transfer reactions catalyzed by the enzyme dehydrogenases. Formazan being non-diffusable, stains the living tissues red. Thus, the living parts of a viable seed should be stained red when incubated in the solution of tetrazolium chloride.

III. Preparation of the solution: 0.2% and 0.5%. The pH should be around 7.0 for proper staining.

IV. Temperature: The test can be performed satisfactorily at room temperature.

*V. Preconditioning period :*18 hours

VI. Preparation after moistening : Remove the lemma and palea

VII. Staining : Seeds are placed in the solution and held at 35°C.

5.3.2.2.3 Seed Treatment

Seeds are subjected to undergo some kind of treatment before sowing for the following purpose.

- To improve germination

- To protect the emerging seedling from insects and disease.
- Induce drought tolerance – ($CaCl_2$ in Cytokinin).
- Induce root growth – (Cytokinin, IAA and GA)
- To supply mineral nutrition – leaf powder, DAP, Micronutrient mixture and MgCl (seed pelleting).
- To fix the atmospheric N by treating the seeds with bio-fertilizers.

5.3.2.2.3.1 Fungicide treatment

Seeds are treated with carbendazim at 2 g/kg of seeds or thiram at 4 g/kg of seeds in a seed-treating drum. Then after 24 hours, the seeds are treated with bio-fertilizers. The quantity of Azospirillum required is 2 pockets i.e. 400 g, cooled rice kanji is used as a binder. First, slurry is made by using rice kanji. Then seeds are added to it and after thorough mixing, the seeds are kept under shade for 30 minutes and on the same day, drying should be done on a polythene sheet or clean concrete surface.

(i) Dry seed treatment

Mix anyone of the following fungicides with the seeds in seed treating drum or polythene lined bag for 15 minutes so that a uniform coating of the fungicide over the seed is ensured. Treat the seeds at least 24 hours prior to soaking.

Carbendazim - 2 g/kg of seeds

Capton or thiram - 4 g/kg of seeds

(ii) Wet seed sprouted seed treatment

If dry seed treatment is not given, mix any one of the following fungicides specified below:

Soak the required quantity of sprouted seeds in the fungicidal solution for 15 minutes just before sowing in the nursery, drain the fungicidal solution and then sow in the nursery.

Fungicide (gm)	Short duration	Medium duration	Long duration
Bavistin	20	13	10
Vitavax	20	13	10
Water	48 liters	32 liters	24 liters

5.3.2.2.3.2 Breaking the dormancy

A desirable dormancy period in a variety is about 2 to 3 weeks. It is short enough to enable the farmers to use the seeds for the succeeding crop and enough to prevent the grain from sprouting in the panicle during a rainy and humid harvest season. A effective way to break seed dormancy is to treat the seeds with diluted nitric acid (0.1 N HN03).

Prepare the seeds. Air-dry the seeds. Then clean them to remove unfilled grains, straw and other foreign matter.

Weigh the seeds : Record the weight in kilograms to know how much solution to prepare (Normal seeding rates range from 20 to 50 kg/ha).

Compute amount of water needed : Always use multiple of 1 liter of water to facilitate liquid measuring. For a large quantity of seed, use 1 liter of water for every kilogram of seeds. For a small amount of seed, use not less than 1 liter of water. Do not use a metal container. A metal container may react with the acid thus reducing the strength of the acid.

Compute volume of concentrated acid. To make a 0.1 N solution, you need 6.3 ml of concentrated nitric acid (purity should not be less than 68%) for every liter of water to be used.

Mix acid and water. Always pour concentrated acid into water. **Caution**: If water is poured into concentrated acid there will be an explosion and splattering. Stir the solution quickly but thoroughly with a bamboo or wood stirrer.

Soak the seeds in acid solution: Pour clean seeds into the acid solution. Remove all floating seeds. Soak the seed for 16 to 24 hours but no longer. Drain off acid solution and dry seeds. Place seeds in the sun. Sundry for 3 to 7 days after soaking. This will give 80 percent germination.

5.3.2.2.4. Testing the Viability of Seed

Testing viability is one of the minor tasks in rice production. It deserves more attention, however, Without knowing the germination percentage, you cannot make a good estimate of how much seed you need to plant a field at a recommended seeding rate.

5.3.2.2.4.1 Selecting Seed by Specific Gravity

Seeds may be separated either mechanically or by the specific gravity method. One mechanical method is to use air as in winnowing to blow the lighter seeds; another is to use of gravity table on which the lighter seeds tend to separate and remain on the side of the slightly sloping table surface, which is jiggled mechanically. You can also use a salt solution of predetermined specific gravity to separate seeds. The light seeds will float and can be removed from the surface.

a. Prepare a salt solution

For traditional tropical varieties (indica), make a solution with a specific gravity of 1.08. Mix 1.65 kilogram of common (table) salt in 10 litres of clean water. Or instead, mix 2.2 kilograms ammonium sulfate in a bucket containing 10 litres of clean water. The desired specific gravity can be checked accurately with a hydrometer or approximately with a fresh chicken egg: The specific gravity is about 1 when one end of the egg surface just breaks the solution surface; the specific gravity is about 1.13 when the surface area of the egg above the solution surface reaches a diameter of 20 mm.

Add the seed sample: Stir the seeds thoroughly in the solution. Allow seeds to settle.

Remove floating seeds : Carefully remove all floating seeds. Avoid spilling them on the ground. Spread them on a piece of cloth or sack, and let them dry.

Wash and dry the good seeds: Drain the bucket and wash the good seeds thoroughly in fresh water to remove the salt. Dry the good seeds and save them for sowing.

b. Soaking in nutrient solution

Soak the seeds in water for 10 hours. Remove the seeds from water and allow the excess water and drain. Place the water soaked seeds in any one of the following solutions for another 10 hours.

Chemical	Soil type	Short (g)	Medium (g)	Long (g)
$MnSO_4$	All type of soils	1000	650	500
KC1	All type of soils	250	160	120
Fe_2SO_4				
$FeSO_4$	Neutral & alkaline soils	1000	650	500
Water	@ 1 litre /kg of seeds	24	16	12

Of the above, KC1 and Fe_2SO_4 are cheaper. Then remove the seed and cover it with gunnies, place it in a dark rooms for another 24 hours for sprouting. Then take out of the seed and before sowing as indicated earlier if dry seed treatment was not given, wet seed treatment may be given and sow in the nursery. Seed thus raised will give "super seedlings" (wet seed treatment gives protection to the seedlings up to 40 days from seedling disease such as blast) If the nutrient treated seeds are not sown immediately, it can be dried in the sunlight to the original moisture and stored even up to a month and then sown. If the nutrient treatment is not resorted to, the required quantity of seeds is tied in a gunny bag, immersed in water for 24 hours and taking out, water is drained off. The bag is kept in a dark room for 24 hours.

5.3.2.2.5. Nursery Area

Recommended area is 20 cents (800 m^2) to meet out the seedlings required for transplanting in one ha of area.

5.3.2.2.6 Manure Application

Before irrigating the Nursery area, FYM or GLM or compost is applied at the rate of 1 t/20 cents of Nursery area.

5.3.2.2.7. Nursery Preparation

After complete spread of the manure in the dry soil, water is allowed 1 or 2 days prior to field preparation. Plough the field under submerged condition until 'Puddled' condition is obtained. Trimming and plastering of the area should be carried out.

5.3.2.2.8 Fertilizer Application

DAP is recommended at the rate of 40kg/20 cents of nursery area. In the case of non availability of DAP, Urea at the rate of 16 kg and Super phosphate at a rate of 120kg/20 cents of nursery area is to be applied. Basal application is given to seedlings when it is ready for transplanting within 20-25 days. More than 25 days of sowing, DAP is applied 10 days before pulling out of seedlings. For clayey soil, 20kg of DAP is applied 10 days after sowing.

Advantages

1. Seedlings absorb phosphorous, store it in the plant and effectively used it in the later stages of plant growth.
2. Reduces P requirement to the main field.
3. Application of Phosphorus to nursery is highly economical

N Deficiency (Yellowing Symptom)

To overcome the N deficiency, top dressing of urea @ 500g/cent of nursery area before 7-10 days of pulling out of seedlings. NaCl is applied when some soil gets harden. Gypsum is also applied @ 1kg/cent 1 or 2 days before pulling out of seedlings.

5.3.2.2.9. Forming beds in the Nursery

Form the Beds with a size of 2.5m with spacing of 30 cm in between them. Soil from the 30cm area are taken and spread over the 2.5m area for raised bed and for easy drainage.

5.3.2.2.10. Water management

Field is irrigated 1 or 2 days before initiating preparation/puddling operation. Thin film of water is maintained at the time of sowing. 18-24 hours after sowing, water is drained from the nursery. Initially 2-3 days alternate wetting and drying is given. Depth of irrigation is increased to a height of 1.5 cm, after wards it is increased to a height of about 2.5cm.

5.3.2.2.11. Weed Management

At the seedling stage, the weeds viz., *Echinochloa crusgalli* and *E. colonum.* cannot be distinguished easily. In order to avoid the weed, pre – emergence herbicide Butachlor at a rate of 1.5-2.0 lit/ha is applied. Herbicides are mixed with sufficient quantity of sand @ 50 kg/acre and should be applied after 8th day of sowing. Keep thin film of water and allow it to disappear. Avoid draining of water after herbicide application. This will control germinating weeds.

5.3.2.2.12. Pest Management

To control the incidence of thrips, green leaf hopper, spray monocrotophos 36 WSC at the rate of 40 ml or endosulfhan 35 EC at the rate of 80 ml for the nursery area. Whenever the dry seed treatment is followed to avoid the incidence of blast disease, apply carbendazim at the rate of 40g for nursery area, 40 lit of water is necessary for spraying the nursery.

5.3.2.2.13. Age of Seedlings and pulling out of seedlings

The normal age of seedlings is 18 to 22 days for short duration, 25 to 30 days for medium and 35 to 40 days for long duration varieties. In order to get high yield, pull out the seedlings at proper stage and transplant it at proper time. Before carrying out the pulling operation, irrigation is to be given. Seedlings are pulled out and bundled at 5 to 8cm dia using banana fibre or straw. Root portion should not be damaged and also should not be dried.

5.3.2.3 Lowland Soil

The major portion of rice crop in tropical Asia is grown under "Lowland" conditions, that is, under "flooded" or submerged conditions (Plate 2). Field prepared in this manner (puddled) are called paddies (puddled soil) and rice grown on paddies are termed "Lowland" rice. Paddies watered by rainfall are termed rainfed paddies. Those watered by irrigation are called irrigated paddies.

A. Characteristics of submerged or flooded soil

A short review of the three major changes (physical, biological and chemical) that occur when a soil flooded will help to understand better the effective fertilizer application of "Lowland" rice crop.

a. Physical changes

Upon flooding, the pore spaces in the soil become saturated with water. The soil swells and hard clods soften and break into small aggregates. Puddling, considered as a necessary operation in land preparation for "Lowland" rice. It drastically reduce water loss by percolation and completely destroys the structural aggregates and pore spaces of the soil. The exchange of air between the atmosphere and the puddled soil is impeded. This slows up the drying of the soil. The flooded puddled soil develops two zones. The upper zone, a thin 1 to 10 mm thick zone, absorbs oxygen from the water, turns brown in color (oxidized state) and reacts to nitrogen like an unflooded upland soil. The lower zone will turn to a dark or blue grey color. This colour change takes place as iron compounds in the flooded soil lose their oxygen. This soil is said to be in a reduced state. The effects of these two zones in flooded soils must be considered for efficient nitrogen fertilization of "Lowland" rice.

b. Biological changes

The absence of soil air (oxygen) in flooded and puddled soils causes a drastic change in the kinds and number of microbes that decompose organic matter. Consequently, the rate of decay of organic matter in a submerged soil is slower than in aerated "upland" soil. Also, nutrient and end products in submerged soil differ from those formed in an aerated "upland" soil. Research has shown that 2 or 3 weeks should be allowed to pass after the paddy is flooded, before rice seedlings are transplanted to save seedlings from the toxic effects of some of these end products. Such temporarily toxic end products, particularly carbon dioxide, soluble iron and soluble manganese, are more likely where large amounts of plant residues (organic matter) are incorporated into the soil. A low land field, previously irrigated to soften the soil, should be completely flooded either just before or immediately after ploughing. Under flooded conditions, the nitrogen in the incorporated plant residues is changed to ammonium form, which is stable under flooded conditions and will later be used by the growing rice plants. If the soil is not flooded, change takes place during which nitrate nitrogen are formed as in "upland" soils. Then when changed to gaseous form, these gas eventually escape into the air. The loss of nitrogen through the transformation of nitrate into nitrogen gas is called denitrification. Since nitrate denitrifies easily in flooded soil, nitrogen fertilizers in nitrate forms are not appropriate for lowland rice.

Plate 3. Main Field Preparation Using Power Tiller Operated Cage Wheel

Denitrification also occurs if basal nitrogen fertilizer (ammonium forms applied at the last harrowing before transplanting) is not incorporated into the soil. The microbes in the oxidized zone change ammonium nitrogen to nitrate. The nitrate then moves down into the lower reduced zone and is transformed into nitrogen gas and lost. However, denitrification of ammonium fertilizer can be prevented by immediately incorporating basal ammonium nitrogen into the lower zone of the soil. Ammonium fertilizer incorporated into the reduced zone is not converted into nitrate and nitrogen gas. This is why ammonium or ammonium forming nitrogen fertilizers such as ammonium sulfate, urea, and ammonium phosphate are used for lowland rice.

Plate 4. Puddling with Tractor drawn cage wheel.

Plate 5. Puddling and levelling with animal drawn Burmese Saturn

c. Chemical changes

Flooding a soil increases the availability of both native soil phosphorus and the phosphorus added as fertilizer. This is one of the advantages of growing rice under flooded conditions. The increased availability of phosphorus takes place because under flooded conditions, iron, aluminium, manganese, and calcium become more soluble and chemically release the phosphorus, which under aerated "upland" conditions, is chemically bound with these elements. The rise in pH of acid soils and lowering of pH of alkaline soils upon flooding likewise increase phosphorus availability.

5.3.2.4 Main Field Preparation

Main field is an area where the seedlings are transplanted or seeds are sown directly. The main field preparation involves the following operations.

5.3.2.4.1 Summer Ploughing

The main field is ploughed during summer months. It has a number of advantages. It conserves the water obtained from rainfall and thus reduces water requirement for main field preparation.

- It exposes the eggs, larva and pupae of insects to direct sunlight. The birds also collect when exposed.
- Destroys the weeds, which are present in the field.
- It helps in proper incorporation of organic matter into the soil.

Ploughing the land during summer to economize the water requirement for initial preparation of land.

5.3.2.4.2. Application of Organic Manure

Before initiating field preparation operations, any one of the organic manure like FYM, compost, press mud or composted coirpith has to be applied at a rate of 12.5 t/ha. They are applied in dry soil itself. Instead of FYM green manure or green leaf manure can also be applied at a rate of 6.25 t/ha.

Computation of Green Manure

1. Green matter or green manure crop is cut from a known area say $1m^2$ area then it is multiplied for 10,000 m^2. At a maximum of 10-15 t of green matter can be obtained.
2. By counting the total number of head loads. Each head load weighs about 20 kg.

3. Direct method-we can access directly from the condition of the crop – for e.g.: Daincha - can supply 10-15t of green matter/ha (Plate 6).

   Sunnhemp - can supply 8-10t green matter/ha.

   Kolingi - can supply 4-6t of green matter/ha

   (i) Addition of green matter to the soil improves the physical properties of the soil.

   (ii) Most green manure crops are leguminous crop having root nodules and can fix atmospheric nitrogen in the soil.

   (iii) By the process of decomposition, it improves activity of microbes present in the soil.

   (iv) Depletes the nutrient in lower layers of soil and conserves in the upper layers.

   (v) They act as cover crops. So they can able to prevent soil erosion.

   (vi) Reclamation of problem soils, mainly sodic soil is possible.

   (vii) Some green leaf manure like neem, pungam etc. have insecticidal property (Plate 7).

5.3.2.4.3. Irrigating Main Field

One or two days before initiation of first ploughing we have to irrigate the field. At the time of ploughing operation a water depth of 2.5cm should be maintained above the soil. The different steps of ploughing are first 1 or 2 ploughings with melur plough (or) country plough (or) mould board plough, then with power tiller as well as tractor attached cage wheel (Plate 3,4 and 5). Ploughing under submerged condition is called as puddling.

Objectives of puddling

- To create impervious surface soil layer and to maintain submerged conditions in the field. It reduces the loss of water through percolation.
- To check heavy loss of plant nutrients – through leaching.
- To create favourably physical, chemical and microbiological conditions for the growth of the rice plant
- To eradicate weeds and to destroy insect eggs, larvae and pupae
- To facilitate the incorporation and distribution of organic manures and inorganic fertilizers in to the soil.
- To create conditions favourable for the seedlings to establish themselves and to encourage the development of root system rapidly.

Plate 6. Green manure- Daincha incorporation using tractor drawn cage wheel

Plate 7. A view of green leaf manure (neem) application in the main field

5.3.2.4.4.Digging corners and maintenance of bunds

- Dig the corners of field, which are not covered by ploughing.
- Cut of 2.5 cm of soil from the top and sides of the bunds to remove the weeds along with their seeds and to destroy the insect pests by using spade/mammutty.
- If the bunds are very broad, trim them to a width of 15 cm and height of 15 cm so that rats do not harbour in the bunds.
- If rat burrows are noticed, insert pellets of 0.5 or 0.6 g Aluminium phosphide and plug the exit holes.
- Apply mud paste to the sides and top of the bund to a thickness of 2.5 cm with a spade/mammutty and plaster it using the flat surface of the spade/mammutty. Plastering the bunds helps in checking weed growth and prevents harboring of insect pests.

Plate 8. Trimming and plastering of field bunds

5.3.2.4.5. Basal Application of Fertilizer

Before completing the last ploughing or puddling operations fertilizers are applied. 50%N, 100% P_2O_5 and 100% K_2O are applied as basal dose.

When green manure or green leaf manure are incorporated, P&K are applied over them in the early morning before incorporation. They stick to plant parts and there will be no loss. They are applied in single dose. For such incorporations, certain implements like Burmese Saturn and green manure trampler are used. After giving basal application, last puddling and leveling is given. Micro nutrient namely zinc sulphate at a rate of 25kg/ha is applied. They should be top dressed in the soil. They are mixed with 25kg of sand for easy application, and are applied before transplanting.

5.3.2.5 How to select good seedlings from nursery

- Uniform height and growth.
- Short leaf sheath
- To have short leaf sheath use, proper water depth (2.5 cm) and
- Good lighting, temperature are needed (25-35°C)
- Free from pest and disease
- Large number of roots and heavy weight.

5.3.2.6 Transplanting

Seedlings can be pulled out without damage to the roots by letting in sufficient water. Wash the roots and made in small bundles. For tying the seedlings only moist straw should be used. If the seedlings are tall, 1/3 to 1/4 of the upper portion of the leaves maybe cut to minimize transpiration and also to destroy the pest's egg masses. Transport the seedlings to the main field in bundles and spread uniformly over the main field and transplant with the required spacing. Necessary seedlings may be kept at each plot for gap filling. If rice root nematode problem is noticed, dip the seedlings root in phosphomidan 0.02% solution for 20 minutes prior to planting.

		Wet nursery		
		Short	Medium	Long
(a)	Age of seedlings (days)	18-22	25-30	30-35
(b)	Number of seedlings per hill	2-3	2	1-2
(c)	Depth of planting (cm)	2-3	2-3	2-3
(d)	Spacing (cm)	15 x 10	20 x 10	20x 15
(e)	Number of hills per sq.metre	66	50	33

5.3.2.6.1. Methods of Transplanting

There are three methods of transplanting.

(i) Random planting.

(ii) Line planting.

(iii) Scattered planting.

In random planting, the seedlings are planted without maintaining spacing and plant population. There will be no optimum population.

In line planting, skilled labourers and ropes are used for planting. Spacing is actually maintained only between the rows and not between the plants. When it is done by using implements or transplanter optimum spacing rows and also plants can be maintained.

a. Use of planting rope and board

b. Use of markers

c. Use of mechanical transplanter

Scattering method of planting is a recent technology of planting. It is just the scattering of seedlings in the main field. It is low cost planting type. But high yield can't be expected. The details are discussed in the subsequent chapters.

Advantages of row planting

- Rotary weeder can be used for weeding
- Easier to apply insecticides, herbicides, fertilizers
- Optimum spacing is attained.

Disadvantages of row planting

- Improper spacing reduces the potential yield by 25-30%
- Planting too close together increase lodging and cost of transplanting
- Planting too apart causes low yields.

Advantages of transplanting

- Plant population becomes more assured
- Since nurseries occupy only 1/10 of area of the main crop, the irrigation, control of insects and diseases and other management practices become easier to reduce the cost of raising seedlings as compared to broadcast wet crop of the same age. Nurseries raised in advance enable the crop to be planted in the field at the right time.

Disadvantages of transplanting

- Labour cost is high
- Seedlings are exposed to possible injury
- Due to transplanting shock, seedlings grow slowly
- Harvesting is delayed

The difference between transplanting and direct seeding is given in Table 12.

Table 12. Difference between transplanting from direct seeding

Criteria	Transplanting	Direct seeding
Growing area	Low land	Low land or upland
Seeding rate kg/ha	30-60	80-100
Seed exposure to predator	Less	Greater
Labour requirements	More	Less
Land preparation	Puddled soil	Both puddled and unpuddled soil
Crop/weed competition	Less	more
Lodging	Less	More
Total water usage	Less	More
Field duration	Earlier	Extended by 15-25 days depending upon the season
Potential yield	Good	Average
Per day productivity	Higher	Lower

5.3.3 Rainfed Rice

Rice yields are generally lower in direct-seeded rainfed lowland than in transplanted irrigated ecology. The major yield limiting factor in the rainfed lowland is high tiller mortality after maximum tillering phase. The tiller mortality is attributed to lag phase, N uptake and low level of leaf nitrogen (<2%). Timely top-dressing to maintain leaf N level above 2% in the post – maximum tillering stage has been found to minimize the percentage mortality and increase thereby the yield level.

5.3.3.1 Issues

Water is a very critical input for the rice cultivation in rainfed uplands. Several agronomic practices have been worked out to conserve water in soil profile and several rainwater harvest structures on farm have been investigated. The on-farm rainwater harvested is used either as life saving irrigation or supplemental irrigation mostly to field area lower in gradient than the water harvest structures. In rainfed lowland ecology, most of the land under a situation, where once accumulated cannot be drained. For such situations, varietal strategy that can withstand submergence for specified duration and can yield well under waterlogged condition and low nutrient application, which is possible only before water stagnated is the answer.

5.3.3.2 Extent of rain fed rice cultivation

The practice of growing upland rice is adopted in the area, which receive an annual rainfall of 800-1,000 mm, and irrigation facilities are not available. The farmer has to depend upon rainwater. A vast tract of paddy in Uttar Pradesh, Bihar, Orissa and parts of Tamil Nadu and Andhra Pradesh is under rainfed upland situation.

5.3.3.3. Land Preparation

Summer ploughing has proved beneficial in increasing water infiltration rate and thus in a higher production of upland rice due to increased moisture availability and less weed infestation. After harvesting the rabi crops in April-May, the field should be ploughed with soil turning plough. The field should be prepared and bunds should be made around the field just after the first shower of monsoon. This will check loss of rain water by run-off.

5.3.3.4 Sowing

Uncertainty of monsoon and unavailability of irrigation water lead to no recommendation about the sowing time, however, it will be worth saying that the crop must be sown after monsoon has set in and about 12-15 cm deep soil layer has become wet. The ideal date for sowing of seeds in northern India is between end of June and first week of July. In some areas dry seeding of rice slightly before onset of monsoon is practiced and the seed remains in the soil, which germinates along with the rains. This type of crop does not have a serious weed problem also but when the sowing is delayed somehow, overnight soaking of seed in plain water before sowing has given good results for quick germination and early establishment of rice plants which has least weed infestation also.

Drilling or sowing rice seed in lines in upland conditions is always better than broadcasting because line sowing of the crop facilitates the cultural operations and it also enables to identify the weed seedlings even at the early seedling stage for weeding purpose. Usually a row-to-row spacing for 30 cm and seeding depth of 5 cm is taken. It is assumed that the moisture stress condition will result in poor germination, therefore, about 25% higher than the recommended seed rate (100-200 kg/ha) will be sufficient for one ha. Drill the seeds with seed drill or behind the plough with the help of funnel in rows after applying fertilizer. The row-to-row spacing should be 20 cm. About 60 kg seed is sufficient for one ha.

5.3.3.5 Application of manures and fertilizers

Manure and fertilizers both play very important roles in rice cultivation. About 10-15 tones of farm yard manure (FYM) or compost should be applied 4-6 weeks before sowing and worked well into the soil. Soil testing is a must to give the balanced fertilizer dose to rice. In the absence of soil testing, apply fertilizers as described below. A good crop of upland rice may be taken up by applying 60 kg N, 30 kg P_2O_5, and 30 kg K_2O/ha. Total quantity of P and K and 25% N should be drilled in the soil at the time of land preparation. Ferti-seed drill may also be used for sowing of seed and placement of fertilizers. After 35-40 days of sowing, top dress 50% of the total N. The remaining 25% N should be top dressed at panicle – initiation stage. Application of Zinc Sulphate @ 15-25 kg/ha as basal or spraying of 0.5%. Zinc Sulphate with 0.25% lime dissolved in water is found to be beneficial for the crop. Use of chelated forms of micro-nutrients is always better for enhancing the crop yields.

5.3.3.6 Weed control

In upland rice weeds are major problem. The upland rice suffers from a very severe weed competition upto 40 days after sowing and if the weeds are not well taken care of the reduction in yield may be 50% or even more depending upon weed composition and its intensity. For an effective control of weeds there are 2 methods mechanical or chemical. The mechanical methods is more time taking, more cumbersome and more expensive, though it is more beneficial. Contrary to this chemical control of weeds is less time consuming, less expensive and more convenient. Thiobencarb (Saturn) @ 2 kg a.i. per hectare should be applied 6-7 days after seeding of rice. It should be followed by manual weeding 40-45 days after sowing. Application of Lasso (Alachlor) or Machet (Butachlor) @ 1.0 litre/ha should be done before crop emergence which controls most of the common weeds in upland rice.

5.3.4 Direct Seeded Semi-Dry Rice

In Tamil Nadu, rice is cultivated in an area of 2.75 million hectares, out of which majority of the area is under transplanting. However, about 3.2 lakh hectares are under semi-dry rice during monsoon season. Further, the practice of semi dry rice culture is fast spreading recently even in canal irrigated areas such as Cauvery delta due to delayed release of water for irrigation from reservoirs. Direct seeding under semi-dry condition is the only possible method in areas of uncertain rainfall and areas with limited irrigation water availability.

In the semi-dry system, rice is brought up like any other upland crop with the receipt of adequate quantity of water either through rains in tankfed areas or through canals in delta regions, at which time, the crop is brought under lowland conditions. Of late, the practice of dry seeding in rice is gaining importance due to delayed onset of south-west monsoon and subsequent availability of water in the canal. Dry seeding facilitates timely sowing and avoid skipping of any season due to late monsoon. Though primarily necessitated by the circumstances, results have shown that this system yields sufficiently high to the tune of 2.5-3.0 t/ha though not at the same level as that of a transplanted crop. It has already been proved that direct seeded rice, if managed properly, can yield almost as high as the transplanted crop, as against mistaken notion about its poor productivity. In areas of labour scarcity and lack of irrigation facilities, direct seeding of rice assumes special importance over and above traditional rice growing areas, due to which there is now a changing trend even among big farmers to shift over to direct seeded rice instead of transplanted lowland rice.

5.3.4.1. Production constraints

Several constraints such as socio-economic, biotic and abiotic are responsible for low productivity of direct seeded semi-dry rice

Socio-economic

1. Poverty: Most of the farmers of upland rice areas being poor are unable to purchase inputs like seeds, fertilizer, pesticides etc.
2. Small and scattered holdings: Marginal and small holdings constitute around 60% of total holdings but account for only one fourth of the total area. Moreover, these small holdings are also scattered which makes adoption of modern technology more difficult.
3. Farmer's unawareness: Most of the farmers are unaware of the benefits of modern technology due to poor extension service.
4. Non-availability of inputs: Seeds, fertilizers, pesticides are not available for farmers easily and in time.
5. Poor mechanization: Improved implements are not used by upland farmers as these are costly, not easily available and need a lot of care.

Biotic

1. Weeds: Direct seeded semi-dry rice suffers severely due to weed infestation. Experimental results revealed that the losses in yield due to weeds may be as high as 60-70 per cent.
2. Diseases: Blast and Brown spot are common diseases causing considerable yield losses.
3. Insects: Termites, stem borer and ear head bug cause substantial loss to crop. Termite attack is high when there is moisture stress.

Abiotic

1. Water stress: Moisture stress is the single most important constraint responsible to low rice yield in the area.
2. Deficiency and toxicity: Soils are poor in organic 'N' and available phosphorus. Acidic nature of soil causes many problems in nutrient supply.

5.3.4.2. Management strategy

Experimental results revealed that it is possible to obtain yield of semi-dry rice upto 3-4 t/ha in the regions if the following agronomic practices are adopted efficiently.

1. Selection of suitable varieties
2. Tillage and land preparation
3. Time and method of sowing
4. Fertilizer management
5. Weed management
6. Water management
7. Cropping systems

5.3.4. 2.1. Selection of suitable varieties

Since variety is a key component of the modern technology, selection of proper variety is essential for realizing good yield. Direct seeded semi-dry rice variety should have the following agronomic and physiological characteristics.

a. High and stable yields (3-4 t/ha)
b. Early seedling vigour and growth with wide droopy leaves
c. Intermediate plant height with stiff straw

d. Early maturity to permit sequential cropping
e. Tolerance to adverse soil conditions
f. Drought tolerance and avoidance
g. Acceptable grain quality
h. Recovery capacity after drought spell

5.3.4.2.2. Tillage and land preparation

Proper and timely land preparation helps in moisture conservation and control of soil borne diseases, insects and weeds. It has been found that deep ploughing of the field in summer effectively reduces the weed population and increases yield of rice (Table 13). Preparatory tillage should be done just after premonsoon rains.

Table 13. Effect of summer ploughing on the yield of semi-dry rice

Treatment	Grain yield (t/ha)
T1 – Summer ploughing alone	1.50
T2 – T1+ country plough	1.92
T3 – T1+ Tractor drawn disc plough	3.11
T4 – T1+ Bullock drawn mould board	2.80

Summer ploughing can greatly reduce the weed menace in the direct seeded rice. Soil compaction is a technique by which soil physical properties can be improved. Moderate compaction of sandy loam soils resulted in increased yield of 5.1 t/ha as that of no compaction (3.9 t/ha) and high compaction (4.1 t/ha). Soil moisture deficit at early stages of growth adversely affect the establishment of semi-dry rice. Seedling drought resistance is important for early establishment of plants.

1. To overcome moisture stress condition
2. For better adaptability of seeds to the environmental conditions.
3. To breakdown the seed dormancy both internal and external
4. To resist salinity, alkalinity and heat due to drought

5.3.4.2.3 Time and method of sowing

Early sowing helps in better establishment of the crop and minimizes the exposure of the crop to the moisture stress. Sowing between 15th September, to 1st October is the optimum period for successful cropping in the rainfed areas of Tamil Nadu. Sowing early in June will produce the highest grain yield while delay in sowing causes decline in grain yield. Experimental results showed that drilling of seeds at 100 kg/ha in rows 20 cm apart provide uniform stand of crop at 2-5 cm depth of sowing as compared to broadcasting under semi dry conditions because sowing in lines facilitates easy weeding and other cultural operations.

5.3.4.2.4 Fertilizer management

The use of both organic and inorganic manures has its advantages as well as disadvantages. So it becomes absolutely necessary to integrate both forms of nutrients in a judicious manner with biofertilizers so as to realize the full potential of applied nutrients, thereby increasing the yield. In case of direct seeding under semi-dry condition, which will be changed into a wet lowland condition by a month or two, the use of integrated nutrient management and cultural practices are of immense value. Use of organic manures like FYM, enriched FYM, composted coir pith, enriched biodigested slurry and inorganic fertilizers of N, P and K along with biofertilizers like Azospirillum are combined with soil compaction under semi-dry condition.

Enriched FYM: Application of FYM increased the utilization of 'P' by the crop when mixed with phosphatic fertilizers and make greater release of phosphorus in available form.

Coir compost: Coir pith contains 0.26% N, 0.9% P and 0.78% K. It also contains 29% organic carbon. It is composted by a specific fungi called 'Pleurotus sajor-caju' and applied to rice may enhance the yield and have better moisture conservation.

FYM : Well rotted FYM contains 0.5% N, 0.2% P and 0.5% K. Based on the analysis, 25 t/ha of FYM supplies 112 kg N, 56 kg P and 112 kg of K. Application of FYM reduced the fixation of applied as well as native 'P' and thereby increased its availability to plants in semi-dry rice and enhanced the growth and yield.

Azospirillum : It is one of the most important associate bacterium found in the rhizosphere of the various grass species, capable of fixing atmospheric 'N' to improve the productivity by supplying nitrogen.

A field experiment conducted at TNAU, Coimbatore during 1987 and 1988 to study the effect of organic manures along with inorganic sources on direct seeded semi-dry rice with ADT 36 revealed that manurial treatment with enriched biodigsted slurry @ 750 kg /ha along with seed treatment of Azospirillum favourably influenced the growth and yield components of rice and hence registered an increased grain yield of 4618 kg/ha and 3686 kg/ha during rabi 87 and summer 88 respectively over FYM alone @ 12.5 t/ha and composted coir pith @ 20 t/ha. Hence integrated use of biodigested slurry with inorganic fertilizers in enriched form in combination with Azospirillum is the most effective nutrient management technology for getting higher yield in rice under semi-dry condition (Shanmugasundaram and Vijayakumar, 1989).

5.3.4.2.5 Weed management

Under direct seeded semi-dry condition, weed problem is more serious due to differences in land preparation, lack of water and proper weed management methods. Under such situations, weeds compete with the rice for nutrients, moisture, light and space and thus creating extremely adverse environment resulting in poor crop growth. Moody (1989) estimated 40 to 85% losses in crop yield due to severe weed competition in semi-dry direct seeded rice. Though manual weeding is an efficient method of weed control, it is still costly, laborious and time consuming in direct seeded rice. Hence application of selective herbicide offer the best alternative for weed management in direct seeded rice.

5.3.4.2.6 Water management

The major problem in semi-dry direct seeded rice cultivation is the availability of water at the right time in required quantity. Since uncertainty is the only certainty of our monsoon rains that too with its erratic and uneven distribution, the soil moisture management practices assumes a greater importance. In semi-dry rice, after the receipt of water, the farmer's practice is to submerge the field to more than 5 cm depth. This was mainly due to the lack of experimental evidence on the irrigation aspects of semi-drying. Since the canal water from big reservoirs for irrigation is going to be a big problem in future, farmers can not depend only on submergence. They should be provided with alternate water management practices which will give sufficient leverage for manipulation of sowing date according to water availability and facilitate restriction of water use to the base minimum with reasonability high yields.

5.3.4.2.7 Cropping systems

Many crops are intercropped with direct seeded rice depending on length of the growing season and farmer's preference. Common systems include rice + maize, rice + cowpea, rice + groundnut, rice + sesamum, rice + soybean, rice + black gram. An intercropping system of short duration rice with long duration pigeonpea at 4:1 ratio could be ideal. Intercropping black gram in between the paired rows of rice recorded higher grain yield than in normal rows of rice.

5.4 RATOONING IN RICE

Ideal situation for ratooning is the single cropped lowland areas in the canal and tank-fed systems where water is available for 190 days depending upon the monsoon rains. Farmers traditionally cultivate a single crop of rice with medium duration (130-135 days) either followed by short duration legumes or millets or leave it as fallow. The surplus water can be economically utilized for ratoon crop after main crop harvest. The major advantage of ratoon crop has shorter duration, lower production cost, higher per day productivity and higher net return in a shorter period of time. But for large scale commercial farming, it is not accepted due to low yield, inferior grain quality, pest problems, non availability of good ratooning variety, and lack of management practices. To make ratoon more realistic, productive and economic, inherent ratooning ability of cultivars must be improved and a suitable package of main and ratoon crop management practices developed (Krishnarajan, 1998).

5.4.1 Crop Improvement

Several studies indicated that intermediate to late maturing varieties perform well as ratoon crop in tropical areas. In four trials conducted at various centres in Tamil Nadu revealed that Bhavani (Petax BPI 76) with a maturity period of 85 days showed the best tiller regeneration ability with 4.0 t/ha ratoon yield during 1985 which is 58.0% of the main crop yield at Madurai centre. In 1984 trial, the same variety Bhavani recorded the highest ratoon yield, which was of 86.6% of the main crop yield. Fifty per cent yield and above yield were also obtained from the other varieties in Madurai and Killikulam centres. The other varieties, which registered more than 50 per cent of the main crop yield, were ACM 8, ACM 10 and Co. 37, which registered 80, 54.3 and 50.20 per cent respectively. Though percentage of ratoon crop yield in ACM 8 is high but both main and ratoon crop yield are low. IR.20 was not promising in its tiller regeneration ability, it yielded only 30.7 and 7.5 per cent over main crop during 1984 and 1985 respectively (Krishnarajan, 1998).

5.4.2 Fertilizer management

Different genotypes responded differently in their tiller regeneration ability in terms of grain yield to different levels of fertilizers N,P&K. Fourteen varieties were evaluated from 1984 to 1994 in Ambasamudram, Madurai and Killikulam. Bhavani, the most responsive variety that gave a ratoon yield of 0.82 to 5.3 t/ha at various levels of N (50 to 125 kg/ha). The maximum ratoon crop grain yield of 5.3 and 5.0 t/ha were recorded at 150 and 120 kg N/ha at Madurai centre. IR.20 did not respond favourably to the N levels, it recorded only 0.19 t/ha of ratoon crop yield at N levels ranging from 50 to 125 kg N/ha. Short duration varieties like vaigai, ADT 36, ASD 16, Co 37 and PMK 1 responded favourably to 75 to 150 kg N/ha with a ratoon crop yield varies from 1.91 to 3.7 t/ha. Regarding application of fertilizer, complete basal application immediately after the harvest of the planted crop registered higher grain yield than two splits i.e. half of fertilizers as basal and remaining fertilizer as top dressing 30 days after ratooning and 1/3 rd as basal, 1/2 each at 15 and 30 days after ratooning.

5.4.3 Stubble height

Stubble heights are important in maintaining good ratoon growth, because enough buds are needed on stems or crown tissue to produce rapid-growing shoots or tillers. Eight varieties were evaluated at Ambasamudram, Coimbatore and Madurai for their tiller regeneration ability at different stubble heights ranging from close to ground level to 30 cm at intervals of 5 and 10 cm. Again, the variety most responsive to different stubble heights was Bhavani and the next best was Vaigai. Most varieties performed well at 20 cm stubble height. IR 8 on the contrary, yielded the maximum (2.9 t/ha) when cut close to the ground. IR.20 and Ponni gave low yields when cut at 15 cm. Tiller regeneration ability in relation to different stubble heights apparently varies with the genotype.

5.4.4 Economics

Economic studies at Agricultural College and Research Institute, Madurai during 1987 showed that between Bhavani and Ponni ratoon rice, Bhavani registered higher net return per rupee invested. A net return of 2.38 per rupee invested was recorded in Bhavani as compared to Ponni, which was recorded 2.04 per rupee invested. In 1989 trial revealed that higher net profit of 12,611 was obtained in White Ponni as compared to Bhavani and Vaigai, which were, recorded Rs. 10,956 and Rs. 7,487 net profit respectively in the planted crop. But in the ratoon crop, 2.12 per rupee invested was obtained in Bhavani as compared to white Ponni and Vaigai which recorded Rs.1.63 and 1.45 per rupee invested respectively. In the fertilizers treatment, a fertilizer recommendation of 100:50:50 kg NPK/ha recorded higher net return which was on per with 100:25:0 kg NPK/ha. The trial at Killikulam showed that in the main crop, the net profit and the cost benefit ratio were maximum in Co.37. The cost of cultivation was comparatively less in ratoon crop than main crop. This is mainly due to no cost improvement for field preparation, seeds and nursery. In the ratoon crop, the varieties Co.37 and PMK-1 performed better.

5.5. OTHER MANAGEMENT PRACTICES

Other management practices like tillage operations, nutrient management, and water management are discussed in detail in the subsequent chapters.

CHAPTER 6

RICE TILLAGE AND RICE MECHANIZATION

Farm implements and machinery are the inputs essential for increasing rice production. Labourer shortage, high cost of labour, low animal power availability and increased works due to the introduction of high yielding varieties have also increased the demand for farm implements and machinery which are used for tillage operations. The word tillage is derived from the Anglo-Saxon words 'tilian' and 'teolian', meaning to plough and prepare soil for seed to sow, to cultivate and to raise crops. **Jethrotull,** who told about thorough ploughing is necessary to make the soil into fine particles, is considered as "Father of tillage".

Manipulation of the soil with tools and implements for obtaining conditions ideal for seed germination, seedling establishment and growth of crops is called tillage or Tillage is defined as a series of mechanical manipulation of the top soil to provide necessary soil conditions favourable for the germination and growth of crops. Tilth is the qualitative characteristics of loose friable and crumby condition of the soil favourable for crop production. Tilth is the physical condition of soil resulting from tillage. When the soil is brought to such a condition ideal for crop growth is called a seed - bed.

Tillage operations are to be done at the right time with the right implements. The tillage operations are carried out with the following objectives:

6.1 OBJECTIVES

- **To improve the physical conditions of soil for good germination and crop growth**
- **To incorporate crop residues and to mix the applied manures and fertilizers with the soil**
- **To break hard soil pans, if any, to increase the soil depth for water absorption and to improve drainage**
- **To ensure adequate aeration to root zone for better microbial and biochemical activities.**
- **To conserve soil moisture and improve moisture retention capacity of the soil by breaking hard pan and crust formed on soil sub surface**
- **To increase the activities of soil microbes**
- **To control and to certain extent pests and diseases which harbor in the soil.**
- **To control soil erosion i.e. soil conservation**
- **To level the field for efficient water management**
- **To control weed and to incorporate**
- **To improve soil porosity**
- **To provide an environment for getting adequate light to the seedling (weed free environment)**

6.2 TYPES OF TILLAGE

Tillage operations can be grouped into on- season tillage, off-season tillage and special types of tillage. For the benefit of readers, the details in short are given as below.

6.2.1 On season tillage

Tillage for a crop from the start of crop season to harvest i.e., cropping season is known as on - season tillage operations. It includes preparatory tillage and inter tillage.

6.2.1.1 Preparatory tillage

Preparatory tillage refers to tillage operations for raising a crop up to sowing or planting as the case may be. It is of two types namely primary and secondary tillage.

Primary tillage starts with onset of the season. Arable soils on an average are ploughed twice, each time in diagonally opposite directions. The objective is to cut, lift, twist and turn the soil while incorporating organic and animal wastes into ploughed layer. It aids in improving the soil moisture regime for subsequent operations even if the rains are delayed.

Secondary tillage includes operations for obtaining seed bed for sowing. Since the soil is in loose condition due to primary tillage there may not be any necessity for ploughing the soil. Implements like Blade harrows are worked once or twice to kill the weeds, break up to clods and to obtain relatively smooth and leveled surface for subsequent sowing.

6.2.1.2 Inter tillage

After the crop is sown, the soil may need further tillage to control weeds, break up crusting or create soil mulch. All such operations up to crop harvest is known as inter tillage or inter cultural operations.

6.2.2 Off season tillage

Tillage operations during uncropped season for special purposes other than that for immediately raising the crop in the season are said to be off-season tillage. Stubble or post harvest tillage, summer tillage, winter tillage and fallow tillage will constitute off-season tillage practices.

a. Stubble or post harvest tillage

Tillage operations are carried out immediately after the harvest of crop to clear off the weeds and crop residues and to restore the soil structure. Thus making the soil ready to store rainwater.

b. Summer tillage

Taking advantages of summer rains the field is ploughed deep to eradicate perennial weeds and to destroy soil borne pathogens, pests and parasites and to check erosion. This tillage is known as summer tillage and this will be carried out in all types of soil.

c. Winter tillage

It is practiced in temperate regions where the winter is severe that makes the field unfit for raising crops. Ploughing or harrowing is done in places where soil condition is optimum to destroy weeds and to improve physical condition of the soil by incorporating plant residues.

d. Fallow tillage

It is also called as mulch tillage or clean tillage. Vertisols are often left uncropped during a part of the rainy season due to unfavourable weather or soil conditions or both. The soil is tilled during this period to control weeds and to create soil mulch for conserving moisture to subsequent crops grown during post rainy season. This practice is called as fallow tillage.

6.2.3 Special type of tillage

The tillage operations carried out to perform specific operations to rectify specific problems are grouped under special type of tillage. They are year round tillage, sub soiling, leveling tillage, wet tillage, strip tillage, ridge, contour tillage and blind tillage.

a. Year round tillage

In dry land agriculture, tillage initiated with the on set of summer showers is continued periodically until sowing, during crop period and even after the crop harvest to avoid weed growth, hard pans and soil water erosion besides conserving soil moisture. Such tillage is referred as year-round tillage.

b. Sub soiling

This type of tillage is carried out to cut open or break the subsoil hard pan or plough pan with the help of sub soil plough or chisel plough. During the tillage, soil is not inverted. Sub soil tillage is done once in 4-5 years in the fields where heavy machineries are used for field operations and where there is a loss of topsoil occurs. Soil type also plays a major role in the formation of hardpan.

c. Leveling tillage

For low land crops like rice, fields require a uniform distribution of water and plant nutrition for better crop growth. This is possible wherein the fields are fairly leveled. Levelers and scrapers are used for this operation. The advantages of leveling are that soil erosion is restricted and other management practices become easy with minimum effort.

d. Wet tillage

Puddling practice for rice is called as wet tillage. Manipulation of soil under water with tools and implements to break down the soil aggregates into colloidal condition is called puddling and the soil is called as puddled soil. The objectives of puddling are bringing the ideal soil condition for planting seedlings, reducing deep percolation losses of water and weed control. Detailed information are given in the subsequent sections of this chapter.

e. Strip tillage

Ploughing is done as a narrow strip by mixing and tillering the soil leaving the remaining soil surface undisturbed.

f. Ridge tillage

It refers to forming ridges by ridge former or ridge plough for the purpose of planting or in site water harvesting. In low rainfall areas, this tillage practice is of much use in moisture conservation.

g. Contour tillage

Tilling of the land along contours and across the slope (contour means lines of uniform elevation) in order to reduce soil erosion and run off referred as contour tillage.

h. Blind tillage

Tillage done after seeding or planting the crop (in a sterile soil), either at the pre-emergence stage of the crop plants or while they are in early stages of growth so that crop plants (cereals, tubers etc) do not get damaged, but extra plants and broad leaved weeds are uprooted and this practice is called as blind tillage.

6.3 RICE - TILLAGE IMPLEMENTS

Implements are those devices used to carry out some work and when the device is used to do some activity before or during or after raising of a crop can be called as tillage implements. Implements can be operated either with human power or animal power or machineries. The earliest tools used by men were digging sticks and simple hoes. As science advances to increase the energy efficiency and cost efficiency modern tools are invented for use. Tillage operations vary according to cultivation practices, water availability, soil texture, topography and source of farm power and equipment used.

6.3.1 Primary Tillage Implements

Primary tillage is the deepest operation performed during the period between two crops. Primary tillage is the initial breaking, ploughing and manipulating of soil from 15-90 cm depth of soil, where as secondary tillage includes seed bed preparation and interculture to a depth of 15 cm. Primary tillage results in burying of weeds and crop stubbles/residues, where as secondary tillage operation causes breaking of clods and mixing of crop residues. The following are the implements used for primary tillage operations in rice field.

1. Country plough
2. Mould board plough
3. Disc plough
4. Reversible disc plough
5. Turn-wrest plough
6. Bose plough
7. Chisel plough and
8. Rotavators and rotary tillers

1. Country (Desi) plough

The indigenous plough consists of a wooden body to which a handle and a shaft pole are attached. The body is made of a bent piece of hard wood with arms making 135° angle thus giving a wedge shape. A small piece of flat iron (share) serves as the piercing point of the plough and is fixed over the plough body with clamps. The one end of the shaft pole is fixed secured with the yoke during working. The working of plough results in the opening of 'V' shaped furrow, the depth and width of which may vary according to size and set up of the plough bottom. The traditional country ploughs are given in Plate 9.

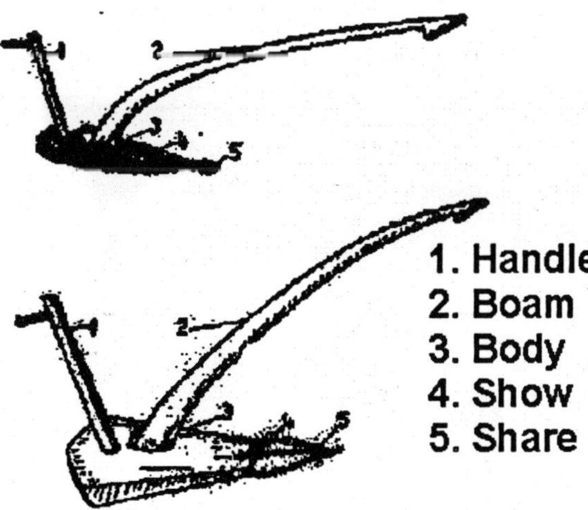

Plate 9. Traditional country ploughs

Adjustments in country plough

The depth of penetration of a country plough can be altered by pulling the implement behind or pushing towards the front thus resulting in deeper or shallow ploughings respectively. First ploughing is done length wise in order to reduce the number of turns at the head. The field is later ploughed cross wise to have thorough coverage of the field.

Ploughing with country plough

The ploughing is started from the right side of the field as the animals are turned towards their left side. First furrow is opened parallel to the end of the field and is taken in the headland and turned at a distance of about 20 feet. The plough is taken along the headlands and second furrow is opened close to the previous furrow. The plough is worked round and round concentrically turning to the left always till nearly half the area of the strip taken for ploughing is completed leaving half the unploughed area in the middle. At this stage, new strip of similar width is taken and proceeded to cover the entire area.

2. Mould board plough

It is a modern tillage implement used to plough deeply and pulverizes the soil. It is more durable, easy to pull and can be adjusted properly. The main parts of the mould board plough are the body, handle, beam, clevis, share mould board, landslide, Wheel and coulter.

Plate 10. Tractor drawn mould board plough.

Body: It is the central piece to which all other parts are attached. The lower portion of the body containing the main working parts is called plough bottom. The side on which the furrow slice laid is called the mould board side.

Handle: It is usually fixed to the body at an angle of 45° and it helps in guiding the plough while ploughing and lifting it while turning at the headlands.

Beam : It is fixed to the plough bottom and connected to the yoke by means of a chain or rope.

Share : It penetrates the soil and makes the passage for the plough bottom. It is made of steel or cast iron. The share makes the horizontal cut and separates the furrow slice from the soil below. The furrow slice moves up the share and gets ready for being twisted and turned by mould board.

The mould board : The furrow slice separated by the share is received by the mould board and is passed over its surface, invert it completely and leaves it resting over the previous furrow. When the furrow slice is being twisted and inverted it, gets broken and pulverized to an extent depending upon the share and curvature of the mould board.

The landslide: It slides along the wall of the furrow and is kept pressed against it by the thrust of the furrow slice on the mould board. It is the support provided by the furrow wall to the landslide that keeps the plough in position and prevents it from falling down under the heavy thrust of the furrow slice. The landslide keeps the plough steady. Landslide is made of cast iron or steel and is usually rectangular in shape and is fixed to the plough bottom. The under side is called "Sole" and the hind part is called "Heel".

Wheel: Wheels are fitted to the end of the beam of heavy ploughs for assisting the ploughman while steadying them, there by ploughing is rendered easier and less tiring. In case of heavier plough, two wheel of different size are fitted, the small wheel resting on the unploughed land is called the "land wheel" and the bigger wheel rests on the previous furrow and is called "furrow wheel".

Coulter: The coulter is passed through a socket in front of the share and clamped to the beam. It is really a knife, which separates the furrow slice from the unploughed land. This type of coulter is called knife coulter. The other type skim coulter is simply a plough body in miniature and is clamped to the beam in front of the coulter where weed growth or manure on the surface require burying.

Ploughing with mould board : By mould board the field is ploughed either by gathering method or by splitting method.

a. Gathering: Ploughing is done from center of the plot and is turned round and round always mould board side. The first furrow is opened in the center and the next furrow is taken by its side over which the first furrow slice had already been laid. The second and third furrows are laid on both the sides of this ridge till the plot is completed. At the end, there will be ridge left over in the center.

b. Splitting : Ploughing is commenced at the two opposite edges of the plot continued inward finally finished with a central furrow. Plough is always turned towards landside. At the end field is left over with a furrow in the center.

Plate 11. Tractor drawn disc plough

3. Disc plough

In the disc plough, the share, the mould board and coulter of the mould board plough are replaced by an inclined concave steel disc of 60-90 cm dia meter set at an angle to the direction of travel. Each disc revolves on a axle; and the angle of the disc to the vertical position and to the furrow wall is adjustable. Lever arrangements are provided to lift the discs clear off the ground and for changing the angle of moulding and adjusting the depth of penetration of the discs into the soil. While working the discs, rotate, scoop out furrows invert the furrow slice and pulverize them thoroughly. Disc plough is especially useful under the following soil conditions.

- Soil with hard pan and sticky soil
- Dry hard ground

- Rough and stony ground
- Ploughing weedy lands and lands with stubbles
- Deep ploughing

4. Reversible disc plough

It is constructed in such a way that the disc can be reversed and the soil is thrown on one side at all sides. The land and furrow wheel adjust themselves properly when the plough is reversed. This saves time taken up by ordinary disc plough. The furrow slice cut at each trip by the reversible disc plough is laid over the previous furrow thus resulting in a leveled field after ploughing.

5. Turn-wrest plough

This is also called reversible plough or one-way plough. The mould board with share is hinged either left or right side of the central body with a hook. When the mould board is needed to shift to another side, the hook is released and the mould board is easily swing to the opposite side. By this when the plough is reached to one end of the land it is not necessary to turn the plough, instead the mould board can be shifted to other side and the ploughing continued.

6. Bose plough (Melur plough)

It is only a wooden plough, which takes a mould board cum share instead of the usual small iron share. It is commonly used in the sandy soils.

7. Chisel plough/Subsoil plough

It is a tractor/bullock drawn implement used to break hard pans that exist in the soil due to continuous adopt of the same type of operation. It consists of curved chisel like type of 37 cm radius of curvature and 3 cm thickness. It is rigidly held in a frame, which is provided with a handle and a shaft pole. It makes a simple vertical cut in the sub soil up to a depth of 45 cm and facilitates the downward movement of water and sub soil drainage. Chiseling becomes necessary in soil with hard impermeable plough pan.

8. Rotavators and rotary tillers

The rotavators are very effective tillage implements for preparing a fine seed bed, weed control and mixing of soil with crop residues. This is also used in puddling operations in some areas. A rotavator consists of many blades, varying in size and shape and mounted on the periphery of a rotating shaft. The shaft gets the drive either from a tractor PTO or from the power tiller transmission unit. A safety guard is provided on the rear to prevent injury to the operator from the rotating blades.

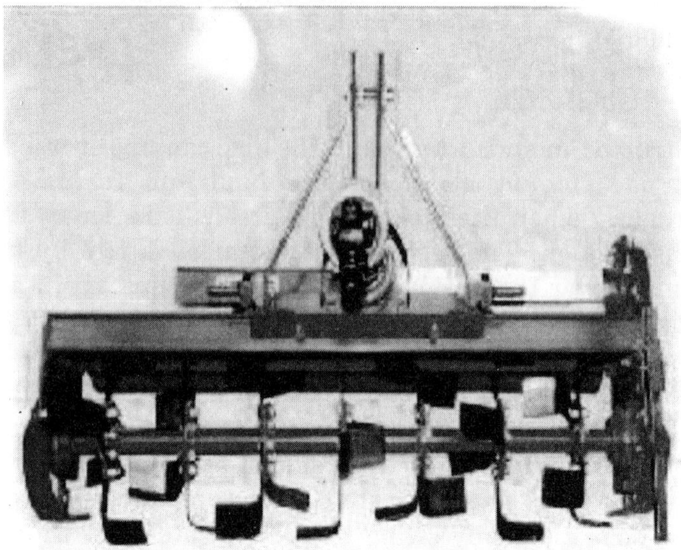

Plate 12. Tractor operated rotavators

6.3.2 Secondary Tillage Implements

The major implements used for secondary tillage operations in rice cultivation include

1. Cultivators
2. Harrows
3. Planker

These implements are used after ploughs for

- Breaking up the furrow slice and working the soil to the required tilth
- Destruction of Weeds
- Stirring the soil and forming mulch
- Mixing of manures and fertilizers with soil
- Covering the seeds

1. Cultivators

Cultivators usually have two on three rows of tynes. The main aim of mounting tynes in two or three rows is to provide a clearance between adjacent tynes about 18-23 cm through which, clods and rubbish can pass freely. Usually tynes are 23-30 cm long, capable of penetrating to a depth of 20 cm. There are many sizes of cultivators in use. The larger ones are designed for tractor mounting. These are used when the soil is ploughed deeply with heavy ploughs to break the big clods that are formed.

2. Harrows

These are also many tyned implements like cultivator, but smaller in size. These are used for breaking the small clods left unbroken by the cultivators and for producing a powdery seedbed. The tynes are set closer to one another and are smaller in size. They open tracks 5 to 8 cm apart and penetrate to a depth of 10 cm. Some of the different types of harrows are. 1. Spike tooth harrow 2. Chain harrow 3. Disc harrow 4. Spring tyne harrow 5. Peg tooth harrow and 6. Blade harrows (Guntakas).

Plate 13. Tractor drawn offset disc harrow

a. Spike tooth harrow

Harrows are teeth shaped having 12.5-25.0 cm length. They are of different shaped such as round, square, rectangular, triangular etc. It can also be called as zigzag harrow.

b. Chain harrow

Chain harrows are used for breaking clods and making smooth surface evenly. It can also be used for covering seeds after broadcasting

c. Disc harrow

They are suitable for hard ground with grassy weed infestation. This is commonly used in India.

d. Spring harrow

These types of harrow are fitted with springs having loops of elliptical shape in the place of rigid tynes. These harrows are suitable for hard and stony fields.

e. Guntaka

It is comparatively large harrows used for stirring the soil and to cover the sown seeds. There are three different types.

i. Meesalu guntaka

The blade is not continuous as that of ordinary ones. The blade has an opening through which the cropped rows pass.

ii. Metla guntake

Six numbers of small blades are fixed to the beam at about 30cm apart by means of a standard. The blades are of 15 cm length pass through the inter space.

iii. Reggala guntaka (R.E. guntaka)

This guntaka can be used for multipurpose operations like harvesting groundnut and other intercultural operations.

6.3.3. Intercultural Implements

1. Japanese rotary weeder

It consists of two small-toothed rollers or drums mounted on a frame provided with handle. Each roller consists of about 5-toothed blades. This implement, while working is pushed and pulled alternatively by the operator in between rows of paddy crop. The float provided will guide the implement smoothly while working and prevent the implement sinking in the puddle soil. The idea of using this weeder is to bury the weeds in the mud so as to decompose them and add organic matter to the soil.

The details regarding cono weeders are given in the subsequent section of this chapter.

Plate 14. Rice rotary weeder

6.3.4. Land Shaping and Levelling Implements

The following land shaping and leveling implements are used for rice cultivation.

a. Wooden float

It is a long sledge like drag with which is used for land smoothening with animal power. Sometimes large size floats are also used with tractors. For better smoothening on the flat fields, it is desirable to float the surface three or more times by working the field lengthwise, crosswise and diagonally. In some areas, farmers are using yoke for levelling main field (Plate 15).

b. Buck scraper

It is a bullock drawn implement used for leveling land. It is made with steel sheets like an open box with a bottom and three sides; the fourth side is left open. The sides are formed by bending the sheet lapping the ends one over the other and riveting them. Two flat steel runners are provided at the bottom to protect the base and prevent the sheet from being worn out by dragging over the land. Two handles are fixed at the sides for assisting in filling the box with earth and for emptying the contents.

Plate 15. Using yoke for mainfield levelling

b. Rollers

Iron rollers are used for breaking clods and compacting the soil. A number of cast iron rings, about 0.6 m in diameter are fixed to an axle and provided with a hooked frame for hitching with a power unit. The surface of the roller is either plain, fluted or ribbed. The fluted rollers concentrate their weight over the projecting ribs and are more efficient in breaking the clods. The plain surface rollers compact the soil evenly. This is used especially in the areas where the clay content is very high i.e., fluffy paddy soils.

c. Leveling board

It is made of wood with 2-2.75 m length and about 20-30 cm in diameter. It is attached to the shaft pole with a hinged hook and used for leveling rice fields, after the final ploughing. When the driver stands over the board, it sinks lightly into to the loose mud and when it moves, the soil in front of it also, is moved but is released when he gets down. For micro levelling, the labourers may be engaged (Plate 16 and 17).

6.3.5. Special Purpose Implements

Special purpose implements for rice cultivation like puddlers, transplanters etc., are discussed in the sections 6.4 and 6.5 of this chapter.

Plate 16. Manual microlevelling in rice mainfield

Plate 17. Main field levelling using wooden levellor

6.4. RICE MECHANIZATION

It is estimated that nearly 145 man-days are required per hectare of rice. Among the various operations, planting accounts for 21 per cent of total human labour requirement. Harvesting and threshing require 43.4 per cent of total labour. Weeding is another operation where human labour input is high (De Datta, 1981). From a study on gender concerns in rice farming, Singh *et al.* (1991) quote that women who constitute a majority of agricultural labour force, perceive drudgery of work as a major constraint for productive efficiency of labour. They have emphasized the need for light tools and implements for planting and weeding and light machinery for threshing. Mechanization can reduce drudgery and enhance timeliness. Women liberated from long arduous hours of work will be able to attend to family needs, household chores and other off-farm income generating activities. With the advent of short duration rice varieties and efficient soil fertility management technologies, crop intensification in rice farming has become a reality. This has led to a reduction in turnaround time between crops. Earliness and timeliness in farm operations have become essential. Exclusive dependence on human labour will become counter productive since it has become scarce and costly in many rice growing regions.

Very often, farm mechanization is equated with just tractorisation or automation of farm operations. In the Indian context, mechanization to fully substitute human labour may meet with social dissensions. But mechanization as a tool to reduce drudgery and to increase labour efficiency and productivity will be welcome. With about 35 per cent of operational land holdings in the size group of lesser than 0.2 ha, the small size and fragmented parcels of land pose operational problems for large machinery in rice farming (Ohja, 1988). Nevertheless, mechanization in rice farming has taken roots in this country at least in those operations in which timeliness and efficiency are hit by high cost and shortage of human and bullock labour and where labour efficiency suffers due to drudgery. Here, mechanization is contemplated to increase production, improve labour efficiency, reduce the cost of operation and to maximize the efficiency of costly inputs like nutrients and water.

6.4.1 Past works on rice mechanization

6.4.1.1 Land preparation

Land preparation is the most important operation in rice cultivation. Power tillers and tractors are extensively employed for land preparation. As farmers were demanding a cheaper alternative, a low cost 5HP garden tractor with a rotary tiller using locally available components, which costs half the cost of a commercial power tiller was developed. The rotary tiller can be operated up to a depth of 15 cm with an effective width of 32 cm. The actual field capacity of the machine was 0.054 ha hr^{-1} and the quantity of fuel required to operate the machine was 0.86 lit hr^{-1}. The cost of the rotary tillage attachment is Rs.1500 and the cost of ploughing with the machine is Rs. 450 per ha. The unit could be used for dry ploughing (Jose, 1990).

Ahamed and Sivaswamy (1994) conducted experiments in farmer's fields with the objective of suggesting technically feasible and economically viable farm mechanization packages for rice farms of Kerala. The trial comprised three mechanization packages, namely low (LMP), medium (MMP) and high (HMP) levels, comprising 24 types of farm implements and machinery, and were, conducted in nine locations. Both the improved packages, namely, HMP and MMP, required lesser labourers resulting in higher net returns than the conventional package. The labour input (man days ha^{-1}) for medium and high level mechanization packages was 127.22 and 107.88 compared to 136.44 for low level package respectively. The corresponding B:C ratios were 1.62, 1.72 and 1.49 respectively.

An operation wise analysis of the effect of selective mechanization on the economics of rice production in three cultural systems viz., transplanted, wet seeded and dry seeded was conducted by James *et al.* (1996). The machines / implements adopted in partly mechanized system of the present study were power tiller (land preparation and transport), IRRI six row manual transplanting machine, power sprayer, manual weeder, IRRI self propelled reaper, paddy thresher and winnower, the total input cost (Rs. ha^{-1}) under non mechanized system for transplanted, wet seeded and dry seeded cultivation was Rs. 22,341, Rs. 20,941 and Rs. 20,053, with the B:C ratios of 1.17, 1.25 and 1.31 respectively. The corresponding values for the partly mechanized cultivation were Rs. 16,757, Rs. 18,756 and Rs.17,462 with B:C. values 1.56, 1.45 and 1.50 respectively. This indicates that only selective mechanization in paddy cropping can sustain the rice cultivation.

6.4.1.2 Seeding and transplanting

In Tamil Nadu and Kerala, mechanical seeding as well as transplanting is not common. A number of research works have been taken for improving the existing machines and developing new machines for seeding and transplanting.

6.4.1.2.1 Paddy dibbler

John and Mohammad (1989) developed a five row bullock paddy dibbler which could drop 4 to 6 seeds per hill spaced at 10 cm apart and cover an area of 0.083 ha hr^{-1}. The labour requirement was 12 man hr ha^{-1}. The cost of dibbler was Rs. 1,800 and cost of operations worked out to be Rs. 203.13 ha^{-1}. A manually operated three row paddy dibbler for dry sowing was developed by Bini (1992), could drop 4 to 6 seeds per hill. The area covered by the dibbler was 0.022 ha hr^{-1} with a field efficiency of 68.68%. The cost of the machine was Rs. 800 and the cost for sowing one hectare of land was Rs.717 compared to Rs. 979 for the manual dibbling method.

Another manually operated 2 row 2 column paddy dibbler designed by Menon *et al.* (1993) dibbled on an average of 2.67 seeds per hill. Hence, only 31.37 kg of seeds were needed to dibble one ha at the placement depth of 5 cm. A 2 row power operated paddy dibbler designed and fabricated by Jayarajan (1996) dibbled 5 seeds per hill. This dibbler required 78 kg of seed ha^{-1}. The depth of placement varied from 3.8 to 4.6 cm at the spacing of 20 to 21.5 cm. The operating cost was Rs. 502.58 ha^{-1}. The field capacity was 0.031 ha hr^{-1} and required 32.34 man hours ha^{-1}.

6.4.1.2.2 Transplanter

Bainu *et al.,* (1994) modified the IRRI six row manually operated rice transplanter for conventional seedlings. After the modification, the average number of seedlings per hill could be reduced from 6 to 2. Missing hills were reduced from 20.83 to 5.5%. Actual field capacity was increased from 0.0139 to 0.0162 ha hr^{-1}. The cost for transplanting by the modified transplanter was worked out as Rs. 432 ha^{-1} while manual transplanting needed Rs.1050 ha^{-1}. Prakash and Sivaswamy (1995) developed a power tiller operated paddy transplanter for conventional seedlings. The machine planted on an average of 55 hills m^{-2} @ 3 seedlings per hill. When the power tiller was operated at a forward speed of 1.2 km hr^{-1}, the transplanter could transplant 2 to 4 seedlings per hill at a spacing of 10 to 12 cm. The average field capacity was found to be 0.13 ha hr^{-1}. The conventional paddy nursery seedlings having 20 days of age, pulled out, root washed and bundled as for manual transplanting was used for this transplanter. The results on the performance evaluation of 8 row mechanical transplanter indicated that in order to raise mat nursery for mechanical transplanting a seed rate between 0.4 and 0.6 kg m^{-2} was sufficient to produce a mat giving seedlings at the rate of 3-4 plants per hill while transplanting. The percentage of missing hills was between 13.5 and 15.6%, and a uniform seedling distribution per hill was observed at lower seed requirement. The variety Jyothi yielded 7678 and Annapurna gave 6797 kg grain ha^{-1} under mechanical transplanted conditions. The corresponding yields under labour transplanted situations were 5913, 6953 kg ha^{-1} respectively. Mechanical transplanting required only Rs. 530 ha^{-1} against the manual cost of Rs. 2300 ha^{-1} (Jaikumaran and Beena, 1996).

6.4.1.3 Weeding

Weed control is one of the important intercultural operations that need mechanization. Kumar *et al.* (1990) developed a power operated paddy weeder for rainfed row crop rice cultivation. The cutting mechanism consisted of 6 wheels mounted on a common shaft and the prime mover is 1.5 HP kerosene engine. The average working speed was 3 m min^{-1} and the area covered was 20 m^2 hr^{-1}.

6.4.1.4 Harvest operations

6.4.1.4.1 Harvesting

Sharma and Mathew (1990) tested the 1.0m IRRI model vertical conveyor reaper. The field capacity of the reaper was 0.135 ha hr^{-1}. The cost of operation of the machine was Rs. 265 ha^{-1}. The cost of harvesting one tonne of grain was Rs. 44.3. Selvan (1995) modified the self propelled vertical conveyer reaper to be operated by a 8-10 HP air cooled Mitsubishi power tiller. The average field capacity of the machine was 0.2 ha hr^{-1}. It was suitable for harvesting paddy both in wet as well as dry fields except the fully lodged crops. A saving of 186 man hrs ha^{-1} was achieved by using this reaper and the cost of operation was Rs. 348 ha^{-1}. A tractor front mounted 2.2m wide paddy reaper windrower was evaluated by Sushilendra (1996) to find out the optimum forward speed for maximum field capacity and field efficiency. The actual field capacity was 0.38 ha hr^{-1} and field efficiency was 54% at an engine speed of 1500 rpm with forward speed of 0.95 ms^{-1}. A simple crop collection unit was developed with the provision for conveying the windrowed crop directly into the collector box. The box was filled within 10 m of travel. The cost of this reaper was Rs. 25,000/-.

6.4.1.4.2 Threshing

John (1991) developed and tested a low cost wire loop type portable power operated paddy thresher. The threshing capacity of the machine was 451.84 kg of paddy hr^{-1}. It was operated by a 0.5 HP motor and two labourers were required for the whole operation. The cost of the thresher and operation was Rs. 3,600 and Rs. 3 per quintal respectively. A hold on type paddy thresher was evaluated by Singh and Ramachandran (1992). The machine was driven by a single phase 2 HP electric motor and 2 persons were handling the whole operation. When it runs at an optimum peripheral velocity of 720 m min^{-1} and at the grain moisture content of 16%, the thresher produced maximum output of 307.16 kg hr^{-1}. The cost of operation per hour was estimated to be Rs.17.5/-. The 8 HP axial flow spike tooth type and 10 HP flow through rasp bar type threshers were evaluated by Sailaja (1993) and observed an average out put of 50kg per HP hr and 107 kg per HP hr respectively. The maximum threshing efficiency of 98.77% and 97.45% was recorded for the flow through rasp bar thresher during *virippu* and *mundakan* seasons compared to only 95.5% and 94.49% respectively for spike tooth thresher for respective seasons. The cost of operation for the commonly available axial flow spike tooth and the flow through rasp bar threshers was Rs. 400 ha^{-1} and Rs. 207 ha^{-1} respectively.

Kumar (1996) developed and evaluated a low cost power operated paddy thresher cum winnower. The power required to operate this wire loop type machine was supplied by a 2 HP electric motor. The capacity of the machine was found to be 300 kg hr^{-1} at grain moisture content of 16%. Two labourers are required for operating the machine and the cost of the machine was Rs. 8,000/-. Further the cost of operation for threshing and winnowing was Rs. 8.5 per quintal.

6.5 RICE MECHANIZATION

6.5.1 Mechanization in land preparation

The source of power for land preparation in rice fields was exclusively bullock power until recently. In recent years, however, the farmers in the background of dwindling bullock power have accepted the use of tractors and power tillers for ploughing and puddling. Although individual small farmers could not own the costly tractors, the practice of custom hiring has made ploughing and puddling with cage wheel fitted tractors an economically viable practice (Plate 18). The power tiller, a smaller machine for land preparation has proved to be an acceptable option for small farmers. It is relatively cheaper than a tractor, its efficiency is reasonably high and it is easily maneuverable from field to field in smallholdings (Plate 19 and 20). Bullock power however, continues to be the main source for land preparation in many rice-growing areas of Tamil Nadu. In such conditions, wetland puddlers can reduce the time and cost of puddling by 50-60 per cent in comparison to conventional ploughs (Srivastava, 1990).

Plate 18. Tractor operated cagewheel and cultivator puddling in rice field.

Plate 19. Pudding rice field with power timer operated cagewheel and rotavator.

Plate 20. Puddling rice field with power tiller operated cage wheel

6.5.1.1 Land preparation equipments

The area covered by Country plough and Helical puddler (Plate 21) will be around 0.016 and 0.06 ha /hr. Heavy machinery in its wake brings new problems. Kuether and Cabrales (1979) observed that in many of the rice growing countries, the fields were reported to develop deepening hardpans as a result of intensive cultivation using heavy tillage equipments. If this condition persists, tractors and draught animals may become very slow or even get bogged down. At the international Rice Research Institute (IRRI), Philipp-ines, a float type tiller called **"Hydrotiller"** has been developed to solve this problem. It can till extremely soft soil twice as fast as hand tractor. The total land preparation process with hand tractor tillage requires about 5-8 passes to finish a hectare of field; the hydrotiller can finish the same area in just two passes (Manaligod and Quick, 1991).

Efforts to improve rice farming machinery for China have resulted in the development of a **"Combined Tillage Equipment'** for paddy fields (Zhu Shen You and Meng All Shen, 1991). It performs the functions of ploughing, harrowing and leveling simultaneously. Chinese farmers have observed that it costs half that of cattle but its efficiency is three times that of cattle.

Plate 21. Helical blade puddler

6.5.2 Mechanization in crop establishment

Advance in varietal development, pest and disease management and nutrient application have increased the yield potential of rice. Realization of this potential depends on crop establishment to ensure optimal population. Transplanting of seedlings is the major means of crop establishment in low land rice. Non-availability of labour in time, high wages and lower efficiency are often quoted as constraints in transplanting with manual labour. Delay in planting, aged seedlings and sub optimal population reduce the yield. Early completion of planting in a command area ensures synchronization of planting in vast area and it favours effective pest management and efficient water management. Non co-operation by planting labourers engaged on contract basis has been quoted by farmers as a constraint for adoption of optimum population in rice in Tambiraparani Command Area (Rangarajan and Veerabadran, 1993). The alternatives for manual transplanting are the mechanization of transplanting and direct seeding.

Plate 22. Direct rice seeder in operation

The IRRI designed **rice transplanter** can be used for transplanting to relieve the drudgery of human labour. This transplanter weighs only 20 kg and needs two men labour. Special type of seedlings received from mat nursery has to be used. The root-washed seedlings from conventional wet nurseries are not suitable. It covers an area of 0.04 ha per day (Tajudeen, 1995). Chinese experience in rice transplanters, which started as early as 1950's indicate that the manual transplanter operated by two persons, can transplant 0.27 to 0.4 ha per day. Its efficiency is reported to be 4-6 times higher than planting by hand (Zhu Shen You and Meng All Shen, 1991). Direct seeding of rice in low land puddle soil is perceived as an alternative for transplanting to overcome labour scarcity and delay in planting. But direct seeding by broadcasting has many limitations such as uneven crop stand, difficulty in thinning and gap filling and inefficient manual weeding. In this context, the **wetland drum seeder** developed by Tamil Nadu Agricultural University based on IRRI model is a handy implement for wet seeding. The manually drawn drum seeder is light in weight and has a simple mechanism. It consists of four sheet metal drums with holes to meter the seeds, two floating skids and a centrally mounted ground wheel that drives the drums. The seeder has a width of 1.8 m and in a single run, it can sow eight rows spaced at 20 cm. It can cover an area of 1.0 ha in a day and requires a seed rate of 80 kg ha^{-1} (Purushothaman *et al.,* 1996). Studies on the use of drum seeder have brought out many advantages : cost reduction, faster growth and establishment, easiness intercultural oper ation, lesser seed rate and higher yield compared to broad casting (Pinar *et al.*, 1994 and Gao, 1995). The seed rate requirement with drum seeder was 80 kg ha^{-1} as against 100 kg ha^{-1} for broadcasting. The best advantage with drum seeder for wet seeding was realized by providing for an intra row spacing of 2 cm by using the appropriate holes in the seed drum. Widening the intra row spacing to 3.5 or 4.7 cm did not confer any significant advantage over broadcasting. The productive tillers at 20 x 20 cm spacing with drum seeder were 413 m^{-2} compared to 265 m^{-2} under broadcasting. Drum seeding resulted in additional grain yield of 640 kg ha^{-1} (16 per cent increase over broadcasting) (Venkatachalapathy and Veerabadran, 1998).

Plate 23. Direct rice cum daincha seeder - 3 rows

6.5.3 Mechanization in harvesting and threshing

This final operation in rice cultivation consumes nearly 40 per cent of total manual labour requirement. The conventional method of harvesting by sickles and threshing by manual labour and animals requires 40 labourers, out of which, 25 are required for cutting and laying the crop (Tajudeen and Job, 1997). In intensive cropping systems, the time between harvest of first crop and land preparation for next crop must be substantially reduced to ensure timeliness. Delay in harvest due to labour scarcity is reported to result in 10 per cent loss of produce. Mechanization of harvest and threshing can therefore help in early and timely harvest, save labour and reduce the yield loss.

Tractor drawn, power tiller operated or self-propelled - reaper harvesters have been developed and proved to be efficient for rice (Srivastava, 1990).

Harvesting method	output	
	ha/h	hr/ha
Sickles (manual)	0.008	125.0
Tractor drawn reaper harvester	0.04	2.5

Based on IRRI model, **a diesel engine operated paddy harvester** was introduced by Tamil Nadu Agricultural University some years ago. This machine did not become popular because of high cost, heavy weight and stain to the operator. A lightweight engine operated paddy harvester with a petrol start kerosene run engine has been developed (Tajuddin and Job, 1997). It is operated by a man at a walking speed of 2.5 km hr^{-1}. The crop is cut and conveyed to the right side of the machine to form continuous windrows. One or two labourers are required to collect the cut sheaves and to form bundles. The harvester is best suited under non-lodging conditions but can be worked under semi lodging conditions also. Its efficiency is one ha per day.

Threshers operated by tractor power or self-propulsion help to improve threshing efficiency and reduce cost and time considerably. In 1973, IRRI developed the axial flow thresher, which is relatively simple and easily maintained. They are much more compact than conventional threshers with comparable output (Manaligod and Quick, 1991). Custom hiring of threshers has caught the attention of the farmers in recent years and threshing by machines is becoming popular. The latest piece of equipment in agricultural machinery, which has caught the imagination of farmers, is the **combine paddy harvester**. It can harvest 10 acres a day and can be operated in fields as small as 30 cents. It needs only a driver and helper to operate the machine. There is no wastage of paddy and the cost of operation is equal to the wages paid for manual labour. This harvester can work in wet conditions and the special rubber track helps free movement in the fields. It stacks the straw neatly on the field as it moves on harvesting the matured grain. The high cost however, makes its use possible only through custom hiring (Srinivasan, 1997).

6.5.4 New Implements for Rice Cultivation

Transplanting of paddy seedlings in wetland is the widely accepted cultivation practice. Labour involved in nursery raising and transplanting constitutes 30 per cent of total labour required in paddy cultivation. During peak transplanting seasons, labourers are not easily available. This leads to delayed transplanting with aged seedlings, resulting in reduction in yield.

For easy understanding of the readers, the details of different implements used for rice cultivation are given below.

6.5.4.1 Puddling Operation

Pudding does the mechanical mixing of soil and water to make soil less permeable. Puddling is effective in soils having a minimum of 20% clay particles and it is the main feature of wetland cultivation. Puddling leads to.:

1. Increase in microporosity of soil, thereby increasing its water holding capacity.
2. Reduced shear strength of soil; makes manual transplanting easier.
3. Reduction in air filled pore volume
4. Increase in moisture suction.
5. Better weed control; weeds die for want of oxygen and weeds will be incorporated
6. Improves soil fertility and fertilizer management.

The main disadvantages of puddling are high water requirement, hindrance to regeneration of soil structure and impeded root development. Different puddlers used for rice cultivation are given below.

1. Animal drawn puddlers

Traditionally, puddling is done by operating a country plough under 5-10 cm of standing water. However, the quality of puddle is poor and its water retaining capacity is low. Therefore, a number of improved bullock-drawn puddlers have been developed. Some of these are

a. Straight-blade puddler.
b. Helical-blade puddler, and
c. Paddle type puddler

In straight-blade puddlers. 4-6 straight blades are mounted vertically on two plates. Although, the quality of puddle is good, this type of puddler results in occasional jerk on the animal. Hence, it has been modified by arranging the blades in a helical form so that at every moment of time, the blades are in contact with the soil. The normal width of puddler available in India varies from 72.5cm to 2m. Generally, two passes are required to obtain a proper field condition. An area of 0.5-0.75 hectares can be covered per day using these puddlers.

2. Power tiller operated puddlers

Power tillers, generally equipped with cage wheels, rotavators and puddlers are used for puddling. Cage wheels are necessarily provided to reduce sinkage of the unit and to develop sufficient traction required for operation of the power unit. An area of 0.6-1.0 hectare day^{-1} can be covered with these. These puddlers are effectively used by the farmers.

3. Tractor operated puddlers

Most of the farmers in Cauvery and Thamirabarani delta areas of Tamil Nadu are using this type of puddlers for rice cultivation. Tractors fitted with cage wheels and puddling unit is used for puddling operation. Puddling units used commonly with tractors are straight-blade puddlers, cultivator or notched-disc harrow. Puddling is normally done at a speed of 3-4 km hr^{-1} under 10 cm of standing water. The effective field capacity varies from 2 to 3 ha day^{-1}.

6.5.4.2 Paddy Seeders

1. Direct paddy seeder for wetland

The direct paddy seeder is useful to sow pre-germinated rice seeds directly in wetland without transplanting (Tajuddin *et al.*, 1994).

Description of the Machine

The paddy seeder consists of two seed drum hoppers, two skids, a 600 mm (60 cm) diameter internally lugged ground wheel and a handle all fixed in a frame work made of 16 mm diameter conduit pipe. The seeder is capable of sowing 8 rows at 150 mm row spacing or 6 rows at 200 mm (20cm) row spacing. The field is to be well ploughed, puddled and leveled at least two days before operating the seeder.

Salient features

- Field coverage : one hectare per day
- Seed rate : 50 kg per hectare
- Yield is on par with transplanted paddy

- Labour for transplanting is saved
- Weeding and intercultural made easy using long handled tools
- Cost of operation : Rs. 450 per hectare (Rs. 2350 for manual transplanting)
- Cost of machine : Rs. 3000/-

Plate 23a. Direct rice cum daincha seeder - 3 rows in operation

2. Direct paddy seeder for dry seedling

The unit is highly suitable for sowing large areas in less time before the moisture is depleted in dry tracts. This is an attachment to the tractor drawn 9 tynes cultivator.

Constructive features

Send box along with cup feed seed metering mechanism are mounted on the cultivator frame and the seeds are dropped in the furrows opened by the cultivator shovels. Detachable side wings are fixed to the existing shovel type furrow openers of the cultivator, which helps in placing the seed at required depth. Power to operate the seed metering discs is taken from the ground wheel through a clutch. A square bar is provided at the back of the unit to close the seeds in the furrows. Marker is provided for easy operation. The operation can stop the dropping of the seeds by disengaging the clutch provided. It is recommended to operate the unit below 4 Km/hr.

Salient feature

Coverage	:	4.0 ha/day
Cost of operation	:	Rs. 250/ha
Cost of conventional sowing	:	Rs. 500/ha
The cost of the unit (attachment alone)	:	Rs. 8,000/-

3. Drum seeder

Paddy transplanting requires more number of labourers and hence, the direct sowing of pre-germinated paddy seeds in puddled soil is now recommended by the Agronomists. The drum seeder is suitable to sow the pre-germinated paddy seeds directly in wetland. Good seeds are pre-germinated by soaking in water for 24 hours and keeping in gunny bags for 24 hours for inoculation and used in the drum seeder. Before using this seeder, the land has to be ploughed, puddled and leveled properly and left for 2 days. Immediate use of this seeder after puddling is discouraged since pulling the seeder will be difficult.

Constructive features

It consists of two drum type seed boxes. Each seed box has series of holes along the circumference of the drum and are kept 15cm apart. By using the seeder, the seeds are dropped on the puddled soil at a row to row spacing of 15 cm. In a single pass, 8 rows are sown. A ground wheel of 60 cm diameter is fixed at the center. A handle is used to pull the unit. For easy pulling in the puddled soil, two floats are fixed. It is suitable for all paddy varieties. The seeder covers an area of one ha in a day. Direct sowing eliminates paddy transplanting operation and the cost of cultivation comes down as given in the Table 14.

Table 14. Comparison of direct sowing with transplanting paddy

S. No.	Particulars	Cost of cultivation (Rs./ha)	
		Direct sowing	Transplanting
1	Nursery preparation	20	140
2	Sowing/transplanting charges	105	1408
3	Price of seeds	210 (84kg)	125 (50kg)
4	Machine cost	17	--
Total cost		352	1673

4. Direct lowland paddy cum daincha seeder

Direct lowland paddy cum daincha seeder is a manually drawn equipment. It simultaneously sows pre-germinated paddy seeds and green manure *viz.*, daincha seeds (*Sesbania aculeata*) in alternate rows on puddled soil. The seeder has two drum hoppers made of galvanized iron sheet, two skids, a ground wheel and a handle. The seeder sows 3 paddy rows and 3 daincha rows, covering 0.75 m in a single pass. The spacing between paddy rows is 250 mm. The hoppers are to be filled with seeds up to 75 per cent to avoid the blockage of free flow of seeds. Using the seeder, paddy can be sown at the rate of 80 kg/ha and daincha at 30 kg/ha. The daincha plants raised in between paddy rows are trampled into the puddled soil in one-month age using long handled cono weeder. The daincha yields green manure approximately 7t/ha. Using 2 to 3 labourers, 0.5 ha can be sown in a day by the seeder.

Plate 24. Direct rice cum daincha seeder - 4 rows

Advantages of paddy cum daincha seeder

- There is scope to save considerable labour and cost of cultivation to a level of Rs. 2500/ha.
- Due to direct sowing, transplanting shock is avoided and the crop matures early.

- Addition of organic matter @ 7t/ha through green manuring which reduce N by 25 per cent
- As the paddy crop is sown in rows, optimum population is maintained, Weeding with cono weeder and other inter cultural operation can be carried out easily.
- Growth of competitive weeds is checked by the shading effect of daincha.
- The seeder is suitable for all types of soil and easy to operate and maintain.

Plate 25. Direct rice cum daincha seeder - in operation

5. Pre-germinated seeders

a. IRRI Multi-hopper seeder

It consists of six hoppers, each having a wooden metering roller. Each roller has six seed pockets on its surface. Roller gets the drive from a single ground wheel provided in the center of the machine. A handle for the operator and a skid are provided. The machine weighs about 17 kg and can cover an area of 0.14 ha in one hour.

b. CRRI seeder

It consists of a seed drum having a number of circular holes along its periphery. Seed rate can be varied by varying the size of openings. The drum is rotated by two lugged ground wheels. A skid to reduce sinkage and a handle for the operator is provided.

c. IRRI drum seeder

It is quite similar to the CRRI seeder. This is an 8-row machine with provision for achieving 3 levels of seed rate. The drums are rotated by a centrally mounted wheel. Two skids are at both ends of the assembly the field. The machine has a field capacity of 1.2 ha per day.

d. IARI seeders

Two designs, a 6–row unit and a 3 row unit have been developed at IARI, New Delhi. The 6 row unit consists of a hopper with a 6 wooden metering rollers. Each roller has six cells, staggered in two rows on its periphery. The rollers get the drive from two lugged ground wheels. An adjustable handle and a float are provided for proper operation. Six furrow openers have been provided below the float for proper placement of sprouted seed. The field capacity of the unit is 0.16 ha per hour. The 3-row unit consists of three individual hoppers with wooden metering rollers having seed carrying cells, two lugged ground wheels, and furrow openers, an adjustable handle and a float. Provision has been made for varying the row spacing and depth of seed placement. The machine can cover an area of 0.08 ha per hour.

6.5.4.3 Paddy transplanters

Traditional method of transplanting by hand is a labour intensive operation. Transplanting of paddy consumes 33 per cent of total labour required in rice production. Transplanting with machine helps in line transplanting of the seedling, which helps in achieving the required plant population per unit area, and makes weeding and intercultural operations easy.

A. Manual Paddy transplanter

The International Rice Research Institute (IRRI) design manual paddy transplanter is suitable for mat type nursery. Normal (root washed) seedlings cannot be planted by this machine. Mat type nursery can be easily grown as discussed in detail in the chapter 5.

Construction of the machine

The machine consists of a seedling tray, forks, handle and skids. By pressing the handle, the forks pick-up the seedlings and plant them in 6 rows. For every stroke of the handle the seedling tray moves sideward for uniform picking of seedlings by the forks. A ratchet and pawl mechanism causes the tray to move laterally when the lever is actuated by planting frame. The planting arm is pushed into the tray and the seedlings are picked up. The seedlings are placed into the soil during the backward movement of the handle. The operator has to pull the machine while punching the handle at the desired spacing. The machine is used to transplant mat type-paddy seedlings in puddle soil (Plate 26).

Plate 26. Manual rice transplanters.

Salient features

Number of rows	: 6
Row to row spacing	: 20 cm
Manpower required	: One operator and one labourer to transport mat seedlings
Field coverage	: 0.25 ha/day
Cost	: Rs. 5000/-

B. Self –propelled paddy transplanter

Self-propelled paddy transplanters have been successfully developed and introduced in Japan and Korea. These are available in walking type as well as riding type models. The transplanters have been developed both for mat type seedling and root washed seedlings.

i. Mat type seedlings-transplanter

This transplanter, consists of a reciprocating tray, ground wheel, flat, a planting unit and a small engine as a power unit. The seedlings in mat form are placed in an inclined aluminum tray and are fed by gravity to the planting fork. Mechanical feeders located at the lower end of the tray intermittently flick the seedling mat and cause it to slide down. A screw mechanism gradually moves the tray from side to side to ensure uniform quantity of seedlings to be picked up by the fingers. The riding type machines are generally of six rows while walking type of 2-4 rows are quite common. Their coverage is from 0.2 to 0.5 ha day^{-1} depending upon number of rows and skill of the operator.

Plate 27. Self-propelled rice transplanter.

ii. Root washed type self propelled paddy transplanter

Root washed type self propelled paddy transplanters are reported to have been developed in China and DPR Korea. The machines use traditional and root washed seedlings. A small engine is used as power unit, which drives the traction wheel through chain and sprocket system. Power for the feeding and planting is transmitted through the propeller shaft from the transmission power take off. The machine covers 0.5 ha day^{-1} depending upon the number of rows.

6.5.4.4. Zero till

Pantnagar zero till is developed to sow wheat directly after harvesting of rice, without performing any tillage operation. It consists of seed and fertilizer box, fluted roller type metering unit, furrow openers and power transmitting unit. The machine can cover an area of 4-5 ha in a day. The machine is reported to give good results in manually harvested plots. Similarly, strip till drills have also been developed for direct sowing of wheat after rice harvest.

6.5.4.5 Liquid urea injector

Most of the urea applied to the paddy fields is evaporated to the atmosphere and hence the plants do not get urea, which results in low yield. To avoid this, application of liquid urea at a depth of 5cm nearer to the paddy plants is done. Urea dissolved in water is pressurized by a peristaltic pump and reaches the soil through furrow openers.

Constructive features

The liquid urea injector has tank, wheel, skid, handle and peristaltic pump. The tank is nothing but a power sprayer pesticide tank. The operator pulls the unit, holding the handle in the field so that the wheels rotate and peristaltic pump pressurize the fluid. The fluid reaches the furrow opener through a plastic tube. The diameter of the wheel is 40 cm. The skid is useful in water stagnated areas for easy handling.

Salient features

The cost of liquid urea applicator per hectare is Rs. 70/-. The field capacity of the machine is one hectare/day. The cost of the machine is Rs. 1200/-.

6.5.4.6 Rotary and Cono weeders

Long handle provided in Rotary and Cono weeders eliminate the drudgery of bending forward while carrying out the weeding operation.

Weeds are removed by rollers and incorporated into the soil. The cost of the machine varies from Rs. 250/- to 1000/ depending upon the model and type of material used. These weeders are now used by the farmers in SRI method of rice cultivation, in Tamil Nadu.

Plate 28. Conoweeder for weeding operation

6.5.4.7. Battery operated sprayer

Agricultural crops are usually sprayed by power sprayers and manually operated sprayers. Power sprayer now costs around Rs. 5000/- and its operational cost is high due to high price of petrol. Repair problems are also more for power sprayers. Manual sprayers require more amount of water and have poor field coverage. Hence a battery-operated sprayer has been developed (Tajuddin, 1998).

Description of the sprayer

The battery operated sprayer consists of a 10 litre capacity plastic tank and 6 V rechargeable battery both fixed in a frame which is carried on the back of the operator. A plastic spinning disc along with a micro-motor is fitted at the end of an aluminium handle. Chemical is taken from the tank to a spinning disc through a plastic hose. A cut – off value is provided in the hose line to stop the flow of spray fluid when desired. The micrometer is wire – connected to the battery through an on-off switch.

Uses

The battery-operated sprayer is useful to spray chemicals on crops like paddy, groundnut, pulses, cotton, vegetables and other crops growing up to a height of one metre.

Salient findings

- Field coverage : 1.5 hectares in a day
- Dry weight : 7 kg
- Water requirement : 45 litres of water per hectare
- Cost of operation : Rs. 60 per hectare (excluding chemical)

6.5.4.8 Battery operated low volume Knapsack sprayer

Generally petrol engine powered power sprayers are used by the Indian farmers. The maintenance cost as well as operational costs is higher for this sprayer. Battery powered sprayer saves these types of expenditure. A6 volt battery supplies the power to the DC motor. The battery can be recharged in the farmer's house itself. It is an improved battery operated low volume sprayer over battery operated sprayer (Tajuddin, 1991).

Plate 29. Battery operated low volume sprayer

Constructive features

A 10 litre capacity plastic tank and 6 volt A battery are fitted in a knapsack sprayer frame. A plastic tube connects the tank and the rotating spinning disc through a cut-off valve. To operate the DC motor, a switch is provided in the handle.

Salient features

This sprayer requires only 50 litres of chemical water mixture per hectare. But when we use the power sprayer and manual sprayer, 150 and 400 liters of water are required respectively. Hence the battery sprayer is recommended in water scarcity areas. The field capacity of the battery-powered sprayer is 0.2 ha/hour. The cost of spraying per hectare is only Rs. 65/-. The operational cost of spraying by power and manual sprayer are Rs. 75/- and Rs. 90/- respectively. Cost of the battery sprayer is about Rs. 1400/-. The repair and maintenance costs are minimum when compared to power sprayer and manual sprayers. The dry weight of the unit is only 7 kg. It is recommended for spraying low height crops like paddy, groundnut, pulses and vegetables.

6.5.4.9. Paddy harvester

Harvesting of paddy is commonly done manually using sickles. Labour requirement is more for manual harvesting of paddy crop. Human stress and drudgery involved in the operation are also more. Timely harvest is not possible in most of the cases because of non-availability of labour during peak harvesting season. A self propelled paddy harvesting machine has been developed based on the design of the International Rice Research Institute (IRRI), Philippines.

Description of the Machine

The harvester is operated by a 5.4 HP diesel engine. The machine consists of gearbox, ground wheels, handle, cutter bar assembly, star wheels and gathering header assembly. The crop should be manually harvested along the four sides of the field for a width of 0.5 m and cleared from the field for providing space to the machine. At one corner an area of 2 m x 1.5 m should be manually harvested to place the machine initially in the field. Since the harvested crop is discharged at the right side of the reaper the harvester should be turned always to the left side.

Plate 30. Rice harvester

Salient findings

Coverage	:	1.5 to 2.0 hectares per day
Shattering loss	:	less than 1 per cent
Height of cut	:	less than 5 cm
Saving in time	:	89 per cent
Saving in cost	:	60 per cent
Cost of the machine	:	Rs. 50,000/-

6.5.2.10 Self propelled paddy harvester

The machine is suitable for non– lodging paddy in both dry and semi – dry field conditions. The field capacity of the machine is 1.5 ha/day. The cost of operation works out to be Rs. 250/ per hectare. The machine costs around Rs. 25,000/-. The machine needs only 3-4 manual labourers for operation including labour for collecting the wind rowed crop. Since the machine is equipped with iron cage wheels, paddy can be harvested even in slushy condition.

Plate 31. Tractor mounted vertical conveyer reaper

Plate 32. Rice vertical conveyer reaper.

6.5.4.11 Combine Harvesters

Both paddy harvesters and paddy threshers used in paddy cultivation require laobourers. In some parts of our country, the farmers have to give 1/6 to 1/8th of the produce as wages for harvesting and threshing. Hence, to overcome the lobour shortage, combine harvesters have been introduced for harvesting and threshing paddy. This machine cuts the crop, threshes and separates the grains from the straw within the field. The threshed paddy straw falls on the backside of the machine. The cleaned paddy grains are stored in the grain tank. For effective operation, the field should not be wet. The rubber tracks are provided to overcome the sinkage problems in wet field. There are two models available in the market.

The following will give an idea about the cost and special features of combine harvesters.

1. Combine harvester- Rubber track

Cost : Rs.13.0 lakhs

Field capacity: one acre/ hour

Hiring charge: Rs.1000-1300/hr

2. Special Features- Std., Balkar B-525 Combine Harvester-Model

- Low diesel requirement
- Low maintenance cost
- No wastage of paddy
- Easy to maintain the tractor
- Easy to lift the vehicle ups and down
- Harvesting more paddy area with less time
- Easy to harvest in high clay and moisture content soils

Specifications

Class: LMV Harvester mounted with tractor

Model: B-525

Type of prime mover: Suitable tractor to be fitted

Blower : 550- dia X 1100 width- 4 blade

Blade: 1075 mm X 128 mm

Drive: V belt and fully drive

Harvester while on harvesting

Overall length	7800 mm
Overall width	3900 mm
Overall height	3480 mm
Harvester while on road	
Overall length	10650 mm
Overall width	2580 mm (with one harvest cutter)
Weight	3980 kg
Cost Approximately	Rs 9.0 lakhs
Hiring charges	Rs. 900-1300 per hour
Field capacity	one acre/hour

The cost of the machine will vary. The hiring charge and field capacity of different implements/machineries will vary depending upon the area, situation, efficiency of operator, field conditions etc.,

6.5.4.12 Paddy thresher

The paddy thresher can be operated by electric motor, tractor, oil engine or power tiller. For easy transportation from one place to another place, the thresher is mounted on tyre wheels. The paddy thresher costs approximately Rs. 60000/-. The field capacity of the thresher is 750 kg/hr. Damage to paddy straw is minimum (Plate 33).

Plate 33. Rice thresher

6.6 MANAGEMENT OF PROBLEM SOILS

Compaction technology for fluffy paddy soils

Compact the fluffy paddy soils by passing 400 kg stone roller or oil drum with stones inside, eight times at proper moisture level (Moisture level at friable condition of soil which is approximately 13.18%) once in three years, to prevent the sinking of draught animals, tractors and labourers during puddling (Plate 34).

Plate 34. A view of the sunken tractor drawn cage wheel in the fluffy soil

6.7 FUTURE THRUST

The investigations so far conducted have contributed much towards mechanization of rice cultivation. Varying agro-ecological situation and size of holdings of rice fields requires suitable modifications of machinery and a research package in this line. Different research programmes on workability of transplanters on different agro ecosystem, main field requirement, mat nursery techniques, varietal preference etc. with respect to mechanical transplanting, modification of reapers to suit varying wet situations and to harvest lodged paddy, workability of combines etc are some of the research priorities. Effective mechanical means of pest control including modification of sprayers for larger area coverage need further attention. Establishment of customary service center and entrepreneurship development programmes for unemployed youth to start these center are a few of frontier areas where sufficient research and development activities are urgently needed.

Considering the prospects for mechanization in rice farming, we can say that mechanization will reduce drudgery, improve productive efficiency of resources, reduce losses and raise the profitability. The various types of machinery suitable for rice farming can be hand drawn, animal drawn, power tiller operated, tractor drawn or self-propelled. While some of them are price reasonable on custom hire basis. For promoting the use of machinery in rice farming, many requirements have to be met. These include creation of awareness among farmers, demonstration and training, infrastructure for manufacture, sales and service and easy credit. More importantly, the mental block that these machines will completely substitute human labourer has to be removed.

CHAPTER 7

IRRIGATION MANAGEMENT

On the eve of World Food Day, October 16, 2003, the World Wide Fund for Nature (WWF) warned that the global farming industry must manage water more efficiently, while still meeting the world's growing demand for food, to avoid a global water crisis. In its latest report, the global environmental body highlights the fact that world food production has to be increased to feed an expanding population with an expected growth of 2 billion people over the next 50 years.

According to the report, many large food-producing countries like the USA, China, India, Pakistan, Australia and Spain have reached or are close to reaching their renewable water resource limits. Water tables are dropping by as much as 10 meters annually in the worst cases, leading to a less reliable supply of water for drinking and sanitation. Agriculture uses 70% of the world's water, rising to 90% in many developing countries. But only 20%-50% of the water drawn actually reaches the crops; most of it is lost during the transfer to fields. "If we do not address the wasteful use of water in farming, this will have serious consequences for achieving the Millennium Development Goal of halving the number of people suffering from hunger by 2015," says Jamie Pittock, Director of the WWF's Living Waters Programme. The report recommends various methods for managing water more efficiently in order to tackle the growing food and water crisis. It highlights the fact that the main causes of water shortage are inappropriate irrigation systems and growing crops that are unsuited to the environment. Thus, the report says, is being driven by misdirected subsidies, low public and political awareness of the crisis and weak environmental legislation. The report identifies cotton, rice, sugarcane and wheat as the `thirstiest' crops in nine large river basins rich in bio-diversity. The WWF believes that growing crops more suited to the location and season would give more 'crop per drop'.

The WWF report also suggests that irrigation systems could be improved through better design, regular maintenance and effective drainage mechanisms. Governments need to allocate water more fairly among farmers where there are shortages. They should also ensure that enough water remains in rivers and wetlands to maintain water supplies, fisheries and wildlife habitats. Governments must do more than make promises. Together with the food industry and consumers they must start a new farming revolution -- one that ensures there will always be enough food and water for everyone,"

Water is a finite resource, which our country has from prehistoric times widely used, misused and abused. Besides, irrigation water is diverted to other uses like industries, domestic purposes etc., at the cost of the share for agricultural purposes. The present scenario at All India level indicated that the supply demand gap amounts to 8.0% and will further increase to 25% in 2020 AD. In Tamil Nadu, the gap is 26% now and further, will be increased to 36% in 2020 AD. Therefore, it becomes, necessary that every drop of water should be used judiciously and economically to realize maximum agricultural production per unit of water used. Among the crops, rice requires water 2-4 times higher than that of other crops of the same duration (Table 15) because the water loss by percolation, seepage and main field preparation are greater in rice as it is cultivated under submerged conditions. The various unavoidable losses in rice cultivation are given in Table 16.

Table 15. Water requirement by different crops

Crop	Economic yield (kg/ha)	Water requirement (cm)	Duration of crop (days)	Daily water requirement (cm)
Rice	5529	120*	115	1.05
Maize	5436	50	105	0.48
Wheat	4573	40	115	0.35
Cotton	1850	65	165	0.40
Sugarcane	112000	235	310	0.76

* Water requirement will vary with varieties.

Table 16. Components of water loss in rice

Sl. No.	Particulars	Water loss (mm)
1	Nursery	40
2	Land preparation	200
3	Evaporation	180-380
4	Percolation + seepage	200-500
5	Main field requirement	200-700
6	Total	1000

Before recommending any irrigation, schedule for any area, one should study some basic things like nature of soil, topography, rain fall, consumptive use of crop, availability of water source, cropping pattern, ground water, contour survey, land use classification etc., Water requirement of rice crop depends on the consumptive uses (Evaporation + Transpiration), percolation, seepage (25-40% of total loss) and the losses that occurs as a result of the way the water is conveyed and distributed (40-50% of total loss) (Surface run off 10%). Water requirement is the quantity of water used regardless of its source for the normal growth and yield of the crop. It includes other special uses like leaching requirement. It also includes other inevitable losses. In most of the places, evapotranspiration requirement during the rainy seasons is approximately 4 to 5 mm/ day. The range of percolation varies from less than 1mm / day in compact soil to several cm/day in light sandy soil. From experiments it was recorded that the optimum rate of percolation varied from 15 mm/day. Average total water requirement also depends upon the method of irrigation practices to be followed. In general it varies from 600 mm to 1000 mm. Hence, In addition to basic water requirement of rice crop, water is required for seedbed preparation and land preparation (puddling). The amount varies depending upon soil texture, soil moisture content prior to land preparation and method of land preparation etc. Where puddling is practiced. In Tamil Nadu, the water requirement for 1st season of rice (kharif/ kar) is very high due to low rainfall, high temperature, high ET etc., During the second season, ET is less due to rainfall, which finally reflects on less water requirement. The irrigation methods vary in different parts of the country due to difference in climate, soil, topography, water requirement and customs.

In general, the following steps may be recommended/ adopted for better water management of rice crop.

- Farmers should have a community nursery
- Reduce the time lag for field preparation, ploughing by tractor in an organized way.

- Proper preparation and leveling the field
- Use of herbicides
- Laying out of inner field channel 30 to 45 cm avoiding field to field irrigation
- Adoption of simple water management practices
- Use of short duration paddy varieties
- The field plot can be 25 to 50 cents, depending on the source of irrigation
- To minimize percolation loss, the depth of stagnated water should be 5 cm or less.
- Care should be taken not to allow development of cracks in the field.

There are different systems of rice cultivation, mostly based on water availability. They are (i) wet rice, grown under total submergence of land, (ii) semi wet or semi dry rice, which are sown under dry or rainfed conditions initially and later converted into submerged condition. (iii) dry rice, grown exclusively under rainfed condition. Due to the availability nature of irrigation water, the wet system rice has two dimensions, viz., direct sown under puddled and transplanting; particularly direct seeding is gaining popularity due to labour scarcity, water constraint and availability of drum seeder for direct sowing in lines as discussed in the earlier chapters.

7.1. WET RICE

7.1.1 Transplanted rice

a. Water management in Nursery

Water management in nursery is important for good germination and healthy seedlings. Before 2-3 days initiating nursery area preparation, water is applied until it attains saturation condition. At the time of ploughing, water is allowed to stand about 2-3 cm followed by puddling and leveling. At the time of sowing, a thin film of water about 2-3 cm is maintained. It will enhance the germination and vigour of the seedling. Water is drained from the nursery field after 18-24 hrs after sowing. Water stagnation in low patches is to be drained into the channel. Water is let in to saturate the field from third to fifth day. From fifth day onwards, the quantity of water applied is increased to a depth of 2.5 cm. Herbicides should be applied on 8th DAS. Application of pre emergence herbicide to paddy nursery is discussed in detail in the subsequent chapter.

b. Water management in main fields

Since the water requirement of rice is higher than that of any other crop of a similar duration, assured and timely supply of irrigation water has a great influence on the yield of the crop. At the time of transplanting, a shallow depth of 2cm of water is adequate. Generally, the water requirement for land preparation varies from 200 to 400 mm depending upon soil type. Summer ploughing helps to conserve the rainwater and reduce water requirement for land preparation. Similarly effective puddling helps to reduce percolation losses, check weed growth; increases availability of plant nutrients (nutrition efficiency), regulate soil and water temperature. The puddling with cage wheel attached to tractor reduces percolation loss, thereby effecting 20% saving in irrigation water and in achieving favourable soil physical conditions.

Since high depth of water will lead to deep transplanting. Maintain 2 cm of water during this period of establishment of seedlings. i.e. the depletion should be recouped as and when it occurs. After the establishment stage, cyclic submergence (as in Table 17) of water is the best practice for rice crop. This cyclic 5 cm submergence has to be continued throughout the crop period.

Table 17. Days after disappearance of ponded water at which irrigation is to be given

Soil type	Summer	Winter
Loamy	1 day	3 days
Clay	Just before / Immediately after disappearance	1-2 days

Experiments conducted at various locations in India and in Tamil Nadu on low land rice revealed that the practice of keeping the soil under shallow depths of submergence i.e., 5 cm of water throughout the crop growth is conducive for higher yields. Maintenance of 5 cm depth of water daily, which includes the water lost due to percolation, ET, seepage and horizontal flow. It is not necessary that 5 cm water should be standing in rice crop. This 5 cm depth of water may be allowed to come down to 0 cm depth after a lapse of a few days as the case may be and the irrigation is given to make the depth to 5 cm. In a very light soil the 5 cm depth may come down in a day or so, in which irrigation is to be given daily. On the other hand, in the case of heavy soils it may take 3 to 5 days or even more depending upon the clay content. Thus in this case the frequency of irrigation would be once in 2 to 5 days or more. Irrigating up to 5 cm depth when the ponded water disappears is able to save irrigation water to the extent of 30 to 50% without affecting the yield.

In Vaigai Periyar Command area, cultivation of early duration rice during June-September and that of the medium duration variety during September – January give higher paddy yields with an irrigation regime of maintaining water levels between 5 cm submergence and saturation resulted in 41% saving of irrigation water.

C. Water management in critical stages

Moisture stress due to inadequate water at rooting and tillering stage causes poor root growth leading to reduction in tillering, poor stand and yield. In the life cycle of rice plant, there are certain critical periods. The water requirement is high during the initial seedling period covering about 10 days. Critical stages of water requirement in rice are (a) Primordial initiation (b) Booting (c) Heading and (d) Flowering. During these stages, the irrigation interval should not exceed the stipulated time so as to cause the depletion of moisture below the saturation level. Seven days after planting, 5 cm of water is necessary to facilitate tiller production and promote firm root anchorage in soil. During vegetative growth phase, water depth should be gradually increased from 3 to 5 cm during the succeeding 8-week period after planting. Draining water during 20-30 days after planting. i.e. active tillering stage is harmful irrespective of the depth of submergence. Water depth is to be maintained at 5 cm level from panicle initiation to flowering stage. Excess water more than 5 cm at booting stage leads to reduction in growth of panicles, delay in heading, weakening of stem and lodging. Rice is most sensitive to moisture stress from 20 days before and 10 days after heading. In contrast to vegetative phase, stress during the reproductive phase causes very drastic effects on yield because the potential recovery of the plant is very low. Water stress during reproductive phase causes serious reduction in yield due to impairing panicle formation, heading, flowering, fertilization, and increased spikelet sterility. At booting stage excess water due to rainfall and canal supply to more than 5 cm inundation leads to delay in heading and reduction of yield. Arrest the irrigation 15 days ahead of harvest. i.e., Water should be gradually drained in the field from 21-25 days after flowering till it is finally drained at about 2-3 weeks before harvest. Undue delay of terminal drainage prolongs ripening and unfavourable harvesting conditions such as lodging. The drainage management in rice fields is generally neglected despite its negative impact on the yield.

Water management practices in main field of rice crop include continuous flooding and intermittent irrigation.

1. Continuous flooding at shallow depth (3-5 cm) has the potential to produce optimum yield. The rice yield is high mainly because of high tiller number. The percolation loss is also low under shallow flooding. Shallow flooding of 5 cm is the best practice since the weeds of grasses and sedges are controlled and increase in fertilizer use efficiency. The distribution and stability of granular pesticides are improved at moderate water depth. Deep continuous flooding (10-15 cm) has the potential to produce optimum yield but may produce less than optimum yield. Net seepage and percolation requirements are increased. The grasses and sedges are controlled effectively. However the broad leaf weeds are relatively unaffected by flooding depth. Midseason drainage may improve the respiratory function of roots, stimulate vigorous growth of roots, checks the development of unproductive tillers and prevents root rot and sheath rot diseases considerably. The field has to be drained completely for the control of brown plant hopper whenever the incidence is noticed.

2. Intermittent irrigation can be practiced in areas with limited water supply and in areas served by pump irrigation. Here, the water losses are minimum and available rainfall is effectively utilized. Under intermittent irrigation good crop yield can be obtained. This is achieved with more productive tillers, short plant height and high lodging resistance. However, scheduling irrigation after disappearance of water in field favours the germination and growth of weeds, which can be controlled by application of herbicide. Poor nitrogen uptake and some losses of nitrogen occur in soils due to alternate aerobic and anaerobic conditions. It is also less profitable to apply high levels of nitrogenous fertilizer with lower level of water supply. Split application of nitrogen and use of slow release nitrogenous fertilizers such as neem-coated urea can be recommended under such conditions to increase fertilizer N use efficiency. In red loamy soils, application of urea super granules and irrigating to a depth of 5 cm water once in 3 days after disappearance of ponded water registered higher yield compared to the application of prilled urea. Intermittent irrigation minimizes formation of toxic substances in the soil and reduces the drainage problem.

Rice can also be grown when soil moisture is maintained at saturation level without serious effect on yield. The irrigation efficiency is generally high when this practice is used on heavy soil. When soil moisture is 70-80% saturation, rice yield starts to decline. When it drops to 50% saturation the yield decreases from 33% to 50% as that of continuous submerged conditions. If water supply is inadequate, irrigation can be given on priority basis depending on the crop. Efficiency of herbicides will be affected either by increasing or decreasing the water level.

d. Water saving methods

There are few water saving methods in rice culture viz.,summer ploughing, quick land preparation, formation of inner bunds near the main bunds, strengthening of field bunds, use of aged seedlings, irrigating the field after formation of hairline cracks, conservation of rain water, terminal drainage prior to harvest and rotational irrigation are recommended as water saving methods in the rice cultivation. The field bunds are strengthened to minimize the seepage loss and to prevent the crab holes. Basal application of fertilizer can be done on the previous day of planting with thin film of water (0.5 cm). At this time there is no need to impound more water. Water can be saved by this practice, instead of flooding, besides increasing fertilizer use efficiency. Rainwater can be conserved by impounding rainwater up to 10 cm depth in the field. Rotation irrigation can be used, when there is good water control through distribution systems and excellent farmers cooperation. By adopting these guidelines in water management practices of rice cultivation, increased rice productivity per unit area per unit volume of water can easily be achieved.

e. Contingent Planning

In command areas, anticipating the release of water, rice crop can be raised under semidry condition up to the maximum of 45 days. Weed control measures should be attended carefully. There after, convert the crop to wet condition on the release of water. This process saves 30 to 40% of water. Spraying cycocel at the rate of 1ml/litre of water under water deficit situation is recommended to mitigate ill effects. Split application of potassium 50% at basal and 25% each of at panicle initiation and flowering alleviates harmful effects of soil moisture deficiency. Foliar spray of kaolin at 3% or potassium chloride at 1% is recommended to overcome moisture stress at different physiological stages of rice. Water management practices in farmer's fields play a pivotal role for maximum yield by effective utilization of the related inputs and environment.

7.1.2 Direct seeded rice

Wet seeded rice requires water less than 20 per cent to that of transplanted rice since wet seeded rice does not require nursery preparation. The wet seeded rice is more water efficient and it requires 20 per cent less water because wet seeding takes only 7 to 10 days of water intensive land preparation compared to 25 to 30 days for transplanted rice. Experimental results on direct seeded rice revealed that about 44 per cent water will be saved for direct seeded rice with saturated soil conditions compared to continuous standing water without significant yield reduction. The total water requirement of wet sown and dry sown rice was 1105 and 1040 mm respectively with water use efficiency of 3.65 and 3.77 kg ha mm^{-1} respectively. For direct seeded rice, irrigating 5 cm depth one day after disappearance of ponded water may be recommended for higher productivity and net returns. But this may not be a viable practice under the present circumstances of water scarcity. So continuous submergence of 2.5 cm throughout the crop period for higher productivity and net returns may be recommended as compared with 5 cm depth of irrigation. Submergence of 2.5 cm results in saving of about 25 per cent in irrigation water requirement over the irrigation need of transplanted rice. In areas of acute water scarcity, irrigating 2.5 cm or 5 cm depth one or three days after disappearance of ponded water may be advocated for efficient use of water with marginal yield loss and considerable net income.

7.1.3 Semidry rice or rainfed lowland rice

Semidry rice cultivation involves growing rice as an lowland crop up to 4th or 5th leaf stage and thereafter bringing it under submerged condition. Generally, the upland condition lasts for 35-45 days after sowing, during which period, the crop experiences moisture stress under varied magnitude. Semidry rice cultivation is mainly practiced where the date of release of canal water gets delayed or become uncertain. In command areas, anticipating the release of water, rice is raised under semi dry condition up to 45 days. Then it is converted into wet condition on receipt of water from canal. This saves about 30-40% of water. In tank commands, the same practice is followed after impounding water in the tanks. Hydrophilic conditioners like Jalsakthi @ 7.5 kg/ha^{-1} as soil application or seed hardening agents like potassium dihydrogen phosphate at 2% solution and succinic acid (100 ppm) would offer good scope for increasing the productivity under semi dry rice cultivation. The water use efficiency (WUE) was higher in 1.40 IW/CPE ratio during summer and 1.10 ratio during wet season. Irrigation at 1.10 IW/CPE ratio was ideal up to 35 DAS which reduced the water requirement by 100 mm of water. The seed hardening treatments with their influence on reducing crop duration and increasing yield, reduced the water requirement and increased the WUE. This emphasizes the prospects of stabilizing and economizing rice production under constraint of irrigation water.

7.2. WATER MANAGEMENT IN COMMAND AREAS

River commands all over the world pose lot of problems due to their vastness. There were international rivers like Nile, Zambazi, Indus and

Ganges. There are inter state (provinces) rivers like Colorado, Narmada, Krishna and Cauvery. All the river commands do have their inbuilt advantages and disadvantages. Water disputes are the integral part of the system command areas. In India, river (delta) systems were existing since historical past. River system shaped the human civilization, land ownership, social and political developments. Absence of land ownership and non-existence of feudal system were due to the need to irrigate the land in India. There was a primitive community land ownership in the historical past due to the need for irrigation. Land could not be sold or purchased by individuals. There was collective mode of production due to soil, climatic and natural calamities like flood, drought and cyclones. In Tamil Nadu "in the early Chola" time, land was communally held by the people in the non-brahmadeya peasant villages as the feeling of solidarity in these village communities was strong. From the view point of social stratification, in relation to land holding, the Ur (village members who appeared as land holders) in early Chola inscriptions were themselves the cultivators and therefore, land owners and cultivators were not separated from each other as different social strata, except in the case of Brahmadeya village where the land owning Brahramin did not cultivate land themselves". Land ownership shaped the history, mode of production and surplus appropriation in river commands.

7.2.1. River Commands

River commands with its river and network of canal system are massive in their physical size, covering large land areas. Cropping system also diverse in nature depending upon location, soil and facilities for drainage. Large quantum of water is to flow to meet the needs of crops. In tropical regions, elaborate infrastructures are to be maintained for irrigation and even for transport facilities, which are massive.

7.2.2. Farmers organization

Water management needs farmer's organization especially in river command areas due to uncertainty of flow in canals and unbounded optimism. Organizations of informal nature up to water resource and Ministry of Federal government including local bodies are to work. Informal organizations are functioning in different parts of Tamil Nadu and other parts of India. Formal organization model has been copied from Philippines. Irrigation Community organizers are the key workers in formal organization. This model was found successful in few commands of Tamil Nadu. Funds are also collected from farmers and matching grant comes from government. The studies proved the possibility of water marketing in lift-irrigated areas for which Farmers Association will go in a long way.

Informal organization in Canal and Tank commands of Tamil Nadu

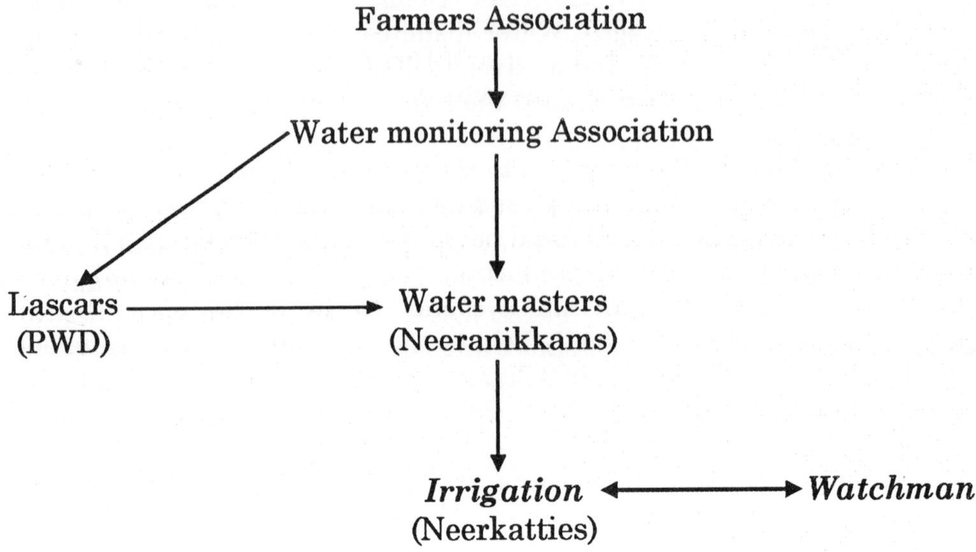

Formal organization in Canal Command areas (Three tier system)

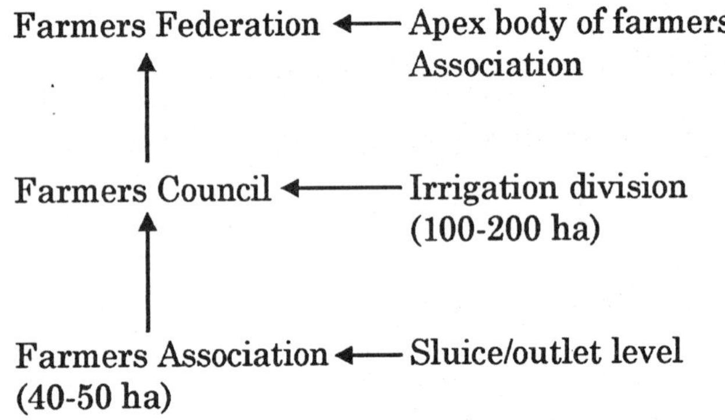

7.2.3 Water Management

Water management in river commands involves catch, collection, timely release conveyance, measurement, distribution, on farm water management and provision of drainage. In field level, efficient water management needs proper land layout, land leveling and water application to the field especially under scarcity. Due to unbounded optimism, water management programmes are suffering in fragmented land holdings and private mode of production. It is estimated that the conveyance loss is about 40 to 50 per cent in canal and tank irrigated areas and 15 to 20 per cent in garden land conditions. The lands in the command areas are piece meal and scattered. In order to take up the above mentioned water management practices, especially forming irrigation and drainage channels land consolidation is a must. But it is difficult under the present land acts and therefore these aspects are to be taken care of in planning water management in the command area.

7.2.4. Modernization

Irrigation canals are to be lined and irrigation structures are constructed for efficient conveyance of water to enable the water to reach the farm head in time. Modernization is useful, purposeful and meaningful for efficient distribution of water. Modernization involves high cost but worth spending to make crop production sustainable.

a. On farm development

As the part of modernization, irrigation infrastructure developed at farm level is termed as on farm development. Minor infrastructures like lined watercourses, Division / Diversion boxes, turnouts, drop structures, etc. are to be formed. Water has to travel on the land surface quickly. There is considerable reduction in conveyance and application efficiencies, Transport facilities through farm roads and formation of thrashing yards are also needed.

b. On farm water management

On farm water management involves application of optimum quantity of water at optimum intervals. Large quantity of water needed for crops at right interval is influenced by crop, stage of the crop, climate and soil. Rice needs almost continuous water supply or at the maximum short period of disappearance of ponded water can be allowed. Rotational water distribution methods have been evolved in different regions of India depending upon the needs of the regions. Irrigation management also needs land consolidation in the present context of fragmentation. Drainage is the integral part of the on farm water management. Similar to irrigation methods, drainage systems have been developed since long time. There is need for improving of the existing irrigation and drainage systems. There is also reuse of drainage water in some parts of canal system.

7.2.5 Water management

As discussed earlier, as rice is a semi-aquatic crop, the standing depth may be ranging from 2.5-5.0 cm. Rice seedlings are transplanted with a standing water of 2.5 cm. On full establishment, the standing depth would be increased. Farmers all over the world believe increased standing would be useful which is correct in terms of weed control. Recent experiments proved that impounding of water to a height of 5 cm one day after disappearance would be useful for water economy without yield reduction or with yield increase of 25 per cent in low-lying areas (Table 18).

Table18. Grain yield in relation to irrigation regimes in Rice-Vaigai (Co.37), water used and water use efficiency

S.N.	Treatment	Grain Yield (kg/ha)	Water Use (cm)* including Rainfall	WUE (kg/ha)
1	Irrigation up to 5.0 cm and replenishing up to 5.0 cm on disappearance of ponded water.	4316	108	40.0
2	Irrigation at critical stages namely maintenance of 5.0 cm at reproductive stages and submergence to saturation during vegetative and ripening phases.	5026	130	39.0
3	Normal method, maintenance of 5.0 cm throughout	4512	166	27.0

* Including 25.0 cm for land preparation

7.2.6 Control of avoidable losses

The avoidable losses are conveyance and delivery losses, that are canal leakage, seepage and farm losses due to uneven distribution resulting from poorly prepared fields and poor handling and management, over irrigation and evaporation. Conveyance and delivery losses account for two-thirds of total water losses. The comparatively high water requirement of rice is attributed to percolation or seepage of water to subsoil and keep the soil saturated with water layers. While engineering devices may reduce the loss due to conveyance and transmission, the farm loss can be minimized by suitable agronomic and water management practices as given below;

(i) Selecting heavy soils or those with hard pans in the subsurface depth to reduce permeability. The soils having percolating rate below 5 mm per day are good

(ii) Growing rice in large blocks instead of in isolated small holdings

(iii) Scrupulous land leveling.

(iv) Reduction of losses by soil filling such as tank silt, fire ash etc.

(v) Use of water saving devices like turn out gates, diversion boxes etc. to have rotational irrigation.

(vi) Compaction of surface soil layer to about 1.60 to 1.88 g/cm^3 depending upon the soil type.

(vii) Use of barriers like polythene sheets, bitumen, asphalt and cement by creating an impervious pan 2 to 3 mm thick at 20 to 30 cm depth, which are very costly under our conditions.

(viii) Manipulation of soil by different puddling implements and machinery

(ix) The local practice of "Kattuthalai" or Kai Varappu in rice fields as to reduce the water loss.

7.3. WATER MANAGEMENT VS WEED MANAGEMENT

Water serves as an effective means of weed control by maintaining continuous shallow submergence especially during the vegetative period, weed growth was effectively suppressed. Control of weeds due to flooding is accomplished by an interaction of the changes in the physical, chemical and biological properties of the submerged soils. In wet seeded rice, weed management is one of the most critical factors because the soil conditions are favourable for simultaneous germination of weed seeds along with rice seeds. Low or high water content prevents the growth of low land rice weeds.

7.4 DRAINAGE

When continuous land submergence of ponding is practiced, it is necessary to drain the soil either through horizontal (surface) or through vertical (subsurface) drainage once or twice during the growth period especially on poorly drained clayey soils having percolation rate less than 2.5 mm/day. This practice is beneficial in removing the toxic substances such as sulphides developed under reduced conditions and in supplying oxygen to the root system. The drainage period will last from 4 to 8 days depending upon the type of soil. In sandy soil the drainage period is about 3 days while in clayey soil, it is about 5 to 7 days. Temporary drainage is beneficial at tillering. Drainage at maximum tillering stage helps to stimulate the vigorous growth of roots and check the development of ineffective tillers. Terminal drainage must be done for two weeks before harvest to hasten ripening of grains and to promote favourable harvesting conditions. Further details are given in the previous sections of this chapter.

CHAPTER 8

NUTRIENT MANAGEMENT

Nutrient is a substance, which is required for the normal growth and yield of the crop. Both the organic manure and chemical fertilizers are essential for rice cultivation. It improves the physical, chemical and biological properties of soil. An integrated approach of plant nutrient system will play an important role in sustainable rice production, especially as the land used for crop production dwindles down due to urbanization and there is a constant stress on the cultivated farm area as more and more areas taken away for non–agricultural purposes and become unavailable for farming. For proper exploitation of yield potentials of high yielding rice cultivars, 13 essential nutrients namely 6 macro nutrients (N, P, K, S, Ca and Mg) and 7 micro nutrients (Fe, Mo Zn, B, Mn, Cl and Cu) are considered generally. Now a days, silicon nutrition of rice is gaining momentum. These elements are essential because deficiency of any one of them adversely affects the physiological functions in plants, resulting in abnormal growth and/or incomplete life cycle.

Nutrient recommendation to a crop is based on three ways

1. Soil test value
2. Blanket recommendation and
3. STCR – Soil test and crop response.

Before cultivating any crop, analyze the soil for its nutrient content and apply the required nutrient. Soil test should be taken for each crop. If it is not possible for soil test, blanket recommendation is a standardized method based on experimental research. Recommendation based on yield target is called as STCR method/approach. Nutrients are applied to rice crop through organic manures, inorganic fertilizers, bio-fertilizers etc.,

8.1. ORGANIC MANURES

Organic manures improve the physical, chemical and microbiological properties of soils and ultimately the crop yields. On an average, organic manures such as FYM and compost contain 0.5-1.0% N, 0.6% P_2O_5 and 0.5% K_2O. Ten tonnes of compost will thus supply 50 - 100 kg N, 20 kg P_2O_5 and 50 kg K_2O, of which about 15 - 30 kg N, 15 kg P_2O_5 and 35 kg K_2O become available during the first season of application. The nutrient status of some common organic manures is given in Table 19.

Table 19. Nutrient status of some organic manures

Category	Source	Nutrient content %		
		N	P_2O_5	K_2O
Animal wastes	Cattle dung	0.3-0.4	0.10-0.15	0.15-0.20
	Cattle urine	0.80	0.01-0.02	0.5-0.70
	Sheep and goat dung (mixed)	0.65	0.05	0.03
	Night soil	1.2-1.5	0.8	0.5
	Leather waste	7.0	0.1	0.3
	Hair and wool waste	12.3	0.1	0.3
FYM composts	Farm Yard Manure	0.5-1.0	0.15-0.20	0.5-0.6
	Poultry manure	2.87	2.90	2.35
	Town compost	1.5-2.0	1.0	1.5
	Rural compost	0.5-1.0	0.2	0.5
	Water hyacinth compost	2.0	1.0	2.3
Oil cakes	Castor	5.5-5.8	1.8	1.0
	Cotton seed	3.9	1.8	1.6
	Pongamia pinnata	3.9-4.0	0.9-1.0	1.3
	Neem	5.2	1.0	1.4
	Niger	4.8	1.8	1.3
	Bassia latifolia	2.5-2.6	0.8	1.8
	Rape seed	5.1	1.8	1.0
	Lin seed	5.5	1.4	1.2
	Sunflower	4.8	1.4	1.2
	Animal meals Blood	10-12	1.2	1.0
	Raw bone	3-4	20-25	--
	Steamed bone	1-2	25-30	--
	Fish meal	4-10	3-9	1.8

Source: Gaur, 1992.

Krishnamoorthy and Kothandaraman (1982) and Gaur and Singh (1982) have provided the potentiality of various organic sources to supplement availability of NPK for the use of crops (Table 20). These organic wastes contribute to about 33.928 m.t of NPK, if their full potential is exploited and used. The summarised results of field experiments conducted all over India have indicated that FYM or compost application, a month prior to sowing, resulted on an average, in yield response of 168 kg/ha in rice (at 12.5t/ha).

Table 20. Potential of organic resources for NPK availability

	Plant nutrients (million tonnes)			
Source	N	P_2O_5	K_2O	Total
Cattle	2.977	0.793	1.332	5.102
Buffalo	0.745	0.276	0.487	1.508
Goat and Sheep	0.214	0.063	0.020	0.297
Pig	0.044	0.027	0.029	0.100
Poultry	0.027	0.020	0.010	0.057
Other livestock	0.079	0.018	0.069	0.166
Human beings	3.228	0.776	0.715	4.719
Farm crop wastes	5.600	2.300	10.700	18.600
Forest litter	0.075	0.030	0.075	0.180
Water hyacinth Compost	0.060	0.033	0.075	0.168
Rural compost	1.130	0.678	1.130	2.938
Urban compost	0.024	0.015	0.030	0.069
Sewage sludge	0.012	0.009	0.003	0.024
Total	14.215	5.038	14.675	33.928

Note: others not included are slaughterhouse waste, bone meal, fishmeal and industrial wastes like pressmud and pyrites.

Utility of other organics has also been tested and found encouraging. Miscellaneous organics like ammonium humate from coal, blood meal manure, water hyacinth, paper mill sludge in rice under coastal saline soils, rice husk, and poultry and pig manures in rice have been reported to supplement 1/3 to 1/2 of nutrients, particularly NPK, required for higher yields, besides showing residual effect and build up of soil fertility in the long run (Gaur, 1982; Maskina *et al.*, 1988, Misra *et al.*, 1989; Sahu, 1990). Similarly sulphitation pressmud (with 1.7% N, 3.6% P and 1.44% K) at 10 t/ha applied 10 days before planting can supplement fertilizer N equivalent to 75 kg/ha (Yaduvanshi *et al.*, 1990). Pressmud increased the yields of rice significantly apart from improving the soil physical properties (Rani perumal *et al.*, 1991), particularly in sodic soils. Based on the results available, calculated fertilizer equivalents of diverse plant nutrient sources are presented in Table 21 by bringing various nutrient sources to a common parameter. INM can be designed from this table for various crops. This should be done within the overall limits of recommended nutrient rates, which should be guided by consideration of nutrient needs and impact on soil properties and, not by waste disposal consideration. Fertilizer equivalents of some organic manures and biofertilizers are given in Table 21.

Table 21. Fertilizer equivalents of some organic manures and biofertilizers

Component	Input level	Fertilizer equivalent of input in terms of yield
FYM	Per tonne	3.6 kg N+P+K (2:1:1)
Green manure (sesbania)	Per tonne	4.4 kg N
Green manure (sesbania)	45 days crop	50-60 kg N for rice
Leucaena lopping	88 kg N in Leucaena	25 kg fertilizer on sorghum
Sugarcane trash	5 t/ha	12 kg N/t
Rice straw + water hyacinth	5 t/ha	20 kg N/t
Rhizobium	Inoculant	19-22 kg N
Azotobacter and Azospirillum	Inoculant	20 kg N
Blue-green algae	10 kg/ha	20-30 kg N
Azolla	6-12 t/ha	3-4 kg/t

Source: Tandon (1992)

8.1.1 Municipal and Sewage Wastes

Municipal and sewage wastes are important sources of organic manure. The total municipal refuse in India is about 12 m.t./annum containing about 0.5% N, 0.3% P and 0.3% K. Sewage sludge availability is about 4.0 m.t./ annum containing about 0.3% N, 2% P and 3% K (Bharadwaj and Gaur, 1985). Sewage farming recycles nutrients for crop production, besides reduing the pollution. Using these wastes improves the soil health and reduces fertilizer (NPK) requirement up to 50% to the succeeding crops (Rani perumal *et al.,* 1991). Thus organic manure can supplement 1/3 to 1/2 of the fertilizer NPK apart from providing other necessary nutrients, improving the physio-chemical properties of soil and having a residual effect of 10-20% on the succeeding crops.

8.1.2. Composting of Organic Wastes

Recycling of organic wastes as manure for sustaining soil health is important in the present context. Burning of rural and urban refuse without energy recuperation or its use for land filling are wasteful processes, which should be avoided (Gaur, 1987). Composting is a microbiological, non-polluting and safer method for disposal and recycling of organic wastes by bioconversion. Efficient celluloytic fungi species such as *Aspergillus, Trichoderma, Penicillium* and *Trichurus* accelerate composting for efficient recycling of dry crop wastes with high C/N ratio and reduce the composting period by about one month. The carbon content decreases by 43-57% and reduction in bulk by 2-8%. Enrichment of partially composted coir wastes can be achieved by *Azotobacter* and phosphate solubilisers (*Aspergillus awamorii* and *Bacillus polymyxa*) to improve the N, P and humus content of the finished compost. Mere addition of low grade rock phosphate increases the P content by 28.3% as compared to control, while inoculation with *Azotobacter* and *Aspergillus* or *Bacillus* increased N content by 32% over control (Gaur, 1982).

Blending 20 Mg of finished city compost with 10 Mg of moist sewage sludge and 200 kg phosphate improved the N content by 73% and reduced the C/N ratio to 8.0 (by 6th week) as compared to city compost with rock phosphate (0.82% N, C/N ratio of 13.8). Similarly blending of 10 kg city compost with 1% rock phosphate and 5 Mg moist sludge raised the N content to 1.45% with a C/N ratio of 15.3 and humus of 6.0 % in the city compost. Thus composts prepared by improved methods of rural and urban composts with improved techniques should be promoted on a large scale for efficient recycling and effective utilization for field crops like rice, sugarcane etc.,

8.1.3 Crop residue management

In India, about 100 m.t of crop residues are available for recycling in agriculture annually. Crop residues are a larger reservoir of plant nutrients (approximately supplementing 0.5, 0.6 and 1.5 m.t. of N, P_2O_5 and K_2O respectively), improve physical and biological properties, and protect the soil from wind and water erosion. Most often, these residues are disposed of by burning *in situ*. Beneficial, harmful and neutral effects of crop residue incorporation in soil on grain yield of succeeding crops have been documented mainly due to differences in crop residue types, amount used, particle size of the residue, time and duration of residue application, soil ecological conditions, nature of succeeding crops, type of soil and amount of fertiliser used. Inoculation of these wastes with efficient cellulolytic microorganisms produces enough simple sugars for growth and multiplication of beneficial microflora like free living N fixers and phosphate solubilizers in soil and increases crop yields (Varma and Mathur, 1990)

In cereal-cereal rotation, incorporation of rice straw (5 t/ha spreaded) inoculated with *Aspergillus* spp in 0-15 cm surface soil one month prior to sowing cereals, contributed equivalent to 50 kg N/ha (Bharadwaj, 1980).

Sidhu and Beri (1989) reviewed the rice straw residue management in 27 field experiments conducted during 1974-85 on Typic Ustochrept soils in rice-wheat rotation. In rice-wheat rotation, residue management of unsoaked rice had significant impact on yield or succeeding crops. It decreased pH (8.2 to 8.0), increased available N (59 to 74 ppm), organic carbon (0.28 to 0.39%), water holding capacity (26.8 to 28.5%) and bulk density (1.56 to 1.49g/cm^3)

Sugarcane trash left out after cane harvest is estimated to be around 19-38 m.t. and is either burnt *in situ* or used as domestic fuel. It can be recycled after composting suitably. Incorporation of soaked (for 120 days) trash (1-2 cm size) at 5 t/ha increased the organic carbon, available K, Zn, Fe and Mn contents but decreased the pH and slightly increased the salt content (Yadav *et al.,* 1987). Favourable effect of pulses (green gram or soybean) stalk incorporation on rice yields (Prasad and Palaniappan, 1987) in rice based cropping systems has been reported (Table 22) besides increasing the organic carbon content of the soil. Thus incorporating residue after harvest of pods is a good practice to sustain the yields of succeeding cereals, besides saving N equivalent to 20-30 kg/ha and improving soil physical and chemical properties. Coir pith, a by-product from coir industries, contains nutrients (N-0.26%, P-0.01%, K-0.78%, C/N ratio- 112:1, Lignin-30% and Cellulose-26%) and its availability is about 1 m.t. in India. Raw and composted coir pith (inoculated with *pleurotus* spp. and 5 kg Urea/t) are presently used for increasing the yield of crops (10-46%) and for enhancing the soil moisture content (Rani Perumal *et al.,* 1991).

Table 22. Effect of residue incorporation

Residue incorporation (t/ha)	Test crops/ sequence	With residue	Without residue grain yield (t/ha)	% change in yield	Soil type
chopped 5t chopped rice straw (at 50 N Kg/ha) 7.5 t	wheat	2.34	1.99	+17.6	loamy irrigated
chopped rice straw 5.0 t	wheat	1.50	1.40	+7.1	silty clay
	rice	1.30	1.10	+18.2	loam, irrigated
chopped wheat straw 7.0 t	rice	6.20	5.90	+5.1	loamy sand irrigated

Adapted from Shaklawat *et. al.,* 1989.

8.1.4. Green manuring

Green manuring is an old practice. In the present context of integrated nutrient supply system, it needs adequate attention since it is a renewable source of input for building of soil fertility and supplementing plant nutrients, especially N. This practice is more appropriate for irrigated agriculture, due to easy mineralization of nutrients from the turned over plants in the soil. Green manures may be either leguminous crops grown in situ (*Sesbania* spp., *Crotalaria juncea*), or leguminous/non-leguminous plants /trees, (*L. leucocephala, Casuarina, Gliricidia maculata, Pongamia glabra, Tephrosia purpurea, T.candica, Cassia tora, Calotropis gigantea*, etc.) grown on bunds of waste lands, for utilising vegetative parts for green manuring of soils. In addition, leguminous crops like cowpea, green gram and black gram are used either as green manure or for stalk incorporation after the harvest of seeds or pods. These leguminous crops fix varying amounts of N from 85 kg in Casuarina to 500 kg N/ha (Leucana). After considering various losses (leaching and denitrification), a modest average estimate of 50 kg N/ha is available to the crop (Gaur and Singh, 1982). Incorporation of 10 t/ha of green manure supplements fertilizer N equal to 40-150 kg/ha and has shown promising results in rice when green manure was integrated with fertilisers and applied a day before planting (Singh *et al.*, 1990). Further integrating P with green manure in the first rice crop produced higher yields of second rice. Green gram haulm incorporation improved the succeeding rice crop yields (Sreekantan and Palaniappan, 1990). N content, biomass production and N accumulation of different green manures are given in the Tables 23, 24 and 25.

Plate 34a. Raising daincha green manure in rice fields.

Plate 34b. Sunnhemp as green manure in rice fields.

Table 23. Biomass production and nitrogen accumulation in green manure crops (at 45 days)

Manure crop accumulation	Fresh biomass (t/ha)	N (kg/ha)
Scəbania aculcata	8.16	37.53
S. rostrata	5.43	38.66
S. speciosa	2.51	15.84
S. grandiflora	3.27	21.25
Tephrosia purpurea	0.95	6.25

(Adapted from Palaniappan *et al.*, 1992)

Advantages of green manuring

- Leguminous green manure crops fix atmospheric N.
- It decomposes easily without leaving much residue in the soil
- Green manures promote the activity of soil microorganisms, CO_2 is liberated and the chemical reaction in the soil are speeded up and more plant nutrients are made available to improve the structure of

the soil. It acts as a binding material in light soil and water holding capacity.

- Green manure crop withdraws plant nutrients from the lower layers and concentrates them in surface soil.
- It is grown as catch crop, shade crop, cover crops or forage crops.
- It supplies nutrients especially N through addition of biomass (120-140 kg N/ha).
- Green leaf is a quick acting as ammonical fertilizers in the tropics.

Application of green manure by way of incorporating and the dual culture of azolla application or BGA to rice were found to save the N requirement. *Sesbania rostrata* and *Azolla microphylla* @ 12.5 t/ha brought a fertilizer N saving of 30% in high yielding varieties of rice(short duration).

Table 24. Nutrient contents and C:N ratios in the green leaf and green manures

	Crop	Total N (%)	C:N	Total P (%)	Total K (%)
Conventional green manures					
I. Annuals					
	Sesbania aculeata	3.97	21:1	0.37	4.80
	Crotolaria juncea	1.90	44:1	0.34	3.60
II. Perennials (leaves and twigs)					
	Gliricidia sepium	2.76	31:1	0.28	4.60
	Pungamia glabra	2.78	33:1	0.27	0.19
Non-Conventional green manures					
I. Trees (leaves and twigs)					
	Azadirachta indica	2.83	70:1	0.28	0.35
	Delonix elata	3.51	27:1	0.31	0.13
	Delonix regia	2.76	32:1	0.46	0.50
	Pheltophorum ferrugenum	2.63	34:1	0.37	0.50
II. Weeds					
	Adothoda vesica	1.32	60:1	0.38	0.15
	Parthenium hysterophorus	2.68	30:1	0.68	1.45
	Eichornia grassipes	3.01	29:1	0.90	0.15
	Trianthema portulacastrum	2.64	32:1	0.43	1.30
	Ipomoea cornea	2.01	43:1	0.33	0.40
	Calotrophis gigantea	2.06	64:1	0.54	0.31
	Cassia pistula	1.60	120:1	0.24	1.20

Source: Pushpavalli *et al.*, 1993.

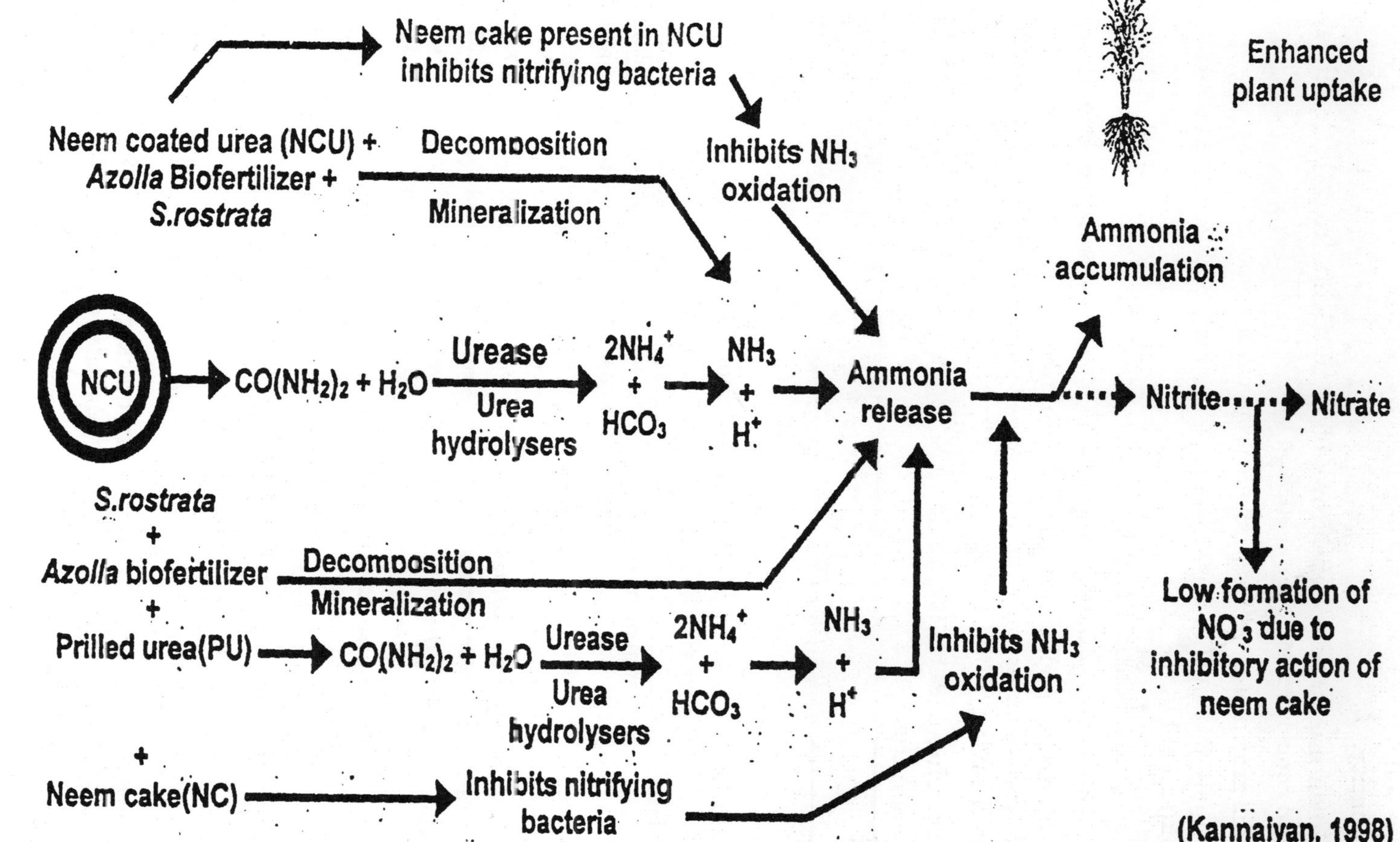

Fig12 INHIBITORY ACTION OF NEEM CAKE ON EFFICIENT UTILIZATION OF NITROGEN IN RICE CROP

Plate 34c. Daincha incorporation using cono weeder.

Table 25. Biomass production and N accumulation in some common green manures for rice fields

Local name	Botanical name	Growing season	output in 46-60 days	
			Green matter (t/ha)	N accumulation (kg/ha)
Sunnhemp	*Crotolaria juncea*	wet	21.2	91
Daincha	*Sesbania aculeata*	wet	20.2	86
Phillipasara	*Phaseolus trilobus*	wet	18.3	201
Green gram	*Vigna radiata*	wet	08.0	42
Cowpea	*Vigna sinensis*	wet	15.0	74
Guar	*Cyamoposis tetragonoloba*	wet	20.0	68
Senji	*Melilotus alba*	dry	28.6	163
Khesari	*Lathyrus sativus*	dry	12.3	66
Berseem	*Trifolium alexandrium*	dry	15.5	67

(Source : Abrol and Palaniappan, 1988)

8.1.5 Non conventional green manures

The leguminous plants, which are quick growing accumulating greater biomass and containing appreciable quantity of N, are generally used as green manures. But non-leguminous annuals, shrubs and trees are capable of providing large biomass and can supply considerable quantity of plant nutrients. Initial set- back seen in certain crops like rice and other crops after the incorporation of organic residues having wide C:N ratio, high lignin content which resist easy decomposition and release of higher combined influence on the establishment of young seedlings might have discouraged the farmers of using those non-conventional green biomass as manures in agriculture. However, such adverse effects could be overcome by proper pre-treatment like composting with suitable microbial inoculants and then applying to the field. Samples drawn from such non-conventional green manures for analysis have brought out their suitability to use as manures. (Palaniappan *et al.*, 1991).

8.1.6 Oil Cakes

Oil cakes of non-edible types like castor, neem and Karanja (Pungam) are widely used as organic manures. The non-edible Oil cakes contain high amount of plant nutrients. Most of the oil cakes are valued much of their alkaloid contents, which inhibit the nitrification process of N transformation in soils. Neem cake contains the alkaloids, nimbin and nimbitidin, which effectively inhibit the nitrification process. The inhibitory action of neem cake on efficient utilization of nitrogen in rice crop is given in Fig. 12. Similarly Karanja seed cake and the bark of the tree possess nitrification inhibitory properties. Karanja, the major furano flavanoid compound present in the seeds of Karanja and pungam, is a potent nitrification inhibitor, equal in efficiency to nitropyrin (N serve) in retarding the nitrification process of ammonical N and increasing the yield, N uptake and grain protein content of rice. (Palaniappan and Siddeswaran, 1994).

8.2. INORGANIC FERTILIZERS

The application of chemical fertilizers is influenced by number of factors.

1. Soil fertility.
2. Nature of crop growing
3. Previous crop and Crop rotation.

8.2.1 Nitrogen

Nitrogen is the major constituent of the cell especially the protoplasm. We can improve the crop growth by applying Nitrogen. If N is applied at higher level, it will reduce the thickness of the cell wall and make the crop to 'lodge'. It will attract the sucking insects. At the lower level of N it will expose the deficiency symptom and finally affect the normal growth of the plant.

Leaves will turn yellow in colour and the crop will become "stunted". At the acute stage, leaves will fall and the plant will die. Hence, nitrogen should be applied at right quantity at right time.

8.2.1.1. Sources of Nitrogen

a. Atmosphere

Atmosphere is the first and foremost source of Nitrogen. 78% of N is in the atmosphere. Due to thunder and lighting, the N in atmosphere will reach the soil / earth surface in the form of nitrate.

b. Soil N

Each soil is having its inherent nutrient source. Nitrogen in soils in mostly present in organic from in soil organic matter (SOM); in most mineral soils, it is 59-95%. About 20 to 40% of N in SOM is present as proteins and another 8-10% as amino sugars. The other N compounds are nucleic acids, amino – lignin complexes etc. Inorganic N is present as NH_4 in the exchange complex (or as fixed NH_4 in clay minerals) and nitrates. In rice fields, nitrates are virtually absent. From the soil, N is available to plants in two ways 1. Mineralisation and 2. Decomposition of the organic matter. Mineralisation is changing the organic N form from parent material into Nitrate form. Decomposition of the organic matter is done by the activities of microbes. Organic N is converted into inorganic nitrogen. From inorganic form, it is converted into nitrite – Nitrate. Crop takes N only in nitrate form. In some cases it absorbs in the form of Ammonia. Flooding resulted in the production of 50-350 ppm ammonium – N in different soils at about 2 weeks. When the soil was pulverized, the rate of mineralisation (N) due to flooding increased and this correlated with the clay humus ratio (1:1 rapid slow 2:1). De Datta (1968) reported that in unfertilized soil, the plant absorbed 74-107 kg N/ha, the amount of soil N mineralized was 57 kg/ha, of which, 49 kg/ha was released until the panicle initiation stage.

Mineralization of soil organic N

Soil organic N compounds are first broken down to simpler compounds such as amides and then these are later converted to inorganic form i.e. ammonium. This process is carried out by a large group of heterotrophs. Ammonium is then oxidised to nitrite by a group of autotroph bacteria belonging to genera *Nitrosococcus* and *Nitrosomonas*. Nitrite is finally oxidized to nitrate by *Nitrobacter* sp., another autotroph. Plants can take up N both in NH_4 and NO_3 form. While most upland plants take bulk of their N in NO_3 form, rice plants particularly those growing under low land conditions take up most of their N as NH_4, Since NO_3 is not stable under submerged soil conditions (discussed later).

Soil organic matter	→	Amide N	→	NH_4	→	NO_2	→	NO_3
	Aminization				Ammonification			Nitrification

Release and availability of native soil N depends upon a number of factors such as C:N ratio of organic residues, soil pH, temperature, moisture and activity of soil micro organisms. Near neutral soil pH, temperatures between 20^0 and 30^0C and field capacity moisture are probably near optimum conditions for nitrification. A C:N ratio wider than 30 in cereal residues can lead to immobilization of inorganic N, because micro organisms use plenty of available energy (C) and inorganic N to multiply their populations. However, with crop residues having a C:N ratio of 15 or less in the case of legume residues, there is enough N to spare and it becomes available to growing crop plants. That is why green manuring with *Sesbania* or *Crotolaria* is done. On the other hand, if large amounts of cereal residues are incorporated, there could be an adverse effect on the young seedlings of the crop sown immediately after incorporation.

c. *Organic manures*

Nitrogen is commonly supplied to the soil through organic and green manures, by application of FYM, compost, with or without addition of N, biological N fixation and incorporation of green manures. A heavier application of FYM or compost improved growth at later stages. The details regarding nutrient content of different organic manures and its fertilizer equivalents are given in section 8.1.

Green manures

Sesbania aculeata (Daincha), (130-185 kg N) *Crotolaria juncea* (Sunn-hemp), (80-130 kg N) *Sesbania rostrata,* (170-220 kg N) *Tephrosia purpurea* (Kolinchi), (70-115 kg N) *Indigofera tinctoria* (Avuri), Philipesara (*Phaseolus* sp.), (85-125 kg N) were found most suitable for sowing and for incorporation within 6-7 weeks after sowing. The application of green leaves collected from various shrubs and trees growing in porombokes road sides, tank bunds and forests is traditional agricultural practice in the rice culture. Rice responds to green leaf better than to FYM or any other manure locally available. Suitable green manure crops can be raised during the fallow period in the single crop wetland regions, which receive some rains in summer. Further details are given in the section 8.1.4.

Decomposition of green manures

When the soil is flooded, the nitrates in the soil are reduced and free N is evolved. There is evolution of CO_2, H, CH_4 etc., from the soil. The CO_2 is used by the algae on the surface for the synthesis of the carbohydrates and concurrent liberation of O_2. The O_2 dissolves in water and moves into the flooded soil slowly, but not effectively beyond 1 cm. Rice roots developed and near the surface may be able to use this O_2. Gaseous O_2 diffuses from the aerial parts to the rice roots through aerenchyma tissues. The green manure incorporated in the soil produces ammonia as an end product and it becomes available for use of rice plants. The production of ammonia is maximum at 4 weeks after transplanting in green manure incorporated field. Low land rice takes N mostly in the form of ammonium together with small quantities of simple N compounds. The nitrates in the soil if any, and in fertilizers are also absorbed but unabsorbed nitrates lost by denitrification and leaching.

d. Bio-fertilizers

(a) Blue algae-10 kg/ha(20-30 kg N/ha) 10 days after transplanting

(b) Azolla – 10 t/ha- 30 kg N/ha (Apply 6 t/ha azolla and incorporate before transplanting)

(c) Azospirillum and Phosphobacteria - seed treatment-seedling dipping–main field application (3+5+10 pockets respectively)

(d) Azolla will be incorporated in rice field during first weeding after draining the water completely. The details are given in the subsequent sections (refer section 8.3).

e. Chemical fertilizers

In recent years, the production of urea and urea based high analysis fertilizers has increased. Ammonium form of fertilizers used is NH_4Cl, NH_4SO_4, NH_4 (PO_4) and anhydrous ammonia.

8.2.1.2. Level of N application

Green manures like daincha should be raised well in advance during fallow period. After 45 days or before flowering, it should be incorporated *insitu* using green manure trampler or tractor drawn cage wheel. If green manures are not raised, apply 12.5 t/ha of FYM or compost. Spread the manure uniformly over the dry soil before letting in water for puddling. If FYM or compost is not applied, apply green leaf manure at 6.25t /ha on the puddled field and incorporate properly to a depth of 15 cm. Maintain 2.5 cm of water and allow 7-10 days before transplanting for decomposition. As far as possible, apply fertilizers as per soil test recommendation, if not followed, adopt blanket recommendation in Tamil Nadu as scheduled here under in Tamil Nadu.

Short duration varieties : 120 : 38 : 38 N P K kg/ha

Medium and long duration varieties : 150: 50: 50 N P K kg /ha

This will vary based on the recommendation being followed in other states. For effective split application of N, maintain thin film of water and allow to absorb and irrigate after 24 hours.

8.2.1.3 Method of Application

Further details are given in the section 8.2.1.6 (A)-split application and 8.2.1.6-(B)-placement. Based on the N requirement of rice crop, it can be applied through foliar spray. Foliar spray of 2% DAP given at flowering and second at 15 days after 1st spray increased grain yield of transplanted and direct seeded low land rice (Annadurai and Palaniaapan, 2000 a and b).

1. Broad casting
2. Placement
3. Foliar application - Urea @ 1-2% Concentration or DAP @ 1-2% Concentration

8.2.1.4 Time of Application

About 50% of N is applied as basal. Remaining 50% is divided into two splits i.e. 25% in tillering stage and the remaining 25% applied at panicle initiation stage or it may be applied as given in the Tables 26, 27 and 28.

Table 26. Way of applying fertilizers to short duration varieties

Light soil	N	P	K
Basal	24	38	19
35 DAS	24	-	-
45 DAS	24	-	-
55 DAS	24	-	19
70 DAS	24	-	-
Total	120	38	38
Heavy soil			
Basal	30	38	19
35 DAS	30	-	-
50 DAS	30	-	19
70 DAS	30	-	-
Total	120	38	38

Table 27. Way of applying fertilizers to medium duration varieties

Basal	30	50	25
AT 40 DAS	30	-	-
55 DAS	30	-	-
70 DAS	30	-	25
100 DAS	30	-	-
Total	150	50	50

Table 28. Way of applying fertilizers to long duration varieties

Basal	25	50	25
40 DAS	25	-	-
55 DAS	25	-	-
70 DAS	25	-	-
95 DAS	25	-	25
115 DAS	25	-	-
Total	150	50	50

8.2.1.5 N dynamics in rice soils and N loss mechanisms

Rice soils since having standing water, have the possibility of N loss by NH_3 volatilization, Nitrification, Denitrification, Leaching and Runoff. The bulk of the loss was found to occur within 5-7 days of application of N.

- Leaching loss : 3-25 % N
- Runoff loss : 19-30 % N
- NH_3 volatilization : 30-50 % N

The surface oxidized layer of soil can vary from 0.2 - 3 cm in thickness depending on the oxygen concentration in the flood water and oxygen consumption. If Urea is applied in low land condition, it will reach the oxidized zone. Amide form is converted into ammonical form and then to Nitrite and Nitrate. Nitrate form in oxidized zone reaches the reduced zone by leaching. Due to activity of Nitrifying bacteria, nitrate form is converted into elemental N_2 and gaseous N_2 escape to atmosphere.

- Loss of N in gaseous form to atmosphere is called as **volatilization loss.**
- Nitrate form of N dissolves in water and reaches lower most soil by **leaching**.

- Nitrogen dissolves in irrigation water and reaches the low-lying area by **runoff.**

1. Volatilization loss

When fertilizer N is applied to soil (urea is taken as the example), it undergoes several changes and on a soil having pH7 or above, a substantial part of applied N could be lost due to ammonia volatilization. Measurements in rice fields showed a loss of 8.3% of applied urea – N due to ammonia volatilization in the seedling stage of rice. Losses were only 1.67% when urea was top dressed at panicle initiation stage. Higher ammonia volatilization losses (20-50%) from surface applied urea were reported by others (Mikkelsen *et al.,* 1978; Vlek and Craswell, 1979).

2. Denitrification

When soil becomes waterlogged, O_2 is excluded and anaerobic decomposition takes place. Some anaerobic organisms have the ability to obtain their O_2 from nitrite and nitrate with accompanying release of N and nitrous oxide. Organism responsible for denitrification is *Psedomone microlocus, Bacillus* etc. Some literatures say that in rice field where the soil remains submerged during most of the growth cycle of crop and therefore reduced due to the lack of oxygen, several species of bacteria e.g. S oxidizing bacteria (*Thiobacillus thioxidans*) use nitrate as a source of O_2 and consequently reduce it to N_2 or N_2O. This process is known as denitrification. It is considered as an important mechanism of N loss in rice filed. Denitrification losses are most severe under alternate flooded and drained conditions, because when drained conditions prevail nitrates formation, which are lost under submergence. Direct measurements of denitrification are quite difficult and in most studies, these are estimated by different method preferably using 15^N (subtracting the N removed by plants, N retained in soil, N lost by volatilization from the N applied).

Nitrification

The biological oxidation of ammonia to nitrate is known as nitrification. Conversion of nitrate (NO_3) is brought about by *nitrosomonas* and *nitrobactor*.

3. Plant uptake

The rice plant can use nitrate N as readily as ammonium N but the proportion of the two forms actually taken up by rice in the field probably depends on the soil water regime. Nitrogen loss mechanism that has so far not received due attention, is N lost through foliage and other plant parts. (Patron *et al.*, 1988) and Mosier *et al.* (1990) suggested that young rice plants may facilitate the efflux of N_2 and N_2O from the soil to the atmosphere.

4. Leaching losses

Of various solid N fertilizers, urea is most susceptible to run off losses when it is surface broadcasted. This is because urea is highly soluble, chargeless organic molecule and cannot be retained by soil particles. As a contrast, NH_4 in ammonium sulphate / ammonium phosphate / ammonium chloride even when applied to most soil surface gets a change to get adsorbed on the exchange complex of clay particles. Considerable N in upland crops could be lost by leaching but again actual measurements are difficult. On the other hand, in lowland rice conditions, leaching losses of N may not exceed 1% (Patil *et al.*, 1990).

Immobilization of applied N

Immobilization of mineral N occurs only when high C:N ratio materials are applied. In general, there is to and fro movement of N between soil organic N and applied mineral N. This has been proved by 15^N studies. In a field study, Goswami *et al.* (1988) found that as much as 31.2% applied N is immobilized. Only a small fraction of immobilized N becomes available to succeeding crop.

8.2.1.6. Nitrogen use efficiency (NUE)

No specific method is available to control N loss. Based on experiences, the following methods may be followed to control N loss and to increase the applied fertilizer N use efficiency. Selection of highly fertilizer responsive varieties and good water and weed management practices will increase NUE.

1. Split application of N
2. Deep placement of N fertilizers (USG)
3. Use of Slow release N fertilizers, nitrification inhibitors and use of N serve

A. Split application of N

Split application of N for increasing NUE is well established. When large amount of N are applied at a time, say all basal, N efficiency is the lowest. Application of N should match the crop needs for N. In rice, 2-3 split doses are recommended for short and medium duration varieties, while 3-4 split doses are recommended for long duration varieties. More splits are also recommended for sandy soils. Split application of nitrogen is also desirable from the viewpoint of nitrogen nutrition of rice plant (Annadurai *et al.*, 2000). Nitrogen absorbed by the plant from tillering to panicle initiation tends to increases the number of tillers and panicles and that absorbed during panicle to flowering increases the number of filled spikelets per panicle. Nitrogen absorbed after flowering tends to increase the 1000 grain weight (De Datta, 1978). In most of areas, K is applied with N in splits. The Tables 26, 27 and 28 given above explain the way of applying the N fertilizers in splits to short, medium and long duration varieties based on the soil types in Tamil Nadu.

B. Deep placement

Placement of fertilizers has received considerable importance especially in the case of phosphatic fertilizers, but it is equally important in the case of N. Ammonia volatilization can be almost prevented by deep placement of N fertilizer, because placement in the reduced zone can prevent or at least retard the formation of nitrates and their subsequent loss by denitrification. In practical, this practice is not followed by the farmers effectively.

C. Use of USG (Urea super granules)

By using USG, NH_3 concentration in floodwater is reduced and prevents rapid conversion of ammonical N to NO_3. Urea is manufactured in the form of higher granules of 1 to 4 gm each. When granules are placed in the reduced zone of rice soils, the losses are substantially reduced. USG response was high in clayey loam soil and not promising in coarse textured soil. Actually, USG's are based on the principles of Chinese mud ball techniques. Urea pellets come in different shapes and sizes. The most popular size is 1g (1cm diameter) granule. Under low land conditions, it will increase NUE by 20-30%. Due to its cost, the problem of placing it at correct place and non availability of low cost USG placement equipments, USG is not commonly used by many of the farmers.

D. Slow release N fertilizers

About 50% of N is lost through denitrification, when urea is applied to the low land rice subjected to alternate flooding and drying. The ammonical form may be in the form of free ammonia or an ammonium salt or in a compound like Urea, which on hydrolysis, in the soil, produces ammonium ion. Nitrate ions are readily absorbed by the plants but at the same time, they are easily leached out particularly in sandy soils. To reduce the loss of N, slow release fertilizers and N inhibitors are found to be useful. Slow – release N fertilizers release N slowly and therefore allow its uptake before it is lost. There are two types of slow release fertilizers.

i. Chemicals

The chemicals with inherent slow N release properties are as follows.

1. Isobutylidene diurea (IBDU)
2. Crotonylidene diurea (CDU)
3. Guanyl urea sulphate (GUS)
4. Urea formaldehyde condensates (UF) and
5. Oxamide

Due to their high cost, these chemicals are not used widely. Urea form is also known as turf fertilizer, and is widely used in USA..

ii. Coated conventional fertilizers

1.Coal tar coated urea

The coal tar at 1 kg / 100 kg of urea is heated to dissolve in a metal container. After slight cooling, 1.5 litre of kerosene is slowly added with constant stirring until a solution is obtained. The coal tar solution is then poured over heaped urea and mixed thoroughly, the finely powdered neem cake is then added and mixed again so as to get uniform coating of neem cake over urea granules.

2. Mud ball urea

- Make mud balls of 1 cm diameter.
- Make a hole in the center so as to accommodate 1 g of urea.
- Place the urea inside and close the opening.
- Place the balls 10 cm deep in the root zone at one ball for every 4 hills.
- By this method, 75 kg N per hectare is sufficient.

3. Paper packed urea

This is similar to mud ball urea. Pack 1 g of urea in piece of paper. Place this packing at 10 cm depth in the zone.

4. Neem treated or coated urea

Blend the urea with crushed neem seed or neem cake 20% by weight. The finely powdered neem cake at 20 kg /100 kg urea is then added to the urea and the contents are mixed again so as to get uniform coating of neem cake over urea granules. Keep it over night before use. The material is then ready for application. Neem coated urea is widely practiced by the farmers in India.

5. Others

Urea can be mixed with gypsum in 1:3 ratio or urea can be mixed with gypsum and neem cake at 5:4:1 ratio to increase the nitrogen use efficiency. Like this, sulphur coated urea and polymer coated urea can be prepared and applied.

D. Use of Nitrification inhibitors

Nitrification inhibitors are chemicals that can retard the nitrification process in soil and thereby reduce the N loss by leaching and denitrification. These chemicals are highly toxic to denitrifying bacteria. It prevents the formation of Nitrite and Nitrate from ammonical form. A number of chemicals have been reported to have nitrification inhibiting properties. The chemicals mostly widely tested are as follows.

1. N – Serve (2-chloro 6 (trichloromethyl) and pyridine),
2. AM (2 – amino – 4- chloro-6 methyl pyrimidine),
3. DCD (dicyandiamide) and
4. ST (Sulphathiazole)

Advantages of slow release fertilizers and nitrification inhibitors

- They improve uptake of N by releasing the nutrient slowly and uniformly and thereby crop growth is improved.
- Due to higher efficiency and uptake of nutrient by the plant, there is scope for reducing the fertilizer dose.
- Leaching losses are minimal and
- The pollution of water and air is low.

Some recommendations on modified N fertilizer for increasing NUE in rice are given in Table 28a.

Table 28a. Some modified N fertilizers to increase NUE for rice

Crop	State	Technique	Application
Rice	Andhra Pradesh	Neem – cake mixed urea	Top – dressing
	Kerala	Neem – cake mixed urea (1:5 ratio)	Top – dressing
	Kerala	Soil – cured urea (6:1 ratio)stored for 24-48 hours in shade	Top – dressing
	Karnataka	USG deep placed 10-15 cm, 7-10 days after planting between alternate hills	Basal
	Karnataka	Soil – cured urea (50 – 100kg soil) kept for 24 hrs.	Top – dressing
	Karnataka	Neem cake blended urea	Basal
	Meghalaya	USG or Rock – P coated urea 48-72 hours	Basal
	Orissa	USG or coated urea	Basal

Tamil Nadu	Neem – cake blended urea (20% cake by wt.)	Top – dressing
Tamil Nadu	Urea mixed with gypsum (1:3)	Basal
Tamil Nadu	Urea treated with coal tar (100 kg urea, 1 kg coal tar, 1.5 litre kerosene)	Basal

Source : Tandon, 1989

8.2.1.7. Nitrogen Assessment

A. Using SPAD Meter

The chlorophyll meter used to asses N status of the plant. The threshold level is 35, it means the leaf nitrogen content is 1.4 gm/m^2 leaf area. The observation are taken from 14 DAT at weekly intervals till flowering at 8-11 am with out any direct sun exposure. The index leaf is 1,2 of 3rd fully expanded top leaf, free from pest and diseases. If the value falls below 35 in short duration rice variety, apply 30 kg of N/ha at 21 DAT, if the value falls below 35 on 38-42 DAT, apply 45 kg N/ha., if the value falls below 35 on 49 DAT, apply 30 kg N/ha.

Advantage

30% saving, 10% more yield, no chemical pollution, increase NUE, N applied according to need of the plant and very easy to operated but very costly.

B. Using leaf colour chart (LCC)

A leaf colour chart developed from Japanses prototype can help farmers to measure the leaf colour intensity which is directly related to leaf chlorophyll content and leaf nitrogen status. It is a simple, easy to use, and inexpensive tool to determine the time of N top dressing in rice crop.

8.2.2 Phosphorus

Phosphorus plays a vital role in grain formation and root development. Phosphorus content in most Indian soils is within 200-800 ppm P; the range of organic P as % of total P is 10-30%. Therefore, most of "P" research in India and elsewhere is directed towards inorganic P, which makes up a major part of total P.

Sources

1. Soil
2. Application of organic manure like FYM, Compost, Press mud and Coir pith.
3. Fertilizer – Single Super Phosphate, DAP, Triple Super Phosphate.
4. Bio – fertilizers: Phospho bacteria, VA Mycorrhizae

Dosage

Short duration varieties : 38kg P_2O_5 /ha.

Medium and long duration varieties : 50kg P_2O_5 /ha.

Method of Application

1. Broad casting

The phosphorus is not easily lost from the system but inputs from sources such as irrigation water and straw are generally very small. The entire dose of P should be given as basal i.e., at the time of last ploughing by broad casting method.

2. Foliar application of DAP at 1-2% concentration to overcome deficiency.

8.2.3 Potassium

Potassium is required for the transport of the products of photosynthesis. K provides strength to cell walls and contributes to greater canopy photosynthesis and crop growth. K increases the number of spikelets per panicle and number of filled grains.

Sources

1. Soil
2. Chemical fertilizer-Potassium Chloride, Potassium sulphate
3. Irrigation water contribute some amount of K

Time of application

Basal application : at 50% with nitrogen as basal

Top dressing : at 25% each at the time of tillering and panicle initiation stages. Or K should be applied along with N as given in section 8.2.1.6 A. When calculating fertilizer K requirements, nearly 80% of K taken up by rice remains in the straw after harvest, making straw an important source to consider.

Dosage

Short duration varieties : 38kg K_2O /ha.

Medium and long duration varieties : 50kg K_2O /ha.

While applying P and K to rice, the following should be followed.

APPLICATION OF PHOSPHORUS AND POTASSIC FERTILIZERS

i. If DAP is applied to the nursery, apply 30% of the recommended dose of P_2O_5 over and above the DAP application to the nursery

ii. When straight fertilizers are applied, mix the phosphate and potassic fertilizers by the quartering method.

iii. When green manure is applied, mix the phosphate and potassic fertilizer by the quartering method. In the place of green manure, pressmud, composted coir pith can also be used.

iv. When the green manure is applied, rock phosphate can be used as a cheap source of phosphorus. It will also hasten the decomposition of stubbles in the second crop.

v. In case of in-situ ploughing of green manure plants into the field, apply the fertilizers on the green manure plants prior to ploughing. Apply in the early morning, so that the fertilizer may stick on the moist leaves.

vi. In case green manure is brought from outside and applied, spread the green manure first on the ploughed field and then apply the fertilisers over the green matter and then incorporate.

vii. Apply 50 kg P_2O_5 / ha as rock phosphate plus 10-12.5 tones of green manure. If rock phosphate is applied, the succeeding rice crop need not be supplied with phosphorus. Application of rock phosphate + single super phosphate / DAP mixed in different proportion (75:25 or 50:50) is equally effective as SSP / DAP alone. Application of phosphate fertilizers with green manure has the following advantages.

   a. The organic acids produced during the decomposing process dissolve the phosphates and make them easily available to the crop.

   b. Potash application along with green manure has no special advantage. But application of fertilizers simultaneously saves labour and is therefore more economical.

   c. Do not apply nitrogenous fertilizers at this time as loss of nitrogen will be inevitable, if the puddle dries out. However, if a sheet of water is maintained over the puddle, there is no loss of nitrogen.

   d. Broadcast phosphobacteria at the rate of 10 packets/ha mixed with sand or soil, with 75% of recommended inorganic P for Kuruvai and Thaladi seasons.

e. Foliar application of DAP: Apply DAP 2% solution as foliar spray to the seed crop at 65th and 80th DAS.

8.2.4 Other nutrients / manures / fertilizers

8.2.4.1. Application of lime

Apply lime to acid soils based on the soil analysis for obtaining normal rice yields. Apply 2.5 t of lime per ha before last ploughing. Apply lime at this rate to each crop up to the 5th crop /year.

8.2.4.2 Basal application of gypsum

Gypsum at 500 kg'/ha is applied basally with NPK in the main field in non-calcareous heavy soils. It will also reduce the methane emission. It is recommended for sodic soil reclamation as an amendment.

8.2.4.3 Application of Zinc sulphate

Among the 7 micro nutrients, rice is exhibiting zinc deficiency symptom. Mix 25 kg of zinc sulphate with sand and to make a total quantity of 75 kg and spread the $ZnSO_4$ + sand mixture uniformly on the leveled field. Do not incorporate in the soil, as there will be Zn phosphate interaction.

8.2.4.4.Silicon nutrition

Silicon is an element absorbed and accumulated by certain plants but its requirement and role are still debated. Silicon is abundantly available in earth's crust and is second only to oxygen. It shares many of carbon, which forms the backbone of most organic molecules but seldom forms an integral component of any biomolecules as the Si – Si bonds are considerably weaker than C-C bonds and many Si-H bonds are relatively unstable and react readily with oxygen. The larger size of Si atom compared to carbon also render it unsuitable as a basic building block despite the fact that it can also form bounds with four other atoms creating a three dimensional net work similar to carbon. Yet a few organisms like diatoms, chrysophytes, silicoflag-llates, some xanthophytes, plants insects and vertebrates absorb and accumulate silicon. Among the plants, silica (SiO_2) concentrations are found to be higher in monocotyledons than in dicotyledons and its level increases from legumes < fruit crops < vegetables < grasses < grain crops (Balasubramaniam, 2003).

Rice accumulates 2-40 percent silica in straw and almost every part of rice contains silica which is not added exogenously as a fertilizer as done in N, P and K and is not readily available in soil. Certain elements like Si etc., are also observed to varying degrees on a few biological forms. Among these, silicon (Si) is considered essential for rice, maize, sugarcane and other graminaceous plants. Rice is known as a silicicolus plant and a Si accumulator (Takahashi *et al.*, 1990) and the plant benefits from Si nutrition (Yoshida, 1975; Takahashi, 1995). Silicon is a tetravalent cation (Si^{4+}) present as the second most abundant element in the earth crust and occurs almost in all minerals. It occurs as oxide of silica. The accessibility of Si to plants greatly depends on how quickly weathering takes place making Si in the solution phases of soil. The prevailing form of silicon in the soil solution is mono silici acid ($Si\ OH)_4$) in which, it is being absorbed by the plants. Si interact with other native and applied nutrients and has the potential to induce resistance / tolerance in the rice plant to biotic (insect pests and diseases) and abiotic stress, (Al, Fe and Mn toxicity, salt injury, lodging etc). (Epstein, 1994). Most likely, the Si content in some regions might be limiting to sustainable rice production. In addition, Si depletion can occur in traditional rice soils from the continuous monoculture of nitrogen responsive high yielding cultivars with intensive cultivation practices. Further with, addition of higher levels of nitrogen in high – tech agriculture, the plants require more silicon to circumvent the succulence and tenderness offered by circumstances. The experiments conducted so far has shown a positive response to silica nutrition. In view of this Si nutrition for rice may be considered important for increasing and sustaining rice productivity.

A. ROLE OF SILICON IN RICE

The functions of silicon on rice are discussed as ameliorative, mechanical and physiological role as under.

i. Ameliorative : When silicate materials are applied to the soils of a large phosphorus fixation capacity, it prevents phosphorus fixation, release a part of fixed phosphorus and then increase the amount of available phosphorus. This fact could be attributed to the release of more soil phosphorus by the replacement of adsorbed phosphate ion by silicate ion. It can also be applied to organic soil, which are deficient in silica. Silicon can also alleviate the aluminium, iron and manganese toxicity in acid soils because of its possible involvement in cation – anion balance in plants. Silicate (S_iO_4) supposedly adds to the excess of anion uptake and therefore expels equal amounts of OH from the roots, which can increase the soil pH at the root surface and eventually decrease the uptake of Fe and Al. However this type of cation – anion balance as a possible mechanism for Fe and Al tolerance in crops grown in acid soils.

ii. Mechanical : On absorption of Si, by rice plant in large amount; silica deposits on the surface of the leaves and stem primarily deposited on the epidermis and vascular tissues and increases the resistance to biotic and abiotic stress. As the sufficient supply of silica permits the rice plant to keep the leaves erect and less curved and thus improves the assimilation system. It also occurs as special morphological skeletal structure like spines, spicules and needles.

iii. Physiological : A silica supply for the rice plant increase the leaf area; prevents the decrease of the photosynthetic activity of the lower leaves and thus enhances the photosynthetic capacity as a whole physiological role of Si in the rice plants (Takahashi, 1995). A silica supply promotes the oxidation power of the plant's root and thus alleviates the toxicity of H_2S. Marshner *et al.* (1990) reported that the application of Si under condition of high phosphorus and insufficient zinc counteracted zinc deficiency induced phosphorus toxicity. Supply of Si increased the physiological available of Zn but the exact mechanism is not known.

The accumulation of Mn and Fe in leaves of rice plants seemed to be most striking, as it showed the brown spotting characteristics (i.e. necrotic and chlorotic areas). If Mn toxicity noticed, Silicon prevented the development of necrotic spots. The effect of the silicon was to decrease the Mn content of the tissues, this might be explained by dilution effect i.e. the much accounts for the lower Mn content. Since silicon caused an increase in yield, beyond that, increase in yield accounted for by overcoming of Mn toxicity a function for Si beyond that of preventing Mn toxicity was postulated (Lewin, 1969). Thus silicon promoted the oxidation power of rice roots with the resulting deposition of Fe and Mn oxides on the roots surface. The effect depended on silicon within the rice plants rather than in the external soil solution. In silicon – deficient plants, Fe and Mn were transported to the aerial parts. To demonstrate this, ^{59}Fe was used. Increased oxygen supply to the roots in the presence of silicon in the plants depended on intake plants. When the tops were excised, the oxygen supply from the top of the root was shut off, and silicon had no effect on the oxidation power of the excised root and root extract. The rate of transpiration of Si – deficient rice plants was increased by about 30 per cent over the rate of control plants. (rates were measured as gram of water lost through transpiration per gram of dry weight per day). These observations suggested a role for silicon in the water economy of the plants. The rate of transpiration is presumably influenced by the amount silica gel associated with the cellulose in the cell walls of epidermal cell. Hence a well thickened layer of silica gel should help to retard water while epidermal cell walls with less silica gel will allow water to escape on accelerated rate.

B. SOURCES OF SILICON FOR RICE

In order to make the silicon fertilization for rice it is essential to trace the sources from which the nutrient can be made available to plants. The sources can be categorized in to inorganic and organic or plant residues.

i. Inorganic

Soil is considered as an inorganic source, if the soil is provided with a rich S_iO_2 content and made available to plants through their transformation reaction or weathering process. Although the soil may be rich in S_iO_2, but their availability may be limited due to various factors. Wollastonite ore (native calcium meta silicate C_a S_iO_3 (24% Si, 34% Ca) and different kinds of slag from iron, ferronickel, and manganese ore smelters have been used as Si sources for rice. These sources vary in their composition and their Si supplying capacity depends on relative amounts of their composition and their Si supplying capacity depends on relative amounts of their basic constituents (soluble in 0.5 m HCI at 30°C) such as monocalcium silicate, di calcium silicate and moticellite. (Savant *et al.,* 1997). Cement has also been used as Si source. Chinese researchers have been testing a new silicate fertilizer containing 230 g water – soluble Si / K fertilizer for rice. Lignite fly ash from Indian thermal power plant containing 23 per cent Si (141 ppm available Si) has been evaluated as a source of Si for rice (Raghupathy, 1993). Rhenania Phosphates prepared by Kali-chemie company of Germany contained 4.2 per cent soluble Si and has been tested as a source of Si for rice.

Basic slag obtained as a byproduct form steel industry, can be considered as a best source. During the manufacture of high quality steel, the impurities like Si, S and P combines with calcium and rise to the top in the furnace and are powdered off. When solidified, they are finely ground and are put on the market as a basic slag. Its chemical formula most commonly accepted is $(CaO)_5$ P_2O_5 S_IO_2. It also contain trace of Zn, Cu, Mn and B. The basic slag so for produced in India is of low grade. The composition of Indian basic slag from different sources are given in Table 29.

Table 29. The composition of Indian basic slag from different source

Composition (%)	Indian iron steel company Burnpur	Tata iron and steel company limited			Hindustan steel limited, Rourkela		Hindustan steel limited, Bhilai	
		Steel melting shop 1	Steel melting shop 2	Steel melting shop 3	OH slag	LD slag	Flusting	Taping
SiO_2	17.21	19.25	15.08	17.30	18.2	15-9	22-27	15-18
FeO	-	12.20	12.20	16.13	10.5-15	12.5-18	15-25	7-12
Fe_2O_3	-	1.47	4.03	2.15	-	-	2-5	2-5
Al_2O_3	3.4	3.27	4.03	1.76	-	-	5-6	5-16
CaO	43-38	45.5	43.20	48.3	43-46	48-52	21-28	58-46
MgO	7-9	9.5	6.87	8.27	5-7	3.5	6-9	10-15
MnO	3.5-5.5	4.8	7.25	3.0	8-12	7-12	7-9	4-6
P_2O_5	4.0-5.5	2.86	7.24	2.74	2.1-2.4	2-4	3-5	2-3

The total annual production of basic slag from different steel mills is about 1.5 million tones. Indigenous basic slag with different particle sizes has been tried at the Agricultural College and Research Institute, Kanke for their respective efficiencies, when applied on an equivalent calcium basis and it has been found that 20-30 mesh size particles of basic slag are considered as best. However, basic slag should be applied in a finely ground form containing 80 per cent material passing through 100 mesh sieve.

ii. Plant / organic sources

Rice straw and rice hull can also be used as Si source. In 1993, world production of unmilled rice (paddy) grain was 519 m.t (FAO, 1993). Assuming an average 1:1.2 grain straw ratio and 4 per cent Si in rice straw, 623 m.t of rice straw containing roughly 25 m.t of Si was available for recycling. Similarly, assuming an average of about 20 per cent rice hull in unmilled rice and nearly 8 per cent Si in rice hull, probably 104 m.t of rice hull containing 8 m.t of Si was available for recycling. Despite the nutrient value of these plant residues, proper recycling of rice straw and rice hull is not common among most rice farmers. Probably for the following reasons (Ponnamperuma, 1984); bulkiness of the materials, additional labour requirement (mainly for transport and spreading) difficulty in incorporating uncut straw, insect, diseases and rat problem associated with rice straw, low benefit: cost ratio and the demand for non agricultural uses. Si content of rice straw shows large variations from 1.7 to 9.3 per cent Si, (Yoshida, 1978) and is influenced by several factors such as soil, irrigation water quality, amounts of fertilizers applied, rice cultivars and season.

C. RECYCLING OF PLANT SOURCES OF SILICON

Rice straw serves as a source of carbon for some microflora, silica and potassium to the plants besides a large number of micronutrients. But farmers seldom resists to apply as a manure because of wider C:N ratio that is likely to lead N – immobilization. However, it has been incorporated into soil as a source of C and / or N for improving the soil physical conditions. Although it contains about 4 per cent Si practically, no reports are seen in the literature on its recycling as a source of Si in the subtropical and tropical rice culture. Although rice straw has a better economic utility as cattle feed, the farmers can make use of the wasted straw for field application as source of C, Si, K, N and P.

In mechanical rice cultivation, the crop is harvested by combine harvester and chopped straw is spread on the land. It is then incorporated in to the soils by disking or puddling, thus recycling nearly all Si, in the straw. Whereas when the crop is harvested manually, only stubbles can be incorporated and the amount of straw thus recycled in this manner depend on the height at which the crop is harvested. For example, in rainfed regions of Indonesia, only panicles are harvested, the straw is slashed, heaped, or spread in the field itself and then ploughed in wet at the beginning of the next season (Ismunadji, 1978). This practice has been potential to recycle nearly 50 per cent of the Si removed by the previous crops. In Philippines, most farmers generally harvest the stalks 60-90 cm from the base of the panicles (De Datta, 1981). This practice also leaves a lot of straw in the fields and helps considerably in recycling of Si. In India, the mechanical harvesting is being introduced where by the recycling of rice straw Si for rice is possible. Hull (Husk) which is almost one fifth of the paddy harvested has an alternative economic use as fuel and therefore not available for field application. But the ash residue after burning which is one fifth of the husk burnt can be successfully utilized as a source of silica and potassium. The husk ash (husk char, black ash) accumulating in rice mills is a rich source of silica (65%) and potash (2%) and is a renewable waste available season after season can be usefully applied to met the silica requirement of rice (Balasubramaniam, 2003).

8.3 BIO FERTILIZERS

An atmosphere around us contains nearly 78% nitrogen, which is in free state. Hence it is not available to plants. In order to meet the food requirements of ever increasing population, the nitrogen fertilizer requirement for crop production by 2010 AD is roughly estimated to be 11.4 x 10^6 tonnes. Increased food grain cannot be produced, unless we carefully make use of biological nitrogen. Biological nitrogen fixation is the key to sustain agricultural productivity through the application of biofertilizers in the field. However, there is an urgent need to transfer this technology on the field of farmers and into an industry by producing these fertilizers on large scale. It is living fertilizer compound of microbial inoculants or groups of micro-organisms, which are able to fix atmospheric nitrogen or solubilize phosphorus, decompose organic material or oxidize sulphur in the soil. On application, it enhances the growth of plants, increases in yield and also improves soil fertility and reduces pollution. Fig. 13 explains the natural bio-resources input for rice crop nutrition. Bio fertilizers are those which contain living organisms like algae, fungi, bacteria etc. these fertilizers contain Nitrogen fixing organisms like photosynthetic bacteria, *Cyanobacteria, Heterotropic Bacteria, Anabaena, Nostoc, Azolla* etc., It fixes atmospheric nitrogen into the soil. This N is available to plants through the activities of microbes, bacteria, algae etc., and through rainfall and lighting.

There are variety of nitrogen fixing micro-organisms present in the nature. These are broadly divided into three categories, viz.

1. Symbiotic micro-organism e.g. Legume –*Rhizobium* Symbiosis.
2. Asymbiotic or free living e.g. *Azotobacter*, blue green algae.
3. Associative Symbiosis, e.g. *Azospirillum*

Inoculants of these micro-organisms have proved their technical feasibility, economic viability and social acceptability. They are therefore called as 'biofertilizer'. Hence the term 'biofertilizer' or microbial inoculants may be defined as preparations containing living or latent cells of efficient strains of nitrogen fixing, phosphorous solubilizing or cellulolytic micro-organisms.

8.3.1 Need for biofertilizers

Biofertilizers have definite advantage over chemical fertilizers. Chemical fertilizers supply over nitrogen whereas biofertilizers provide in addition to nitrogen certain growth promoting substances like hormones, vitamins, amino acids, etc. Crops have to be provided with chemical fertilizers repeatedly to replenish the loss of nitrogen utilized for crop growth. On the other hand, biofertilizers supply the nitrogen continuously throughout the entire period of crop growth in the field under favourable conditions. Continuous use of chemical fertilizers adversely affect the soil structure whereas biofertilizers when applied to soil improve the soil structure. The deleterious effects of chemical fertilizers are that they are toxic at higher doses. Biofertilizers, however, have no toxic effects (Kannaiyan, 1993).

It may be borne in mind that biofertilizers are no substitute for chemical fertilizers. At present, the use of chemical fertilizers is far below the recommended level. Therefore, the aim and object of spread of biofertilizers technology as a Industry has to build up efficiency in use of chemical fertilizers supplemented by low cost inoculants to the extent possible. Main constraints in spread of biofertilizers as an Industry are production and supply of efficient cultures of micro-organisms to farmers well before sowing and secondly, Quality control aspect and lack of publicity, etc. At present, in Maharashtra there are roughly 45 entrepreneurs located mostly in cities, of which, very few have qualified staff and facilities for production of biofertilisers. In order to cover the vast areas of legumes, pulses, oil seeds, cereals and horticultural crops, we need more manufacturers who can prepare the biofertilizers in particular locality and supply to farmers as per their need. Besides, whatever biofertilizers are prepared are not reached to farmers prior to sowing. At many places they are given free of cost, which lowers down the importance of product and farmers don't use them carefully. Sometimes, biofertilisers are supplied after expiry date and hence expected results are not obtained. A quality control aspect is the most important in biofertilizers. Now-a-days most of the products sold in the market are below standard. A few entrepreneurs possess ISI mark for their products. The government have no control over manufacturers of biofertilizers in any of the states of India. There are no rules and regulations related to production and sale of biofertilizers in Maharashtra. Hence farmers are confused about rates, quality and expiry dates of biofertilizers. Similar is the case of prices; they are varied from place to place. Besides nitrogen fixation, they can also synthesize carbohydrates, amino acids and peptides.

8.3.2 Bio Fertilizers for Rice

8.3.2.1. Azotobacter

Isolation of the strains of Azotobacter and other free living nitrogen fixing bacteria from different crop plants grown under various agro climatic regions of Tamil Nadu were made and tested for their efficiency in economizing the use of fertilizer nitrogen without affecting the yields of different crops. Such efficient strains of Azotobacter are being supplied as peat soil based azotobacter inoculants to the farmers. Azotobacter inoculants are to be applied to the rice crop by adopting the following methods, as suggested by Kannaiyan, 1993.

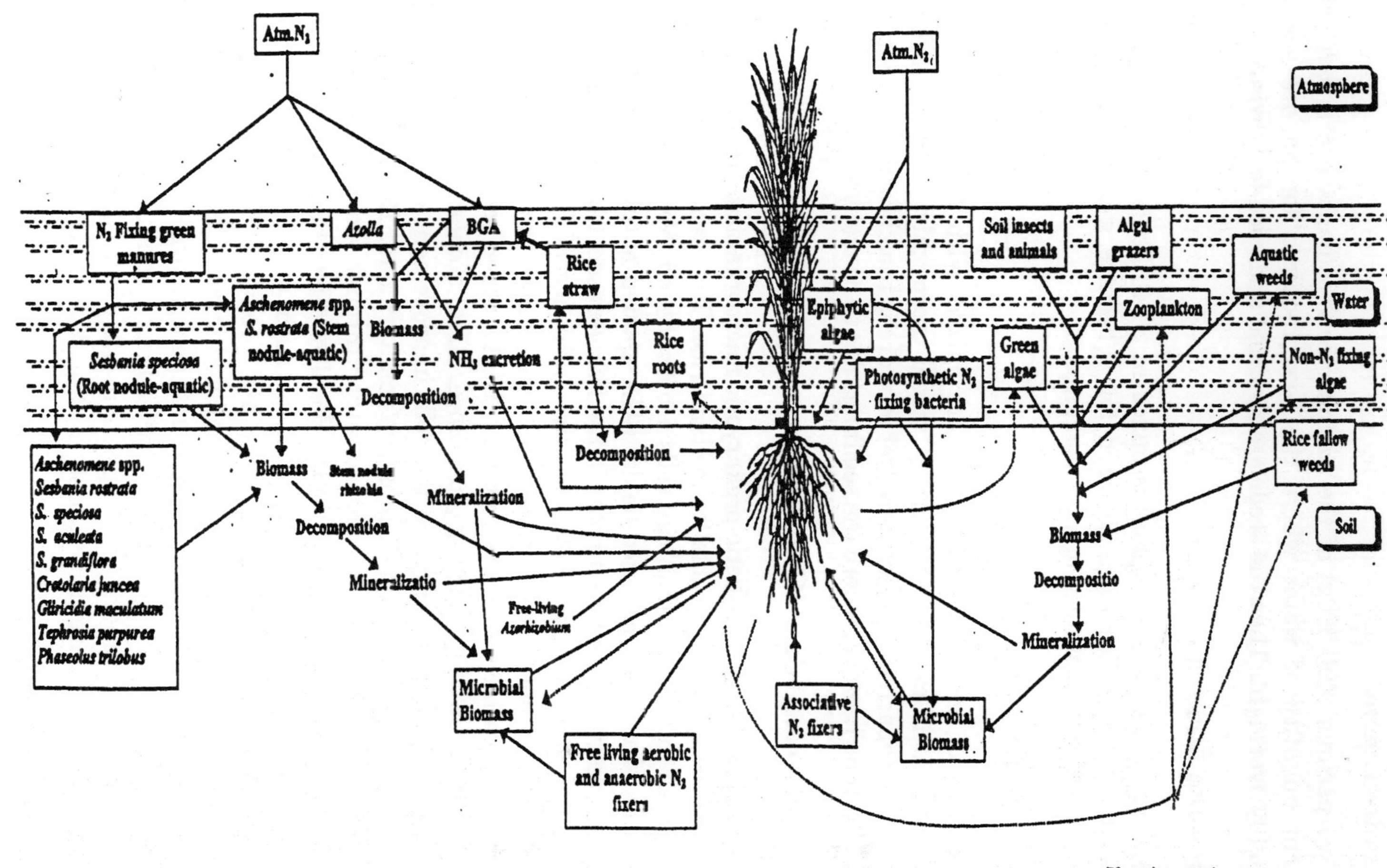

Fig. 13 NATURAL BIORESOURCES INPUT FOR RICE CROP NUTRITION

(i) Seed treatment

Two packets (400-500g) of the azotobacter inoculants are mixed with sufficient quantity of water where the seed required for ha are soaked thoroughly overnight. Then the seeds are sir dried and used for sowing.

(ii) Seedling dipping

One kg of inoculants (5 pkts) is mixed in 30-40 litres of water and the root portions of the seedlings required for a hectare were dipped in this suspension for 10-15 minutes and transplanted.

(iii) Soil application

Two kg of inoculants (10 pkts) are mixed with 20-25 kg of well decomposed cattle manure and further mixed thoroughly with 20-25 kg of soil. The Azotobacter cattle manure soil mixture is then broadcasted into the main fields (one hectare) where the seedlings are to be transplanted.

8.3.2.2 Blue-green Algae (Cyanobacteria)

The algae is photosynthetic prokaryotic microorganism capable of fixing atmospheric N and are suitable for lowland rice. The predominant genera of N-fixing BGA in 32% of Indian rice soils are *Anabaena, Nostoc*, *Calothrix* and *Tolypothrix*. In submerged rice fields, BGA applied at 10 kg N/ha, a week after transplanting, contribute 20-30 Kg N/ha/season (Aiyer *et al.,* 1972). Various field studies have shown positive influence of BGA on rice under different agro-climatic conditions. In some cases, increased yields of subsequent crops have also been noticed. BGA's other beneficial effects are liberation of growth promoting substances like auxins and amino acid compounds, addition of humus to the soil, improvement of aeration to rice roots and reducing the sulphide injury (Venkataraman, 1982). The nitrogen fixed by the algal biomass is released either through exudation or through microbial decomposition of the cells. Besides nitrogen, blue-green algae synthesizes a number of other substances such as carbohydrates, amino acids, peptides etc.; these substances are produced to a lesser extent extracellulary, while large portions assimilated interacellularly are liberated into the surrounding medium. Important genera are *Nostoc, Anabaena, Tolypothrix*, and *Cylindrospermum*. Further large quantities of nutrients are released through microbial decomposition after the death of the algal biomass. Some algae are decomposed in 2 or 3 days, but others withstand microbial degradation for more than 4 weeks depending on the physiological state of the algae, the composition of the associate microflora, the suitability of the cell wall as a substrate for micro-organisms and the relative biodegradability of specific compounds of the cell walls. If cost of 10 kg/ha BGA is only Rs. 30 and BGA inoculation produces on an average increase in grain yield by 300 kg/ha. BGA is not very popular among farmers due to its low growth rate, contamination and poor competitiveness. There is need to develop technology acceptable to the farmers and to popularise its usage in rice fields, saving 25-30 kg fertiliser N/ha. This will be equivalent to 1.5-1.7 Mt. urea used to rice in India (Venkataraman and Till, 1990).

Mass scale multiplication

The field selected for mass multiplication of blue-green algae is prepared well and leveled uniformly. The field is divided into small plots of size 20 m x 2 m (1 cent), by providing suitable bunds and irrigation channels. Water is allowed into a height of 5 cm. Super phosphate at 2 kg and lime at 100 – 200 g are added to each plot and inoculated with 2 kg starter culture of blue green algae. Carbofuran (Furadan) – 200 g is also added to prevent the activities of soil insects. Water is let in at periodical intervals, so that the height of water level is always maintained at 5 cm. The growth of algae is rapid and in about 10 days, a thick mat is formed which floats on the surface of the water. After 15 days of inoculation, the plots are allowed to dry up in the sun and the algal flakes collected and stored. From each harvest 30 to 40 kg of dry algal flakes are obtained from each bed.

Application or inoculation to the main field

Inoculation of BGA is highly suitable during kharif season. Application of blue green algal flakes @ 10 kg/ha is recommended for rice crop. The algal inoculant is applied 10 days after rice transplantation. At the time of BGA inoculation, 500g of super phosphate may be mixed in the BGA flakes which encourages better growth and establishment in the rice fields.

8.3.2.3. Azolla

A floating water fern having symbiotic relationship with N fixing cyanobacterium, Anabaena azollae, produces 2-4 kg biomass/ha/day and incorporation of 10 t fresh Azolla/ha is equivalent to addition of 25-40 kg N/ha (Table 30). Azolla is a free floating water fern that fixes atmospheric nitrogen in association with N fixing BGA, Anabena azollae and contributes 40-60 kg N/ha to rice crop. Azolla is primarily grown as a green manure for rice.

Table 30. Effect of Blue green algae and Azolla on rice yields in different places of Tamil Nadu.

Treatments/location	Rice yield (t/ha)		
	Vellore	Trichy	Cuddalore
BGA (N-P-K Kg/ha)			
a) 50-25-25	4.24	4.56	3.79
b) 50-25-25+BGA (10 kg/ha)	4.46	4.79	3.88
c)75-37.5-37.5	4.40	4.49	3.95

Azolla	**Bhuvaneswar**		**Thanjavur**	
	kharif	Rabi	kharif	Rabi
(a) Control- 0N Kg/ha	2.50	1.13	2.89	2.39
(b) Azolla 5 t/ha (incorporated)*	2.96	1.22	3.43	2.45
(c) Azolla 10 t/ha (incorporated)*	3.18	1.57	4.05	2.69
(d) Azolla 5 t/ha (inoculated)+ 30 kg N /ha	3.10	1.46	4.42	2.69
(e) Azolla 10 t/ha (inoculated) + 30 kg N /ha	3.53	1.90	4.60	3.33
(f) 30 kg N /ha as Urea	3.03	1.38	3.82	2.89
(g) 60 kg N /ha as Urea	3.42	1.69	4.97	3.43
(h) 90 kg N /ha as Urea	3.38	2.04	4.79	3.55

kh - kharif * incorporated before planting,
+ inoculation followed by incorporation at 20 days after planting.

Azolla as green manuring

Azolla is primarily grown as a green manure for rice crop. The nitrogen accumulated in the azolla body is considered available to rice plants only after decomposition of Azolla. Azolla is decomposed in 8-10 days. A layer of azolla covering a hectare of rice field contains about 15 tones of green manure and ensure about 25-30 kg N/ha and the amount can be doubled by growing a second azolla layer after the first crop has been incorporated.

Dual cropping

Azolla can be used effective in dual cropping with rice to supply nutrients after the formation of mat either by decomposition after over growth or by incorporation into the soil. Azolla inoculated at 0.125 kg/m^2 on 7-10 days after planting is found to establish and cover the area in 15–30 days. Azolla was incorporated in the field during first weeding after draining the water completely.

On dry weight basis the analysis of Azolla showed 85 – 94% moisture, 3-4% N, 0.5 – 0.8% P_2O_5, 2-4% K_2 O, 0.45% Mg, 0.35 – 0.90% Ca and 0.11 – 0. 16% Mn (Kannaiyan *et al.*,1998). Azolla is found to be supplement source of nitrogen to the rice crop. There is scope to reduce 15 to 25% of N fertilizer when Azolla is incorporated in the field. So far, 6 extants species and 25 fossil species of Azolla have been identified. Some Azolla species are *A. caroliniana, A. filiculoides, A. mexicana, A. mexicana, A. rubra, A. microphylla, A. nilotica and A. puniata.* The results of International Network on Soil Fertility and Fertilizer Evaluation for Rice (INSFFER) have shown that incorporation of one crop of Azolla before or a little after transplanting rice was equivalent to a split application of 30 kg fertiliser N/ha. It can be applied along with fertiliser N to step up rice yields by 13-54%, besides building up organic carbon, total N and available P in the soil from season to season (Satapathy and Singh, 1980). However, limitations in wide usage of Azolla are less water availability and low survival rate during summer with temperature over 35-45°C, high phosphate requirement of 60-80 kg/ha for good Azolla growth, problem of transportation of inoculum of 10-12 t required for one ha and arthropods menace checking the growth of Azolla. There is need to have less thermosensitive strains with high multiplication rate and low phosphate requirement. This could save fertilizer N equivalent to 25-40 kg/ha (Verma and Bhattacharyya, 1990).

8.3.2.4. Azospirillum

(a) Seed treatment : **(3 pockets – 600 gm)**

Six hundred gm of peat based Azospirillum should be mixed with seed required for sowing one hectare along with rice kanji as a binding agent. Afterwards it should be dried under shade for 30 minutes. The treated seeds should be packed in gunny bag and soaked in water for 24 hours in a water trough. The sprouted seeds should be sown in the nursery bed. The water used for soaking should also be poured in to the nursery plot itself before the seeds are sown.

(b) Soil inoculation

Two thousand g (10 pockets) of peat based inoculants along with 25 kg of powdered FYM and 25 kg of soil should be thoroughly mixed and broadcasted in the mail field (one ha).

(c) Seedlings root dipping

Trench of 4 x 4 m with 10 cm depth should be made and water slurry with 1000 g peat based inoculation should be prepared. The root portion of the seedlings should be dipped for 10-15 minutes in the slurry and then transplanted.

8.3.2.5. Phosphobacteria

The activity of this is completely different from others. Phosphobacteria and fungi solubilize the residual and insoluble phosphates available in the soil and convert them into orthophosphates in the soil-water solution, very near to root surfaces. The phosphate solubilizing power is further enhanced by addition of organic matter. The efficient isolates are *Pseudomonas steriata*, *P. rathonis*, *Bacillus polymyxa* (bacteria), *Aspergillus awamorii*, *A. niger*, *Penicillium digitatum* and *Schwanniomyces occidentalis* (fungi). These cultures increase the efficiency of rock phosphate in neutral and alkaline soils and improve the quality of compost applied (Gaur, 1982; Verma and Bhattacharya, 1990). Mere inoculation of seeds or seedlings with phosphate solubilisers can make phosphorus available equivalent to 30 kg P_2O_5 as SPP by solubilising soil and applied P (Gaur, 1990). The adaptability of these organisms on large scale under field conditions needs to be explored. Phosphorus material is fixed in soil or converted to unavailable form. They mobilize the applied and available phosphorus in the soil, when compared to other nutrients. It is applied by following methods (as in the case of Azospirillum application).

It can be applied as:

i. Seed treatment – 3 packets

ii. Seedling dipping – 5 packets and

iii. Soil application – 10 packets

8.3.2.6. Vesicular Arbuscular Mycorrhiza – VAM

It is a fungal colony. It normally forms a mat around the root portion of higher plants. The roots of most of world's plants are colonised by specific soil fungi to form, vesicular arbuscular mycorrhizae (VAM), a mutualistic symbiosis that can be considered an exploitation of soil phosphate and more efficient use of added fertilisers, particularly slowly mobile ions like some forms of N, Cu and Zn. VAM can be harnessed to improve the productivity in agricultural and horticultural crops and forestry by increasing the efficiency of applied fertilisers and enhancing plants' survival in nutrient deficient soils or degraded habitats (Barea, 1991). Its use on commercial scale is yet to be stabilised. It is recommended at a rate of 50g/m^2, only for paddy nursery.

8.3.2.7. Rhizobium

In rice, it is not directly involved in any activity but can be applied along with green manure seeds. This will induce the formation of root nodules in green manure crop and thus, fixes nitrogen. This has symbiotic association with higher plants. Biological nitrogen fixation by leguminuous green manure crops in symbiotic association with rhizobium is a low cost input for rice crop. Research findings show that the strain Sesbania rostrata was associated with two potential rhizobial strains. The strain ORS-571 nodulates both roots and stem of *S. rostrata*. The utilization of N fixing leguminous green manure crops for rice would be a viable component of integrated nutrient management.

Some of the promising strains of rhizobium / azorhizobium for inoculation of N fixing green manure are SRS-SK-2, SRR-SK-1, SAR-SK-5, SSR-SK-4, TPR-SK-5, PTR -SK 1.

8.4 INTEGRATED NUTRIENT MANAGEMENT (INM)

Integrated use of organic, inorganic and biofertilizers for crop production is called as INM. Plant nutrients can be supplied from different sources viz. organic manures, crop residues, bio-fertilizers and chemical fertilizers. For better utilization of resources and to produce crops with less expenditure, integrated nutrient management is the best approach. In this approach, all the possible sources of nutrients are applied based on economic consideration and the balance required for the crop is supplemented with chemical fertilizers.

- Apply 12.5 t/ha of FYM or compost.
- Spread the manure uniformly over the dry soil before letting in water for puddling.
- If FYM or compost is not applied, apply green manure @ 6.25 t/ha on the puddle field and incorporate properly to a depth of 15 cm.
- Green leaf manure can also be applied @ 6.25 t/ha.
- Maintain 2.5 cm of water and allow 7-10 days before transplanting for their decomposition.
- Apply biofertilizers by all means.
- As per as possible apply fertilizers as per soil testing recommendation. If it is not available adopt blanket recommendations as scheduled here under.
- Short duration varieties : 120 : 38 : 38 N P K kg/ha
- Medium and long duration varieties : 150 : 50 : 50 N P K kg/ha.
- For effective split application of N, maintain thin film of water and allow to absorb well and irrigate after 2 hours.

8.4.1 INM in Cropping system

Work on INM in intensive cropping system is rather recent and results are available only for a few selected cropping systems. These studies are of long term in nature as sustainability aspects and cost effectiveness have to be carefully assessed before making recommendations to the farmers.

8.4.1.1 N management in rice based cropping system

Rice occupies a major portion of the national food grain production. Rice-rice-pulse, rice-wheat, rice-oilseed, rice-green manure/fodder and rice-cotton are common intensive cropping systems followed by farmers in irrigated and rainfed lowlands in different parts of the country. Long term studies carried out under AICARP have indicated that it is possible to substitute a part of fertilizer N needs of Kharif crops by FYM without any adverse effect on the total productivity of the system. Appreciable yield increase in rice with green manuring has been reported from almost all rice growing countries. The flood tolerant green manure crop of *Sesbania rostrata* could be successfully raised prior to rice and in between two rice crops in a rice-rice-green gram system (Siddeswaran, 1992). The additional rice and wheat yields that could be harvested in a rice-wheat system due to green manuring with different species ranged from 731 to 1602 kg ha^{-1}. Green manuring with *Sesbania* was found to bring larger benefits than other species. Intercropping of the flood tolerant *S. rostrata* in transplanted rice in 12:1 ratio was successfully demonstrated. Growing and *in situ* incorporation of *S. rostrata* at preflowering stage registered a grain yield of 6.3 t/ha as against 5.8 t/ha in sole cropping of rice. Transplanting 30 days old *S.rostrata* seedlings for every 10 rows of rice (1.5 m) and incorporating biomass produced at 30 days after transplanting increased the rice yield by 15% when compared to no green manuring (Palaniappan *et al.,* 1991). Multipurpose green manures like *S. rostrata* and *S. aculeata* can also be intercropped in green gram and pigeon pea during the pre-rice summer season in 2:4 row ratio. Incorporation of sole *S.rostrata* increased the subsequent rice yield by 33 to 36% over the use of sole green gram residues and combined use of *Sesbania* with grain legume residue by 20 to 22%.

In rice-rice-pulse rotation, Srinivasulu Reddy (1988) reported that green manuring alone to either of the rice crops registered a grain yield similar to 50 kg/ha fertiliser N application. This trend of equivalence was noticed consistently, even at higher levels. Combined application of sunnhemp and rock phosphate in the rice-rice-green gram system enhanced the P use efficiency. Inclusion of green manures in rice-based cropping systems saved the cost of fertiliser nutrients to the tune of Rs. 350 ha^{-1} in an year

(Siddeswaran and Palaniappan, 1990). Grain legumes are grown in rice-based cropping systems for protein, fodder and green manure purpose. It is found that about 18 grain legume species are important in various rice farming situations of Asia. In the irrigated environment of tropics and subtropics, green legumes can be grown in rotation with one or more rice crops per year (Buresh and De Detta, 1991). The productivity of rice where legume residue was incorporated was superior to no legume. Higher agronomic efficiency and apparent recovery were observed with cowpea residue than green manure or urea N (John *et al.,* 1989). Inclusion of grain legumes like cowpea and black gram during the pre-rice summer in the rice-rice-green gram system yielded 5 to 7 q/ha of grain and addition of haulms enhanced the SWM rice grain productivity of the system, in terms of rice grain equivalent, was enhanced with the inclusion of black gram and cowpea in the system (Siddeswaran, 1992). The N fertilizer substitution in rice by incorporating green gram haulms was 20-30 kg (Palaniappan *et al.,* 1990). Incorporation of cowpea haulms could save 35-50 kg N/ha. Use of biofer-trilizers like BGA, azolla or azospirillum has been reported to increase the productivity of crops and reduce the fertiliser N requirement substan-tially in rice-based cropping systems. Choice of appropriate biofertilisers and applic-ation at the recommended dose and time would go a long way in saving of fertiliser nutrients. In cereal-cereal rotation, incorporation of wheat or rice straw at 5t/ha inoculated with *Aspergillus* sp. in the top 15 cm soil one month prior to sowing cereals contributes equivalent to 50 kg N/ha (Gaur *et al.,* 1970, Tiwari *et al.,* 1987). Sidhu and Beri (1989) reviewed the wheat and rice straw residue management in 27 field experiments conducted during 1974-85 in rice-wheat and maize-wheat rotations. While incorporation of chopped wheat straw was effective on maize, unchopped wheat or rice straw had no significant impact on the yield of succeeding crops. Under INSURF network, it was found that rice straw (5t/ha) incorporation along with prilled urea was better than prilled urea at equal N level of 58 or 87 kg at Coimbatore during 1984-87 (IRRI, INSURF, 1991). However, incorporation of straw at Pant-nagar and Faizabad did not give encouraging results.

8.4.1.2 Phosphorous and Potassium management in rice based cropping system

The applied phosphatic fertilisers undergo changes in the soil and non-labile pool of phosphate benefits the succeeding crops. The residual effect of applied P differs with the capacity of plant species to use the native and applied P. In a rice-grain legume cropping system, the total response to P was higher when P fertiliser was applied to grain legume crop and the residual P utilised by the succeeding rice crop, than direct P application to rice crop and the residual P utilised by grain legume. Tiwari *et al.* (1980) reported that rice-wheat system recorded higher grain yield when 60 kg P_2O_5/ha was applied to wheat only in comparison to its application to rice only. Further it was concluded that among the crops of wheat, potato, maize and rice in a system, P should be applied to potato, which is more responsive to P than other crops.

In rice-based cropping systems upland crops respond to K application more than rice. Experiments conducted at Kanpur indicated that application of K at 30 kg/ha to both the crops in rice-wheat system produced additional yields of 496 and 612 kg/ha of rice and wheat respectively (Tiwari *et al.*, 1980). The results of AICARP experiments conducted on cultivators' fields indicated that rabi crops were more responsive to K than Kharif crops.

8.5 NUTRITIONAL DISORDERS

Physiological or nutritional disorders are abnormalities in rice growth caused by soil conditions. When studies do not show conclusively that pathogens or pests cause certain plant abnormalities, scientists classify such abnormalities as physiological disorders pending verification of their true causes. Presently, 25 physiological disorders are observed in 16 countries. About half of these diseases, have been studied rather extensively.

The main causes of physiological or nutritional disorders are: -

- Deficiencies of nutrients.
- Excesses of elements, either essential or non-essential; and disturbances of root metabolism caused by some compounds produced in soils under sub-merged conditions.

8.5.1 Deficiencies

A. N,P,K Deficiencies

Nitrogen and phosphorus deficiencies in rice are common in Southeast Asia. Potassium (K) deficiency is often found in some calcareous and lateritic soils whose K supplying power is quite low. Potassium deficiency is also associated with bronzing (iron toxicity). The visual symptoms of N, P, or K deficiency manifest themselves in the colour of the leaves and their effect on plant development.

a. Nitrogen deficiency: Plants become stunted and yellow in appearance first on lower leaves. In case of severe deficiency the leaves will turn brown and dies. Deficiency symptoms first appear at the leaf-tip and progress along the midrib until the entire leaf is dead. Either urea or DAP at 1-2% is recommended to correct N deficiency.

b. Phosphorus deficiency: Stunted dark green plants with erect leaves and reduced tillering. To correct P deficiency, 1-2% Dap spray is recommended.

c. Potassium deficiency : Dark green leaves with yellowish brown leaf margins or dark brown necrotic spots first appear on the tips of older leaves. Grains formation is poor. Weakening of the straw, which results in lodging. 1-2% KCl foliar spray is recommended to correct K defiency.

B. Deficiencies of other plant nutrients

a. Iron Deficiency

Entire leaves become chlorotic (deprived of chlorophyll) and then whitish. If iron supply is cut suddenly at a certain time, the newly emerging leaf subjected to chlorosis.

b. Zinc Deficiency

The midribs of the young leaves, especially the basal leaves, become chlorotic. The more general symptoms are appearance of brown blotches and streak in lower leaves, followed by stunted growth.

Correction of zinc deficiency : Basal application of zinc sulphate at 25 kg/ ha or 0.5 per cent zinc sulphate foliar spray on 30,40 and 50 DAP for medium duration and long duration varieties and, 20, 30 and 40 DAP for short duration varieties.

c. Sulfur Deficiency

The symptoms are very similar to those of nitrogen deficiency. Consequently, it is almost impossible to distinguish by visual symptoms alone between sulfur and nitrogen deficiencies.

d. Calcium deficiency

The tips of upper leaves become white. In an extreme case, growing point dies. The leaf colour remains green. Root elongation is retarded, and the tip of root becomes brown.

e. Magnesium deficiency

Interveinal chlorosis in lower leaves.

f. Manganese deficiency

First, interveinal chlorosis appears in younger leaves, the older leaves remaining relatively yellowish green. The chlorotic streaks spread downward from the tips to the base of leaves and later on some dark brown necrotic spots appear on these leaves. The newly emerging leaves become short and narrow with severe chlorosis.

g. Boron deficiency

The plant height is stunted. The emerging leaf shows white tip and dies in a severe case.

h. Copper deficiency

The leaves appear bluish green, and then become chlorotic near the tip. The chlorosis develops downward along both sides of the midrib followed by dark-brown necrosis of the tips. The new emerging leaves fail to unroll and maintain a needle like appearance for the entire leaf, or occasionally half the leaf. The basal end develops normally.

l. Silicon deficiency

Please refer the section 8.2.4.4.

8.5.2 Excesses of Non-essential Elements

a. High Salt injury

Excess salts in the soil is a hazard to rice plants. This problem is found in coastal areas where seawater affects the soil and in inland areas, where there is little rainfall. In coastal areas, high salinity is often associated with acid sulfate soils, which have low pH (based on air-dried soil sample) and high sulfur content. Iron toxicity is frequently observed in this kind of soil. In inland areas, high salinity is often associated with high alkalinity or high pH. Iron deficiency is often observed in alkaline soils.

Symptoms of salt injury

Plants are stunted and produce less tillers. The tips of leaves become whitish. Frequently, some parts of leaves become chlorotic.

b. Bronzing (iron toxicity)

Rice plants need iron. However, the presence of too much iron in the soil can cause iron toxicity or bronzing. Measurements have shown that the soluble iron content in paddy soils increases to a maximum within 2 to 3 weeks of flooding and then declines. Iron toxicity or bronzing has been observed on lateritic sandy soils and peaty soils.

Bronzing symptoms

The symptoms manifest themselves during the tillering stage of the rice plant. First, tiny brown spots appear in lower leaves starting from the tips and spreading toward the base of leaves. Leaf colour usually remains green. In severe cases, the entire leaves look purplish brown. Affected plants also show dark-brown, coarse, scanty root system. Excess iron may induce phosphorus and potassium deficiency.

CHAPTER 9

WEED MANAGEMENT

A weed is a plant growing where it is not desired or a plant out of place. At world level, 30,000 species are identified as weeds. Nearly 18,000 species are causing very high damage to crops. 350 plant species, 100 plant genera, 60 plant families are identified as weeds for paddy.

1. Major family is poaceae having more than 80 species. Best example is *Echinochloa crusgalli.*
2. Family cyperaceae having 50 species e.g. *Cyperus difformis, C. rotundus, C. iria* 10-15 species belong to other families.

Yield loss in rice varies from <10 to 90% depending on the rice culture and management practices. Direct seeded upland rice is more susceptible to weeds than low land rice. Since puddling itself is an excellent means of controlling weeds. Out of 350 weeds species of rice <20 are major importance. Critical period or crop weed competition period is the period up to which the weeds are kept under control. For transplanted rice, the critical period is 25-35 days after transplanting. For direct sown paddy, it is 60 DAS. Occurrence of important weed in the field is depending upon the duration of the variety grown. Weed species that cause problem in the rice vary with soil, temperature, altitude, longitude, rice culture, seeding method, water management, fertility level and weed control technology (Subramanian *et al.,* 1991).

9.1 HARMFUL EFFECTS

- Weeds compete with the crop for the soil moisture, soil applied nutrient, space, sunlight, etc.

- Weeds can able to remove 1250 MT of water/ha and 1/3rd of nutrients applied within a period of 2-3 weeks. Weeds have smothering effects over crop. During early stage, weeds grow very well and prevent sunlight to crop.
- Weed reduces the yield of the crop and affects the quality of the produce e.g.: *Echinochloa* in rice.
- Weeds are alternate host for insects and pathogens.
- Weed causes health hazards to human beings and livestock. e.g. *Parthenium hysterophorus.*
- Due to presence of perennial weeds it reduces land value and productivity of land. E.g. *Solanum elaegnifolium. Cyperus* sp.
- Presence of weeds also blocks the irrigation and drainage channel.
- Weed creates problem in land preparation and increases the time for preparation of land.
- Weeds reduce harvesting and processing efficiency.
- Energy consumption requirement is more to control weeds.

9.2 ADVANTAGES

- Fodder for cattle e.g. *Cynodon* dactylon.
- Leguminous weeds fix atmospheric N in the soil e.g. *Tephrosia purpurea.*
- Some weeds act as very good medicinal plant. e.g. *Eclipta alba.*
- Weeds and Weed residues act as component in compost. e.g. *Ipomea cornea.*
- In hilly areas, weeds conserve soil and prevent soil loss from erosion. e.g. *Cynodon dactylon.*

9.3 CLASSIFICATION OF WEEDS

Based on various characters, weeds are classified.

a. Based on life cycle

1. Annual
2. Biennial
3. Perennial

Annual: Weeds complete its life cycle within a season or one year.

Biennial: Weeds complete its life cycle within two season or two year. First season is adopted for its vegetative growth and second season is for reproductive growth.

Perennial: Life cycle of the weeds is more than two seasons.

b. Based on growth habit

1. Grasses
2. Sedges
3. Broad leaved weeds

I. Grasses

Grasses are monocots having narrow leaves with parallel venation, round hollow stem with distinct nodes and inter nodal regions. Leaves are arranged in alternate rows.

II. Sedges

Sedges are monocots with narrow and parallel venation. Stems are triangular stems, which are solid and pithy. There is no distinct nodal and inter nodal regions. Leaves are arranged in whorl fashion.

III. Broad leaved weeds

Board leaved weeds are dicots. They have very broad leaves with different venation and the stem of the broad leaves is round and solid.

c. Based on habitat

1. Wetland Weed – Hydrophytes
2. Garden land Weed – Mesophytes

d. Based on edapic factor

1. Red soil weeds
2. Black cotton soil weeds
3. Laterite soil weeds etc.

e. Based on nativity

1. Indigenous (Indian origin)
2. Exotic weeds: Weeds are introduced from other countries. e.g. *Parthenium hysterophorus,* which is introduced from USA.

f. Based on economic value

1. Aquatic weeds
2. Problematic weeds
3. Parasitic weeds
4. Poisonous weeds

9.4 FACTORS INFLUENCING THE WEEDS GROWTH IN RICE YIELD

Many ecological and crop production principles influence the presence and abundance of group of weeds in rice field which are given below.

- Soil
- Seeding method
- Soil moisture regime
- Crop rotation
- Air and soil temperature, altitude and longitude
- Land preparation
- Fertilization and fertility level
- Rice cultivars
- Weed control technology
- Interaction of above factors.

9.5 WEEDS UNDER DIFFERENT ECO SYSTEMS

9.5.1. Rain fed Upland Weeds

Under rain fed upland conditions, the problem due to weeds is very severe and the yield is drastically reduced. The following Table 31 gives the list of dominant weed species reported from different states of India.

Table 31. Dominant Weed species in upland rice culture reported from different states of India.

State	Dominant Species	Reference
Andhra Pradesh	*Amaranthus spinosus, Euphorbia heterophylla, Celosia argentina, Panicum* sp.	Pillai *et al.* (1976), Singh and Pillai (1996)
Assam (also Meghalaya, Tripura and Nagaland)	*E. Colonum, Boerhaavia hispida, Bidens, pilosa, Galinsoga parviflora, Eupatorium odoratum, Phyllanthus niruri, Mimosa pudica, Amaranthus spinosus, Setaria pumila*	Singh and Patel (1989) Singh (1988) Gogoi and Sarma (1993)
Bihar	*Rungia repens, Physalis, minima, Caesulia axillanris, Digeria alternifolia*	Pandey and Thakur (1988) Pandey *et al.* (1991)
Delhi	*Amaranthus gracilis, Digeria arvensis, Alternaria sessila, Cyperus diformis, Leptochloa chinesis, Justica* sp.	Kaushik and Mani (1978) Bhol and Singh (1987ab) Chander & Pandey (1996)
Karnataka	*Paspulum Conjugsatum Dinebera retroflexa, Portulaca oleracea, Spilanthus alba*	Mahadevaswamy and Nanjappa (1991)
Tamil Nadu	*Dinebra arabica, Chloris barbeta, Croton, bonplandianum, C. spersiflurus*	Mohamed Ali and Sankaran (1984); Ravichandran (1993)
West Bengal	*Digitaria longifolia, Elytrophorus arteculata, Eragrostis interrupta, Setaria glauca*	Mukhopadhyay (1974,1978).

9.5.2. Rain fed Low Land Rice

In low-lying areas, rice is generally grown/ direct seeded before the onset or after the onset of monsoon. Initially, the conditions resemble those of rainfed upland direct seeded rice, but, 25-30 days after sowing, rain comes and the conditions change to wetlands. In these areas, the yield is very low due to the presence of many terrestrial, semi aquatic or aquatic weed species (Table 32).

Table 32. Weed species in rainfed lowland rice.

State	Predominant weed species	Reference
Assam	*Scirpus* sp., *Fimbristylis miliacea, Monochoria vaginalis, Jussiae sufforocticosa, Ludwigia octovalvis, Echinochloa crusgalli*	Gogoi and Kalita (1992)
Bihar	*Chara* sp., *Spirogyra* sp., *Paspulum scorbuiculatum, Ipomea reptants, Echinochloa colonum, Echinochloa crusgalli, Cyperus difformis*	Pandey (1984)
Tamil Nadu	*Fimbristylis miliacea, Scirpus mucronatus, Cyperus iria, C. difformis, Eclipta alba, Marsilea quadrifoliata, Ammanea bacifera, Echinochloa colunum, Sphaeranllus indicus, Monochoria vaginalis*	Mohamed Ali and Sankaran (1984), Ramiah and Muthukrishnan (1992), Ramkrishnan *et al.* (1992)

Himachal Pradesh	*Echinochla colonus, Cyperus rotundus, Paspalum parpalodes, Panicum* sp., *Setaria gluea*	Thakur and Bassi (1994)
U.P.	*Cyperus iria, C. difformis, Cynodon dactylon, Nymphaea stellata, Hyndnilla verticillata, Typha* sp., *Protomegton* sp., *Eleocharis dulcis, Salvania* sp., *Chara zeylanica*	Varshney and Singh (1976).

9.5.3. Wet Seeded Puddled lowland Rice

Under these conditions, the problems due to weeds are lesser than dry direct seeded rice. Both under puddle direct sown and transplanted conditions, the weed species are almost common, but the weeds dominance and densities may vary largely. The important weed species in puddle direct seeded and transplanted rice are given in Table 33.

Table 33. Weed species in puddled direct seeded and transplanted rice.

Dominant weed species Monocot	Dicot	State	Reference
Panicum colonum, Paspullum conjugatum	*Fissendocarpa linifolia, Hydrolea zeylanica, Cuphea balsamona, Rotala indica*	Assam	Gogoi (1995)
E. colonum, Paspasllum scorbiculatum	*Oldenlendia dichotoma, Sphenocdlea zeylanica, Monochoria vanglis, Ludwigia perviflora*	Bengal	De and Mukhopadhyay (1983)
Cynodon dactylon, Cyperus esculentus, Aeshymemone, armericana, Commelina benghalensis	*Marselia quadrifoliate, Oxalis carniculata, Ipomea aquatica*	Orissa	Misra and Lenka (1973)
Panicum repens, Dactyloctenium aegypticum, Cyperus species	*Rungia repense Caesulia axillaris, E. hirta, Eclipta alba*	Bihar	Pandey and Sukla (1990) Singh *et al* (1992)
Fimbristylis sp., Paspulum fasciculatum	*Phyllanthus niruri, Lippia nodiflora*	U.P.	Rao and Singh (1994)
Leptochloa chinensis	*Eclipta alba*	New Delhi	Chander and Pandey (1996)
Commelina benghalensis Cynodon dactylan Eleusine indica, Paspulam distichum, Cyperus sp., *Commelina benghalensis*	*Chlorisd barbata, Dinebra arabica, Ammania baccifera, Caesulia axillaris, Euphorbia sp., Cynotis axillaris, Langascea mollis, Cephalandra indica*	M.P.	Jain *et al.*, (1984) Kulmi (1991)

Panicum sp., Fimbristylis miliacea Cyperus difformis	*Jussia suffructicosa, Marsilia minuta*	A.P.	Raju and Reddy (1995)
Panicum repens Cyperus difformis	*Lindernia parviflora*	Karnataka	Reddy *et al.* (1995)
C. difformis Digitaria sanguinalis Fimbristylis, C. iria, Paspalum distichum	*Eclipta prostrata Ammania baccifera Marselia minuta*	Tamil Nadu	Srinivasan *et al.* (1992)
Panicum elatum, Setaria glauca	*Ageratum conyzoides*	H.P.	Shah *et al.* (1990) Singh and Kharawara (1984)
Ischaemum rugosum	*Caesulia axillaris*	Punjab	Chela & Gill 1980
Panicum colonum Sporobolus indicus	*Caesulia axillaris, Cynotis* sp., *Ammanaia baccifera, Ludwigia octovalvis*	Gujarat	Patel *et al.* (1988)

9.5.4. Transplanted Rice

Transplanting of rice is done in the areas where irrigation water is available during rice growing period. When compared to direct seeded rice, the problems due to weeds are lesser. The weeds occurring in the transplanted rice fields are called as wetland weeds. Wetland weeds are hydrophilic or mesophytic. They require less O_2, low temperature, and less higher intensity. They are heavy seed producer and they are adapted to anaerobic condition and waterlogged conditions. The list of wetland weeds in Tamil Nadu, with their family and common names is given in the following Table 34. The characteristics of wetland weeds are given in Table 34a.

Table 34. List of wet land weeds with their family, botanical and common names

Botanical name	Family	Common name (Tamil)
I. Grasses		
Echinochloa colonum	Poaceae	Jungle rice or Kudiraivali / varsanampul
E. crusgalli	,,	Barn yard grass
Panicum repens	,,	Injipul
Brachiaria mutica	,,	Water grass
Cynodon dactylon	,,	Devil grass / Arugampul
II. Sedges		
Cyperus rotundus	Cyperacea	Korai, purple nut sedge, nut grass.
C. irria	,,	Vayal oosi korai
C. esculantus	,,	Yellow nut sedge

C. difformis	"	Vayal vatta korai
C. bulbosus	"	Vayal korai
Fimbristylis mileacea	"	Keerai
III. Broad leaf weeds		
Marsilia quadrifoliata	Marsileaceae	Arai keerai
Ammania baccifera	Lythraceae	Neermel nerappu
Asteracantha longifolia	Acantheaceae	Neer mulli
Eclipta alba	Compositae	Karishilanganni
Phyllanthus niruri	Euphorbiacea	Keela nelli
Phyllanthus maderaspatensis	Euphorbiacea	Mella nelli
Centella asiatica	Umbeliferae	vallarai
Mimosa pudica	Mimosideae	Thottachuringi
Ludwigia parviflora	Enagraceae	Seelai poondu / nirkkrambu
Lippia nodiflora	Cruciferaceae	Poduthalai
Monochoria vaginalis	Pontederiaceae	Vayal kalai
Commelina benghalensis	Commelinaceae	Near thamarai
Cyanotis axillaries	Commelinaceae	Neer palli
Sphaeranthus indicus	Compositae	Kotta kananthai

Table 34a. Characteristics of wet land weeds

Weed	Habit	Habitat	Propagation	Characters
Cyperus rotundus	Perennial	Wet and dry land	Underground tuber, seed	Triangular stem
Enhinochola colonum	Annual	Wet land	Seeds	Looks like rice seedling with prominent ear head
Panicum repens	Perennial	Requires moist condition	Under ground stem, seeds	Branches are covered with pointed scales, inflorescence is much branched
Ammannia baccifera	Annual	Requires moist condition	Seeds	Erect glabrous succulent stem, dense red flowers.
Ludwigia parviflora	Annual	Wetland	Seeds	Auxiliary cluster spines, erect glabrous, yellow flowers
Eclipta alba	Annual	Requires moist condition	Seeds	Prostrate and erect, leaves are lanceloate, white or yellow flower
Lippia nodiflora	Annual	Wetland	Seeds	Leaf at the base are smooth serrated the tip
Marsilia quadrifoliata	Annual	Wetland	Runners	Four lobed leaf with long petiole
Astercantha longifolia	Annual	Requires moist condition	Seeds	Dense cluster spine in the leaf is axis

Among the weeds in rice, *E. crusgalli* is the most trouble some weed of transplanted rice in the world. *E. colonum* is second in importance. It tends to grow all around the equator but *E. crusgalli* has a greater range from north to south. Considering the importance of these two weeds in transplanted rice, the differences between *E. crusgalli* and *E. colonum* and the differences between grasses and sedges are given in the following Tables 35 and 36.

Table 35. Difference between *E. crusgalli* and *E. colona*

Characters	*E. crusgalli*	*E. colonum*
Panicle	Usually nodding	Erect
Branching	Close on rachis more or less branched	Distant from rachis simple
Spikelet	Often not arranged in rows	Usually arranged in four rows.
Awn	Variable in length	Awnless
Stigma colour	Red or white	Blackish purple
Anther colour	Brown / yellow	purple

Table 36. Difference between grasses and sedges

Grasses	Sedges
1. Stem is round, cylindrical	Stem is triangular
2. Ligular process in prominent	Absent
3. Stems are hollow	Pithy in nature
4. Internode is prominent	It is not prominent
5. Alternate leaf arrangement	Whorl leaf arrangement

9.6 WEED MANAGEMENT

In general, the weeds are controlled by the evolution of

1. Weed eradication
2. Weed control
3. Weed Management and
4. Integrated weed management.

9.6.1 Weed Eradication

It is the complete removal of weeds in a locality or in a particular area. It is possible only in the case of introduced weeds or problem weed in a very limited area. This is not possible in rice cultivation to eradicate a particular weed completely.

9.6.2 Weed Control

Controlling the weeds throughout the crop growth period. This method is uneconomical. It will increase the cost of cultivation and thus reducing the profit.

9.6.3 Weed Management

Managing the weeds upto crop weed competition period and it is highly economical in rice cultivation.

9.6.4 Integrated Weed Management (IWM)

Integrated Weed Management is a combination of mechanical, manual, biological, cultural and chemical methods or IWM is integration of several weed control methods to achieve effective and economical wed control. IWM also includes preventive measures like.

1. Use clean seeds and weed free seeds
2. Should not irrigate or drain the water from the weed-infested field to clean field. After application of herbicide, don't drain and disturb the water for 3 days.
3. Summer ploughing and grazing of lands during off season
4. Proper puddling and leveling the land
5. Pull out the seedlings before transplanting
6. Prevent transport of machinery / implement from weed infested field to clean field.
7. Application of well decomposed organic manure
8. Good and efficient water management practices.

Advantages

- It prevents weed shift
- Reduces crop – weed competition
- Prevent resistance of weed species for a particular herbicide
- Reduces the herbicide residual effect
- Prevents environmental pollution
- It is suited for high cropping intensity
- Most economical and effective methods.

9.6.4.1 Mechanical Methods

Various tools and implements are used for weed control. Rotary weeder and cono-weeder is mostly used. It uproots and incorporates the weeds in the same field, thereby increases organic matter content of soil.

9.6.4.2 Manual methods

Farm labourers or members of farmer family are normally engaged for hand weeding operation. Two manual weeding is recommended for rice. First weeding is given on 15-20 DAT and second weeding is given on 30-35 DAT. This method is a common practice, being followed by most of the farmers in India (Plate 35).

Plate 35. Hand weeding in rice field.

9.6.4.3 Cultural Method

It involves operations from ploughing up to weeding.

9.6.4.4 Biological Methods

Bioagents like insect pest and pathogens are used. E.g. Parthenium is controlled by using Zygogramma. It is not adopted under paddy condition.

9.6.4.5 Chemical Method

Herbicides are used for controlling weeds. Chemical method has a number of advantages like .

1. Quick and efficient in weed control.
2. Larger area can be covered in a shorter period.
3. Highly economical.
4. Highly selective towards weeds.

The important chemicals and its trade name for controlling weeds in rice fields are given below for easy reference.

CHEMICAL NAME	TRADE NAME	CONCENTRATION
Butachlor	Machete or Delchlor	50% EC
Thiobencarb	Saturn	50% EC
Pendimethalin	Stomp	48% EC
Anilophos	Aniloguard	50% EC
Fluchloralin	Basalin	45% EC
2,4 D Na Salt	Fernoxone	50% WP
2,4 DEE	2,4 DEE	50% WP

Disadvantages

1. Farmers are lacking in knowledge in use of herbicides.
2. It is uneconomical where labourers are available cheaply.

The dose and method of application of herbicides are given below in detail for the benefit of the readers.

A. Methods of Application

Herbicides are applied either to the soil (soil application) or to the foliage (foliage application). Depending on the properties of herbicides, mode of action and selectivity, different methods are adopted. Environmental factors, convenience and cost are other factors that influence the choice of correct method of application. An improper method of application can result in poor weed control and / or severe crop injury.

I. Soil application

1. Soil surface application - Herbicides are sprayed on the soil surface to form a uniform herbicide layer. The herbicides thus sprayed, due to their low solubility may penetrate only few centimeters into the soil. Weeds germinating in the top layer are killed due to incidental absorption of herbicides. Most of the triazines and amides are applied by this method. The herbicides for surface soil application has to be soil active, less soluble and less volatile. After application, the surface soil should not be disturbed.

2. Soil incorporation - Some of the herbicides belonging to aniline and carbamate groups are volatile. When they are applied to the soil surface, they are lost by volatilization. When applied to soil surface, a one third of fluchloralin is lost within three days. Volatile herbicides are incorporated into the soil to reduce losses. Generally, these herbicides are applied before planting as it is difficult to incorporate the herbicide after sowing.

3. Sub-surface application - The perennial weeds like *cyperus rotundus* and *cynodon dactylon* are to be controlled, the herbicides are applied to the lower layers of the soil. This is done by injecting herbicides into the soil at several points.

4. Band application - Herbicides can be applied as narrow bands over or along the crop row. The weeds in between the rows can be controlled by intercultivation or by non-selective herbicide. This method is useful where labourer is expensive and intercultivation is possible. Weeds in maize can be controlled effectively by spraying atrazine on seed row at the time of sowing and non-selective herbicides in between the rows after weeds have emerged.

II. Foliar Application

1. Blanket application - Blanket application is the application of herbicides over the entire leaf area. Only selective herbicides are applied by this method.

2. Directed application - Application of herbicides in between the crop rows directing towards weeds is known as directed spray or directed application. Care is taken to avoid spray fluid falling on the crop.

3. Spot application - Herbicide solution is poured on weeds in cropped and uncropped field infested with obnoxious weeds in isolated patches. This method of application of heavy doses of herbicides on isolated patches is called spot application.

4. Basal application - Brushwood and unwanted trees are treated with herbicides by different methods depending on the situation. Generally, foliar application is the most common method of treating brush. However, if the trees are taller, say more than 2.5 m, it is not possible by ground sprays. Under such a situation, basal bark application is resorted to. The bark of the trees at the base of the stem up to 30 cm height is removed and a drenching spray of herbicide is given to the base.

5. Cut – surface application - The most difficult brush weeds or trees are cut of the ground surface, concentrated herbicides or paste of herbicides is applied to the stump or cut surface.

B. Time of Application of Herbicides

For efficient and effective control of weeds, time of application of herbicide is important. It depends on herbicidal properties like selectivity, solubility and volatility. In addition, weed dynamics and crop tolerance also decide the time of application.

a. Fallow application - Application of herbicides well in advance of sowing i.e. more than 10 days before sowing is known as fallow application. Fallow application of herbicides are resorted to control problematic weeds with higher does of non-selective herbicides, when there is no crop. It is generally done during the fallow period to control weeds like Cyperus and Cynodon with high doses. In temperate countries, application of higher dose of 4.5 - 8 kg a.i./ha of atrazine has given higher maize yield only when applied in autumn or early spring.

b. Pre-plant application - Application of soil active herbicides 2-10 days before sowing or planting is called pre-plant application. It is resorted to when the herbicides are to be incorporated or when the toxic effects on crop are more initially. Volatile herbicides like fluchloralin, trifluralin are incorporated into the soil to avoid volatilization losses. As it is not possible to incorporate herbicides into the soil after sowing of the crop, they are applied before planting and this process is known as pre-plant incorporation (PPI). Sometimes, it is necessary to apply slightly higher doses of herbicides to kill problematic weeds. In this situation also, pre-planting application is resorted to.

c. Pre-emergence application - Application of herbicides after sowing of the crop but before emergence of crop and weeds is called pre-emergence application. Pre-emergence application of herbicides is done from one day after sowing to four days after sowing. Only selective herbicides are used for pre-emergence application. Highly soluble herbicides are not advisable because these herbicides may leach, if heavy rain occurs after application. Generally, germinating weeds are killed by pre-emergence application and this gives competitive advantage to the crop.

d. Post-emergence application - Application of herbicides after the emergence of crop is called post-emergence application. It is generally resorted to when the crop has grown sufficiently to tolerate herbicides. Post-emergence application is resorted to kill weeds that appear late in the crop. Generally, post-emergence application of herbicides is done about 30 to 40 days after sowing.

C. Calculation of Dose

The commercial products of herbicides contain the actual herbicide, adjuvant and other filling material. The percentage of adjuvant and other inert material used as fillers depends on formulation and the company, which is manufacturing the particular herbicide. Therefore, it is necessary to make recommendations based on the active ingredient or acid equivalent.

Active Ingredient: The chemical in commercial product that is directly responsible for the herbicidal activity is called active ingredient (a.i). Active ingredient of each herbicide formulation is displayed by the manufacturer on the container. It is generally expressed as per cent by weight or by volume. For e.g., Atrataf 50% weight / weight indicates that the commercial product of atrazine contains 50% of the active ingredient. If the recommendation is 0.5 kg/ha, 1.0 kg of commercial product is to be used.

$$\text{The amount if commercial product to be used} = \frac{\text{Recommended dose}}{\text{a.i.}} \times 100$$

Acid Equivalent (a.e): Some herbicides are active organic acids such as phenoxy acetic acid, picloram and chloramben. But often many of these are in the form of their salts and esters. For instance, instead of 2, 4-D in acid form, sodium salt or amine salt or ester form is used. The theoretical yield of parent acid from such a herbicide formulation is called its acid equivalent. The acid equivalent of sodium salt of 2,4-D is 92.5%. It indicates that 2,4-Dichlorophenoxy acetic acid is 92.5% in sodium salt of 2,4-D. However, it has become customary with the herbicides in relation to their final product. Acid equivalent is, therefore, less than active ingredient.

9.6.5 IWM under different ecosystems

9.6.5.1. IWM for rainfed upland rice

IWM is suggested for weed control in upland rice. Combining 2 or more weed control measures helps in realizing higher grain yield and reduces built-up of perennial weeds owing to continuous use of herbicides, or any other type of weed control measure, growing of same crop and following same cropping system year after year. Raising of competitive rice cultivars attenuates the virulence of prevalent weeds to considerable extent. Inter-cropping of rice also reduces the weed threat and helps in reducing the persistence of herbicides (Kolhe and Mittra, 1981). The following different methods may be followed depending the situations/regions for controlling weeds in upland rice.

a. Manual methods

Hand weeding is the most widely used method for controlling weeds in uplands rice. Normally two weedings are done. Time of weeding slightly varies in different regions but weedings are done within 15 to 45 days after sowing; these also coincide with critical period of crop weed competition. In general, two early hand weedings at 2 weeks intervals after sowing are better than one handweeding either at 15 or 30 or 45 days after sowing.

b. Mechanical methods

Only limited number of implements such as hand hoes and bullock drawn desi ploughs are used to control weeds in upland direct seeded rice.

c. Chemical methods

Use of herbicides is gradually increasing in upland rice culture. Herbicides are expensive to small farmers but not to the large farmers who face problem of labour shortage (Behra *et al.,* 1997). A large number of herbicides have been tested by agronomists and these are: Anilophos, Avirosan, Bentazan, Benthiocarb, Bifenox, Butachlor, Butralin, Chlorsulfuroan – methyl, Dichlofop methyl, Dinitramine, Dichlomate, Drepaman, 2,4-D, Fluchloralin, MCPA, MCPB, Molinate, Nitrofen, Pendimethalin, Pretilachlor, piperphos, Propanil, Paraquat, Oxadiazan and Oxyflourfen, The quantity of chemical required for controlling weeds in upland rice is as follows, as suggested by Rajendra Prasad (1999).

Thiobencarb	- Pre – emergence application at 1.5 - 2.0 kg ha^{-1}
Anilofos	- Pre – emergence application at 0.3-0.4 kg ha^{-1}
Avirosan	- Pre – emergence application at 0.75 –1.0 kg ha^{-1}
Butachlor	- Pre – emergence application at 1.5 -2.0 kg ha^{-1}
Nitrofen	- Pre – emergence application at 2.0-4.0 kg ha^{-1}
Oxadiazon	- Pre – emergence application at 0.5- 2.0 kg ha^{-1}
Pendimehalin	- Pre – emergence application at 1.0 - 2.0 kg ha^{-1}
2,4 D	- Post – emergence application at 0.5-1.0 kg ha^{-1}
Propanil	- Post – emergence application at 2.0-4.5 kg ha^{-1}

9.6.5.2 IWM for rainfed lowland rice

a. Manual methods

Hand weeding is not possible owing to stagnation of water in fields during critical period crop – weed competition.

b. Mechanical methods

It is the most common method of controlling weeds. A mechanical operation called beushening or biassi or bidahan is widely practiced in north India. It is the practice of cross – ploughing the field which generally has 5 to 10 cm standing water with a desi plough a month after sowing once or twice depending upon weed density and rice plant population. The ploughing is followed by laddering the same day or a few days after. This aims at bringing an uniform plant population in the field. This operation results in marked decreases in weed density and increase in rice grain yield (Sahu and Jena, 1968).

c. Chemical methods

The information on chemical weed control in rainfed lowland rice culture is scanty. Herbicides tested include: anilofos, bentazone, butacholr, 2,4,-DEE, bensulfuron methyl, quinclorac, nitrofen, oxadiazon, oxyfluorfen, pendimethalin, thiobencarb, pretilachor, copper sulphate, propanil, bifenox and fluchloralin. The results available for controlling weeds in rainfed low land rice due to different herbicides are given in Table 37.

Table 37. Effect of some important herbicides on grain yield of rainfed lowland rice.

Herbicides	Dose (a.i. kg ha^{-1}	Time application DAS/ DAT	Grain yield q ha^{-1}			Reference
			Herbicide	Weedy check	Yield increase	
			A	B	A-B	
Anilofos	0.40	4	42.1	24.6	17.5	Tuteja *et al.* (1995)
Butachlor	1.5	5	21.7	16.7	5.0	Pandey (1984)
Butachlor Bifenox	1.50	4	46.5	14.6	21.0	Tuteja *et al.* (1995)
Nitrofen	2.0	5	29.7	16.7	13.0	Pandey (1984)
Pendimethalin	1.30	5	23.3	16.7	6.6	Pandey (1984)
Pendimethalin	0.75	6	13.0	8.0	5.0	Gogoi & Kalita (1989)
Thiobencarb	1.50	5	25.1	16.7	8.4	Pandey (1984)
2, 4-D IPE	1.0	5	27.7	16.7	11.0	Pandey (1984)
Anilofos + 2, 4-DEE	0.4 + 0.4	4 & 35	47.9	24.6	23.3	Tuteja *et al.* (1995)
Butachlor + 2, 4-D Na	1.5 + 1.0	8 & 25	22.2	16.9	5.3	Ramiah and Muthukrishnan (1992)
Butachlor + 2. 4-DEE	1.5 + 0.4	4 & 35	46.5	24.6	21.9	Tuteja *et al.* (1995)
Pendimethalin + 2,4-D Na	1.25 + 1.0	8 & 25	29.0	16.9	2.1	Ramiah and Muthukrishnan (1992)

Pendimethanlin + HW	1.0 + one	pre s.	21.5	11.5	10.0	Kathiresan *et al.* (1997)
Thiobencarb + 2, 4-D Na	1.5 + 1.0	8 & 25	20.5	16.9	3.6	Ramiah and Muthukrishnan (1992)
Quinclorac	0.20	6	14.0	8.0	6.0	Gogoi & Kalita (1989)

Pre s. * = Pre sown sand mixed followed by hand weedling 20 DAS.

In Tamil Nadu, for controlling weeds in semi dry rice, Pre-emergence application of herbicide viz., thiobencarb, 1.5 kg/ha and pendimethalin 1.25 kg/ha on 8th day after sowing is recommended. Any one of the following herbicides viz., Saturn 50 EC (thiobencarb) 2.5 lit/ha or stomp 30 EC (pendimethalin) 4.2 l/ha in 625 l of water is to be sprayed on 8th day after sowing with the help of hand operated sprayer fitted with fan or deflector type of nozzle. One irrigation given within 5 days after application of herbicides will improve efficiency of herbicides. The herbicide layer should not be trampled until irrigation. The pre-emergence herbicides can be mixed with 50 kg sand and applied uniformly over the leveled field without pits followed by immediate irrigation. Depending on the weed flora second weeding by hand or herbicide can be given. If dicots such as *Eclipta alba* are dominant, 2,4 D sodium salt (Fernoxone 80% WP) at 0.8 kg/ha can be sprayed on 35 day after sowing. If grasses are dominant manual weeding is recommended for the second weeding.

9.6.5.3 IWM for direct seeded rice

a. Manual methods

Manual weeding has been found very effective in arresting weed growth and augmenting crop yield and is widely practiced. This method is followed where women laborers are available.

b. Mechanical methods

Mechanical weeding is also effective. Finger hoe is employed in light soil.

c. Chemical methods

Herbicides have been found effective in controlling weeds but their efficacy is slightly lower than that in transplanted rice. In transplanted rice, standing water of a few millimeter favours uniform distribution and enhances reactivity and results in grater herbicidal activity. The results available for controlling weeds in direct seeded rice due to different herbicides are given in Table 38.

Table 38. Effect of some of important herbicides on grain yield of puddled direct seeded rice.

Herbicides	Dose (a.i. kg ha^{-1}	Time application DAS/DAT	Grain yield q ha^{-1}			Reference
			Herbi-cide	Weedy check	Yield increase	
			A	B	A-B	
Anilofos	0.45	5	20.3	1.2	19.1	Nandal and Singh (1994)
	0.500	5	11.0	7.0	4.0	Chander and Pandey (1996)
Butachlor	1.50	6	24.2	18.0	6.1	Singh and Biswas (1981)
	1.50	6	33.8	18.6	15.2	Kumar and Gautam (1986)
	1.0	5	10.4	7.0	3.4	Chander and Pandey (1996)
Oxadiazon	0.50	5	21.6	1.2	20.4	Nandal and Singh (1994)
Oxyfluorfen	0.15	6	39.0	18.6	20.4	Kumar and Gautam (1986)
Pretilachlor	0.75	5	19.3	1.2	18.1	Nandal and Singh (1994)
Thiobencarb	1.0	6	38.0	18.6	19.4	Kumar and Gautam (1986)
2, 4-DEE	0.8	6	30.7	18.6	12.1	Kumar and Gautam (1986)

9.6.5.4 IWM for Low Land Rice (Transplanted)

a. Manual methods

It is an effective method in the areas where lobourer availability is not a problem and if hand weeding is done before critical period of crop – weed competition, it results in marked increase in rice yield.

b. Mechanical methods

Now-a-days, This practice is not in vogue owing to closer planting. - However, in light textured soil, finger hoes can be used. In the case of line planting, rotary weeder or conoweeder are used for incorporation of weeds in the field itself.

c. Chemical methods

It is the most acceptable and widely used measure in transplanted rice. Herbicide application brings about marked increase in crop yield. A large number of herbicide have been tested and these widely differ in selectivity and herbicidal activity. Tested herbicides include;

Anilofos, Avirosan, BAS 3510, Bentazone, Bifenox, Butachlor, C288, Chlorsuluron, Cinmethyline, Dinitramine, Dichloromate, 2,4-D, Fluchloralin,

Fluroxypyr, MCPA, Metsulfuron, Molinate, MT–B-3015, Nitrofen, Pendimethalin, Pretilachlor, Piperophos, Propanil, Purazosulfuran ethyl, Paraquat, Oxadiazon, Oxyfluorfen, Thiobencarb, Tridiphane, WL 29226. The effect of important herbicides on grain yield of transplanted rice is given in Table 39.

Table 39. Effect of some Important herbicides on grain yield of transplanted rice.

Herbicides	Dose (kg a.i./ ha^{-1}	Time appli-cation (DAT)	Grain yield q ha^{-1}			Reference
			Herbicide	Weedy check	Yield increase	
			A	B	A-B	
Anilofos	0.4	4	54.7	41.9	12.8	Nagaraju (1994)
	0.4	4	63.0	47.0	16.0	Pandey (1993)
	0.5	3	68.70	5.70	63.0	Singh *et al.* (1990)
Bifenox	2.0	4	41.93	30.08	11.85	Pandey and Singh (1981)
	3.0	7	45.14	34.05	11.09	Ray and Mustafee (1980)
Butachlor	1.5	4-6	49.0	42.0	7.0	Pillai *et al.* (1976)
	2.0	3	65.0	32.8	37.2	Singh *et al.* (1990)
	1.5	6	32.3	29.8	2.5	Gogoi and Kalita (1992)
Chlorimuron	0.012	4	37.4	29.6	7.0	Chander & Pandey (1996)
ethyl 2, 4-D (EE)	3.2	6	42.54	22.38	20.16	Singh and Biswas (1981)
	1.2	4	56.60	41.90	14.7	Nagaraju (1994)
	0.75	4	44.10	38.50	11.17	Singh and Biswas (1981)
Nitrofen	1.5	6	34.17	23.00	11.17	Singh and Bhandari (1985)
Pendimethalin	1.0	3	69.50	32.80	36.7	Singh *et al.* (1990)
	1.0	6	26.88	23.659	3.29	Kulmi (1992)
Pretilactor	0.75	3	28.00	22.10	5.90	Gogoi (1995)
	1.00	4	61.49	41.88	19.16	Naqaraju (1994)
	1.0	3	27.92	23.59	4.33	Kulmi (1989)
	0.75	3	37.86	21.78	16.07	Nandal and Singh (1994)
Propanil	3.0	2	19.80	14.60	5.20	Mishra and Singh (1989)
Oxadization	1.0	4	34.70	21.80	12.90	Pandey and Sukla (1990)
	0.5	4	33.05	23.40	9.65	Singh *et al.* (1994)
	0.75	3	64.38	52.81	11.57	Brar *et al.* (1997)
Oxyfluorfen	0.3	3	69.00	32.80	36.20	Singh *et al.* (1990)
	0.075	4	54.08	41.88	12.20	Nagaraju (1994)

Thiobencarb	1.5	7	47.20	28.8	18.40	Sharma *et al.* (1994)
	1.0	3	30.7	22.1	8.60	Gogoi (1995)
	2.0	4	35.20	21.80	13.4	Pandey and Shukla (1990)
	1.5	6	26.82	23.59	3.23	Kulmi (1989)
	1.5	4	30.44	19.26	11.18	Singh and Singh (1988)
Thiobencarb +.05 2, 4D (EE)	+1.0	+5	38.87	22.92	15.95	Subramanian and Mohamed Ali (1987)
Anilofos +0.3 + 0.6 2, 4-DEE	+0.6	4	62.71	41.88	20.83	Nagaraju (1994)
Anilofos +0.2 + 0.5 2, 4-DEE	+0.5	4	44.48	25.58	18.9	Rao and Singh (1997)

DAT = Days after transplanting

In addition to chemical methods, the following points are to be considered for efficient weed management in transplanted rice.

- Use of weed free seeds
- Summer ploughing and grazing of lands in the off season
- Proper puddling and leveling of the land
- Pull out the weed seedlings before transplanting

In transplanted rice, weed management practices should be done both in nursery and main field, which are as follows.

a. Low land nursery

Weeds are a major problem in raising healthy nursery. Weed species like *E.colonum* and *E. crusgalli* are two major weed species in the nursery. They are morphologically akin and therefore their removal becomes difficult. Further their growth is slow and they do not attain sufficient height within a week and results in problem of their removal by hand. Any one of the herbicide viz., Machete 50 EC or Delchlor 50 EC (Butachlor) 160 ml/20 cents or Saturn 50 EC (Thiobencarb) 160 ml/20 cents, Ronstar 25 EC (Oxadiazon) 160 ml/20 cents or stomp 30 EC (Pendimethalin) 260 ml / 20 cents has to be mixed with 4.0 kg sand on the day of application and applied uniformly over the nursery on 8th day after sowing. After the application of herbicide, water should not be drained for 2 days from the nursery. Subsequent irrigation has to be given before hairline cracking (Mohamed Ali and Sankaran, 1984).

b. Main Field

In Tamil Nadu, the following herbicide recommendations are being followed for controlling weeds in transplanted rice.

i. Pre emergence herbicide

Pre – emergence herbicide should be applied on 3-5th DAT. 50 kg of sand is mixed with any one of herbicide given below and broadcasting is done. It should be followed by one hand weeding on 30-35th DAT.

1. Butachlor (Machete) — 50% EC @ 2.5 lit/ha.
2. Thiobencarb (Saturn) — 50% EC @ 2.5 lit/ha.
3. Pendimethalin (Stomp) — 48% EC @ 4.5 lit/ha.
4. Anilophos – 50% EC @ 1.5 lit/ha.

ii. Post Emergence Herbicide

Post emergence herbicide should be applied in the later stage of the crop. It is available in the 50% WP. Deflector, fan type nozzles are recommended for application. Before post-emergence herbicide application, one hand weeding is to be done.

1. Butachlor – 1.2 lit/ha
2. 2-4 D EE ---1.5 lit/ha.
3. Thiobencarb — 1.2 litre/ha
4. Pendimethalin – 1.5 lit/ha + 2.4 D EE 1.5 lit/ha.
5. Anilophos — 1 lit/ha + 2.4 D EE 1.5 lit/ha.

2-4 D Na salt @ 1.25 kg/ha is recommended for upland paddy.

iii. Herbicide mixtures

Pre-emergence and post emergence herbicide mixtures are also recommended for rice crop to have broad spectrum weed control.

Thiobencarb - 1.0 kg + 2-4 D EE 0.5 kg/ha

Butachlor - 1.0 kg + 2-4 D EE 0.5 kg/ha

Fluchloralin - 0.5 kg + 2-4 D EE 0.5 kg/ha

Oxadiazon - 0.5 kg + 2-4 D EE 0.5 kg/ha.

CHAPTER 10

PEST AND DISEASE MANAGEMENT

Pest and disease management practices should be done both in nursery and main field.

10.1 PEST AND DISEASE MANAGEMENT IN NURSERY

10.1.1 Pest Management

An area of 800 m^2 (20 cents) nursery is required for planting one ha or main field. Forty litres of spray is necessary for spraying the nursery area.

i. Army worm

Drain water from the nursery and spray chloropyriphos 20 EC 80 ml or Endosulfan 35 EC 80 ml during late evening.

ii. Thrips

Sampling : Pass a wet table tennis bat (painted with white paint) over the foliage in 12 places in the nursery. If thrips population exceeds 60 numbers in 12 passes or if rolling of ½ area of first and 2nd leaves in 10% of seedlings is noticed, apply one of the following insecticides. Phosphomidan 85 WSC 25 ml or Monocrotophos 36 WSC 40 ml or Endosulfan 35 EC 80 ml.

iii. Green leaf hopper

Sampling : Take 25 net sweepings in the nursery area. If the population exceeds 60 for 25 sweepings or 20/ m^2 by actual counting, spray any one of the following insecticides.

Fenitrothion 50 EC	-	80 ml Phosphomidon 85 WSC-25 ml
Fenthion 100 EC	-	40 ml Quinalphos 25 EC – 80 ml
Phosalone 35 EC	-	120 ml Endosulfan 35 EC – 80 ml
Monocrotophos 36 WSC-		40 ml

Maintain 2.5 cm of water in the nursery and broadcast anyone of the following Carbofuran 3 G 3.5 kg or Phorate 10 G 1.0 kg or Quinalphos 5 G 2.0 kg.

iv. Case worm

Mix 250 ml of kerosene to the standing water, dislodge the cases by passing a rope and drain water, collect the cases and destroy, and spray Monocrotophos 36 WSC 40 ml or Quinalphos 25 EC 80 ml.

v. White tip nematode

Presoaking overnight and sun drying for 12 hours at 6 hrs/day for two days, two or three days prior to sowing to denematize the seeds.

vi. Rice root nematode

Application of carbofuran 3 G at 3.5 kg/20 cent nursery.

10.1.2 Disease management in nursery

i. Blast / Brown spot

If dry seed treatment alone is done, spray the nursery with any one of the following fungicides on observing the initial infection of blast or brown spot in the nursery along with insecticides. Carbendazim 40 g; copper oxychloride 100 g, Mancozeb 80 g; Edifenphos 40 ml for blast and any one of the above fungicide except carbendazim for Brown spot.

ii. Tungro disease

Apply Carbofuran 3 G at the rate of 3.5 kg ten days after sowing or spray two rounds (10 and 20 days after sowing) with anyone of the following insecticides: Monocrotophos 36 WSC 40 ml, Phosphamidon 85 WSC 40 ml, Fenthion 100 EC 40 ml for 20 cents nursery.

iii. Blast

Seed treatment with Carbendazim 4 g/kg of seed and spraying of Edifenphos at 1%, immediately on noticing blast. The spray formulation (500 g) of *Pseudomonas fluorescens* may be dissolved in 500 litres of water and used for one ha.

10.2 PEST AND DISEASE MANAGEMENT IN MAIN FIELD

10.2.1 Pest Management in main Field

A. Stem Borers

Yellow stem borer *Scirpophaga incertulas* (Pyraustidae: Lepidoptera)

Dead heart in young seedlings and tillers 'white earhead' in panicle stage.

B. Caterpillar

Gall fly or Gall midge- O*rseolia oryzae* (Cecidomyiidae : Diptera)

Silver shoot (or) Anaikomban or onion leaf, which is a tubular gall in the place of normal shoot in tiller.

C. Leaf feeders

a. Whorl maggot - Hydrellia sasakii – (Ephydridae : Diptera)

Yellowish white longitudinal marginal blot with holes in a few places on the emerging leaves. Leaves shriveled, plants stunted and maturity delayed – maximum damage 30 DAT – MAGGOT.

b. Swarming caterpillar or Army worm or cutworm- Spodoptera mauritia (Noctuidae : Lepidoptera)

Cutting of seedlings in large scale similar to cattle grazing by nocturnal feeding, peduncles of ears bitten through immaturing crop CATERPILLAR.

c. Case Worm – Nymphyula depunctalis (Pyralidae : Lepidoptera)

Leaf tips with cut ends, larval damage is by scrapping of chlorophyll in alternate line of darker and light steaks. Leaf tubes found floating on water. LARVA.

d. Leaf folder – Cynaphalocrocis medinalis, Marasmia patnalis, M. ruralis (Pyralidaea; Lepidoptera)

Longitudinal folding of leaves and scrapping of green tissues, drying of leaves.

e. Skipper butterfly – Pelopidas mathias (Hesperidae : Lepidoptera)

Edges of leaves are fastened with webbing. Backward rolling of leaves, feeding from margin inwards. LARVA.

f. Green horned caterpillar-Melanitisleda ismene (Satyridae: Lepidoptera)

Leaves defoliated from the margin or tip irregularly LARVA

g. Yellow hairy caterpillar- Psalis pennatula (Lymantridae : Lepidoptera)

Irregular feeding of leaves, Similar to horned caterpillar damage. LARVA.

h. Hispa or spiny bettle - Dicladispa armigera (Chrysomelidae : Coleoptera)

White parallel streaks or white patches along the long axis of leaf.

Blister like spots towards leaf tip. Adults scrape green matter – GRUBS AND ADULTS.

i. Grasshoppers-Oxyanitidula, Hieroglyphus banian (Acrididae: Orthoptera)

Irregular feeding on seedling and leaf blades. Cutting of stem at panicle stage NYMPHS & ADULTS.

D. Sap Feeders

a. Thrips - *Stenchaetothrips biformis* (Thripidae : Thysanoptera)

Yellow or silvery streaks on leaves of young seedlings. Terminal rolling and drying of leaves from tip to base. NYMPHS & ADULTS.

b. ***N. vriescens*** (Cicadellidae : Hemiptera)

Yellow of leaves from tip downward. *N. cinticeps and N. impicticeps* are sectors of rice tungro, Yellow dwarf transitory yellowing and other viruses. NYMPHS & ADULTS

c. ***Zig zag leafhopper***- *Recilia dorasalis* (Cicadellidae : Hemiptera)

Top drying and orange discolouration of both margins of leaves. NYMPHS & ADULTS

d. Blue leafhopper- *Empoascanara maculifrons* (Ciciadellidae : Heiptera)

Hopper burn in seedlings in the form of whitish waxy lines on leaf blades in the initial stage of attack and subsequent drying NYMPHS and ADULTS.

e. ***White backed plant hopper***-*Sogatella furicifera* (Delphacidae: Hemiptera).

Hopper burn symptom and drying of young plants NYMPHS and ADULTS.

f. Brown plant hopper- *Nilaparvata lugens* (Delphacidae : hemiptera)

Hopper (Pugaiyan in Tamil) drying and browning of young plants, circular patches of drying, lodging of matured plants. Vector of grassy stunt, ragged stunt and wilted stunt. NYMPHS and ADULTS.

g. Earhead bug *-Leptocorisa acuta, L. oratorius* (Alydidae : Hemiptera).

Yellowing of leaf tip. Individual grains become chaffy. Black spot on the grain at the site of feeding puncture. Obnoxious odour emanates on disturbing the bugs. NYMPHS and ADULTS

h. Black bug*- Scotinophara Iurida* (Podopidae: Hemiptera).

Presence of brownish black bugs at the base of plants; plants stunted with reduced number of tillers, leaves turn reddish brown or yellow and dry (bug burn)

i. Stripped bug or Nama vandu in Tamil - *Tetroda histeroides* (Pentatomiodae: Hemiptera)

Leaves become yellow.

j. Mealy bug *– Brevennia rehi* (Pseudococcidae : Hemiptera)

Plants weak, yellowish and very much stunted in patches ; presence of white waxy fluff in leaf sheaths.

k. Green leafhopper*-* Nephotettis nigropictus (-apicalis) N. cincticerps, N. impicticeps

E. Economic threshold level (ETL) of important insect pests

Sl.No.	Pest	ETL
1.	Thrips	25/5 passes
2.	Stem borer	10% dead hearts-- 2% white ear
3.	Gall midge	10% silver shoot
4.	Whorl maggot	25% damaged leaves
5.	Leaf folder	10% vegetative stage
		5% boot leaf stage
6.	Green leaf hopper	60/25 net sweepings
		5/hill vegetative stage
		10/hill flowering stage
		2/hill Tungro endemic area 20/m^2 (Nursery)
7.	Brown plant hopper	1/tiller
		2/tiller when spider is present at 1/hill
		5 bugs/100 panicles at flowering stage.
8.	Earhead bug	16 bugs/100 panicles at milky stage.

F. Pesticides Recommended

Main field (Per acre)

Apply any one of the insecticide as given below, for each insects (pesticide per acre) and the details are given in Table 40.

1. Thrips

Endosulfan 35 EC 300 ml, Monocrotophos 36 WSC 200 ml.

2. Stem borer, Gall midge, Green Leaf hopper

Fenthion 100 EC 200 ml, Phosalone 35 EC 600 ml. Quinalphos 25 EC 300 ml, Endosulfan 35 EC 300 ml, Fenitrothion 50 EC 250 ml, Phosphomidon 85 WSC 110 ml.

3. Whorl maggot

Endosulfan 35 EC 300 ml, Fenitrothion 50 EC 250 ml.

4. Brown Plant hopper

- Avoid use of excess nitrogen
- Use resistant varieties like PV 3, CO 42 and moderately resistant variety like ADT 36.
- Provide rogue spacing at every 2 m, to facilitate insecticide application.
- Control irrigation by intermittent draining the water
- Set up light traps to monitor population and control.
- Avoid use of insecticides causing resurgence such as synthetic Pyrethroids, Methyl Parathion, Fenthion and Quinalphos.
- Pre-flowering stage : Phosphomidon 85 WSC 200 ml, Monocrotophos 36 WSC 500 ml, Phosalone 35 EC 600 ml, Chloropyriphos 20 EC 500 ml. Carbofuran 3 G 7 kg. Neem oil or eluppai oil at 6% (60 ml/1)
- Post flowering stage: 20 kg Carbaryl 10% D 10kg, drain water before using insecticides. Direct the spray / dust towards the base of the plant.

5. Case Worm

Drain water from the field. Dislodge the cases by running a rope. Apply any one of the following insecticide Monocrotophos 36 WSC 200 ml, Endosulfan 35 EC 400 ml, Phosphomidon 85 WSC 200 ml.

6. Leaf folder

Avoid use of excess nitrogen. Set up light traps to attract moths. Use any one of the following; Carbaryl 50 WP 1 kg. Quinalphos 25 EC 400 ml, Monocrotophos 36 WSC 200 ml, Phosalone 35 EC 600 ml, Fenthion 50 EC 500 ml, Chloropyriphos 20 EC 500 ml.

7. Stem borer

Set up light traps to attract moths, Use any one of the following; Phosphamidon 85 WSC 100 ml, Fenitrothion 50 EC 250 ml, Endosulfan 35 EC 300 ml, Fenthion 100 EC 200 ml.

8. Mealy bug

Use any one of the following; Phosphomidon 85 WSC 200 ml, Fenitrothion 50 EC 400 ml, Phosalone 35 EC 600 ml, Dimethoate 200 ml.

9. Earhead bug

Apply any one of the following insecticides in the early morning or late evening. BHC 10% D 10 kg, Carbaryl 10% D 10 kg, Fenthion 100 EC 200 ml. Monocrotophos 36 WSC 200 ml, Fenitrothion 50 EC 400 ml, Malathion 50 EC 400 ml.

10. White tip nematode

Spray Fenthion 100 EC 200 ml, Monocrotophos 36 WSC 100 300 ml or Phosphomidon 85 WSC 100 ml at 45 days after transplanting.

11. Rat

Use poison bait, zinc phosphide 1 part with 49 parts propped corn /rice or dry fish or warfarin 0.5% one part with 19 parts of bait material (0.025%) or Bromodiolone 0.25 w/w (1.49) at 0.005%.

12. Crabs

Poison bait with warfarin 0.025%.

G. General Instructions

- Do not apply the insecticides as a matter of routine. Apply insecticides only when the target insect population is above economic threshold level (ETL).

- Do not mix two insecticides as they will hasten the development of resistance in insects.
- Avoid insecticides, which induce vegetative growth and resurgence of sucking pests.
- Do not apply the same insecticide repeatedly
- The concentration given against each insecticide is on active ingredient basis (a.i.) for high value spray application.

In short, the important insects, their nature of damage and control – chemical control measures are given in Table 40.

Table 40. Important Insect pests or rice, their nature of damage and chemical control

Pest	Nature of damage	Economic Thres-hhold Level(ETL)	Chemical control
Brown plant hopper (BPH)	Both nymphs and adults suck sap mainly from basal portion of plants. Damaged plants turn yellow exhibiting hopper burn' symptoms. Circular patches of `hopper burn' are seen in the field under severe infestation. Pest transmits grassy stunt, ragged stunt and wilted stunt diseases.	5-10 hoppers per hill	Applycarbofuran 3G 20 Kg ha^{-1} or phorate 10 G 12.5 kg ha^{-1} or spray carbaryl 50 WP 0.15% or monocrotophos 36WSC 0.04% or chlorpyriphos 20EC 0.05%. Direct the spray towards the base of the plants.
White backed plant hopper	Both nymphs and adults such sap mainly from the basal portion of the plants. Attacked plants turn yellow and rust red from tip downwards exhibiting `hopperburn' symptoms.	5-10 hoppers per hill	-do-
Green leaf hopper (GLH)	Both nymphs and adults such sap from foliage. The infested leaves turn brown from tip downwards exhibiting 'hopperburn' symptoms. The pest transmits tungro virus disease.	2 hoppers per hill in tungro endemic areas.	Apply carbofuran 3G 20 kg ha^{-1} or phorate 10 G 12.5 kg ha^{-1} or spray carbaryl 50 WP 0.15% or monocrotophos 36 WSC 0.04% or chlorpyriphos 20EC 0.05%.
Rice mealy bug	Both nymphs and adults suck sap from stems of plants. Plants remain stunted and even dry under high incidence. The infestation is localised in patches.	NA*	Spray endosulfan 35EC 0.07% or fenitrothion 50 EC 0.1% or dimethoate 30EC 0.03% or malathion 50EC 0.1%. Direct the spray towards the base base of plants.

Paddy aphid	Both adults and nymphs such sap from foliage and panicles and cause yellowing of leaves. Severe infestation results in chaffy earheads.	NA*	Spray endosulfan 35EC 0.07% or fenitrothion 50 EC 0.1% or dimethoate 30EC 0.03% or monocrotophos 36 WSC 0.04%.
Yellow stem borer	Caterpillars enter the tiller to feed and cut the central shoot at the base producing 'dead hearts' in young plants. After panicle initiation the damage by caterpillars results in chaffy panicles called 'White ears'.	5% dead hearts during vegetative. 2% white ears during panicle stage.	Adopt seedling root dip treatment in 0.05% chlorpyriphos emulsion for one minute before transplanting in endemic areas. Apply carbo-furan 3G 20 kg ha^{-1} or phor-ate 10G 12.5 kg ha^{-1} or spray malathion 50 EC 0.1% or phosalone 35EC 0.1% or endosulfan 35 EC 0.05% or quinalphos 25 EC 0.04% or fenitrothion 50 EC 0.1%. Spray should coinc-ide with hatching of borer eggs.
Leaf fodder	Larvae fold the leaves by fastening leaf margins with stitches and then feed on green tissue inside the fold. The infested leaves become white and membranous.	4% folded leaves	Spray carbaryl 50 WP 0.25% or quinalphos 25 EC 0.05% or monocrotophos 36 WSC 0.04% or chlorpypyriphos 20 EC 0.05% or fenitrothion 50 EC 0.06%
Climbing cutworm	Young larvae feed on foliage. Mature larvae climb the plants and cut off the earheads or their rachies during night. During day larvae hide at the base of the plants. Panicles and grains are seen lying in the field.	4-5 larvae m^{-1} crop row	Dust carbaryl or malathion 25-30 kg ha^{-1} or spray chlorpyriphos 20 EC 0.05% or endosulfan 35 EC 0.07% or quinalphos 25 EC 0.075%. Direct the spray towards the base of the plants.
Swarming Caterpillar	Larvae damage the crop both in nursery and field. They feed voraciously and leave behind only the stalks of the plants. The crop looks as if grazed by cattle. The caterpillars migrate in large nos. from field to field	NA*	Dust carbaryl or malathion 25-30 kg ha^{-1} or spray endosulfan 35 EC 0.1% or chlorpyriphos 20 EC 0.05% or quinalphos 25 EC 0.075%. Direct the spray towards the base of the plants and soil crevices. Carry out control measures on community level.
Green horned caterpillar	Larvae defliate the plants by either completely eating the leaves or spare the mid rib after eating the leaf lamina. Damaged crop presents grazed look.	NA*	Dust carbaryl or malathion 25-30% ha^{-1} or spray chlorpyriphos 20EC 0.05% or endosulfan 35 EC 0.01% or quinalphos 25 EC 0.075%.

Rice Caseworm	Young larvae feed on green tissues on the leaf. Mature larvae cut leaf tips & roll them to form tubular cases in which thy live. Larvae feed during night by protruding their heads from the cases. Cases can be seen attached to plants at water level or floating in water.	NA*	Spray endosulfan 35EC 0.07% or monocrotophos 36 WSC 0.04% or carbaryl 50 WP 0.25%
Rice hispa	Both grubs and adults damage the crop in nursery as well as field. Grubs feed on the Chlorophyll by mining into the leaves. Beetles feed on chlorophyll between the veins of the leaf lamina producing parallel white streaks.	2 adults or 2 damaged leaves per hill	Dust carbaryl or malathion dust 25-30 kg ha^{-1} or spray chlorpyriphos 20 EC 0.05% or quinalphos 25 EC 0.075% or monocrotophos 36 WSC 0.04%.
Rice root weevil	Grubs feed on roots of the plants. Infested plants appear stunted, yellowish and lodge. Translucent white grubs are found clinging to the roots of the plants.	NA*	Apply carbofuran 3G 20 kg ha^{-1} or phorate 10G 12.5 kg ha^{-1}.
Gall midge	Maggots enter the stem and feed on growing point of the cetral shoot. As a results of feeding the central leaf of the attacked tiller is converted into shining tubular structure called 'silver shoot'. The affected tiller does not produce any panicle.	5% silver shoots	Carryout seedling root dip treatment in 0.05% chlorpyriphos emulsion for one minute before transplanting in endemic areas. apply carbofuran 3G 20 kg ha^{-1} or phorate 10 G 12.5 kg ha^{-1} or spray carbaryl 50 WP 0.15% or monocrotophos 36 WSC 0.04% or phosalone 35 EC 0.1% or fenitrothion 50 EC 0.1%.
Rice gundhi bug	Both nymphs and adults suck juice from developing grains during milky stage and as a result some of the grains in the panicle remain chaffy. Black or brown spots appear aroung the puncture made by bugs in the grain.	1 bug per hill	Dust carbaryl or malathion or endosulfan 25-30 kg ha^{-1} or spray endosulfan 35 EC 0.07% or malathion 50 EC 0.1% or fenitrothion 50 EC 0.1% in early morning or late evening.
Panicle thrips	Both nymphs and adults damage the crop by desaping emerging panicles which results in chaffy grains. In severe infestation entire panicle may remain chaffy.	NA*	Spray malathion 50 EC 0.1% or carbaryl 50 WP 0.1% or endosulfan 35EC 0.07% or monocrotophos 36 WSC 0.04% or dust malathion or carbaryl 25-30 kg ha^{-1}.

*NA - Information not available in literature

Adopted from Subhash Chander, 1999

10.2.2 Diseases Management In Main Field

10.2.2.1 Fungal Diseases

A. Blast

The fungus affects leaf, node and neck of the panicle. On the leaves, the spots are typically spindle shaped with dark brown margin and grey centre. Spots are encircled by yellow pale, coalescence of spots is observed resulting in bigger lesions and leading to drying of leaves which cause weakening of node. The neck region of affected panicle becomes black, brittle necrotic and break. The mycelium is inter and intracellular and septate. Conidiophores emerge through stimulate opening (or) pierce through epidermal cells.

Management

- Growing resistant varieties – Ponmani, ADT 40 CO 25 in samba CO 37, Co 43, CO 44, IR20, ADT 36, ADT 39 in Thaladi seasons in Tamil Nadu.
- Moderately resistant varieties - CO 45, ADT 37, IR 64, ASD 18
- Avoid cultivation of highly susceptible varieties viz., IR 50 and TKM 6
- Removal and destruction of weed hosts in bunds / channels
- Seed treatment with Captan/Thiram or Carbendazim or Carboxin at 2g/kg
- Seed treatment with *Trichoderma viride* @ 4g/kg or *Pseudomonas fluroscens* @ 10 g/kg
- Avoid closer spacing
- Spray nursery with carbendazim 25 g / Edifenphos 25 ml for 8 cent nursery as discussed earlier
- Spray main field with Edifenphos 250 ml/ha, Probenphos, Kitazin – 500 ml /carbendazim 250 g/Pyroquilon 500 g/ha.

B. Brown leaf spot

On leaves, the spots are rectangular or oval or elongated or sesame seed shaped. They are uniformly brown or reddish brown. Symptoms are observed on leaf sheath and also on ear heads. The affected leaf sheath turns brown. On the surface of grain, brown lesions either partially or fully observed.

Management

- Field sanitation – removal of collateral hosts and infected debris
- Crop rotation, adjustment of planting time and proper fertilization
- Use slow release nitrogenous fertilizers
- Tolerant varieties viz., CO 44, Cauvery , Bhavani
- Use of disease free seeds
- Seed treatment with Thiram/Caption @4g/kg
- Spraying the nursery with Edifenphos 40 ml/Mancozeb 80g /Captafol 40 g for 8 cents of nursery
- Spraying the main field with edifenphos 500 ml, Mancozeb 1 kg captafol 625 g/ha.

C. Narrow brown leaf spot - *Cercospora janseana*

Symptoms : It produces short, linear brown sports appear on leaves and also on sheaths, pedicels and glumes. The spots appear large numbers during later *stages of crop growth.*

Management : Spray carbendazim @ 250g / Mancozeb 1 kg/ha

D. Sheath blight

Black irregular lesions formed on leaf sheath, which is outside near water level. Infection will proceed to culm region. Inside the culm and on leaf sheath large number of small spherical brown sclerotia are formed. Primary spread is by soil borne sclerotia. Secondary spread is by a air bone basidial spores.

Management: Avoid excess N fertilizer/optimum spacing/eliminate weeds, apply organic amendments. Avoid flow of irrigation water from one field to another, summer ploughing/burning of stubbles spray carbendazim 250 g or Chlorothalonil 1 kg/Edifenphos 1 lit/ha. Seed treatment with *P. fluroescens* @10 g/kg followed by seedling dip @2.5 kg of products / ha dissolved in 100 lit and dipping for 30 minutes. Soil application of *P. fluroescens* @ 2.5 kg/ha after 30 days after transplanting (mixed with 50 kg of FYM/Sand). Foliar spray at 0.2 concentration commencing from 45 day after transplanting at 10 days Interval for 3 times depending upon the intensity of disease.

E. Foot rot

Abnormal elongation of affected seedlings is the conspicuous symptoms. The affected seedlings and plants are lean and lanky. Adventitious roots are produced on two or three nodes, which are above soil level. Mycelium is septate and hyaline. Two types of conidia are produced (i.e.) Micro conidia : single celled, Hyaline (or) rarely one septate. The chlamydospores may be terminal or intercalary.

Management : Treat the seeds with Thiram or Captan or Carbendazim @ 2 g /kg seeds.

F. Sheath rot

Leaf sheath enclosing the ear head exhibits dark brown or black circular to irregular patches. The panicle does not emerge fully from the flag leaf. The glumes are discolored. White powdery growth of mycelium is seen inside the leaf sheath and also on the panicle. Grains are discolored. In advanced stages it causes earhead chocking. Mycelium is colorless and septate. Conidiophores are branched once or twice and conidia are on tips. Conidia are unicellular hyaline.

Management

- Apply recommended doses of fertilizers
- Adopt optimum spacing
- Spray carbendazim 250 g / Edifenphos 1 lit or Mancozeb 7 kg or Chlorothalonil 1 kg/ha at boot leaf stage and 15 days later.
- Soil application of gypsum in 2 equal splits (500 kg/ha) reduces the incidence.

G. Stem blight

The fungal affects leaf sheath near water level. Brown spots are visible on leaf sheath. They become greenish white with brown margin. The fungus produces sclerotia on affected lesions.

H. Grain discoloration

The grain discoloured either reddish brown or black either in patch or in full. Affected grains are shriveled. Helminthosporium : Conidia coloured, many septate, straight or slightly curved.

Curvularia : Conidia colured; second cell bigger 4 celled.

Management

- Spray the crop at boot leaf stage with Mancozeb 1 kg or carbendazim 250 g/ha
- Store the grains at 13.5% moisture content.

I. False smut (Lakshmi disease)

Few grains in the ear head exhibit the symptoms. Affected grains are converted into grain velvety mass, which will be double the size of normal grain. The glumes are not affected. Conidia are oval green, single celled and oval.

J. Oodupathi disease

The panicle becomes as slender, dirty, grey coloured cylindrical, rod-like structure. The glumes are attached with mycelia to the surface of central rachis. The disease is common in Cumbum valley area of Tamil Nadu.

Management

- Hot water treatment at 54°C for 10 minutes
- Removal of collateral hosts like *Cyanodon dactylon* in the bunds and field.

10.2.2.2 Bacterial Diseases

A. Bacterial Blight (BB)

Symptoms are chiefly confined to leaves. Initial infection is seen as water soaked spot at the tip of leaf. Infection proceeds along the margin of the leaf in an irregular heavy manner. The affected portion turns straw coloured. The central portion remains green for some time. The entire leaf finally becomes blighted. The bacterium is rod shaped with a single polar flagellum.

Management

- Burn stubble and use optimum dose of fertilizers.
- Avoid clipping of tip of seedlings at the time of transplanting.
- Avoid flooded conditions and remove weed hosts.
- Grow resistant varieties viz., IR 20 and TKM 6.

- Spray streptomycin sulphate and tetracycline combination 300 g + copper oxychloride 1.25 kg/ha.

10.2.2.3 Viral and Mycoplasma Diseases

a. Rice Tungro Disease (RTD)

The chief symptoms are stunted growth of plant, discolouration of leaves in various shades of yellow to orange. The discolouration starts from the tip and proceeds down wards. Older leaves exhibits lusty coloured spots or dots of different sizes. The ear heads are small and grains are ill filled. The virus is transmitted by *Nephotettix* spp. (Green Jassid), as discussed in the previous sections of this chapter.

Management

- Destroy weed hosts of the virus and vectors.
- Grow tolerant varieties like IR 50 and CO 45.
- Control the vector in the nursery by Carbofuran 170 g/cent at 10 day after sowing.
- In main field, use Phosphomidan/Fenthion 500 ml/Monocrotophos 1 lit /ha or neem oil 3 per cent at 15 and 30 day after transplanting.

b. Rice Yellow Dwarf (RYD)

Plant become chlorotic and stunted. Large number of thin and pale tillers with leaves yellowish green are seen. The affected plant will look like a clump of grass. The disease is transmitted by *Nephotettix* spp.

Management

- Destroy the weed hosts and vectors.
- Spray Phosphomidan or Fenthion 500 ml or Monocrotophos 1 lit /ha.
- Resistant varieties are IR 62 and IR 64.
- Burning of stubbles.

The botanicals i.e. materials of plant origin for controlling rice pest and diseases are given in the chapter 18 (section 18.5.6.3 and 18.5.6.4).

CHAPTER 11

GROWTH ANALYSIS

The term growth is familiar and generally we equate growth with an increase in size. Growth is usually accompanied by changes in form or shape. We call it as differentiation. The term development is used to encompass the activities resulting from growth and differentiation. The terms growth differentiation and development thus encompass the events related to the progressive unfolding of the plants genetic thoroughly known. Growth can be measured in many different ways, depending upon the kind of information required for the different types of experiments. Some of the examples of growth parameters are plant height measurement, plant fresh weight and dry weight, individual leaf size (length, width, area), dry weight partitioned among organs such as roots, stem, leaves and fruits, other parameters like cell number in tissues and organs, concentration of specific chemical constituents such as nucleic acids, soluble nitrogen, proteins, lipids, carbohydrates in tissues and organ. The time factor will be considered to get the rate of growth information.

Growth is the most impressive character of living organism. The formation of new cells during germination and seedling emergence at the expense of metabolites and growth substances are from the seed food reserves. The biological growth requires orderly assimilation of substances from the environment and their conversion into the organization of plant body. A crops economic viability can be understood only when we have the knowledge on the potential yield of the crop and the practical limitations that occur in the field. Identification of morphological and physiological characters (parameters), which are closely associated with higher crop productivity and how it is affected by both environment and management practices in a crop regime, are to be thoroughly known.

11.1 MATHEMATICAL LAWS

Mathematics provides the experimental agriculture with a powerful and necessary tool to help to attain the objectives or a basic understanding of the main determinants of crop growth and production and its prediction. The two major requirements of this type are; simple and direct application to the agricultural research problem in hand and to use the application and to integrate our understanding of the fundamental physiological properties of plants in relation to the performance at the agricultural field level. Growth is measured quantitatively and its course traced in the form of growth curves. These curves may be fitted into concepts or meaningful growth models utility their respective growth equation. The growth curve is a graphical representation of some plant growth characteristics (weight, height, leaf area, growth rate etc) plotted against quite slow referred to as lag phase of growth; then there is a period of rapid increase in size called lag phase of growth or exponential growth phase and this is followed by a period of decreased growth called senescence phase of growth. The curve is sigmoid shaped and lends itself for mathematical manipulation.

The classic works of Lei beg, Sacks, Blackman and Mitscherlich are of several theories on growth factor limitation and plant response. Some of these concepts have been elevated, erroneously to the status of the laws. Lei beg's law of minimum states "A deficiency or absence of one necessary constituent, all others being present, renders the soil barren for crops for which the nutrients is needed". Blackman's theory of "optima and limiting factors" (1905) was stated as follows: "when a process is conditioned as to its rapidity by a number of separate factors, the rate of the process is limited by the pace of the slowest factor". Light and carbon dioxide are required for photosynthesis. The Blackman theory suggests that there is an abrupt cessation of the process", The "Blackman response", if one of these factors becomes limiting. In Mitscherlich's "law of diminishing return", the increase in any crop produced by an unit increment of a deficient factors is proportional to the decrement of that factor from the maximum". The response is curvilinear instead of linear, as Blackman had suggested.

11.2. GROWTH ANALYSIS

Growth analysis is defined as the analysis of yield influencing factors or parameters and plant development as net photosynthate accumulation. Growth analysis does not require any sophisticated instruments. Only two measurements, made at frequent intervals are required for growth analysis and for example, the parameters are leaf area and dry weight. Dry weight is determined by standard procedures. Leaf area (one side only) can be determined currently by a photoelectric device that needs leaf area directly as individual leaves are fed into the instrument. Another common method is using linear regression analysis : area = b (l x w), where b = slope, l=length and w = leaf width. Growth analysis can be made of individual plant parameters or of plant communities. Analysis of individual plant growth, generally made at early stages, including the following.

1. Relative and absolute growth rates
2. Unit leaf rate of net assimilation rate
3. Leaf area ratio
4. Specific leaf area
5. Specific leaf weight and allometry in growth (e.g. S-R ratio)

Agronomists generally analyze plant community growth, since it represents the accumulation of economic yield. Quantities used in growth analysis of plant communities include (I) leaf area index (ii) leaf area duration (III) crop growth rate of total biomass (usually the above ground parts) and the economic biomass. The growth correlation between two parameters are given below.

a. Allometry : The relationship between the growth rates of individual parts of an organ or organism is referred to as allometry.

b. Shoot -root ratio: The allometry of top growth to root growth, usually expressed as shoot root ratio, has its own physiological significance. Since it can reflect one type of tolerance to drought.

c. Apical and lateral growth: Vegetative and reproductive growth are other allometric relationship in growth correlation.

d. Harvest index: Harvest Index (HI) reflects the proportion of assimilate distribution between economic and total biomass. HI is similar to the partitioning coefficient.

11.3 BIOMETRIC OBSERVATIONS/ GROWTH CHARACTERS

Proper sampling is a must to get reliable results. An appropriate sample is one that provides an estimate or a sample value that is as close as possible to the value that would have been obtained in the plants in the plot. The difference between the sample value and the plot value constitutes the sampling error. A good sampling technique is one that gives a small sampling error.

11.3.1. Components of a plot sampling technique

To develop a plot sampling technique for the measurement of a character in a given trial, the researcher must clearly specify the sampling unit, the sample size and the sampling design.

a. Sampling unit: The sampling unit is the unit on which actual measurement is made. Some commonly used sampling units in replicated field trials are a leaf, a plant, unit area etc.

b. Sampling size: The number of sampling units taken from the population is sample size. The required sample size for a particular experiment is governed by the size of the variability among sampling units within same plot and the degree of precision desired for the character of interest.

c. Sampling design: The manner in which 'n' sampling units are to be selected from the whole plot. Ex: Simple random sampling, multistage random sampling, stratified random sampling etc. Gomez & Gomez (1984) prescribed that sample size is purely based on the variability and the degree of precision required for the character of interest. The sampling designs described by those authors are often impractical in experiments with frequent samplings.

d. Modified sampling procedure- Two border rows at the periphery of the sampling area are avoided for sampling. A hill in any other row is first located randomly. From that hill, nine other consecutive hills in the same row, away from the nearest border, are observed. Tiller numbers are noted serially (1-10). In the same sampling area on the diagonally opposite side, tiller observations are done in 10 other hills (11-20). Then, the mean tiller number for the 20 hills is calculated. If the mean tiller number is 7.2, for example 5 hill with tiller numbers 6,6,7,8 and 8 (i.e. one hill with mean tiller number, 2 hills with (mean-1) tiller number and 2 hills with (mean + 1) tiller number are to be sampled. The 5 hills with required number of hills may be found among 20 hills already counted or from the neighbouring area. The 5 selected hills are to be carefully pulled out (for DMP and plant analysis) (*Thiagarajan et al., 1995*).

11.3.2. Observations on growth characters

a. Plant height: The height of the plant is measured from the base of the plant to the tip of the longest leaf stretched and up to the tip of panicle at flowering and expressed in cm.

b. Dry matter production : Five plant hills selected at random from each plot outside the net plot, are cut close to the ground level, dried in hot air oven at 80^{0}C for 72 hours. Dry weight of the samples is recorded and expressed as kg or t/ha.

c. Tiller number : In each net plot, four quadrate of 0.25 m^2 area are selected at random and total tillers are counted and expressed as number m^2.

11.3.3 Observations on growth analysis

a. Leaf area index (LAI)

Area of leaf blade per unit area of land surface. The data are generally collected by the following method. Two replications of 4 hills each are taken. The four hills are considered as a sample. From a sample hill, eight fresh leaf blades are taken at random, their total area is measured by a blueprint method. The sample leaves are dried in the oven and the dry weight per cm^2 is calculated. The leaf area of the four hills is computed from the total leaf blade weight and expressed as the weight per cm^2. The LAI is then calculated by dividing the leaf area per four hills by the field area occupied by the four hills. Leaf measurement is taken at flowering stage and LAI is worked out as suggested by Palanisamy and Gomez (1974).

$$LAI = \frac{L \times W \times K \times \text{Number of leaves per tiller}}{\text{Area occupied by the plant}}$$

Where L = length of the third leaf from the top (cm)

W = width of the third leaf from the top (cm)

K = constant factor (0.75)

b. Crop growth rate (CGR)

CGR is calculated as suggested by Watson (1948) using the formula

$$CGR = \frac{W_2 - W_1}{t_2 - t_1} \text{ g/day}$$

Where, W_1 and W_2 are the shoot dry weights recorded at time t_1 and t_2.

c. Relative growth rate (RGR)

It is defined as the growth rate per unit plant weight. RGR is calculated as suggested by Williams (1946) using the formula.

$$RGR = \frac{\log_e W_2 - \log_e W_1}{t_2 - t_1} \text{ g/day}$$

where, W_1, W_2 and t_1 and t_2 are the same as given for CGR.

d. Net assimilation rate (NAR): Rate of plant weight increase per unit area .

$$NAR = d_w / d_t \times LAI$$

In practice, it is calculated by the equation as suggested by Williams (1946) for estimating NAR.

$$NAR = \frac{(W_2 - W_1)\ (\log_e L_2 - \log_e L_1)}{(t_2 - t_1)\ (L_2 - L_1)} gdm^{-2}\ day^{-1}$$

Where, L_1 and L_2 refer to leaf area (in dm^{-2})at t_1 and t_2; W_1 and W_2 refer to plant weight at t_1 and t_2.

11.3.4. Observations on other Growth Parameters

a. Apparent assimilation: The balance between photosynthesis and respiration.

b. Extinction coefficient (K): Defined as

$$K = \frac{-\log_e \frac{I_0}{I}}{LAI}$$

when I_o and I are the light intensity at the top and bottom of a population with leaf area of LAI. It is derived from I = c= K. LAI

c. Grain – straw ratio: Ratio of grain weight to straw weight at harvest.

d. Growth efficiency : Defined as W/S, where W is the plant weight increase and S is the weight of substances used for growth. S can be subdivided into the substances used for respiration (R) and those that become constituents of newly formed organs or accumulated substances (W).

e. Growth duration : Days required from sowing to 50 per cent flowering. Generally, it takes another 30 days to harvest.

f. Nitrogen response : Changes in magnitude of grain yield with increase in nitrogen application.

g. Optimum leaf area index: LAI at which maximum dry matter production takes place.

h. Photosynthetic rate : Unless specifically mentioned, refers to the rate of CO_2 absorption of a population in the field in light plus the respiratory rate. The rate of CO_2 exchange by a population is measured by the following method: Hills free from border effect are enclosed in a transparent vinyl chamber whose base area is 60 x 60 cm (in the case of 30 x 30 cm spaced population, four hills are enclosed). Air is passed through the chamber from a blower and the CO_2 concentration of the passing air at the inlet and outlet of the chamber is measured by a Hitachi infrared gas analyzer. The airflow rate adjusted to maintain the difference in CO_2 concentration between inlet and outlet is less than 50 ppm.

i. Plant weight : Weight of dried plant expressed in g per m^2 field area. It is frequently referred to as dry matter. Generally four hills are sampled from each of two replications, roots are removed, washed, dried in an air-forced oven at 70-80^{0}C for more than three days, and then weighed.

j. Plant Population : Group of rice hills. In this study, hills are always regularly spaced.

k. Respiratory rate : The rate of CO_2 release from a population in darkness. Generally, measurement is made after 10 PM.

l. Thickness of leaves : Weight of leaf blade per leaf is expressed in g/100 cm^2.

m. Translocation percentage : The percentage of straw weight decrease from flowering to harvest, to the panicle weight increase during the same period. The growth analysis and its parameters are very useful in obtaining a better understanding of yield influencing development during the crop growth cycle.

11.3.5. Observations on yield characters

The following yield characters are to be studied to know their relationship with yield.

a. Number of productive tillers per sq.m -The total number of productive tillers on 60th day after planting and at harvest were recorded in all the hills of two quadrates of 1 sq. m fixed randomly in two places of plot.

b. Length of panicle and No. of grains per panicle- The panicle length and number of filled grains and unfilled grains are recorded from 40 primary panicles in each plot.

c. 1000 grain weight- Weight of 1000 grains was weighed and recorded in 10 samples from each plot.

d. Yield Attributes on Economic Yield-Number of panicles / m^2, number of grains / panicle, % of ripened grains and weight of 1000 grains are called as yield attributes on economic yield.

e. Yield component analysis

The grain yield can be divided into several components. The equation to estimate the grain yield is.

$$Y = N \times W \times F \times 10^{-5}$$

Where,

Y = grain yield (t/ha)

N = spikelet numbers/m^2

W = 1000 grain weight and

F = filled spikelets (%)

1. Relative Importance of yield components

A study on the relative importance of each yield component for the grain yield computed by means of correlation revealed that the combined yield components accounted for 81% of the yield variation, number of spikelets/m^2 alone explained 60% of the variation and filled spikelet per cent and grain weight together accounted for 21% (Table 40a). If the contribution of all the yield components were taken as 100, which would have been true, had there been no error in measurement, the contribution of number of spiklets/m^2 would have been 74% and the combined contribution of filled spikelets per cent and grain weight would have been 26%.

Table 40a. Contribution of each yield component to grain yield

Yield component	Contribution to total variation in yield (%)
1. Spikelet number/m^2	60.2
2. Filled spikelet (%) and 1000 grain weight (g)	21.2
3. Spikelet number/m^2 and filled spikelet	75.7
4. Spikelet number/m^2 and 1000 grain weight (g)	78.5
5. Spikelet number/m^2 and filled spikelets and filled spikelets (%) and 1000 grain weight(g)	81.4

Clearly, spikelet number/m^2 was the most important component limiting the yield. Spikelet number per unit land area is affected by

- Cultural practices – spacing, no. of seedlings/hill and nitrogen application
- Growth characteristics – tillering (seedling emergence)
- Climatic conditions – solar radiation and temperature

Under subnormal cultural practices, the spikelet number per unit land area can be increased by increasing the plant density and nitrogen application. A higher plant density is required for low tillering, early maturing varieties and low nitrogen levels. Rice is most sensitive to high temperatures at heading and next most sensitive at about 9 days before heading. One or two hours of higher temperature at anthesis leads to sterility. High temperature before or after anthesis has much less effect on sterility. There is a good correlation between number of spikelet/m^2 and total nitrogen uptake at heading: as the amount of nitrogen absorbed by the crop increased, the number of spikelets/m^2 increased. Nitrogen fertilization affects sterility caused by low temperatures at the reduction division stage. When temperatures are above or far below the critical, nitrogen supply has little effect on sterility. At moderately low temperatures (16^{0}C), however, the percentage of sterility increases with an increasing nitrogen supply. The adverse effects of high nitrogen under low temperatures during the reproductive stage can be alleviated by increasing the amount of phosphate applied.

2. Filled Spikelet Percentage

Factors such as weather, soil, fertilizer application and incidence of diseases and insects affect filled-spikelet or sterility percentages. Sterility percentage refers to the percentage of unfertilized spikelets plus the percentage of partially filled spikelets. About 9% was the average sterility percentage. Crops under shade during ripening had a low percentage of filled spikelets because of increased numbers of partially filled spikelets. Some varieties may have lower filled spikelet percentages at high levels of nitrogen. At high levels of nitrogen, lodging or culm bending decreased filled spikelet percentage. Lodging reduces the cross sectional area of vascular bundles and disturbs the movement of assimilates and absorbed nutrients via the roots. It also disturbs the leaf display and increases shading. The percentage of filled spikelets often decreases as the number of spikelets/m^2 increases. When solar radiation is low, the source activity may be insufficient to produce enough carbohydrates to support the growth of all the spikelets. As a result, the number of unfilled spikelets may increase. Air temperatures below 20^{0}C may cause high percentages of sterility if they persist for more than a few days during booting or heading. High temperature shortens the grain filling period. Persistent cloudy weather will be more detrimental to grain filling under high temperature because of a shorter ripening period. Temperature higher than 35^{0}C at anthesis may cause high percentage of sterility. Strong winds can cause sterility at flowering by desiccating the plant, impair grain growth by causing mechanical damage to the grain surface and cause the crop to lodge. A hot air frequently causes white head, particularly when it occurs during panicle exertion. In the coastal areas, strong winds from the coast often contain broken water and cause sterility. High salinity in the soil causes a high percentage of sterility. Drought at flowering often causes high percentages of sterility in rainfed lowland and upland rice.

11.4 WAYS TO INCREASE THE YIELD

A. Shallow Planting

This method enhances greater number of tillers. The growing point comes to the ground surface under shallow planting and the number of tiller is increased. In shallow transplanting, seedlings can easily absorb nutrients and consequently the number of tillers increases.

B. Prevention of root damage after transplanting

Apply 1 kg of gypsum or DAP/cent of nursery one or two days prior to pulling out of seedlings to avoid root damage. The optimum water temperature, low air temperature, windless cloudy and rainy conditions favour rooting. During tillering stage, 35°C in the day time and 15°C in the night seem to be nearly optimum for increasing the number of tillers. Normal spacing of transplanted hills increase the number of panicles per unit area. However spacing varies with duration of the variety. Suppressing nonbearing and weak tillers: It can be achieved by complete draining of the surface water just after time of the final emergence of bearing tillers, to the extent that cracks will appear on the surface.

C. Steps to increase the number of panicles

- Healthy seedlings should be short and thick in nature, which make the plant to develop more roots.
- An optimum spacing should be adopted
- To increase the number of panicles, the vegetative growth period (from transplanting to the panicle initiation) should coincide or start with low temperature, which favours lengthening of vegetative growth period of varieties irrespective of duration.
- Adequate basal fertilizers
- For increasing the number of tillers a nitrogenous top dressing needs to be given only during the early growth stage i.e. within 20 days after transplanting.

D. Steps to increase the number of grains / panicle

- Controlling the development of an excessive number of panicles.
- Raising strong and large tillers by the time of neck node differentiation.

Plate 36. Rice panicles - a closer view

The number of differentiated spikelets (grains) is mainly determined by the amount of assimilated products for 10 days from the neck node differentiation to the initial stage of spikelet differentiation. Hence, careful management practices such as adequate water supply, free from insects and diseases and other favourable conditions up to the time of neck-node differentiation have to be followed. For increasing the number of differentiated spikelets, the most important condition is to nourish the plant well in the 7th day to 10th day period from neck node differentiation stage to the spikelet differentiation stage. The neck node differentiation stage generally occurs 7 to 10 days earlier than the spikelet differentiation stage (panicle formation stage). Top dressing with N fertilizer is essential at spikelet differentiation stage than that applied at the neck node differentiation, which occurs before maximum tillering stage in many cases. If neck node differentiation stage occurs after maximum tillering stage, then it is better to apply nitrogenous fertilizer at the neck node differentiation stage itself to increase the number of spikelets. For top dressing at the neck node differentiation stage, potassium fertilizers can be effectively used along with N fertilizers in sandy soils with low fertility. The percentage of degenerated spikelets will be of more than 30% in an ordinary crop. This is due to nutrient deficiency in the period of reduction division. Hence, top dressing just before the reduction division stage for decreasing the number of degenerated spikelets and increasing the number of grains per panicle is most important (Plate 36).

E. Steps to increase the percentage of ripened grain

- Encouraging early seasonal cultivation:
- Prevent the production of excessive number of spikelets:
- Top dressing at the time of neck node differentiation stage is apt to make the plant increase the number of grains excessively and to reduce the percentage of ripened grains.
- Top dressing just before reduction division i.e., 7 days before heading does not increase the number of grains, but increase the number of ripened grains, making the percentage number of ripened grains as well as the yield is increased.
- Make the rice plant reaches it's heading time in a healthy physical conditions.
- Ripening period of the variety must coincide with sunniest season.
- Top dressing with N fertilizer at full heading stage.
- Promote the kernels development by making the abortive kernels develop fully and increasing the percentage of ripened grain.
- Reduce damage caused by disease and insect pests and after heading.
- The heading and flowering stage of rice should not coincide with peak period rainfall and wind to prevent the plant from lodging.

F. Steps to increase the weight of grains

When a quick acting fertilizer such as ammonium sulphate is used, the top dressing just before the initial stage of reduction division (panicle initiation) (7 days before the heading) enables the plant to enlarge the hull size. The kernel ripening period should coincide with the sunniest season in each location. The kernels develop better on clear day than on a cloudy day, and much better during the day time on a clear day than during night, and further they develop most remarkably at the time between 12 and 4"O" clock on a clear day. The rate of translocation decreases markedly when temperature goes below 20°C.

11.5 ROGUEING AND FLOWERING INDEX STUDY

11.5.1 Rogueing

Rogue is a plant of the same crop but with different varietal character and these plants have to be removed to maintain the varietal purity. The rogueing at pre-flowering or at flowering stage and at maturity stage is purely based on the varietal characters.

11.5.2 Flowering Index

There are two stages in flowering. The first stage is the flowering of the main shoot and second stage is general flowering of the side shoots. The flowering of the main shoot is often hastened by transplanting and more so by delayed transplanting. Side shoot takes its own time to flower. When young seedlings planted flowering starts early but the period is protracted. If aged seedlings are transplanted, flowering is delayed but synchronous. If flowering is not uniform due to various reasons like age of seedlings, application of N at basal and duration of the varieties, a slighter dose of N can be applied to get uniform flowering. The progress of flowering can be indexed as following.

Calendar date (1)	Count no. (2)	No. of shoots in 5 plants at random at each count (3)	No. of panicle emerged (4)	Ratio of col. 4 to col.3
5th Nov.	1	40	4	0.100
8th Nov.	2	38	9	0.236
11th Nov.	3	42	14	0.333
14th Nov.	4	42	32	0.780
17th Nov.	5	40	88	0.950
Mean of flowering index				0.480

CHAPTER 12

POST HARVEST TECHNOLOGY

Post harvest technology is as old as agriculture. However, during the recent part, it has emerged as an independent discipline or science and technology. This term encompasses an array of handling and processing system from the stage of maturation till consumption of the produce and includes harvesting, threshing, cleaning, grading, drying preservation, storage, processing, packaging, transportation, marketing and consumption systems. An estimated Rs. 1.56 crores worth of rice is lost every year in Tamil Nadu because of inefficient post harvest practices according to a study organized by ICAR (Alagusundaram, 2004).

12.1 HARVESTING

Harvesting is the process of collecting the mature rice crop from the field. Harvesting of paddy includes cutting, stacking, threshing, cleaning and hauling of paddy. The goal of good harvesting methods is to maximize grain yield, and to minimize grain damage and quality deterioration. Harvesting can be done manually using sickles and knives, or mechanically with the use of threshes or combine harvesters. Regardless of the method, a number of guidelines should be followed that will ensure that harvest losses are kept of a minimum and grain quality is preserved during harvest operations. In short, guidelines for proper harvesting include

- harvest at the right time and moisture content
- avoid delays in threshing after harvesting

- use the proper machine settings when using a threshing machine
- clean the grain properly after threshing

To get the highest head rice recovery, the crop could be harvested at 21-24 per cent moisture level. Short duration varieties are harvested between 25 and 35 days after 50 per cent flowering. Medium and long duration varieties harvested after 27 and 39 days at 19-23 per cent moisture level will give highest head rice recovery.

Following are the few pre-harvest operations and checks for obtaining maximum head rice yield.

i. Drain the water from the field 7 to 10 days before the expected harvest date as draining hastens maturity and improves harvesting condition.

ii. When 50 per cent of the panicles turn straw colour the crop is ready for harvest. Even at this stages, the leaves of some of the varieties may remain green. The crop should be cut, when 80-85% of the grains are straw coloured.

iii. Confirm maturity by selecting the most mature tiller and dehull a few grains. If the rice is clear and firm, it is a hard dough stage (Plate 37).

Plate 37. Physiological maturity symptoms of rice

Plate 38. Kuruvai rice is ready for harvest and nursery is ready for next samba crop

Plate 39. Manual rice harvesting with sickle.

At this stage harvest the crop, thresh and winnow the grains. Dry the grains to 14% moisture level and store. The crop is harvested as soon as the grain reaches the physiological maturity. The moisture percentage of grains at this stage may be 25 to 30 per cent. The recommended optimum moisture content is 21-24 per cent. It comes after 28-32 days of 50 per cent flowering. Harvest the crop by cutting close to the ground, transport the harvested crop as small bundles to the threshing floor. Threshing can be done either by machine or by human labour. Then initial cleaning is given and the produce is stored. Maturity may be hastened by 3-4 days by spraying 20% NaCl, a week before harvest to escape monsoon rains (Plate 38 and 39).

Harvesting systems

1. Manual harvesting and handling, followed by threshing and cleaning by hand or using manually operated tools.
2. Manual harvesting, followed by machine threshing.
3. Machine harvesting with a reaper followed by machine threshing.
4. Combine harvesting (Plate 40 and 41)

Plate 40. Loading harvested rice bundles in bullock cart.

Plate 41. Transporting harvested rice bundles to threshing floor.

12.2 THRESHING

The threshing floor available in villages is muddy. Lot of grains stick to the soil rendering the marketability of grains very difficult. Under such circumstances, good threshing floor is a must.

12.2.1 Manual Threshing

The common method for threshing by hand is separating the grain from the panicle by impact. This is done by hand beating, treading, or by holding the crop against a rotating drum with spikes or rasp bars. Hand beating methods are normally used for threshing rice that easily shatters. Manual threshing methods include :

1. Foot threshing or trampling
2. Beating against a threshing rack
3. Flail – use of stick for threshing
4. Using a pedal thresher or treadle thresher (Plate 42 and 43).

Plate 42. Manual rice threshing

Plate 43. Winnowing manually threshed rice

12.2.2 Machine threshing

Given the high labourer requirements of manual threshing, in many countries threshing of paddy is now mechanized by use of small stationary machine threshers. Depending on farming systems, post-harvest practices and infrastructure, threshers come in different sizes and range from small portable units without cleaner (Philippines), to large truck mounted units (Thailand). Threshing is either done in the field, near the field or at the nearest road (Plate 44 and 45).

Plate 44. Paddy thresher - Machine threshing

Plate 45. Paddy thresher - Machine threshing

12.3 Drying

Use of driers found to minimize the loss of grains during drying.

12.3.1 Principles in Grain Drying

Threshed rice with hull (lemma and palea) is called paddy, rough rice, or paddy rice. Rice grain dehulled by milling is termed milled rice. It is more advantageous to store paddy than milled rice since the hull, which covers the palay, protects the kernel from storage pests. However, paddy storage requires special precautions. Paddy is not an inert product; it is a living respiring biological product. Unfortunately, this very important property of paddy is often overlooked during grain drying and storage systems unknowingly. Drying is the process that reduces grain moisture content to level where it is safe for storage. It is the most critical operation after harvesting a rice crop. Delays in drying, incomplete drying or ineffective drying will reduce grain quality and result in losses. Drying and storage are related processes and can sometimes be combined in piece of equipment (in-store-drying). Storage of incompletely dried grain with a higher than acceptable moisture content will lead to failure regardless of what storage facility is used. In addition, the longer the desired grain storage period, the lower the required grain moisture content must be. The purpose of drying is to reduce the moisture content of rough rice to a level safe for storage. As even short term storage of high moisture paddy rice can cause quality deterioration, it is important to dry rice grain as soon as possible after harvesting – ideally within twelve hours. The following Table 41 shows the recommended moisture content (MC) for storage of paddy grain and seed, and potential problems when the moisture content exceeds these limits.

Table 41. Moisture contents required for safe storage for different storage periods

Storage period	Required MC for safe storage	Potential problems
2 to 3 weeks	14-18%	molds, discoloration, respiration loss
8 to 12 months	13% or less	Insect damage
More than 1 year	9% or less	Loss of viability

The purpose of storage is to provide the dried grain with protection against insects, molds, rodents and birds, and to prevent moisture from re-entering the grain. The respiration process of the grain is externally manifested by decrease in dry weight, utilization of oxygen, evolution of carbon dioxide and release of energy in the form of heat. The respiration rate is influenced by the degree of hydration of tissues. Thus, one of the methods of preserving or suppressing biological deterioration of paddy is reducing grain moisture by drying. At low moisture content (13 to 14 per cent), the grain is relatively dormant and can be stored for long periods under proper conditions. Under the traditional rice production systems with yields averaged 30 to 40 calams per hectare, the old methods of harvesting, threshing, sun-drying, and sack storage of grain are adequate to local farmers. However, the scenario changed due to the introduction of high yielding varieties and the irrigation programs, which led to extraordinary increase in yield. Today, farmers may have to depend on grain driers to dry their high grain production.

12.4 CLEANING

Threshed grain contains all kinds of trash. This trash can be vegetable such as chaff, straw, empty grains and foreign seed, as well as mineral materials such as earth and stones. Grain should be cleaned as soon as possible after harvesting and certainly before storage. The simple traditional cleaning method is winnowing, which uses wind or a fan to remove the light elements from the grain. Mechanical winnowers that incorporate a fan and several superimposed reciprocating sieves or screens are now used in many countries. These can be manually powered or motorized and have capacities from 100 kg per hour to 2-3 tons per hour. Where combine harvesters are used, there is a trend towards using large capacity centralized seed cleaners. These are normally equipped with a series of vibrating sieves and are capable of 10-30 tons per hour.

Plate 46. Winnowing and cleaning of cattle threshed rice

Winnowers, cleaners and graders can be used by replacing the conventional method of winnowing, which largely depends on the wind velocity and manual grading (Plate 46).

12.5 STORAGE

Rice storage facilities take many forms depending on the quantity of grain to be stored, the purpose of storage, and the location of the store. In general, it is recommended that rice for food purposes be stored in brown rice or husked rice form rather than milled rice as the husk provides some protection against insects and helps prevent quality deterioration. However, when rice can be stored as brown rice, 20% less storage capacity will be needed.

The requirements for a good storage system include:

- Prevention of moisture re-entering the grain after drying
- Protection from insects, rodents and birds
- Ease of loading and unloading
- Ease of maintenance and management

Rice grain is hygroscopic and in open storage systems the grain moisture content will eventually equilibrate with the surrounding air. High relative humidity and high temperatures contribute to high equilibrium or final moisture content. In many tropical countries, the equilibrium moisture content is above safe storage moisture levels.

12.5.1 Safe storage conditions for grain

Safe storage of rice for longer periods is possible if three conditions are met:

- Grain is maintained at moisture levels of 14% or less and seed is stored at 12% or less
- Grain is protected from insects, rodents and birds
- Grain is protected from re-wetting by rain or imbibing moisture from the surrounding air.

The longer the grain needs to be stored, the lower the require moisture content will need to be. Grain and seed stored at moisture contents above 14% may experience the growth of molds, rapid loss of viability and a reduction in eating quality.

12.5.2 Minimize damage during storage

Little can be done to prevent or reduce the invasion of crops in the field by fungi. However, the following recommendations should help to prevent storage fungi problems or minimize damage from storage fungi in stored grains.

1. Harvest as soon as the moisture content allows for minimum grain damage.
2. Adjust the harvesting equipment for minimum kernel or seed damage and maximum cleaning.
3. Clean all grain harvesting and handling equipments thoroughly before beginning to harvest. Clean bins or storage facilities thoroughly to remove dirt, dust and other foreign material, crop debris, chaff and grain debris.
4. Clean grain going into storage to remove lightweight and broken kernels or seed as well as foreign material and fines
5. Moisture content is far the most important factor affecting the growth of fungi in stored grain. After harvest, grain should be dried to safe moisture contents as quickly as possible.
6. Aerate grain to safe and equalized temperatures through the grain mass.
7. Protect grain from insect and mite damage.
8. Check stored grain on a regular basis and aerate as needed to maintain low moisture and proper temperature.

The storage should also be done using the storage structures like bins, gunny bags, impregnated with chemicals in order to protect it from weather elements like temperature, moisture and aeration and also from insects, rodents, fungi and other micro organisms. The recommended moisture content of paddy is 12% for safe storage.

12.5.3 Optimum moisture content for storage and milling

In Tamil Nadu, the moisture content of newly harvested rice usually ranges from 20 to 26 per cent. At this moisture content, grains are too wet for milling and storage. Optimum moisture content for milling is about 11 percent presently used in the Tamil Nadu. Nevertheless, some rice millers mill paddy with 16 per cent moisture leading to considerable grain damage. A moisture content of 12 percent is considered the threshold value for safe storage. To prolong storage over six months, the grain moisture should be reduced to 13 per cent. Here again the need for using artificial driers is amplified. Researchers estimate that about 10 per cent of the crop is lost if the grains are left in the field to dry down to 16 to 14 percent before they are harvested. (Alagusundaram, 2005). Thus, farmers may lose a considerable portion of their produce, unless they use more efficient post harvest practices.

12.5.4 Use of grain driers

Using mechanical driers has the following advantages.

1. The drying operation can be controlled more accurately. Thus, the moisture content of the grain can be reduced more closely to the levels necessary for milling with maximum "head rice" recovery and effective preservation of seed viability.

2. Drying can be done in foul or fair weather.

3. The farmer can plan the harvest season for more efficient use of available labour.

The moisture content of the grain is one of the most important indices for determining whether the grain should be harvested or is dry enough for safe storage and milling with maximum recovery. Therefore, it is essential to understand the common methods of representing moisture. It is also important to know how to determine the moisture content accurately. The rice industry expresses moisture content (M.C.) as a percentage of the total weight.

$$\text{w.b.} = \frac{\text{Wm}}{\text{Wt}} \times 100 = \frac{\text{Wm}}{\text{Wm} + \text{W d.m}} \times 100$$

where,

w.b.	=	wet basis moisture content (%)
Wm	=	weight of moisture in the grain (g)
Wt	=	Total weight of the grain (g)
W d.m.	=	weight of dry matter in the grain,(g)

For expressing moisture content on wet basis, it is incorrect to state that one has removed 7 per cent of 100 kg or 7 kg of water from 100 kg of grain if moisture content of the 100 kg of grain is lowered from 20 to 13 percent. This is because the denominator of the equation changes in relation to the amount of moisture removed. For example, at 20 percent moisture content, 100 kg of wet grain contains 20 kg of water and 80 kg of dry matter. If the same amount (100 kg of wet grain) is dried, the resulting weight of moisture is calculated using the following equation.

$$\text{Wm} = \frac{(0.13)(80)}{0.87} = 11.96 \text{ kg}$$

Therefore, the amount of water removed from the 100 kg of wet grain when the grain is dried from 20 to 13 percent is 20.00 minus 11.96 or 8.04 kg. Since it is necessary to reduce the moisture content of the grain to 13 per cent for safe storage, a table has been prepared to facilitate the computation of moisture load. For example at an initial moisture content of 20 per cent, the moisture load is 8.04 kg per kg of grain. Therefore, to dry 100 kg of wet grain to 13 per cent moisture level, it requires evaporation of 0.804 x 100 or 8.04 kg water.

12.6 GRAIN QUALITY

Grain quality is not solely a varietal characteristic but also depends on the crop production environment, harvesting, processing and handling systems. Therefore, maintaining good grain quality is the concern of all disciplines including breeding, agronomy, entomology, chemistry and engineering. The quality of grain is best, when it reaches physiological maturity. Management from that point forward will determine the rate of decline in quality but is unable to stop some decline in quality from occurring.

Rice grain quality represents a summary of the physical and chemical characteristic, which may be genetic or acquired properties.

The genetic properties include

- Chemical characteristic (gelatinization temperature, apparent amylase content, gel consistency, alkali spreading value and aroma)
- Shape
- Size
- Color of grain
- Chalkiness
- Bulk density
- Thermal conductivity
- Equilibrium moisture content and
- Flowability

The acquired properties or environmental factors are either additional to the normal complement of genetic qualities or are the consequence of certain genetic qualities being lost or modified. The important acquired properties are:

- Moisture content

- Grain purity
- Physical and pest damage, cracked grains,
- Presence of immature grains
- Milling-related characteristic (milling and head rice recoveries, grain dimensions, whiteness, milling degree and chalkiness)

12.7 PARBOILING

Parboiling is a hydrothermal treatment followed by drying before milling for the production of milled parboiled rice. It originated in India. Half of India's rice crop is parboiled. Parboiling is particularly good in the case of coarse and medium rice of soft structure because such rice suffer excessive breakage when milled raw. Parboiling involves soaking paddy in water for a short time followed by heating once or twice in steam and drying before milling.

Advantages

- Dehusking of parboiled rice is easy.
- Grain becomes tougher resulting in reduced losses during milling
- Milled parboiled rice has greater resistance to insects and fungus.
- Loss of nutrients due to the removal of husk and bran in milling are decreased. During parboiling, the vitamins and minerals present in hull (outer covering of the paddy) and bran coat are dissolved and seeped into the endosperm. Part of the scutellum and germ, which are rich in B vitamins, get fixed to the grain and hence losses of B vitamins are less.
- Loss of water soluble nutrients due to washing of rice is less in parboiled rice compared to raw rice.
- Parboiling improves digestibility and Protein Efficiency Ratio when compared to raw rice.
- Parboiled rice will not turn into glutinous mass when cooked.
- Parboiled rice swells more to the desired softness when cooked.
- Parboiling stabilizes the oil content of the bran.

Disadvantages

- Sometimes it develops an unpleasant smell and hence not preferred.
- Change of colour may occur during parboiling.

12.7.1 Conventional process

This process consists of the following steps:

1. Steeping paddy in cold water for 2 or 3 days in large cement tanks
2. Steaming of the soaked paddy for 5-10 minutes and
3. Drying in the sun.

During soaking of paddy in cold water for 2 or 3 days, fermentation sets in and off flavour develops in the grain. The moulds present in the grain also grow and may produce toxic metabolites. For example, afflotoxin may be produced from *Aspergillus flavus.*

12.7.2 Hot soaking process

The hot soaking process, developed at the Central Food Technological Research Institute, Mysore consists of the following steps:

1. Soaking of paddy in water at 65-70°C for 3-4 hours.
2. Draining of water and steaming of soaked paddy in the same vessel for 5-10 minutes and
3. Drying of the paddy in the sun or in mechanical driers.

12.7.3 Modified parboiling process

Several modifications of the parboiling process have been developed in different countries. These are briefly described below.

Avorio process : In this process developed in Italy, paddy is kept in perforated baskets and moved in a hot water tank for soaking and later steamed under pressure in a rotating cylinder, followed by drying in hot air driers.

Conversion process : In this process developed in U.S.A., soaking of paddy is carried out in cold water. The air in the soaked paddy is removed by vacuum steamed under pressure and dried in a vacuum drier.

Malek process : The process, developed in U.S.A., consists of soaking paddy at 100°F for 3-6 hours followed by steaming for 15 minutes. The steamed paddy is dried in a hot air drier at a low temperature.

12.8 MILLING OF RICE

Milling or removal of the outer covering (hull) and seed coats (bran) of the paddy grain involves two basic operations, which could be done by one machine or by a combination of two or more separate units. These operations are necessary to convert rough rice into a form suitable for human consumption.

Before milling, paddy is cleaned to remove the large, heavy impurities. During dehulling and milling, coarse outer layer of bran and germ are removed. Paddy on milling yields hull 20%, bran 8%, polishing 2% and rice 70%. In India, paddy is milled either by home pounding or by mechanized rice mills. In home pounding, there are more broken rice than mechanized miller. Storage life is short as the fat in the bran develops rancidity.

The process of milling involves the following steps.

- Rice is passed through two stone or rubber discs rotating at different speeds and by shearing action on the grain, the hull is pulled away. The whole kernel from which the hulls have been removed is known as brown rice.
- This is then milled in a machine called pearler to remove coarse outer layers of bran and germ by a process of rubbing, resulting in unpolished milled rice. Some amount of breakage of rice occurs at this stage.
- Unpolished rice is liable to develop rancidity and so it is next polished in a brush machine, which removes the aleurone layer and yields "polished rice".
- Sometimes the polished rice is further treated in a device known as trumbol to give a coating of sugar and talc to increase shining brightness of the grains.
- Rice is separated from the broken kernels. Large kernels are called as second heads, medium ones are called screenings, smallest ones are called the brewers rice.

Most varieties of coarse rice are not highly polished. As per the Indian Government regulations, the extent of polishing should not exceed 5%. If rice is milled beyond 10% then most of the thiamine is lost. The percentage of losses of different nutrients during milling are; protein 15%, fat 82%, thiamine 85%, riboflavin 70%, pyridoxine (vitamin B6) 50%. The degree of milling determines the amounts of nutrients removed.

Losses during milling can be compensated by the following process.

- By under milling / unpolishing rice the loss of nutrients can be reduced. Rice that is under-milled do not have the customary white luster and they are more subject to insect infestation and flavour deterioration than white rice.

- A second method is by processing the rough rice prior to milling to increase retention of vitamins. This is done by parboiling which is commercially known as converted rice.

Another way of arresting the losses occurring in the milling of rice is the ratified enrichment of the grain. A premix has been developed in which the rice is wetted with a solution of thiamine and niacin, then dried. A second coating of iron pyro phosphate is distributed on the rice. The rice premix is highly resistant to washing, cooking and storage losses.

12.8.1 Three General types of rice mills

12.8.1.1. "Kiskisan" or Hulling – scouring mill

This machine removes the hull and bran in one operation in a solid fluted cylinder rotating within a hollow cylinder. The hulling process is effected by the shearing action of the rotating solid cylinder ribs against a stationary blade whose clearance can be varied according to the variety of the rice being milled. The mixture of crushed hull, bran and broken drop through the perforated lower half of the hollow cylinder. The milled rice is discharged at the outlet end. This machine gives a total milling recovery of 64 to 66 per cent (Arboleda *et al.*, 1964).

12.8.1.2. Under – run Disk Sheller or "Cono" Type Mill

In this type of mill, shelling or hulling and scouring are accomplished in separate operations. The hulling process is done in a twin horizontal disks coated with emery stone. The upper disk remains stationary while the lower disk rotates. Centrifugal force and abrasive action effects the shearing of the hull. Scouring is done in a frustum of a cone coated with emery stone rotating inside a screen mesh of the same shape provided with three rubber brakes. The rubber brakes choke the movement of the grain and due to induced attrition, the bran is removed and expelled through the screen mesh while the pearled rice is discharged at the bottom outlet. Total milling reconvert in this mill is from 66 to 68 per cent (Brown, 1966).

12.8.1.3. Rubber – roll Type

This mill has separate units for hulling and scouring. The hulling process is an application of the principles of granular dynamics, which is achieved through the frictional action of two rubber rolls that differ in their circumferential speeds. Fractional force between the rubber rolls separate the hull from the grain and produces brown rice. Scouring and pearling is then accomplished in units similar to that of the "cono" mill type. The rubber – roll mill may give a total milling recovery of 71 to 74 per cent (Brown, 1966).

12.8.2 Modern rice mill

The milling of rice in a modern mill consists of the following steps: (1) cleaning and grading; (2) dehusking (shelling); and (3) polishing (milling)

A. Cleaning and grading

The paddy as received will contain foreign matter such as stones, clay particles, straw, chaff and dirt that have to be separated in paddy cleaner.

B. Dehusking (shelling)

The shelling is carried out normally using two different types of shellers; (a) Disc sheller and (b) Roller type sheller.

a. Disc sheller : This machine is also known as "Under Runner sheller". This consists of two discs. The inside surface of the discs is covered with a mixture of emery and hard cement and the clearance between the discs is adjusted close to the length of the paddy grain to be shelled. One plate is stationary, while the other plate revolves. The husk is removed by aspiration.

b. Roller type sheller: Commercial units of this type are manufactured in some countries. It consists of two horizontally set rubber rollers rotating in opposite directions, the differential rotation between the two being kept at about 200 per minute. Dehusking is effected by the grain hitting the rotating rollers. The resulting brown rice contains the pericarp and germ almost intact. The breakage of rice is minimum with this machine.

C. Polishing (milling)

The brown rice obtained by shelling is subjected to polishing (milling). Two types of equipments viz., (1) Cone type polisher and (2) Horizontal type polisher are commonly used.

a. Cone type polisher : The cone type polisher consists of an inverted truncated cone coated with a mixture of cement and emery, rotating inside a wire basket. The distance between the wire screen and the cone is adjustable and the degree of polishing is determined by this adjustment.

b. Horizontal type polisher: The horizontal type polisher consists of a horizontally set straight or slightly tapering metal or emery coated cylinder fitted inside a fixed outer cylindrical framework. The working of the horizontal polisher is similar to that of the cone polisher. As a result of milling, rice polishing and germ are removed and white rice is obtained.

The average yields of various products of milling of paddy in modern rice mill are given in Table 42.

Table 42. Percentage yields of various products during milling of paddy by Modern mill

Cleaning losses		2.0 – 3.0
Hulls		19.0-21.0
Brown rice		76-79
Milled products:		
	White head rice	60-70
	Broken rice	2-4
	Bran and polishing	6-8

12.8.3 Old milling process

Engelberg type huller (sheller-cum-polisher)

The Engelberg type huller is the simplest and cheapest among paddy milling machines. This machine is a sheller-cum-polisher. It is used widely in India and other Asian countries. The huller consists of a solid heavy fluted cylinder, which rotates at 500-600 rpm inside a hollow stationary cylinder with uniform perforations. The machine is completely filled with paddy and the packing force is regulated by a blade that runs along the length of the machine and protrudes between the upper and lower halves of the perforated cylinder. The dehusking and polishing (removal of outer bran layer and germ) are effected by friction. Skill and close attention are required for efficient operation to prevent excessive breakage.

12.9 RETENTION OF VITAMINS AND MINERALS

The retention of vitamins and minerals in various forms of rice are given in Table 42a.

Table 42a. Retention of vitamins and minerals in various forms of rice

Nutrient	Brown rice	Polished	Hand pounded rice	Parboiled rice
	Quantity µg/ 100g	Expressed as percentage of the quantity present in brown rice		
Thiamin	320	15	66	72
Riboflavin	156	35	87	66
Niacin	4600	37	85	83
Folic acid	36	22	60	30

Source : Acharya, 1998

12.10 DEFINITION OF TERMS

The readers are requested to refer Glossary for detailed information. For easy understanding of this chapter, the important terms are given.

Hulling, husking or shelling – refers to the removal of the outermost covering of the palay (paddy) grain.

Scouring or pearling – means the removal of the bran.

Polishing – means removal of the inner layer of the seed coat (very fine rice bran) from the milled rice to make the surface glossy and clean.

Total milling recovery – refers to milled rice recovered from the original rough rice (paddy) in percent, which is determined by the following equation.

$$\frac{\text{Milled rice}}{\text{rough rice}} \times 100 = \%\ \text{total milling recovery}$$

Head or whole grain recovery – refers to ¾ to whole grain recovered from the original rough rice in per cent.

Grading- The separation of broken from the head rice

Brokens – are known as white rice grains less than ½ the original length.

Brown rice – are rice grains with only the hulls removed.

Whitening- The process of removal of silver skin and the bran layers of the brown rice

Glazing- Process of coating the surface of the rice with thin layer of talc and glucose to have shiny and more transparent appearance.

12.11 ERADICATION OF FUNGAL, INSECT AND RODENT DAMAGE ON FOOD GRAINS

Methods commonly employed for the eradication of damage to food grains by fungi, insects and rodents are outlined below.

12.11.1 Prevention of damage by fungi

Drying the grains in the sun or in driers to an optimal moisture content will help to prevent fungal damage. The dried grains should be stored in properly constructed warehouses to prevent adsorption of moisture. Fumigation with methyl iodide, methyl bromide or a mixture of ammonia and phosphine controls the growth of fungi in stored grains.

12.11.2 Eradication of insect damage

For the eradication of insects from food grains, traditional methods are being used by the farmers in rural areas. These, in general, are not very effective. Modern methods using insecticides as fumigants are highly effective in controlling insect infestation and are also safe for use. Some of them are described below:

A. Insecticides

I. Contact insecticides : Only a few contact insecticides, *viz.*, pyrethrin and malathion are suitable to use on stored products. The other commonly used insecticides such as DDT, Lindane, BHC and Deldrin are highly toxic and hence unsuitable for mixing with food commodities.

II. Pyrethrins: Pyrethrins are safe for use as insecticide. However, the active principles are unstable and the insecticidal activity is lost in a short period.

III. Malathion : This is a 'safe' insecticide and many countries allow direct admixture of malathion in foodstuffs provided the residual malathion in the food does not exceed 8 ppm.

B. Fumigants

Fumigants are highly volatile chemical compounds which, when applied to foods, volatilize, penetrate and kill the adult insects, immature stages of insects and eggs. A large number of fumigants have been suggested by various workers, e.g., carbon disulphide, carbon tetrachloride, ethylene dichloride, methyl bromide, ethylene dibromide, phosphine, etc. Of these, only phosphine (in the form of aluminium phosphide), methyl bromide and a mixture of methyl bromide and ethylene dibromide are widely used.

I. Phosphine (aluminium phosphide): The tablet or pellet form of aluminium phosphide makes the application simple to grain stored in bags or in bulk. The fumigant is highly toxic to man. As it requires exposure of 3 days, precautions must be taken to avoid poisoning of persons handling this fumigant.

II. Methyl bromide: This has been widely used in many countries for the fumigation of food grains, legumes, oilseeds, etc. It is highly effective in destroying different stages of insects and eggs.

III. Methyl bromide and ethylene dibromide mixture : A mixture of methyl bromide and ethylene dibromide (Duro-fume process) has been developed at CFTRI, Mysore, for the fumigation of food grains and legumes.

This mixture is more suitable for use in tropical climate than methyl bromide.

The Botanicals for rice storage pests are given in detail in Chapter 18-section 18.5.2.3.

12.11.3 Insect proofing of bags

A process for making insect-proof gunny bags has been developed at CFTRI, Mysore by spraying a solution containing insecticides on the outside surface of the bags. Insects do not infest grains packed in such bags for a period of 6-8 months.

12.11.4 Rodent control

Methods used for minimizing the damage caused by rodents include proofing, repelling, trapping and poisoning.

A. Proofing **:** The storage bins and warehouses should be made rodents proof.

B. Repelling **:** Various types of chemicals have been used as rodent repellents, but some have given satisfactory results completely.

C. Trapping **:** Various types of traps have been in use for trapping the rodent, but it is only partly successful.

D. Poisoning **:** The use of poisons incorporated in baits is generally confined to control storage warehouses. When used correctly, poison baiting is one of the cheapest and most effective methods of controlling rats on a large area. The poisons used are arsenic oxide, zinc phosphide and sodium fluoracetate.

CHAPTER 13

RICE PRODUCTS AND BY PRODUCTS

Using rice, several products are prepared after fermentation and the by products of rice are used in many ways.

13.1 FERMENTED PRODUCTS

13.1.1 Idli

It is an important food for the south Indian people. Idli is a parboiled rice and black gram dhal based fermented and steamed food. It is easily digestible and used as food even for infants and invalids.

Parboiled rice and whole black gram after dehusking are soaked separately in proportion of 3:1. They are ground separately. Black gram has to be ground into a fine paste and rice has to be slightly coarsely ground. The ingredients are mixed together, salt is added. The mixture must be allowed to ferment overnight. Idlis are made by steaming in perforated containers.

Parboiled rice, which is relatively hard in texture is found to be more suitable than raw rice for making soft and spongy idlis. The specialty of black gram in idli preparation is owing to the mucilaginous material present in it. This mucilaginous principle helps in the retention of carbon dioxide evolved during fermentation. The protein surfactant forms the foam and the spongy network, which is stabilized by the viscous polysaccharide, thus preventing the collapse of the foam. The microorganisms present in the ingredients and those from the air bring about the fermentation. The bacteria recognized in idli fermentation are *Lactobacillus delbrueckii*, *L. lactis*, *Streptococcus lactis*,

S. faecalis, Leuconostoc mesenteroides and Pedicoccus cerevisiae. A number of wild yeasts are also implicated in fermentation. The acid and gas production are mostly related to the growth of bacteria. Microorganisms developed during the presoaking of black gram have a high autofermenting ability and also more potentiality to act on rice semolina. The optimal condition of fermentation is 30^{0}C for 16 hours. With the progress of fermentation, there is an increase in batter volume, acidity and nonprotein nitrogen. Fermentation increases the content of B vitamins.

13.1.2. Dosa

It is an another important food for the south Indian people. Parboiled rice and whole black gram are soaked together in the proportion of 4:1 to 6:1. Ingredients are ground fine and salt is added. It is allowed to ferment overnight and then dosas are made.

13.1.3. Aappam

To the rice flour, yeast, kanji or toddy is added to facilitate fermentation. It is allowed to ferment overnight. The batter is poured in a kadai and the kadai is tilted. (it is not handled like dosai). It is closed with the lid and with the steam that is evolved, it is cooked.

13.1.4 Kali

Ragi flour is soaked in water. Next day, it is cooked along with rice (broken). It is allowed to ferment overnight and on third day, it is consumed.

13.1.5. Dhokla

Generally, it is made with a combination of pulses like red gram dhal, green gram dhal, bengal gram dhal or Bengal gram flour. Another way is that rice flour and Bengal gram flour are taken in the proportion of 3:1. It is mixed with sour butter milk and allowed to ferment for six hours. It is steamed in a plate, cut into pieces and seasoned.

13.1.6.Rice Vadam

The left-over cooked rice is soaked in excess water and allowed to ferment overnight. Water is drained off. Rice is mashed and to it, is added finely cut onions, chillies, and lime. Made into small vadams and kept in the sun for dehydration.

Advantages of Fermentation

- It improves the flavour and texture of the product.

- Nutritive values like Vitamins B and C are increased
- Fermented products are easily digestible.
- It decreases the pH and inhibits the harmful microorganisms
- Certain harmful substances like trypsin inhibitors and phytins are decreased during fermentation

13.2 COOKED RICE

Well cooked rice is fluffy and dry with each rice separate and firmer but tender. Occasionally, rice with more moisture is preferred because it packs together easily and holds its shape. It is necessary to use just the right amount of water in cooking rice because when too little water is used, the cooked rice is likely to be hard, brittle and small in volume, but when too much water is used the rice sticks together in a pasty mass formed by the starch of the rice grains and the excess water. Usually two cups of water are added for a cup of rice though it may be variable. Pressure cooking is the best method of cooking rice as it retains maximum nutrients and gives proper cooking quality of the grain. Because it is a blonde food, rice can be prepared in many ways and with many flavours. The cooking quality of rice can be expressed in terms of consistency and swelling number.

$$\text{Swelling number} = \frac{\text{Volume of cooked rice}}{\text{Volume of uncooked rice}}$$

13.2.1 Ready to cook cereals

Ready mixes have been developed for a large number of traditional foods such as idli, dosa, idiappam, sevai etc. Macaroni foods also have become popular. Nowadays wet mixes, ready–to-cook idli and dosa batter are available in the market. Frozen samosas ready for frying are also available. Traditional foods like parched cereals are easy to cook. Dehydrated pulao, kichidi and upma and ready – to – cook chapattis are used in armed forces.

13.2.2 Ready –to-eat-cereals

Parched cereals like puffed rice and puffed wheat can be consumed without much cooking. Bread has become part of our daily life. Chapattis stored in cold storage are popular among Indians living in USA. Defense Food Research Laboratory, Mysore has developed a technology by which chapattis can be stored for more than six months.

13.2.3 Quick cooking rice

The rice is cooked to gelatinize the starch and subsequently dried. The processed rice has a porous structure that will permit rapid rehydration, usually within 5 minutes.

13.3 RICE PRODUCTS

13.3.1 Rice starch

Rice starch granules are quite small and are embedded in a protein matrix. To separate them from protein, broken rice is steeped for 24 hours in 5 times its weight of 0.3% caustic soda. The caustic soda treated granules are washed, dried and ground into flour. The flour is then mixed with about ten times its weight of caustic soda solution. This removes gluten. After 24 hours, the starch that settles down is removed, washed and dried. Rice starch is used in puddings, ice-creams and custard powder. It forms a tender opaque gel.

13.3.2 Rice bran

Breakage of the white rice kernel during milling also results in small fragments of the endosperm becoming part of the bran fraction. These broken fragments are primarily starch and normally contains 10-12% of bran. If the bran is subjected to a short term high temperature treatment after milling, the lipase activity is destroyed and stabilized bran of edible grade is produced. Bran produced from parboiled rice is already stable and does not require any further stabilization treatment. Stabilized or parboiled food grade rice bran is normally finely granulated light tan in colour and it has blonde flavour and can be used in preparations like bread, snacks, cookies, and biscuits. In addition, rice bran is a very rich source of dietary fibre so it is an effective stool bulking agent. The nutritive value of rice bran is given in Table 43.

Table 43. Nutritive value of rice bran per 100 g

Nutrient	Amount	Nutrient	Amount
Energy, kcal	393.0	Riboflavin, mg	0.48
Protein, g	13.5	Niacin, mg	29.8
Carbohydrate, g	48.4	Calcium, mg	67.0
Fat, g	16.2	Iron, mg	35.0
Thiamine, mg	2.7	Fibre, g	4.3

13.3.3 Rice bran oil

Central Food Technological Research Institute, Mysore, has developed low cost chemical methods to extract edible grade oil from rice bran. National Institute of Nutrition, Hyderabad has certified that this oil is toxicologically safe for human consumption. This oil is rich in Vitamin E, which gives oxidative stability to the oil. In addition, it has higher cholesterol lowering effect than other oils. The keeping quality of this oil is high when compared to other oils. Foods deep fried in rice bran oil showed that this oil is less absorbed or consumed in fried foods when compared to food fried in groundnut oil. This oil is available in the market, in the name of "Harvest".

13.3.4. Parched- processed rice products

About 4-5% of total supplies of rice in India are converted into rice products – parched rice, parched paddy and rice flakes.

A. Parched rice

Parboiled rice is used in making parched rice. Parboiled rice is soaked in salt water till its moisture content increases to about 20 per cent. Parched rice is prepared by throwing rice in sand heated to a high temperature in iron or earthen pan. On stirring, rice begins to crackle and swell. Then the contents of the pan are removed and sieved to separate the parched rice from sand. Parched rice is a crisp product with a grayish to brilliant white colour and is sold either salted or unsalted. It is eaten as such or mixed with buttermilk and then consumed.

B. Parched paddy or puffed rice

Sun ripe paddy is filled in earthen jars and is moistened with hot water. After 2-3 minutes, the water is decanted and the jars are kept in an inverted position for 8-10 hours. The paddy is exposed to sun for a short time and then parched at 190-210°C for 40-45 sec. During parching, the grains swell and burst into soft white product. The parched grains are sieved to remove sand and winnowed to separate the husk. Puffed rice prepared in the above manner is made into balls with jaggery and used as a snack.

Advantages of parched rice products

- Easily digestible and hence good for children and old people.
- Readily available to eat due to faster cooking.
- It adds variety in the diet.
- Since iron pans are used, iron content is added.

- Improves flavour and texture.

C. Flaked rice (Beaten rice)

Flaked rice is made from parboiled rice. Paddy is soaked in water for 2-3 days to soften the kernel followed by boiling in water for a few minutes and the water is drained off. The paddy is heated in a shallow earthen vessel or iron pan till the husks break open. It is pounded by an iron pestle or using iron roller, which flattens the kernel and removes the husk. The husk is separated by winnowing. Flaked rice is thin and papery and has white colour. Rice flake is a precooked product and can be readily reconstituted by soaking in warm water for 20 minutes. It is widely used in India for the preparation of snacks.

13.3.5 Aromatic rice

Aromatic rice varieties contain acetyl 1-1 pyrroline that is the major principle for aroma while cooking. These are characteristic of the rice variety. Non-aromatic rice varieties can be flavoured by adding one or two fresh leaves of *Pandanus latifolius* while cooking as they contain 2-acetyl 1-1 pyrrole. Further details about aromatic/non aromatic basmati/scented rice are given in the subsequent chapters.

CHAPTER 14

SOIL SAMPLING AND SOIL TESTING

Most of the agricultural soils in their virgin state had accumulation of 'N' from trees and grass residues. 'P' is limiting because of the parent soil material, which is low in this nutrient. Hence for increasing production, 'P' was supplied through organics and through mineral sources. As yields increased from use of 'P' fertilizers in legumes and later 'N' fertilizers, soils were depleted of other nutrients and a need to potassium (K) and some of the macro and micronutrients was discovered. Now, for economic production of most crops in developed countries depend on the use of fertilizers. Similar is the case in developing countries with the advent of new high yielding varieties. As yields increased and soil fertility levels were depleted by crop removal, erosion and low fertilization, general fertilizer recommendations based on soil series were no longer satisfactory for many farm situations. Then came the chemical soil testing. Soil testing should answer the following two questions.

1. Is the nutrient supply in the soil is adequate for the crop to be grown?
2. If not, how much should be applied for this crop or for the rotation to be followed?

In soil testing, it is the 'available' portion of nutrients to be determined. For this, the amount of nutrients that is directly proportional to that taken up by growing plants in a growing season is considered. The soil samples should be collected from the field for soil testing and each sample collected must represent the area being sampled. Utility of the results obtained from the laboratory analysis depends very much on the sampling precision. Hence, collection of large number of samples is advisable so that sample of desired size can be obtained by sub sampling.

14.1 SOIL SAMPLING

Collection of soil samples is one of the most important steps in a soil testing program. Because of the inherent variation in soils, along with variations resulting from management, one must recognize that even with the greatest care at this step undoubtedly, there will be possibility for a larger error from sampling than from the laboratory analyses. Hence the objective of sampling is to select the area and collect a sample that will most adequately represent the average fertility of the sampled area. Fertility variation will be more in soils with a history of fertilization than that on unfertilized soils. In unfertilized soils, inherent differences among soil types results from differences in weathering products from parental materials. In case of fertilized soils, the changes in soil fertility are directly related to fertilizer applied, method of application (broadcasting-uniform changes in fertility, Band placement-broad differences in fertility) and cropping systems. Briggs during 1974, found that from 150 soils samples collected from 105 m^2 area, the N content varied form 9 to 87 kg/ha, 'P' content from 24 to 74 kg/ha and 'K' content from 610 to 1099 kg/ha. For practical purposes, the objective of soil testing is to estimate the average fertility level of a field. These are two methods by which this could be done

(i) By taking the average results of analysis of the core soil samples taken

(ii) Prepare a composite soil samples by mixing all the core samples taken and then do the analysis.

To understand the complete picture on the variation within a field, individual cores have to be sampled. For practical and economic reasons it may be just as satisfactory to take one composite soil sample of mixed core samples for one analysis.

14.1.1 Guidelines for soil sampling

1. The sample should represent only one condition. Using a soil survey map, fields should be divided into sampling units based on soil series. If there is more than one soil in the fields, it should be subdivided and a separate sample collected for each soil.

2. Avoid unusual area. Take the composite sample from the most representative areas of a field. Avoid eroded areas, filled in areas, edges of field, turn around areas where fertilizers are spread more heavily and fertilizers spill areas.
3. Sample row cropland in middle, away from rows where fertilizer might have been placed in band.
4. Late summer or fall following summer crops is a good time to collect samples because the general variations in crop growth can be seen and may indicate areas within the field that need a special sample.
5. Subdivide the field into sampling units based on cropping or fertilization histories. Each area that has been cropped or fertilized differently from other areas should be sampled as a separate unit.
6. Subdivide large fields into smaller units, possibly upto 5 hectares, and take a sample from each. If the field is uniform, then larger areas can be included in the composite sample in future years.
7. The depth of sampling should be to the effective root zone. Sampling depth for paddy crop should be 15 cm.
8. Annual sampling at least for a few years will show some variability but should give a good indication of the general fertility level, when several years are averaged. Also this should show trends in fertility changes in the field. After a few years of sampling, annually the general condition of the field should be established and sampling every two or three years would be satisfactory.

14.1.2. Methods of soil sampling

Different methods are employed in soil sampling. The most common methods are use of scope to remove the soil from the required depth for sampling. For the analysis of total nutrients, soil sampling should be done before ploughing. For the analysis of available nutrients, wet soil samples should be collected before puddling. The other methods of soil sampling are as follows.

14.1.2.1. Rectangular core sampler

In this, a pre-fabricated steel rectangular container of the dimension 30 cm x 15 cm x 40 cm of length, breadth and height respectively is used. This is driven into the soil using an iron hammer of 5 kg to a depth of 30 cm. This is then pulled out using a iron rod by two persons standing on either side. The container after removal is shade dried and opened to get soil samples from different layers (0-30 cm). The soil from different layers can be cut and dried, and used for analysis.

14.1 2.2 Core Auger

This could be used in upland paddy soils for collecting soil samples upto 30 cm depth for analysis.

Materials required - Mammuty or spade or auger, Polythene or cloth bags, A pan or bucket, Khurpi (for collecting profile samples)

Procedure

Each field is to be treated as a separate sampling unit, fields with varying soil types, slope and those, which were under different past management practices such as fertilization, liming and cropping pattern needs separate sampling units. It is advisable to sample after the harvest of the crop and if sampling is to be done in standing crop, collection should be done between the lines. Collection of sample in a Zig – zag pattern across the field ensures homogeneity. During sampling, dead furrows, wet spot areas near main bunds and trees, manure heaps and irrigation channels need to be avoided. The surface litter is to be removed at the sampling spot. Then the auger is driven to plough depth of 15 cm and soil sample drawn. At least 10 to 15 samples are to be collected from each sampling unit and placed in a bucket or container. If auger is not available, a 'V' shaped cut to a depth of 15 cm is to be made at each sampling spot. About 1-5 cm thick slices of soil from top to bottom of exposed face of the 'v' shaped cut is to be removed and placed in a clean container. The samples thus collected are to be thoroughly mixed and foreign materials like roots, stones, pebbles and gravels are to be removed. Then, the bulk is to be reduced to about half to one kilogram either by quartering or compartmentalization. Quartering is done by dividing the thoroughly mixed sample into four equal parts. The two opposite quarters are discarded and the remaining two quarters are remixed and the process repeated until the desired sample size is obtained. In the case of compartmentalization, the thoroughly mixed soils is uniformly spread over a clean hard surface and divided into smaller compartments by drawing lines along and across the length and breadth with fingers or small twigs. Then from each compartment a pinch of soil is obtained. The sample so collected is put into a clean numbered cloth / Polythene bags.

14.1.2.3 PVC tube sampling method

In this, PVC pipes of 20 cm in length and 2 inch diameter is cut and driven into the soil using wooden mallet. This pipe along with the soil is then removed by gentle slope and the soil sample is taken out from the pipe using a flat end stick for further analysis.

14.1.2.4.Other methods

In recent times, the researchers have found that for paddy crop under submergence, drying of soil sample will vitiate the available nutrient status of the soil. Hence, they go in for wet soil analysis or by extracting the wet soil using 2 molar KCl and this extract could be analyzed. The scientists at IRRI have developed a technology for estimating the available nutrients *in situ* when there is crop in the field using Piezometers.

PIEZOMETERS: In this, a PVC tube of 1-inch dia and 75 cm long is taken. One end of the pipe is filled with a ceramic cup of 10cm length. To the other end, it has a rubber stuff with two holes through which two glass rods are inserted. The instrument is placed at the root zone of the crop (up to 30cm). The soil solution at the root zone gets collected in the ceramic cup. This is then pumped out into a flask using a suction pump. This soil solution is analyzed to EC, pH and available nutrients in paddy field under submerged condition.

14.1.3 Processing and Storing

The sample collected from the field is dried in shade by spreading on a clean sheet of paper after breaking the large lumps if present with a wooden mallet. Then the soil material is sieved through a 2 mm sieve. The material passing through the sieve is collected and stored in a clean container with proper labeling for laboratory analysis. For the determination of organic matter it is desirable to grind a representative sub sample and sieve it through 0.2 mm sieve.

If the samples are meant for the analysis of micronutrients viz., copper and zinc, great care is needed in handling the sample. Brass sieves should be avoided and aluminium or stainless sieves are to be used. It is better to use stainless steal or Polythene materials. Air drying of soil must be avoided if the samples are to be analyzed for nitrate and ammonia contents and bacterial count.

14.2. SOIL TESTING

14.2.1 Methods of soil analysis

14.2.1.1 Soil Reaction

The pH is a measure of the hydrogen or hydroxyl ion activity of the soil - water system. It indicates whether the soil is acidic, neutral or sodic (alkaline) in reaction. Since the crop growth suffers much both under very low pH (strongly acidic) as well as very higher pH (sodic) conditions, appropriate reclamation measures becomes essential.

Different methods for pH measurement

1. Saturated paste method
2. Soil: Water suspension method
3. Soil:$CaCl_2$ suspension method
4. Universal indicator method
5. Using pH indicator papers

1. Saturated paste method

This method is more precise, accurate and universally accepted method. This method is used for identifying specific soil problems like acidity or alkalinity and salinity.

2. Soil: Water suspension method

This method is used for making fertilizer recommendations; besides it is quite rapid.

3. Soil:$CaCl_2$ suspension method

This method is used to make the variability in salt content of soil. The results are independent of dilution over a wide range of soil: solution ratio and more reproducible than those obtained in water universal indicator, pH indicator paper are much more rapid, though not accurate.

A. Regents

Commercially available buffer tablet of pH 4.0, 7.0 and 9.2

or

Buffer solution pH 4.0 : Mix 5.0 ml of 0.2 M potassium acid pithalate and 0.4 ml of 0.2m M Sodium hydroxide and dilute to 200 ml

Buffer solution pH 7.0 : Mix 50 ml of 0.2 M Monopotassium phosphate and 29.54 ml of 0.2 M Sodium hydroxide and dilute to 200 ml.

Buffer solution pH 9.2 : Prepare 100 ml 0.05 M Borax solution.

B. Instrument

PH meter: pH meter is known as potentiometer or multimeter

The determination of pH amounts to the determination of the potential developed when a hydrogen electrode is dipped in a solution of hydrogen ions. The determination of potential of a single electrode (half cell) is not practicable and therefore, it is customary to couple such a half cell with another half cell of constant and accurately known electro motive force so that the total potential difference of the two cells when suitably connected can easily be measured by means of a potentiometer. That electrode called as reference electrode. The commonly used reference electrode is a calomel electrode.

Rating of Soil with reference to pH	pH
Acidic	< 6.5
Neutral	6.5-7.5
Alkaline	>7.5

14.2.1.2 Electrical conductivity

A. Principle

The amount of soluble salts in a solution increases, the electrical conductivity also increases. This electrical conductivity is measured in forms of the resistance offered to the flow of current using a conductivity bridge.

B. Reagents

0.01N Potassium chloride solution (or) saturated solution of Calcium sulphate.

C. Instrument

Conductivity meter or Solu-bridge

The conductivity determination is based on "Ohm's law" which states that for the same type of material, the resistance is directly proportional to the length of the conductor and inversely to the cross section or diameter.

On conductivity measurement, it is aimed to measure the intensity of current flowing between two electrodes of known and constant surface area through the electrolyte solution, keeping the distance between the electrodes constant.

Total soluble salts in soil and soil solution play a major role in the uptake of nutrients and water by the plants. In low soluble salts concentration, the osmosis of the plant is normal but in highly saline soils, the revere osmosis or plasmolysis leads to drying of plants. This EC measurement is particularly important in the case of crops sensitive to salt and can help the farmers in certain problematic soil areas in deciding which crops could be grown successfully or what reclamatary measures may be required to increase the yield.

Rating of Soil with reference to EC	EC
Harmless	0-1.0
Injurious	1.0-3.0
Critical	>3.0

14.2.1.3. Estimation of avaliable Nitrogen in Soil (by Alkaline Permanganate Mcthod)

A. Principle

A known weight of soil is mixed with excess of alkaline permanganate and distilled. Organic matter present in soil is oxidized by the oxygen liberated by $KMnO_4$ in the presence of NaOH and thus ammonia is released. This released ammonia is absorbed in a known volume of boric acid (2%) containing double indicator and converted to ammonium borate. This ammonium borate is titrated against standard H_2SO_4

$2KMnO_4$	-	$NaOH + K_20 + 2MnO_2$
Organic matter	-	Release of R NH_2
$RNH_2 \quad + H_2O$	-	$ROH \quad + \quad NH_3$

B. Reagents

1. 0.32% $KMnO_4$ Solution (3.2 g of $KMnO_4$ dissolved in litre of distilled water)
2. 2.5% NaOH solution (25 g of NaOH dissolved in 1 litre of distilled water)
3. 2% boric acid (20 g of boric acid dissolved in 1 litre of distilled water)
4. N/50 H_2SO_4 (3 ml of conc. H_2SO_4 is diluted to 1 litre with distilled water and standardized by titration with N/10 Na_2CO_3. This gives N/10 H_2SO_4(from this N/50 H_2SO_4 is prepared by dilution).

5. Double indicator (Bromocresol green (0.5g) and methyl red (0.1g) dissolved in 100 ml ethyl alcohol)

C. Procedure

- Weigh and transfer 20 g of soil into a distillation flask.
- Add 30 ml of distilled water just to moist the soil and 1 ml of liquid paraffin or 1 g of paraffin wax (to avoid frothing)
- Add few pieces of glass heads also to avoid frothing and pumping. Add 100 ml of freshly prepared 0.32% $KMnO_4$ and 100 ml of 2.5% NaoH to the soil in the distillation flask.
- Place 100 ml beaker containing approximately 20 ml of 2% boric acid with double indicator below the delivery end of the condenser in the distillation set.
- Distill the contents at a steady rate and collect the liberated ammonia in boric acid.
- Continue the distillation until the release of ammonia from distillation take place or about 30ml of distillate is collected (bringing moist red litmus paper near the outlet of condenser which turn blue as long as ammonia is being evolved.)
- Titrate the ammonia collected in boric acid with N/50 H_2SO_4.

D. Calculation

Weight of the soil taken = 20 g

Volume of the N/50 H_2SO_4 consumed = X ml (titre value)

1 ml of N/ 10 H_2SO_4 = 0.0014 g N

Therefore 1 ml of N/50 H_2SO_4 = 0.00028 g N

X ml of N/50 H_2SO_4 = 0.00028 X 'X' g N

This is present in 20 g of soil (i.e. 0.0028 X 'X' g)

Therefore, available N present in kg/ha = 0.00028('X'/20)2 X 10^6

Rating of Soil with reference Organic matter content (OMC)	OMC
Low	<0.5 %
Medium	0.5-0.75 %
High	>0.75

Rating of Soil with reference to available N	Available N (Kg)
Low	Up to 280
Medium	280-450
High	>450

14.2.1.4 Estimation of available Phosphorus (Olsen's method for neural, alkaline and calcareous soils)

A. Principles

The carbonate (Co_3^{2-}) ions from sodium bicarbonate ($NaHCo_3$) will react with Calcium (Ca^{2+}) and Calcium carbonate (Ca Co_3) is precipitated, thus allowing the P to come into solution. The amount of P extracted is determined colorimetrically.

B. Reagents

1. 0.5 M $NaHCO_3$ (pH 8.5) (42 g of $NaHCO_3$ dissolved in 1 litre of water)
2. Darco G.60 or activated carbon (free to phosphorus)
3. Ascorbic acid

Reagent A

A. Dissolve 12 g of ammonium molybdate in 250 ml of distilled water.

B. Dissolve 0.291 g of antimony potassium tartrate in 100 ml of water.

C. Prepare 100 ml of 5 N H_2SO_4 (138 ml conc. H_2SO_4 made up to one litre) Mix the above three reagents well and make up the volume to two litre.

Reagent B

Dissolve 1.056 g of Ascorbic acid in 200 ml of Reagent A.

C. Procedure

- Weigh accurately 5 g of soil and transfer to a polythene-shaking bottle
- Add 50 ml of 0.5 M $NaHCO_3$ (pH 8.5) and a pinch of Darco G.60 (to make the extractant colorless)
- Shake the contents in a reciprocating mechanical shaker for 30 min.
- Filter the contents through Whatman No.40 filter paper (the filtrate should be clean and colourless)
- Pipette out 5 ml of the aliquot into 25 ml volumetric flask and add 4 ml of reagent B.

- Make up the volume to 25 ml with distilled water
- Shake the contents and allow for few minutes for colour development
- Measure the intensity of blue colour in photoelectric colorimeter using a red filter at 600 nm
- Using the standard curve, find out the concentration of P (ppm) and calculate the available P content.

D. Calculation

Weight of soil taken = 5 g

Volume of $NaHCO_3$ used for extraction = 50 ml

Volume of extracted solution used for P estimation (aliquot) = 5 ml

Calorimeter reading = T

Concentration of P read from standard graph for the reading = 'x' ppm

= 'x' microgram/ml

= 'x' / 10^6 g/ml

Therefore, in 25 ml of solution = 'x' / 10^6 x 25 g

This is present in 50 ml of the extract solution and 5 g of soil

Therefore, available P_2O_5 in kg/ha = $\frac{\text{'x' } 10^6 \times 25 \times 50 \times 2}{5 \times 5 \times 10^6}$

Rating of Soil with reference to available P_2O_5	Available P_2O_5 (Kg)
Low	<11
Medium	11-22
High	>22

14.2.1.5 Estimation of available Potassium (Neutral normal ammonium acetate method using Flame photometer)

A. Principle

The K ions in the exchange sites are replaced with NH_4^+ and K^+ is released. The concentration of K ions in the solution is then determined using flame photometer.

B. Reagents

1 N ammonium acetate (Neutral in pH): Dissolve 77 g of AR grade ammonium acetate (CH_3 $COONH_4$) in 1000 ml of distilled water. Adjust the pH to 7.0 with ammonia solution or acetic acid.

C. Procedure

Transfer 5g of soil into a polythene-shaking bottle. Add 25 ml of 1 N ammonium acetate extractant and shake the contents in a mechanical reciprocating shaker for 5 minutes. Filter the contents through whatman No.40 paper. Feed the filtrate in to the flame photometer and record the reading. Calculate the available K content using a standard curve.

D. Calculation

Weight of soil taken = 5g

Volume of extractant used = 25 ml

Flame photometer reading = T

Concentration of K in the standard Curve = X ppm

= X micro gram / ml

= $X / 10^6$ g/ml

Therefore, in 25 ml of solution = $X / 10^6 \times 25$ g

This is present in 5 g of soil

Therefore, available K in soil in kg/ha = $\frac{\text{“x”} \times 25 \times 2 \times 10^6}{5 \times 10^6}$

$$\frac{\text{“x”} \times 25 \times 2 \times 10^6}{\frac{5 \times 10^6}{5 \text{x} 5 \times 10^6}}$$

Rating of Soil with reference to available K	Available K (Kg)
Low	<118
Medium	118-280
High	>280

14.2.1.6 Available micronutrients

Extremely small quantities of micronutrients needed by the plants but, now a days, deficiency of micro nutrients in crops like rice at an alarming stage noticed due to the use of high analysis inorganic fertilizers, little or no use of organic manures and heavy uptake of micronutrients by high yielding cultivars.

Number of methods, mostly colorimetric has been widely used for majority of the micronutrients. Neutral normal ammonium acetate and chelating agents like EDTA, have been used as extractants for Fe, Zn and Mn, and the extracted amount determined colorimetrically. Zinc determination by dithiozonate method has been very popular. Also alternate method as described by Jackson (1962) is still employed. Available soil 'Boron is generally extracted by hot water and molybdenum by ammonium oxalate methods. The Boron in soil extracts is determined by Azomethine-H and molybdenum by Dithiol method. However, after Atomic Absorption Spectrophotometer discovery, for a much and accurate analysis of available Zinc (Zn), Copper (Cu), Iron (Fe) and Manganese (Mn), the DTPA – method (Lindsay and Norvell, 1978) as given below is most widely used.

14.2.1.6.1. Determination of Zn, Cu, Fe and Mn using DTPA method

Principle

DTPA, a chelating agent, combines with feee metal ions in solution and forms soluble complexes. Due to the reduced ionic activity in solution desorption takes place, bringing in some more ions from solid phase. DTPA offers the most favourable combination of stability constants for the simultaneous complexing of Zn, Cu, Fe and Mn. Since Fe and Zn deficiency is frequently experienced in calcareous soils, the method is designed to avoid excessive dissolution of $CaCO_3$ with the release of occluded mincronutrients which are normally not available to plants. This is achieved by (i) the inclusion of soluble Ca^{++}, and (ii) buffering the reagent at pH 7.3 with TEA (which burns cleanly during flame atomization). When the extractant is added to soil, additional Ca^{++} and some Mg^{++} enter the solution. This is largely because the protonated TEA exchanges with these ions from the exchange sites and this leads to the increased ionic concentration of Ca^{++} in the solution, which in turn helps in suppressing the dissolution of $CaCO_3$, DTPA extractant has the ability to chelate Zn, Cu, Fe and Mn in competition which Ca^{++} and Mg^{++} and unlike most other chelating agents, it applies a moderate stress to solubilize soil Fe at a pH where $CaCO_3$ is not dissolved. Suitability of this method has been proved through excellent relationships between the test values and plant utilisable nutrients under pot and field studies concludd the world over.

I. Determination of available zinc

Instruments

(i) Atomic absorption spectrophotometer (AAS)
(ii) Mechanical shaker

Reagents

- **Dilute HCI :** AR grade HCI diluted 5 times with double distilled water.
- **DTPA extractant :** Dissolve 1.967 g of AR grade diethyline-triamine-penta acetic acid (DTPA) and 1.470 g of $CaCl_2$. $2H_2O$ (AR grade) in about 25 mL of double distilled water (DDW) by adding 13.3 mL of triethanolamine (TEA), followed by 100 mL more of DDW. Transfer the solution to one litre volumetric flask giving 4 to 5 washings. Just before making up the volume, adjust pH to 7.3 with dilute HCI. This reagent has 0.005 M DTPA, 0.1 M TEA and 0.01 M $CaCl_2.2H_2O$.
- **Standard stock solution `A' (1000 mg Zn L^{-1}) :** Weigh exactly 1.0 g of pure Zn metal (AR grade) and dissolve it in minimum volume (about 10 mL) of dil. HCI (1:1) and make the volume to one litre.
- **Standard solution `B' :** Dilute 5 mL of solution A to 100 mL to get solution B containing 50 mg Zn L^{-1}.
- **Standard working solutions :** Dilute 0.5, 1.0, 1.5, 2.0, 2.5 and 5.0 mL portions of solution B to 50 mL to get working standards containing 0.5, 1.0, 1.5, 2.0, 2.5 and 5.0 mg Zn L^{-1}. The working standards should be prepared in the medium of the extracting solution after every few days as it cannot be preserved for long.

Procedure

1. Weigh 10 g of soil sample in 100 mL conical flask.
2. Add 20 mL of the DTPA extractant and shake for 2 h on a mechanical shaker.
3. Filter through Whatman No. 42 filter paper, discarding first few drops. For quick filtration, Whatman No. 1 filter paper can also be used if the filtrate is clear.
4. Use the filtrate for Zn measurement using AAS.
5. Freed the standard working solutions and prepare a standard curve by plotting AAS readings against Zn concentrations.

Calculation

$$\text{Available (DTPA-extractable) Zn in soil (mg kg}^{-1}) = \frac{A \times 20}{10} = A \times 2$$

Where A stands for the Zn concentration in aliquot as read from X-axis of standard curve against the sample reading.

II. Determination of available copper

Available copper can be determined in the DTPA extract similar to Zn, using AAS. For this, the standard stock solution can be prepared as given below;

Accurately weight 1.0 g AR grade copper metal wire or turning and dissolve it in 50 mL of diluted HNO_3, (1:1 with DDW) and finally make the volume to one litre. This is solution A containing 1000 mg Cu L^{-1}. Prepare solution B containing 50 mg Cu L^{-1} by diluting appropriate volume of solution A. Finally prepare working standards containing 0.25, 0.50, 1.0, 1.5, 2.0 and 2.5 mg Cu L^{-1}.

III. Determination of available iron

Iron in the DTPA extract can also be determined with the help of AAS exactly in the same manner as Zn and Cu described above. However, the working standard solutions of Fe should be prepared for higher concentrations as the DTPA-extractable Fe content of soil is generally more than both Zn and Cu. Thus, the Fe standards may be prepared as given below:

Prepare standard stock solution (solution A) by dissolving exact 1.0 g of AR grade Fe mental in about 50 mL of 1:1 diluted HNO_3 and dilute the contents to one litre with DDW. Prepare solution B by diluting 50 mL of solution A to 500 mL to get 100 mg Fe L^{-1}. Finally prepare working standard solutions containing 1.0, 2.0, 3.0, 5.0 and 10.0 mg Fe L^{-1} by diluting appropriate volumes of solution B with the medium of extraction (DTPA).

IV. Determination of available manganese

DTPA extractable Mn is also determined following the same technique as adopted for Zn, Cu and Fe. For this prepare the standard solutions as follows :

Weigh 1.583 g of AR grade MnO_2 or 1.0 g of pure Mn metal and dissolve it in 50 mL of dillute HNO_3 (AR grade). Make the volume to 1 litre with DDW to get solution A having a Mn concentration of 1000 mg L^{-1}. From solution A, dilute 25 mL to 250 mL with DDW to get solution B having 100 mg Mn L^{-1}.

Finally prepare working solutions of 0.5, 1.0, 2.0, 2.5 and 5.0 mg Mn L^{-1} concentration by diluting 0.5, 1.0, 2.0, 2.5 and 5.0 mL portions of solution B to 100 mL.

14.2.1.6.2 Determination of available boron (Azomethine- method)

Principle

Boron in soil exists in organic and inorganic frame. The compounds of boron, which have high availability, are water soluble. Boron dificiency is encountered in highly calcareous and acid soils, while toxicity in salt-affected soils as well as soils irrigated with high B water. Soil extraction with cold or hot water continues to be the most acceptable method for determining the available B in soils. In soil extract, B can be determined by azomethine-H method (John *et al.,* 1975). The method employs azomethine-H as the reagent to form a stable coloured complex of H_3BO_3 at pH 5.1 in aqueous media, which retains proportional absorbance-concentration properties for several hours independent of the presence of a wide variety of salts. The maximum absorbance occurs at 420 nm. This technique is rapid, reliable and more convenient to use than traditional procedures employing carmine, curcumin or quinalizarin.

B. Instruments

(i) Refrigerator

(ii) Colorimeter or spectrophotometer

(iii) Polypropylene test tubes of 10 mL or 15 mL capacity

C. Reagents

- **Azomethine-H reagent :** Dissolve 0.45 g of azomethine-H in 100 mL of 1% L-ascorbic acid solution in polypropylene reagent bottle. Fresh reagent should be prepared weekly and stored in refrigerator.
- **Buffer solution** : Dissolve 250 g of ammonium acetate (NH_4 OAC) and 15 g of ethylene diamine tetra acetic acid (EDTA disodium) in 400 mL of distilled water. Slowly add 125 mL of glacial acetic acid and mix.
- **0.01M $CaCl_2$ solution :** Dissolve 1.47 g of $CaCl_2.2H_2O$ in distilled water and adjust the volume to 1000 mL.
- **Boron standard solution :** Dissolve 0.114 g of boric acid (H_3BO_3) in distilled water and adjust the volume to 1000 mL. Each mL of this

solution contains 20 mg of B. Dilute 0.5, 1, 2, 3, 4, 5, 10, 20, 30, 40 and 50 mL of stock solution to 100 mL with distilled water to have solution with concentrations of 0.1, 0.2, 0.4, 0.6, 0.8, 1.0, 2.0, 4.0, 6.0, 8.0 and 10 mg of B L^{-1}, respectively.

D. Procedure

1. Weigh 20 g of air-dry and 20 mesh sieved soil sample in a 250 mL. B-free conical flask.
2. Add 40 mL of 0.01 M $CaCl_2$ solution.
3. Attach a water-cooled reflux condensor.
4. Heat the flask until first sign of boiling, and then reflux the contents for exactly 5 minutes.
5. Allow to cool without removing the condensor.
6. Transfer a 1 ml aliquot of blank, diluted B standard or sample solution into 10 or 15 mL polypropylene tube.
7. Add 2 mL of buffer and mix.
8. Add 2 mL of azomethine-H reagent, mix and after 30 minutes read the absorbance at 420 nm.
9. Refer these readings to that of standard curve prepared with 0, 0.1, 0.2, 0.4, 0.6, 0.8, 1.0, 2.0, 4.0, 6.0, 8.0 and 10.0 mg of B L^{-1} solutions for converting readings to B concentrations in soil extracts.

14.2.1.6.3 Determination of available molybdenum (Dithiol Method)

A. Principle

Amongst the essential nutrients for plants, Mo is required in the smallest quantity. It is used by plants in the form of MoO_4 ions and its availability, unlike other elements, is high in the alkaline pH range. As it is required in very small amount, tests for availability of Mo are rarely carried out, particularly in alkaline soils. Ammonium oxalate method is, however, the most popular for extraction of available Mo from soil. The extractant is advantageous as it is easy to prepare, does not allow a significant change in pH during shaking due to buffering action and molybdenum forms stable complex with oxalic acid. The extraction is based on the replacement of MoO_4 ions from exchange sites, which is irreversible due to formation of strong Mo-oxalic acid complexes.

B. Instruments

(i) Colorimeter or spectrophotometer

(ii) Hot plate

(iii) Refrigerator

(iv) Water bath

C. Reagents

- **50% Potassium iodide solution:** Dissolve 50 g in 100 mL of doulbe distilled water (DDW).
- **50% Ascorbic acid solution :** Dissolve 50 g in 100 mL of DDW.
- **10% Sodium hydroxide solution** : Dissolve 10 g of NaOH in 100 mL of DDW.
- **10% Thiourea solution :** Dissolve 10 g in 100 ml of DDW and filter. Prepare fresh before use.
- **Toluene-3, 4-dithiol solution (commonly called dithiol) :** Weigh 1.0 g of AR grade melted dithiol (51°C) in a 250 mL glasss beaker. Add 100 mL of the 100% NaOH solution and warm the contents upto 51°C with frequent stirring for 15 minutes. Add 1.8 mL of thioglycolic acid and store in a refrigerator.
- **10% Tartaric acid :** Dissolve 10 g in 100 mL of DDW.
- **Iso-amyl acetate**
- **Ethyl alcohol**
- **Ferrous ammonium sulphate solution :** Dissolve 63 g of the salt in about 500 mL of DDW and then make the volume to one litre.
- **Extracting reagent :** Dissolve 24.9 g of AR grade ammonium oxalate and 12.6 g oxalic acid in water and make the volume to one litre. Adjust the pH at 3.3 with dil. HCl/dil. ammonia solution.
- **Standard stock solution (100 mg Mo L^{-1}) :** Dissolve 0.150 g of AR grade MoO_3 in 100 mL of 0.1 N NaOH, make slightly acidic with dil. HCl and make the volume to 1 L.
- **Working standard solution (1 mg Mo L^{-1}) :** Dilute 10 mL of the stock solution to 1 L.

D. Procedure

a. Soil extraction

1. Weigh 25 g of air-dry soil sample in a 500 mL content flask. Add 250 ml of the extracting solution (1:10 ratio) and shake for 10 h.
2. Filter through Whatman No. 50 filter paper. Collect 200 mL of the clear filtrate in a 250 mL glass beaker and evaporate to dryness on a water bath.
3. Heat the contents in the beaker at 500°C in a furnace for 5 h to destroy organic matter and oxalates. Keep overnight.
4. Digest the contents with 5 mL of HNO_3-$HClO_4$ mixture (4:1), then with 10 mL of 4N H_2SO_4 and H_2O_2, each time bringing to dryness.
5. Add 10 mL of 0.1N HCl and filter. Wash the filter paper with separate 10 mL of 0.1N HCl and DDW until the volume of the filtrate is 100 mL.
6. Run a blank side by side (without soil)

b. Colour development

7. Take 55 mL of the filtrate in 250 mL separately funnels and add 0.25 mL of ferrous ammonium sulphate solution and 20 mL of DDW and shake vigorously.
8. Add excess of KI solution and clear the liberated iodine by adding ascorbic acid drop by drop while shaking vigorously.
9. Add one mL of tartaric acid and 2 mL of thiourea solution and shake vigorously.
10. Add 5 drops of dithiol solution and allow the mixture to stand for 30 minutes.
11. Add 10 mL of iso-amyl acetate and separate out the contents (green colour) in colorimeter tubes/cuvettes.
12. Read the colour intensity at 680 nm (red filter).

C. Preparation of standard curve for Mo

1. Measure 0, 2, 5, 10, 15 and 20 mL of the working standard Mo solution containing 1 mg Mo L^{-1} in a series of 250 mL separatory funnels.
2. Proceed for colour development as described above for sample aliquots.

3. Read the colour intensity and prepare the standard curve by plotting Mo concentration against readings.

d. Calculation

$$\text{Available Mo in soil (mg g}^{-1}\text{)} = A \times \frac{250}{200} \times \frac{1}{25} = \frac{A}{20}$$

Where, A stands for Mo concentration in mg as obtained on X-axis against a sample reading.

14.2.1.6.4. Alternate method for determining, available molybdenum (Ammonium oxalate method)

Mo can also be determined by using graphite furnace based atomic absorption spectrophotometer. For low Mo concentration in soil extract (< 0.5 mg g^{-1} soil), the method is specially useful. Anion-resin method is also used for this purpose but the ammonium bicarbonate – DTPA method suggested by Soltanpour *et al.* (1982) and described below appears to be the better alternative (particularly for alkaline soils) if ICP-AES is available.

A. Principle

The NH_4HCO_3 reagent used in the method not only acts as an extractant but also as a buffer. During the extraction, the pH of soil-reagent mixture increases to 8.5, dissolving the Ca-, A1- and Fe molybdates (Soltanpour, 1991). The HCO_3 ions also help in displacing MoO_4 ions from the colloidal surface. The extracted Mo can be measured by using ICPS.

B. Instruments

(i) Mechanical shaker

(ii) Inductivity coupled plasma spectrophotometer

C. Reagents

- **AB-DTPA extracting solution** : Add 1.97 g of AR grade DTPA to 800 mL of DDW. Add 2 mL of 1:1 diluted NH_4 OH to aid dissolution and prevent effervescence. When most of the DTPA is dissolved, add 79.06 g of NH_4HCO_3 and stir gently to dissolve. Adjust the pH to 7.6 with dil. HCI or dil. NH_4OH. Make the volume to one litre. Store the reagent under a 2-3 cm layer of mineral oil, as the pH of the reagent changes spontaneously.
- **Standard Mo solution** : Dilute 1 mL of standard Mo solution (containing 1 mg Mo L^{-1}) to 500 mL with AB-DTPA reagent. Because

of linearity obtained with ICP-AES, a control and one standard are generally enough to prepare the standard curve.

D. Procedure

1. Take 10 g of 2 mm sieved soil sample in 100 mL conical flask.
2. Add 20 mL of the AB-DTPA reagent and shake for 15 minutes.
3. Filter through Whatman No. 42 filter paper.
4. Take 0.25 mL of conc. HNO3 in a 10 mL beaker and then carefully add 2.5 mL of the extract to it.
5. Mix on a rotary shaker for 15 minutes and determine Mo using ICP-AES.

CHAPTER 15

RICE BASED CROPPING SYSTEM AND CONSTRIANTS IN RICE PRODUCTION

Lowland cropping systems of Asia are diverse and the cropping pattern ranges from continuous flooded rice production with upland crops including non-food crops. The selection of particular system depends on many factors relating to national goals, edaphic and climatic factors as well as the experience and tradition of the local farmers. When water is sufficient, generally farmers give priority to lowland rice based cropping systems with two rice crops followed by a non-rice food crop. Wherever water is limited, one rice crop followed by an upland crop either pulses or oilseeds is the predominant system of cropping in India.

15.1 RICE BASED CROPPING SYSTEMS

15.1.1 Rice based cropping systems in India

In general, the cropping systems followed in India are as follows

Double cropping

Rice - Wheat

Rice - Potato

Rice - Gram

Rice - Rice

Triple Cropping

Rice - Wheat - Rice

Rice - Potato - Rice

Rice - Rice - Rice

Rice - Wheat - Mung

Jute - Rice - Potato

15.1.2 Rice Based Cropping Systems in Tamil Nadu

The following cropping systems are being followed in Tamil Nadu, depending upon the soil type, irrigation source etc.

1. Rice - Fallow
2. Rice - Rice
3. Rice – Pulses (Blackgram / Greengram / Soybean)
4. Rice – Oilseeds (Groundnut / Sesame)
5. Sesame / Groundnut - Rice
6. Rice – Rice – Blackgram/ greengram
7. Rice – Rice - Groundnut
8. Rice – Rice - Sesame
9. Rice – Rice – Rice
10. Green manure- Rice- Pulses/Oilseeds
11. Rice- Rice – Pulses (greengram/ blackgram)- sesame

15.1.3 Cropping systems in problem soils

Sodic soils area characterized by high pH values, excessive exchangeable sodium, poor physical condition leading to poor aeration, very low infiltration rate and hydraulic conductivity. They are also deficient in available Ca, N and Zn. Growing rice as the first crop in Kharif season and wheat crop during winter season and green manuring with Daincha in the summer season was found to the best at CSSRI, Karnal. In another system, berseem is grown in the place of daincha. Growing of grasses like Para grass and Karnal grass has also shown good promise in sodic soils, because these grass are tolerant and found to reclaim the soil slowly.

The state wise cropping systems for coastal saline soils are given in Chapter 19.

15.2. MANAGEMENT PRACTICES FOR RICE BASED CROPPING SYSTEMS

a. Water Management

Production potential of rice based cropping systems is dependent upon the effective utilization of water, one of the most critical factors in determining crop yields. Efficient use of water by alternate wetting and drying or rotational system of irrigation result in water saving. This can be better utilized to bring more area under rice. In case of water deficit, changing the cropping pattern from rice-rice sequence to irrigated / dry crop either groundnut or sesame followed by medium duration rice variety will benefit the farmers with higher monetary returns besides saving water. Further details are given in Chapter 20 section 20.3.

b. Nutrient management

i. Nitrogen

i. Continuous cropping with rice results in loss of nitrogen

ii. Inclusion of legumes in the system increases the productivity of the systems. e.g. Residual effect of mung, groundnut, lentil increased the grain yield of following rice.

iii. One green manure in the system adds N equivalent to 80 kg / ha to the following rice and residual effect in succeeding crop.

iv. In Northern India, higher grain yield of rice and follow up wheat could be obtained by the use of slow release N fertilizers and nitrification inhibitors in rice crop. Further details are given in Chapter 8 – section 8.4.1.1.

ii. Phosphorus

In a cropping system, residual effect of phosphorus is given much importance. Crops differ in their ability to utilize residual phosphorus, which gets converted into different reaction products. For example, flooded rice is found to make better use of residual phosphorus than wheat and therefore in a rice-wheat sequence, P application is advocated preferably to wheat. In potato based systems viz., Potato-mung-paddy, potato-paddy-paddy and potato-jute-paddy, P should be applied to potato. In case of rice-pulses / oilseeds systems followed in Tamil Nadu, P should be applied to rice and the residual effects were found to be significant in follow up crops of sesame, mungbean and soybean. Further details are given in chapter – section 8.4.1.2.1.

iii. Sustaining the productivity of rice-wheat cropping system in north India

- Inclusion of green manure in the system
- Including a legume crop in the system
- Application of FYM to rice crop
- Application of slow release N fertilizers to rice crop will increase the availability of N to Wheat.

C. Inclusion of rice fallow pulses

In the Cauvery and Thamirabarani delta regions of Tamil Nadu, rice fallow pulses like green gram, black gram and soybean are grown as relay crop to utilize the residual moisture and nutrients available in rice field. No tillage operation is done. The seeds are broadcasted in the standing crop of rice 7-8 days before harvest. The broadcasting of pulses in no till system saves time; avoids soil crusting; reduces weed growth and utilizes the residual soil moisture effectively. The pulse crop covers the land surface in 2 to 3 weeks. Pulse seeds sown close to rice stubbles will penetrate the soil and grow into the decomposing root channels of the rice plant. Additional nutrients may be available for the pulse from the decaying rice roots. Some times, the seeds are dibbled in 30 cm x 10 cm spacing after the harvest when the soil is in waxy condition. To compensate the poor germination of pulse seeds under high moisture conditions, 25% extra seeds can be used for sowing. Seeds are sown from mid of January to mid of February in Tami Nadu and Andhra Pradesh. In Northern and Central India, due to severe winter, rice fallow can be utilized by sowing short duration rabi pulses like lentil and lathyrus and rice is grown during kharif / South West Monsoon season. The agro techniques for rice fallow pulses grown in Tamil Nadu, are given below.

1. Crops and Varieties of Rice fallow pulses

Crop	Varieties				
Blackgram	ADT 2,	ADT 3,	ADT 4,	ADT 5,	TMV 1
Duration (days)	70-75	70-75	65-70	62	60-65
Yield (kg/ha)	970	720	600	1320	1320
Greengram	ADT 2	ADT 3	VBN 1		
Duration (days)	70-75	66	65		
Yield (kg/ha)	800	500	770		
Soybean	C01	ADT 1	C02		

Duration (days)	85-90	75-80	85-90
Yield (kg/ha)	1270	1340	1250
Lentil*	Pusa 1,	Pusa 6	
Lathyrus*	Pusa 24	Nirmal	

* grown in north and central India

2. Seed rate

Blackgram and greengram : 25 kg/ha

Soybean : 75 kg/ha

3. Spacing **: 30 cm x 10 cm**

Plate 47. Rice fallow pulse - blackgram

4. Foliar spray

Foliar spray of 2% DAP (2 kg/ 100 lit.of water) at the appearance of the first flowers and the second spray 15 days after. Vermi pelleting at 50 g/kg of seed is also recommended for higher yield.

5. Plant Protection

For Protection against inflorescence and pods borers, Dust any one of the following insecticides @ 25 Kg/ha Endosulphan 4% dust, Quinolphos 1.5%

dust, Phosalone 4% dust, Carbaryl 15% dust, or spray Endosulfan 35 EC 1.0 lit./ ha, or Monocrotophos 500 ml/ha.

6. Harvesting

Harvest the plants when 80% of the pods are matured. The colour of the majority of the pods at harvest will be black in black gram, and brown in green gram. The pod moisture content will be 17-18%. Dry the pods immediately to 12-13 per cent moisture content along with the plant and then thresh to separate the grains.

15.3 CONSTRAINTS IN RICE PRODUCTION

It is now estimated that 56% yield potential alone is realized. The contributing factors to yield gap are estimated to 40-45% for plant density, 35-45% for fertilizers and 15-20% for land preparation, insect and disease control, weed control and seedling age.

Post harvest technologies should be more emphasized since 7 to 26% of crop is estimated to be lost due to grain shedding, improper storing and processing.

The constraints that limits rice production can be summarized as follows.

- Reduction of cultivated area due to urbanization, industrialization and population explosion
- Inadequate supply of quality certified seeds
- Delayed and prolonged transplanting and sub optimal plant population
- Decreased use of organic manures like FYM/compost resulting in reduction of native fertility of soils.
- Imbalanced and inefficient use of fertilizers
- Wide spread zinc deficiency
- Biophysical factors such as delayed Monsoon
- Lack of assured water supply in time
- Scarcity of underground water available for irrigation
- Impeded drainage and salinity in coastal regions
- Increase in problem areas due to discharge of industrial effluent into channel rivers.

- Lack of high yielding varieties ideally suited to relatively harsh sub ecologies in the rainfed ecosystem, which accounts for 58 per cent of the total area.
- Diseases and pest epidemics
- Problems due to water logging and drainage in tail end areas of delta.

The problems due to water logging and drainage are given below

15.3.1 Water Logging and Drainage Problems

Prolonged presence of excess water in a place where there is inadequate drainage results in water logging. The excess water may originate from direct irrigation, seepage from reservoirs, canals and ditches or it may come into the area through subsurface flow from other areas where there is excess precipitation or irrigation. Thanjavur, Thiruvarur and Nagai districts in Tamil Nadu are facing water logging problems due to heavy rainfall during north east monsoon periods.

15.3.1.1 Effect of water logging

Excess water in agricultural field is detrimental to crop production as it restricts aeration, affects soil temperature, biological activity and structural stability of soil, hinders with tillage operations resulting in poor root development and ultimately poor crop production. Excess water slowly builds up the water table, which ultimately comes to the root zone and even up to the ground level. The rising water table will not only affect the crop growth but also brings the harmful salts from deeper layers/ root zone in some areas where the soils are highly sensitive to irrigation. This water logging and salt build up is commonly occurring phenomena in several command areas/ localities where drainage is a problem. One can often see stunted growth of crops, with yellowing of leaves in low areas of field when the soil has become saturated and water logged. Such condition results in root damage caused by reduced supplies of oxygen and accumulation of carbondioxide with the related effects on the soil plant relationship.

a. Soil aeration

Soil aeration is a function of the soil particles size and their arrangement, and the degree of saturation of the soil moisture content. If the larger pores are free of water so that the moisture level is below field capacity, gas diffusion can process satisfactorily. When the profiles becomes saturated, however, the rate of diffusion declines. As the decomposition and other biotic activity continue to take place in the soil, the oxygen level drops and carbon dioxide content increases. The lowered oxygen content of the soil atmosphere under saturated condition affect different crops in varying degree. The plants that withstand saturation, have specialized air-conducting tissues in their stem and roots (e.g. rice crop).

b. Uptake of nutrients

Adequate supply of oxygen to the roots is essential for efficient uptake of nutrients. High moisture levels also reduce the uptake of plant nutrients. Restricted root growth limits the volume of soil, from which the plant may draw nutrients. Also when the aeration is poor, this may accumulate sufficiently high concentration of reduced iron and manganese to become toxic to the roots. Decomposition of organic matter, when aeration is poor may also produce hydrogen sulphide, a gas that is toxic to the roots. Accumulation of organic acids in flooded soils also leads to the inhibition of growth of rice roots.

c. Denitrification

Denitrification also occurs rapidly in saturated soils, apparently because of the competition for nitrogen by soil micro-organisms that thrive in the saturated soils and the reduction in number of nitrifying organisms due to lack of aeration. Tissues of plants grown in soil that has been maintained near saturation contain less crude protein indicating that nitrogen balance in the soil has been upset.

d. Redox Potential

If the soil is flooded with water the oxidation-reduction potential gradually falls to a limiting value. In the absence of oxygen, the oxidation-reduction potential falls more rapidly especially in surface soils. At low oxidation reduction potential, reduced compounds especially of ferrous iron and sulphide/ sulphur are produced.

e. Biological activity

Lack of aeration impedes biological activity. The process of decomposition of organic matter is slowed down and the rate of mineralisation of organic nitrogen compounds is decreased. Some pathogenic organisms becomes more active under anaerobic conditions, which may cause problems causing soil borne diseases. Poor aeration and a high moisture content, which usually occur together in poorly drained soils, directly affect the occurrence and severity of plant diseases. The rate of mineralization of nitrogen is also affected by soil temperature. Poorly drained soils are slow to warm up in the cold seasons. Low soil temperature restrict root development and depress biotic activity.

f. Soil structure

The arrangements of particle is affected by frequent and prolonged saturation. The poor structure found under such condition may be attributed mostly to the lack of biotic activity, reduced root activity and lack of wetting and drying cycles.

g. Salt problem

Extensive areas of land in the arid and semi arid regions of India have gone out of cultivation due to the rise of water table and accumulation of salt. Excessive irrigation and poor water management are the chief causes of water logging and salt build up.

15.3.1.2 Drainage

The purpose of drainage is primarily to remove excess water from the soil. The provision of drains is a must in every irrigated and rainfed farm for sustained farming. Excess water even for a brief period may cause severe damage to crop and soil. Some times drainage becomes a greater problem than irrigation especially in low-lying areas. The beneficial effects of drainage in improving the productive capacities of soil are.

- Removal of excess water
- Improvement of soil structure
- Increasing in depth of plant root zone
- Improved air circulation, warmed soil
- Improved organic matter decomposition and nitrification, reduced erosion
- Increased leaching of salt and prevention of its rise in to soil surface

Good drainage, both surface and internal is essential. Provision should be made to drain excess rainfall promptly and safely. Surface drainage may be needed to prevent saline – alkali conditions in a soil by leaching. If the land is not naturally drained, artificial drainage must be established. Integrated irrigation and drainage planning is often necessary for laying out a farm area for efficient water use.

15.3.1.2.1 Types of Drainage

Drainage can be either natural or artificial. Most lands have some natural drainage evacuating excess water over land surface, towards depressions like lakes, swamps or rivers. In many cases, however, natural drainage is inadequate and artificial or man-made drainage is required. The artificial drainage is classified into two major systems known as (i) surface drainage and (ii) sub surface drainage, accordingly the system drains off excess runoff or sub surface water.

a. Surface drainage

Surface drainage is the removal of excess water from the surface of the land. This is normally accomplished by shallow ditches, also called open drain, discharging in larger and deeper collector drains. In order to divert the surface excess water towards the drains, the field surface is given an artificial slope by means of land preparation or land grading. In flat areas, the main problem is the timely removal of stagnated water in shallow topographic depressions. In sloping areas, it is usually the removal of excess water without causing erosion.

The need for surface drainage is the result of a combination of certain climatic, hydrologic and soil condition, and of topography and land use. Excess precipitation, surface run off from upland areas, and overflow from rivers may result in the ponding of water in the land. This is especially so if the soils have low infiltration rate and poor internal drainage. In considering the relation between precipitation, soil conditions and surface run off, two concepts are recognized viz. (i) The threshold concept i.e. the rate of rainfall does not exceed the basic infiltration rate. The soil will absorb water till it is saturated only then the water will start to flow over the ground surface and (ii) The infiltration concept i.e. the rate of rainfall exceeds the infiltration rate and stagnant water occurs at the surface before the soil is saturated.

The surface drain for flat land is done by the following way.

1. Bedding system
2. Parallel field drain system
3. Random system
4. Parallel open ditch system

For sloppy area,

1. Cross slope ditch system
2. Standard erosion control terrace

b. Sub surface drainage

These drainage systems are not followed under Indian conditions. How ever, the following details are given for the benefit of readers to understand the possibilities of sub surface drainage.

In sub surface system, the removal of ground water or excess surface water penetrating into the soil is effected by providing under ground outlets at a component depth to drain off free subsoil water and to convey it to natural outlets so as to lower the water table sufficiently. The different types of subsurface drainage (1) Open trench (2) Pipe drains (3) mole drains and (4) perforated pipes.

1. Open trench

Open trenches are narrow but deeper open excavations provided in the field to convey the sub surface water. As deeper trenches are not stable and cause hindrance to ploughing and tillage operations, other under ground drains are preferably used.

2. Pipe drains

The drains include concrete and burnt clay tile, corrugated plastic tubing or other perforated conduit. Corrugated steel pipe with a high structural strength is suitable to withstand high soil loads, to cross-unstable soils that require the rigidity of a long pipe and to provide a stable outlet into open ditches. Tile for lateral drains (drains intended to receive water directly from the soil as contrasted with main lines which received most of their flow from laterals) are usually 0.3 m lengths, 75 to 150 mm inner diameter. The tiles are laid end to end in the bottom of a trench that is then back-filled.

3. Mole Drains

Mole drains are cylindrical channels artificially produced in the sub soil without digging a trench from the surface. They are similar to pipe drains except that they are not lined with tile or other stabilizing material. Moling is a temporary method of drainage where soil conditions are suitable. Moles function efficiently for the first few years and then gradually deteriorate. The maximum life is 10 to 30 years.

4. Perforated pipes

Instead of tiles, perforated pipes or tubes are also provided where the clay or concrete tiles are not stable due to the surface loads. Perforated corrugated steel pipes are suitable where many surface loads are anticipated. These pipes are laid in open trenches and back filled.

CHAPTER 16

HYBRID RICE

Rice provides about 2/3rd of the calorie intake of about one billion people in humid and sub humid regions of Asia and 1/3rd of the calorie intake of about one billion people in Africa and Latin America. Achieving self sufficiency in rice production and maintaining price stability are important issues in low income countries because of the importance of this crop in national food security and generating employment and income for low income people. The yield ceiling of rice varieties of green revolution era must be lifted further because by 2020 A.D., world must produce 350 m.t. in addition to the produces available today. This production increase must be achieved on less land, labour, water and plant protection inputs and it must be sustainable also. The increase in rice production can be achieved either by improving existing plant type or by developing heterotic rice hybrids. Development of heterotic rice hybrids seems to be appropriate and more promising.

16.0 NEW TOOLS TO MEET CHALLENGES OF WTO REGIME

a. Hybrid Rice

In the immediate future, hybrid rice is the practically feasible and adoptable technology for enhancement of rice production and productivity. Hybrids have recorded 15-20 per cent higher yield than the highest yielding inbred varieties in the farmers' fields. Adoption of hybrids at present is limited to only 2 lakh hectares annually, due to grain quality consideration and higher seed cost among other constraints. But these and other constraints are likely to be overcome shortly, on account of intensive research underway on these aspects. The potential area in the country for hybrid rice is anywhere between 8-15 million hectares and adoption in this large area will increase the rice production by 10-20 m.t.

b. New Plant Type and Super Hybrid Rice

The other possibilities for increasing rice production are through development of super hybrid rice, with a yield potential of 100kg/ha/day, development of new plant types capable of yielding 10-12 t/ha and adopting integrated crop management strategies. Research on all these aspects has been intensified and is likely to pay rich dividends in near future.

c. Bio- technological tools for yield stability and stress tolerance

Biotechnology can help in improving rice breeding through transfer of economically important traits across genus/ species barrier into the rice gene pool. (i.e. broadening the genetic base), manipulation of target trait without disruption to the non target regions of the genome (i.e. Increasing efficiency in selection) and shortening the breeding cycle.

The three broad applications of rice biotechnology that are expected to contribute both directly and indirectly towards rice improvement efforts in India are DNA marker technology, Genetic engineering and Application of genomic tools. DNA marker technology: Among the different type of markers used in rice genome analysis and tagging different traits, PCR based DNA markers like RAPDs, micro-satellites, Inter-Simple Sequence Repeats (ISSRs), Sequence tagged sites (STSs) are being widely used due to their simplicity, robustness and speed of assay. Three important applications of molecular markers to rice breeding and genetics are: determining the allelism of gene(s) conferring identical pheno-types, use in marker assisted selection and map based cloning of genes. The immediate impact of marker aided rice breeding will be seen in the success of breeding for durable pest resistance such as that against BLB, Blast and Gall midge where gene pyramiding is required; breeding for multiple pest resistance where simultaneous screening of segregating populations against various stresses is difficult and in breeding for nutritionally enriched rice such as golden rice where a combination of exotic genes confer the trait of interest.

d. Genetic transformation

Genetic transformation is another tool that promises to revolutionize Indian rice improvement programme. The most important advantage of transgenic technology is the capacity to mobilize useful genes from non-rice gene pool to rice with least disruption to rice genome. Transgenic rice with resistance to insect pests and pathogenic viruses, fungi and bacteria have been developed and are being tested under limited field evaluation in India.

These transgenic rice mainly include Bt genes like CrylAb, CrylAc conferring resistance against stem borers and leaf folders, pr GNA (*Galanthus agglutinin*) gene conferring resistance against sucking pests and conferring resistance against bacterial leaf blight. Rice transformation with chitinase gene Chi 11 against sheath blight and potato protease inhibitors for insect resistance has also been reported. However, the main issues confronting use of transgenic approach for rice improvements are of bio-safety and public concern. Further a plethora of IPR issues are to be resolved before the transgenic rice can be released for commercial cultivation. Legal wrangles through which the discoverers of 'Golden Rice' technology went through before releasing it free to the poor developing nations is well known. While the limited release of Bt cotton for commercial cultivation by the Govt. of India may pave the way for easy clearance for other transgenic crops, at least in cases where private sector is taking initiative.

e. Other promising technologies

Similar to DNA marker technology and transgenics, genomics is another area full of prospects. With the task of structural genomics of rice nearly complete with release of DNA sequence data for the entire rice genome, stage is now set for comparative and functional genomics, wherein the known sequence are being deciphered for their function and commonality in other cereal crop genomes. Geneticists and breeders can acquire an almost unlimited number of DNA markers and for crop improvement. Quantitative geneticists can gain instantaneous access to the genes in their segregating population that may be responsible for the traits like yield, abiotic stress tolerance etc. that are controlled by multiple loci. Breeders can aim at several traits simultaneously for the rice improvement programme. These possibilities are bringing other goals such as nutritional fortification of rice with enriched vitamin, micro-nutrients within the breeding objectives. Breeding for abiotic stress tolerance such as drought and adverse soil conditions are now proving to be simpler targets. Researches are seriously considering to transfer C_4 pathway of photosynthesis, similar to the one found in crops like maize, into rice to make it more efficient of harvesting solar energy and converting it into higher grain yield.

Another great challenge being seriously addressed is the development of aerobic rice-the rice that can be cultivated with less water with no yield reduction. However, it is a long way before these technologies are transformed into products. The details regarding aerobic rice are given in Chapter 21- section 21.5.

Table 44. Comparison of highest yielding rice hybrids and inbreds at IRRI during 1986-1995

Season	Trial	Hybrid	Yield (t/ha)	Difference from the best check	% of Check	Growth duration (Days)
1986DS	I	IR54754A/IR66R	7.4	1.2*	119	126
	II	IR54754/ARC 11353R	7.9	2.3*	142	133
1986 WS	I	IR4752A/IR64 R	3.9	1.0*	134	126
	II	IR19728A/IR25167-9-2 R	3.6	0.6*	120	122
	III	IR 46830A/IR 50R	4.1	0.7*	120	110
1987 DS	I	IR46830/IR29723-143-3-2-R	6.4	1.8*	139	112
	II	IR54755A/IR2797-125-3-3-2 R	7.8	1.8*	130	130
	III	IR 54752A/ARC 11353 R	6.8	1.0*	117	128
1988 DS	II	IR46830A/IR9761-19-1R	4.8	0.0	100	108
	IV	IR5752A/IR64R	5.3	0.9*	120	120
	V	IR54752A/IR15325-13-3-3-2R	5.9	0.8	116	122
	VI	IR54752A/IR13146-45-2-3	6.8	1.1*	119	116
		IR46830A/IR9761-19-1R	3.2	1.0*	145	105
1988 WS	I	IR54752/IR9761-19-1R	6.3	1.5*	131	111
	II	IR54752A/IR28228-119-2-3-1-1R	6.5	1.0*	119	124
1989 WS	I	IR54752A/IR2686-C2-118-6-2R	3.1	0.4	114	120
	II	IR54752/IR54752-19-3R	3.5	0.8*	131	136
1990 DS	I	IR62829A/IR9761-19-1R	4.0	0.6	117	110
	II	IR62829A/IR31805-20-1-3	4.5	-0.6	088	116
	III	IR58025A/IR10198-66-2R	4.8	1.5	112	116
	IV	IR5825A/IR9723-143-3-2-1R	5.6	1.0*	121	128
1990 WS	I	IR58025A/IR54752-22-19-3R	3.0	0.0	100	135
	II	IR5825A/IR40750-82-2-2-3R	3.2	0.2	107	126
1991 DS	I	IR62829A/IR35366-62-1-2-2-3 R	4.7	0.7*	118	112
	II	IR58025A/IR54745-2-45-3-2-4R	5.4	1.2*	128	122
1991 WS	I	IR58025A/IR 19058-107-1R	6.4	1.2*	123	113
	II	IR62829/IR47310-94-3-1 R	5.1	1.1*	128	120
1992 DS	I	IR62829/IR54883-4-1-3 R	6.4	0.5	108	112
	II	IR62829A/IR46 R	7.2	0.7	111	115
	III	IR58025A/IR66R	6.3	0.8*	114	117
1992 WS	I	IR64608A/IR42686-C2-118-62	3.7	-0.6	086	111

	II	IR58025A/IR48725-B-B-1412 R	4.3	0.2	105	116
	III	IR58025A/IR37712-90-3-3-3-2 R	4.1	0.4	111	119
	IV	IR58025A/IR54056-64-2-2-2R	4.4	0.7*	119	126
1993 DS	I	IR5025A/IR50404-57-2-2-3R	6.4	0.7	112	110
	II	IR58025A/IR34686-179-1-2-1R	7.4	0.8*	112	128
1993 WS	I	IR58025A/BG915R	4.0	0.8*	125	105
	II	IR58025A/IR1078-33R	4.3	-0.7	086	114
1994 DS	I	IR58025A/IR2221-14R	7.3	0.6	109	111
	II	PM8A/IR29723-143R	7.5	0.5	107	126
1994 WS	I	IR5805A/IR9606-119-3R	5.8	1.0*	121	105
	II	IR6875A/IR46R	6.6	1.6*	132	122
1995 DS	I	IR58025 A/IR58103-6-3R	9.5	1.6*	120	114
	II	IR58025A/RP633-76-1R	9.6	1.6*	120	121
		MEAN				
		DS	6.5	0.95	117	
		WS	4.2	0.58	117	
		GRAND MEAN	5.6	0.80	117	

DS : Dry season, WS : Wet season, * Significant 5% level

Rice research has to be geared up to surmount the technological challenges in breaking the genetic yield barriers, improving input use efficiency and developing environmentally acceptable strategies for alleviating losses due to pest and diseases.

16.1 HISTORY

Although heterosis in rice is known to rice researchers, since 1926, its commercial exploitation was demonstrated by Chinese scientists by developing heterotic rice hybrids yielding 15-20% more than the available best varieties (Yuan *et. al.,* 1994). The increase in yield was harnessed without any significant difference in input use. Successful commercial exploitation of heterosis in rice by Chinese scientists has encouraged the rice scientists all over the world to initiate the research on exploring potential and problems of hybrid rice production technology (Heu *et al.,* 1984). In general, commercial rice hybrids have shown 20-30% advantage for grain yield. Hybrid varieties gave 29-45% better yields than the pure line varieties in China during 1981-1990. Standard yield heterosis in rice hybrids evaluated along with pure line varieties in Huinan province, China during 1979-81 was found to range from 11.1 to 20.4%. Hybrids have proved their superiority over pure line varieties at IRRI (Table 44) as well as in other countries like India, Indonesia, Philippines, Korea, USA and Egypt. Studies had further shown that under high yielding environment, yield advantage for hybrids over pure line varieties was higher in low yielding environment.

Further studies have shown that the increase in yield in rice hybrids due to their increased dry matter resulting from higher leaf area index, higher crop growth rate and higher harvest index resulting from their increased spikelet number and increased grain weight. Expansion of leaf area due to increased number of tillers is the main factor for achieving heterosis in crop growth rate of rice hybrids during the first month after transplanting. Degree of heterosis decreased thereafter, it again increased after heading thus, contributing to the higher grain yield.

16.2 PRESENT STATUS

Commercial success of hybrid rice in China has clearly shown potential to meet the ever-increasing demand for rice world over. Efforts to develop and use this technology in India were initiated in 1970s, these were systematized and intensified since December, 1989 with launching of a Project "Promotion of research and development efforts on Hybrids in rice" with UNDP/ICAR assistance. During 1997, hybrid rice was planted in about 1,00,000 ha area. Now this project is being operated as a National Research Network with 12 centres across the country with Directorate of Rice Research, Rajendranagar, Hyderabad as coordinating centre. The target area of hybrid rice for cultivation includes Punjab, Haryana, Western Uttar Pradesh, Coastal Karnataka for South India. There are three strategic research centres, conducting basic research located at DRR, Hyderabad, CRRI, Cuttack and IARI, New Delhi. Seven associate centres viz. Coimbatore, Karnal, Pantnagar, Chinsurah, Mandya and Karjat are engaged in conducting region specific research.

16.3 MAJOR OBJECTIVES OF HYBRID RICE RESEARCH

1. Development of hybrids with 15-20% yield advantage over the best check.
2. Optimization of seed production package.
3. Standardization of package of practices for cultivation of hybrids, and
4. Conducting basic research relevant to the project goals.

As a results of concerted efforts so far 13 public bred rice hybrids viz. APHR 1, APHR 2, KRH 1, KRH 2, CORH 1, CORH 2, ADTRH 1, DRRH 1,

CNRH 3, Pant Sankar Dhan 1, Sahayadri, Narendra Shankar Dhan 2 and Pusa RH 100 have been released for general cultivation in different parts of the country. The salient features of these rice hybrids are given in Table 45. Besides, four rice hybrids viz., RH 371, AA 6201, HR 7120 and RH204 have been developed by private sector and are being marketed in the country. Hybrids are also being developed and evaluated for shallow lowland ecosystem, for grain quality characteristics, for pest and disease resistance and of basmati qualities. Strategic research centre IARI, New Delhi is exclusively engaged in developing basmati type hybrids. This centre has developed basmati CMS lines Pusa 3A and Pusa 2A and identifying suitable restorers.

Table 45. Rice hybrids released in India

S.No.	Name of hybrid/ (Year of Release)	Cross	Duration (days)	Yield in OFT@ (t/ha)		Yield adv. over check (%)	Released for the state of
				Hybrid	Check		
1.	APHR-1 (1994)	IR58025A/ Vajram R/	130-135	7.14	5.27 (Chaitanya)	35.4	Telangana & Rayalaseema regions of Andhra Pradesh
2.	APHR-2 (1994)	IR 62829A/ MTU 9992 R/	120-125	7.52	5.21 (Chaitanya)	44.2	Telangana & Rayalaseema regions of Andhra Pradesh
3.	CORH-1 (1994)	IR 62829A/ IR 10198 R/	110-115	6.08	5.23 (IR 50)	16.2	Irrigated regions of Tamil Nadu
4.	KRH-1 (1994)	IR 58025A/ IR 9761 R/	120-125	6.02	4.58 (Mangala)	31.4	Irrigated regions of Karnataka
5.	CNRH-3 (1995)	IR 62829 A/ Ajaya R/	125-130	7.49	5.45 (Khitish)	37.4	Boro season of West Bengal
6.	DRRH-1 (1996)	IR 58025A/ IR 40750 R/	125-130	7.30	5.50 (Tellahamsa)	32.7	Telangana & Rayalaseema regions of Andhra Pradesh
7.	KRH-2 (1996)*	IR 58025A/ KMR-3 R/	130-135	7.40	6.10 (Jaya)	21.3	Irrigated regions of Karnataka
8.	Pant Sankar Dhan-1 (1997)	IR 58025A/ UPRI 93-133 R/	115-120	6.80	6.20 (Pant Dhan-4)	9.7	Western and plains of Uttar Pradesh and Uttaranchal.
9.	CORH-2 (1998)	IR 58025A/ C20 R	120-125	6.25	5.20 (ADT-39)	20.2	Irrigated regions of Tamil Nadu

10.	ADTRH-1 (1998)	IR 58025A/ Ir 66 R/	115-120	7.10	4.90 (ASD-18)	44.9	Irrigated regions of Tamil Nadu
11.	Sahyadri (1998)	IR 58025A/ BR 827-35 R/	125-130	6.64	4.89 (Jaya)	35.8	Konkan region of Maharashtra
12.	Narendra Sankar Dhan-2 (1998)	IR 58025A / NDR 3026 R	125-130	6.15	4.94 (Sarjoo-52)	24.5	Eastern parts of Uttar Pradesh
13.	PHB-71 (1997)*	—	130-135	7.86	6.14 (PR 106)	28.0	Haruana, UP, TN
14.	PA 6201 (2000)*	—	125-130	6.18	5.03 (Jaya)	22.9	Eastern and some parts of Southern India.
15.	HRI-120 (2001)*	—	135-140	6.11	4.91 (Jaya)	24.4	Telangana region of A.P., Karnataka, Konkan region of Maharashtra, Plains of Uttaranchal, Eastern U.P., Orissa and Tripura.
16.	Pusa RH-10 (2001)*	Pusa 6A/ PRR 78 R/	120-140	4.35	3.11 (Pusa Bas. 1)	39.9	Haryana, Punjab, Western, UP, Delhi
17.		—	125-130	6.89	5.60 ()	22.7	Western India

@OFT = On-farm trials * hybrids released by CVRC

Experiments conducted on economics of hybrid rice cultivation during last several years have shown that under good management conditions on an average, an extra net profit of about Rs. 3000-3500 per hectare can be earned. The only additional cost in cultivation of hybrids is the seed. All other cultivation practices at present are same as those for high yielding rice varieties of the area. High cost of seed is due to low production of hybrid seed per unit area. Seed yield is generally 1-1.5 t/ha.

Seed yield obtained on seed parent (CMS line) in a hybrid rice seed production plot is function of :

1. Yielding ability of CMS line as determined by its fertile counterpart.
2. Proportion of male sterile line in relation to pollen parent and
3. Out-crossing rate of the male sterile line.

Extensive research in China, IRRI and in our own country has led to the identification of the guidelines for successful seed production of hybrid rice (Yuan, 1985 and Mao, 1988). These guidelines include.

i. Selection of seed and pollen parent with synchronized anthesis.

ii. Selection of seed parent with long stigma, longer duration and wide angle of flower opening.

iii. Selection of pollen with a high percentage of residual pollen per anther after anther dehiscence.

iv. Synchronization of flowering.

v. Use of optimum row ratio between A and B/R line.

vi. Use of seed and pollen parent with short, horizontal flag leaf or long erect flag leaves.

vii. Use of GA_3 to improve panicle exsertion and prolong anthesis.

viii. Planting of seed parent and pollen parent rows across the prevailing wind direction and use of supplementary pollination.

ix. Selection of optimum time of flowering of parent lines in seed production plots.

Hybrid rice seed production using CMS system involves two steps viz., multiplication of A line and production of hybrid (A/R) seed. However, production of hybrid seed with the EGMS system involves one step only i.e., production of hybrid seeds. In this system male sterile line can be multiplied by growing EGMS line under suitable day length and /or temperature situation. By adopting the standard practices recommended for hybrid rice seed production, seed yield of 1.5-2 t/ha can be obtained. These practices are constantly reviewed and changed from time to time. This way Chinese have improved seed yield from 0.67 t/ha in late 1970s to 2.6 t/ha at present. However, in a country like India seed yield is still very low. It is around 1.0t/ha. However, seed yield of 1.5-2.0 t/ha has been obtained from experimental plots at certain locations by following the recommended practices. Recently the Chinese have switched the emphasis to 2-line hybrids from 3-line hybrids. They have developed several 2-line commercial hybrids. In China, 2-line indica hybrids like liangyou pelte and liangyou pelshan and 2-line japonica hybrids *viz.* N5088S/R 187 and 7001 S/Xiushul 04, 7001 S/wanhul etc. have been developed and are being grown commercially. Outside China, the 2-line hybrids are till in experimental stages. In India, thermosensitive genetic male sterility (TGMS) system is being used for developing 2-line rice hybrids.

16.4 HETEROSIS

Hybrid rice breeding programme exploits the phenomenon of heterosis. In very broad sense, heterosis can be defined as superiority of F_1 when two genetically dissimilar parents are crossed. Though heterosis has been observed for various morphological, physiological and biochemical characters, in an applied breeding programme, the concern is primarily with the economic yield potential. In practical breeding programmes, usually standard heterosis is considered, which is defined as superiority of the F_1 hybrid as compared to the highest yielding released check variety or hybrid. Heterosis breeding differs from the inbred breeding programme. Inbred breeding programme accumulates productivity genes that perform well under homozygous conditions, while heterosis breeding assembles genes that perform well under heterozygous conditions.

Standard heterosis can be estimated as

$$\text{Heterosis (\%)} = \frac{\text{F1} - (\text{standard check})}{\text{Standard check}} \times 100$$

At present, two approaches viz., (i) three line system of heterosis breeding or CGMS system and (ii) two line system of heterosis breeding are being followed in rice. Heterosis in rice for grain yield ranges from 15-25% over the check varieties and was first observed and exploited commercially by the Chinese scientists from 1976 onwards. Rice strictly being a self pollinated crop requires a stable economically viable system for hybrid seed production. And with the discovery of male sterile cytoplasm wild Abortive – WA and from wild species of *Oryza sativa spontanea* in China, the development of several stable male sterile lines like V20 A, IR 58025 A, IR 62829 A, by backcross method for transfer of sterile cytoplasm directly or indirectly from WA source was made. The man responsible for the success story of hybrid rice in China is ***Prof. Yuan Long Ping,*** so he is called as ***"Father of Hybrid Rice"***.

Today, in China nearly 60 per cent of the total rice area is under hybrid rice and increased substantially the rice production and productivity. Successful development and extensive adoption of hybrid technology in China, promoted augmentation of hybrid breeding since, 1990. The network approach with generous support from the ICAR, UNDP/FAO and Mahyco Research Foundation has led to the development of as many as 13 hybrids (APRH 1, APRH 2, DRRH 1, KRH 1, KRH 2, CNRH 3, CORH 1 (MGR), sahyadri, NSO2, PSO1, PUSA RH 1, CORH 2 and ADTRH 1 etc.) which yield 1.0-1.5 t/ha more than the best varieties. With the hybrids planted on a commercial scale over 60,000 ha last year, India has become the second country after China to make hybrid rice technology a field reality. In India, TNAU released CORH-2, the first salt tolerant rice hybrid especially for sodic soils. This hybrid gives one tonne per ha more than the conventional varieties like TRY 1 and CO 43 under sodic soils.

In this approach, Three line *viz.*, (1) Cytoplasmic Genetic Male Sterile line or A line (2) Maintainer line or B line and (3) Restorer line or R line are used.

16.4.1 Cytoplasmic genetic male sterile system (CGMS)

In CGMS system, the sterility / fertility is due to interaction of nuclear gene with the cytoplasm. In line, the sterility inducing cytoplasm is usually from the related wild or weedy species and the nucleus contain recessive fertility restorer genes, hence the line is sterile. The maintainer line or 'B' line contains normal cytoplasm and recessive fertility restorer genes. Hence, it is fertile. The cytoplasmic male sterile line and maintainer line are exactly same genetically. They differ only in cytoplasm content. CMS line has sterility inducing cytoplasm whereas maintainer line has normal cytoplasm. The restorer line possesses fertility restoring genes in dominant conditions. The cytoplasm may be either sterility inducing or normal one. The R line is fertile. In CMS multiplication, A line is crossed with B line. Since both possess fertility genes in recessive condition, and the cytoplasm contributed by 'A' line is sterility inducing one, the resultant F_1 is sterile (*i.e.*) it is nothing but the 'A' line. This is the method for multiplication of 'A' line.

In hybrid seed production A line is crossed with 'R' line. Since 'R' line possesses fertility restorer genes in dominant conditions, through the cytoplasm contributed by 'A' line is sterility inducing one, the resultant F_1 that is used as commercial hybrid is fertile one. The parental lines should posses the following desirable features.

- Good general combining ability
- Desirable floral traits promoting cross pollination and hence, higher seed set.
- Adaptability to local conditions.

16.4.2 Restorer improvement

Desirable features of restorer lines are :

- Strong and stable restoration over seasons and locations/regions

- Highly vigorous growth habit with large number of spikelets containing big and plumby anthers filled with large number of pollen grains.
- High amount of residual pollen
- Complete dehiscence of anthers
- Profuse tillering, synchronous heading and concentrated blooming of spikelets.
- Multiple disease and pest resistance
- Good general and specific combining ability
- Acceptable grain quality

16.4.3 Two line approach of heterosis breeding

Crop scientists have been engaged in developing an effective and viable two line approach to exploit hybrid vigour in rice since more than a decade. Two major approaches have emerged as a result and they are being commercially exploited now for rice improvement in China. They are (i) Use of Environment Sensitive Genetic Male Sterility (EGMS) and (ii) Use of Chemically Induced Male Sterility (CIMS)

16.4.3.1 Environment Sensitive Genetic Male Sterility (EGMS)

Alteration of fertility / sterility in response to environmental factors particularly with respect to temperature and photoperiod is utilized in two line heterosis breeding. At a particular critical range of temperature, certain genotypes become male sterile and those can be used as female parent in development of hybrids, while at some other range of temperature or photoperiod, same lines become fertile and hence can be multiplied without a maintainer line unlike in three line system of heterosis breeding. This type of male sterility is referred to as environmental sensitive genetic male sterility (EGMS). Those genotypes which exhibit sterility / fertility with respect to temperature are called Thermosensitive Genetic Male Sterile (TGMS) lines, and those showing similar response with respect to photoperiod are termed as Photoperiod sensitive genetic male sterile (PGMS) lines. In the development of hybrids, only two lines *viz.,* female and male are needed and hence this system is called "Two line Heterosis Breeding".

Advantages

- There is no need for maintainer line; hence procedure for development of hybrids is simpler
- Any genotype can be utilized as male parent; hence, the chances of development of heterotic hybrids are greatly increased.
- Negative effects associated with sterility inducing cytoplasm are avoided.
- Seed production programme is simple and more efficient. The field area ratio between CMS multiplication, hybrid reproduction and commercial cultivation of F_1 hybrid is 1:100 : 1000 as against 1: 50 : 5000 in three line system.

A. Desirable characteristics of TGMS / PGMS lines

- Population of EGMS lines should consist of more than 1000 individual plants and their morphological characters should be uniform.
- The percentage of male sterile plants should be 100% and the degree of male sterility (percentage of abortive pollen grains should be over 99.5%.
- The change of fertility to sterility and vice-versa should be quite clear and distinct.
- Male sterile period should be last for more than 30 consecutive days.
- Seed setting during fertile period should be over 30%
- Out crossing percentage during sterile phase should be lower than check 'A' lines.

B. Development of two-line experimental hybrids

The development of two-line experimental hybrids is similar to the ones in three system of heterosis breeding. It essentially consists of the following.

- Making several test crosses between the TGMS / PGMS line and several promising genotypes. Here, it may be emphasized that any promising genotypes can be used as a male parent. Unlike in three line system, where only those genotypes carrying dominant restorer genes should be used as male parent.
- Evaluation of the test crosses to identify promising two line hybrid combinations
- The subsequent procedures of observational yield trials, regional or national trials, multilocational trials, on –farm verification trials etc.,

for identification of best possible hybrid remain the same, as in three-line system.

16.4.3.2 Commercial utilization of two-line hybrids

Two-line hybrids have been grown commercially in China for the past couple of years. It is estimated that during the last three years, the two-line hybrids were grown in an area of more than 0.1 m.ha. and it is expected that by this time, more than 1 m.ha will be covered by two-line hybrids in China. Several promising TGMS / PGMS lines have been developed in China and many two-line hybrids are under extensive large scale testing programme. Some very promising inter-subspecific (indica x japonica) hybrids have been developed with an yield advantage of 11-25% over the highest yielding and extensively grown three line hybrid Shan-You 63. Research on two-line breeding has been initiated in some other countries such as Japan, India, Philippines (IRRI) and it may take couple of years before two line hybrids are commercialized. For tropical countries, such as India and Philippines, TGMS system may be more suitable, since wide temperature ranges exist at different altitudes in different seasons and the required day length differences are not found. Several TGMS lines have been identified in India and at IRRI, presently the temperature ranges inducing sterility / fertility are being standardized and few experimental hybrids produced are being evaluated in the initial hybrid rice trials at the research farms.

16.5 HYBRID SEED PRODUCTION

Hybrid vigour in rice has been first reported by Jones (1926). This has led to speculation regarding the production of hybrid rice by utilizing cytoplasmic male sterility. Most Japonica rice has normal cytoplasm, but indica varieties with sterile cytoplasm and fertility restoring system have been identified. But difficulties have been encountered in obtaining sufficient seed set by cross pollination to make hybrid rice seed production economically feasible. After the implementation of UNDP/FAO project entitled "Development and use of hybrid rice technology in India", the hybrid rice production in India has become a success story.

Hybrid rice seeds were produced using (cytoplasmic genetic male sterility) three line system. The two genes Rf1 and Rf2 are the genes for fertility restoration. The process of hybrid rice production involves continuous supply of agronomically improved cytoplasmic male sterile rice (A), maintainer line (B) and fertility restorer (R) line in system. Maintainer and restorer lines are maintained by selfing, while CMS line and F_1 seeds are produced with efforts to enhance cross pollination in field. F and S refer to fertile and sterile cytoplasm. Rf and Sf are fertility restoring and non restoring genes respectively.

16.5.1 Techniques of hybrid rice seed production

The following points are to be taken in to account for a successful hybrid rice production.

1. Choice of field **:** Fertile soil, protected irrigation and drainage system, sufficient sunshine, no serious disease and insect problem.

2. Isolation **:** To ensure purity of hybrid seed and avoid pollination by unwanted pollen, isolation is a must.

Space isolation : No other rice varieties should be grown except pollen parent with a range of 100 m distance.

Time isolation : A time of over 20 days is practiced (The heading stage of other variety with in 100 m range should be 20 days earlier or later over the MS line).

Barrier isolators : Topographic features like wood lot, biological barrier crops like sugarcane, sorghum, coconut etc., tall green manure crops like sesbania and daincha crops to a distance of 30 m / artificial obstacles of (plastic sheet) above 2 m height.

3. Optimum time for heading and flowering

Favourable climatic conditions for normal flowering are

(i)	Mean temperature	-	24 – 28°C
(ii)	Relative humidity	-	70 – 80%
(iii)	Day and night temperature difference	-	8 – 10°C
(iv)	Sufficient sunshine		
(v)	Sufficient breeze or air movement		

4. Synchronization of flowering

As the seed set on MS line depends on cross pollination it is most important to synchronize the heading date of the male and female parents especially for the hybrid combination having the parents up to quite different growth duration. In addition, in order to extend the pollen supply time, the male parent is usually seeded twice or thrice at an interval of 5-7 days.

5. Row ratio, row direction and planting pattern

Row ratio refers to the ratio of number of rows of the male parent to that of the female parent in the hybrid seed production field. The layout of row ratio depends on the growth duration of the R line, growth vigour of the R line, amount of pollen shed and plant height of the R line.

Row ratio and spacing of A and R lines in the main field

R	R	A	A	A	A	A	A	A	A	R	R
0	0	*	*	*	*	*	*	*	*	0	0
0	0	*	*	*	*	*	*	*	*	0	0
0	0	*	*	*	*	*	*	*	*	0	0
15 cm											
0	0	*	*	*	*	*	*	*	*	0	0

30 cm 20 cm 15 cm (male : female ratio = 2:8)

- R line should have enough pollen to provide
- The row direction should be nearly perpendicular to the direction of winds prevailing at heading stage to facilitate cross pollination. Practically, a row ratio of 2:8 is currently widely used in India for hybrid seed production. Generally, the R line is transplanted with two to three seedlings per hill and separated by a spacing of 15 cm from plant to plant, 30 cm from one row of restorer to another and 20 cm from CMS line. The MS line is transplanted with one to two seedlings per hill with a spacing of 15 x 15 cm. A good population structure to get more seed yield is given below :

(a) Seeding / hill	(b) Hill / sq.m	(c) Effective tillers / sq.m
A line = 1 – 2	A line = 30	A line = 300
R line = 2 – 3	R Line – 5	R line = 120

(6) Prediction and adjustment of heading date

Even if the seeding interval between both parents is accurately determined, the synchronization of their flowering might not still be attained because of variation in temperature and difference in field management. By examining the primordial initiation of panicle, the flowering date can be made by applying quick releasing nitrogen fertilizer on the either developing parent and the later developing parent with 2% solution DAP spray. By this measure, a difference of 4 to 5 days may be adjusted.

(7) Leaf clipping, gibberellin application and supplementary pollination

These techniques are very effective for increasing the out crossing rate.

(a) Leaf clipping: The leaves taller than the panicles are the main obstacles to cross pollination and, therefore, should be cut back. Generally leaf clipping is undertaken 1-2 days before the initial heading stage, and more than 2/3 rd of the blades of flag leaves are cut back from the top.

(b) Application of gibberellin (GA_3): GA_3 can adjust physiological and biochemical metabolism of rice plant and helps in hybrid seed production by stimulating the elongation of young cells. In most of the CMS lines, about 20-30% of spikelets of a panicle are inside the flag leaf sheath (exsertion is only 70%). GA_3 effects exsertion of panicle completely out of half leaf sheath. In India, recommended dose of GA 3 is 50 g/ha using knapsack sprayer and 25 g/ha with ultra low volume sprayer.

Advantages

- Enhances panicle and stigma exsertion
- Speed up growth of late tillers and increase effective tillers
- Flag leaf angle is increased
- Reduces unfilled grains
- Enhances seed setting and seed yield

Spraying stage : 5 per cent of panicle emergence

Spraying time : 8-10 AM is the best time

(c) Supplementary pollination

Shaking the R lines panicles by rope-pulling or rod driving during anthesis can enhance the crossing rate. This is carried out during peak anthesis (10-12 AM) period.

8. Rogueing

To get 98% purity of CMS lines and R lines, in addition to strict isolation, a thorough rogueing is also necessary.

9. Harvesting and processing

- The male parent harvested first
- Care should be taken to avoid admixture of male and female lines

- Female line should be threshed separately in a well cleaned threshing floor
- Seed dried in shade to 12% moisture content
- Packed in suitable, cleaned gunny bags after grading

16.6 FUTURE PROSPECTS

From the above, it is now clear that bulk quantities (1-2 t/ha) of hybrid rice seed can be obtained in tropical, subtropical and temperature countries by using suitable seed and pollen parents and by adopting correct seed production technology. Further, the seed yield can even be increased beyond 2.9 t/ha by developing seed and pollen parents with higher out crossing potential and higher labour inputs. An economic evaluation of hybrid rice seed production has been reported by several workers, (He *et al.,* 1988). Studies have shown that hybrid rice seed production is profitable, if the commercial hybrid rice seed yield is around 2.0 t/ha. From the economic analysis made on hybrid rice in China, the following conclusions were made with regards of prospects of hybrid rice in future.

i. This technology would be relevant for countries where the rice supply is projected to fall short of the demand.

ii. It would be acceptable for irrigated rice ecosystems where there is no more increase in the yielding ability of pure line varieties.

iii. The technology likely to succeed in the countries like India, Indonesia, Sri Lanka, Vietnam and Philippines where wages are low due to high ratio of labour : land ratio.

iv. Region specific research and seed production infrastructure of hybrid rice can be economically feasible to set up in large countries like India and Indonesia, those are having significantly large irrigated areas.

v. Before making a large scale commitment to hybrid rice related infrastructure, a detailed assessment of land and labour requirement ought to be made.

vi. Unit cost of GA_3, an important component of hybrid rice seed production technology, would decrease with an increase in the scale of production.

According to Dr. M.S. Swaminathan (1993), the hybrid varieties have three advantages, first, they increase average yield in general; second, they provide more skilled jobs in seed production and third, they also provide a tool to improve the crop management practices at farm level. Given the prominence of rice in Asia, rice hybrids can have same catalytic effect on the development of region's seed industry as that of hybrid corn seed industry in North America. There is an urgent need to effectively transfer the technology already generated by conducting on-farm and frontline demonstrations. There is a need to motivate, mobilize, activate and coordinate the seed production personals. Large scale adoption and spread of this technology will primarily depend upon the economics of seed production. Rice hybrids with higher magnitude of heterosis, coupled with better grain quality and resistance to diseases/pests need to be developed. The hybrid rice seed production technology is to be further fine-tuned to obtain an average seed yield of about 2.0 t/ha commercially. The current level of heterosis (15-20%) in indica rice hybrids is economically viable, but a higher level observed in indica-japonica hybrids will be more attractive. This would require tackling more associated problems like hybrid sterility, tall stature, late maturity and poor grain quality. Some indica-japonica rice hybrids have been developed and are under testing in countries like China, Korea and Japan.

Rice hybrids with desired disease/insect resistance and acceptable grain quality can be developed by an appropriate choice of parental lines. Therefore, while selecting parental lines, their level of disease/insect resistance and grain quality should be monitored critically. In order to develop rice hybrids possessing premier grain quality such as basmati, both parents of hybrid must possess basmati qualities viz., long slender grain, high grain elongation ratio, intermediate amylose content, soft gel and strong aroma. The low frequency of restorers among basmati lines is an important handicap, which can be overcome by using the TGMS system to develop basmati rice hybrids. Experiments have shown that genetic improvement for flowering behaviour and floral traits of seed and pollen parents and modification in seed production practices are useful in increasing seed yield beyond 2.0 t/ha. Therefore, incorporation of the elongated uppermost internode gene in male sterile lines should be useful in improving their panicle exsertion. Seed born diseases are major problems in hybrid rice, therefore, they must be tackled. In future, A, B and R lines can be purified by deployment of biotechnological tools through anther culture, transfer of CMS into elite breeding lines, tagging of WC genes, characterization of genetic diversity of parental lines and transfer of genes for resistance to diseases/pests etc.

In the area of seed production, efforts must be directed to increase the out crossing potential of CMS lines and suitable management practices to increase out crossing rate in seed production should be vigorously investigated. Various measures to bring down the cost of hybrid seed production search for alternatives to GA_3 use etc. must be vigorously explored. Possibilities are to be explored to spread this technology in favourable rainfed shallow lowland ecosystem and for irrigated winter season (boro) in Eastern India.

CHAPTER 17

RICE BASED INTEGRATED FARMING SYSTEM

Population has been increasing by leaps and bounds. India's population is expected to be 1370 and 1660 million in 2030 and 2050 AD respectively. The Food production to the level of 289 m.t and 349 m.t. are needed to satisfy the above population. It becomes necessary to increase production level per unit area almost to double from what we is to-day. It is not always possible to increase the productivity vertically beyond certain limit. Horizontal expansion also has come to zero level. In the years to come, it is expected that the cultivable area is going to be on the decreasing trend by virtue of diversion of part of it for industrial purposes. Evidently we have to look forward an alternative to cope up the nutritional requirement of ever increasing population. In this context, talking merely on quantity of grains of cereals, pulses, oilseeds and their productivity will not solve the problem. Ultimately, how best the nutritional value of grains can be thought of to be satisfied with the alternates like egg, meat, milk, vegetables, fruits, edible mushroom etc. These enterprises can very well be taken up for productivity with an area, which is quite unfit for crop activity viz., elevated patches in the field, around fence line, adjacent to residence, store room, thrashing floor etc. Compounding with increasing population in most of the tropical countries with a (Subbian and Palaniappan, 1992) warrant to produce more food from the limited land available for cultivation, the agricultural systems have encountered numerous ecological and environmental constraints. The goal of sustainable agriculture can be achieved by seeking the optimal use of internal production inputs in a way that provide acceptable levels of sustainable crop-productivity and live stock production resulting in economically profitable return. It would be possible to reach this level with proportionately less use of inputs through integrated farming systems. The choice of components in the farming systems and their management must ensure that the interaction between components should be the most complementary with least competitiveness. Integrated farming system approach is not only a reliable way of obtaining fairly high productivity, but also a concept of sustainable agriculture (Swaminathan, 1987).

17.1 WHY INTEGRATED FARMING SYSTEM?

In an agricultural country like India, the average land holding is very small. There is no scope for increase in farm size. As mentioned above, The population is steadily increasing without any possibility of increase in land area. The income from cropping of an average farmer is hardly sufficient to be assured of regular income for a decent living. He has to think of other enterprises, which will complement his cropping activity. This is made possible through optimal crop and livestock mix consistent with the farm resources immediately available. Farming includes crop, livestock, poultry, fishery and tree culture. A judicious mix of one or more with cropping complements, the cropping enterprise through effective recycling of residue/ waste. A carefully chosen and planned farming system,. keeping in view the soil and environment conditions, users in greater dividend. Farm as a unit is to be considered and planned for effective integration.

The future of Indian agriculture depends heavily on the development of appropriate farming system as it fits to resource poor farm families of different agro-ecological zones. The endowment of abundant sunshine, long growing season, responsive soil type, combination of surface water, ground water and seasonal rains and above all a progressive peasantry offer vast scope for an intensive farming system through multiple cropping and diversified farming including animal husbandry, forestry, sericulture, fisheries and the like. The challenge is to upgrade technological and social disciplines on a continuous basis and integrate these disciplines to suit the region and the farm families in a manner that may ensure increased production with stability, ecological sustainability and equitability. Research studies carried out in lowlands, uplands with supplemental irrigation and dry lands have demonstrated the technical feasibility and economic viability of the integrated farming systems attempted. Besides facilitating cash income, these models generate additional employment for family labourer and minimize the risks associated with mere crop production. If the farming system is linked with agro-processing and other post harvest technology, the farm economy could be further stabilized and strengthened. This requires a massive extension approach and institutional support.

17.2 DEFINITION

Farming system is a complex inter-related matrix of soil, plants, animals, implements, power, labour, capital and other inputs controlled in part by farm families and influenced to varying degrees by political, economic, institutional and social forces that operate at many levels. The farming system, therefore, refers to the farm as an entity of interdependent farming enterprises carried out on the farm. The farm is viewed in a holistic manner. The farmers are subjected to many socio-economic, biophysical, institutional, administrative and technological constraints.

Integrated Farming is defined as biologically integrated farming system which integrates natural resources and regulation mechanisms into farming activities to achieve maximum replacement of off-farm inputs; secures sustainable production of high quality food and other products through ecologically preferred technologies; sustains farm income; eliminates or reduces sources of present environmental pollutions generated by agriculture, and sustains the multiple function of agriculture (IOBC, 1993). Thus, integrated farming systems must be built on the sound foundation of ecological science. It emphasizes a holistic approach. Such an approach is essential because agriculture has a vital role to play that is much wider than the production of crops, including providing diverse, attractive landscapes and encouraging biodiversity and conserving wild life. Sustainable development in agriculture must include integrated farming systems with efficient soil, water, crop and pest management practices, which are environmentally friendly and cost effective. The future agricultural system should reorient from the single commodity system to food diversification approach for sustaining food production and income. Integrated farming systems, therefore, assume greater importance for sound management of farm resources to enhance farm productivity, which will reduce environmental degradation and improve the quality of life of resource poor farmers and to maintain agricultural sustainability. The aim of the integrated farming system can be achieved by:

- efficient recycling of farm and animal wastes;
- minimizing the nutrient losses;
- maximizing the nutrient use efficiency;
- following efficient cropping systems and crop rotations; and

- complementary combination of farm enterprises

The farming system conceptually is a set of elements or components that are interrelated which interact among themselves. At the centre of the interaction, the farmer is exercising control and choice regarding the type and results of interaction. Any farming system, however, is subject to what is potentially possible in technical terms. It is the human environment that provides sufficient condition for development and utilization of a particular system. A farming system obviously is very complex. Therefore, any agricultural technology well suited to a particular agro-ecological situation and socio-economic environment may not be adopted by other farmers. Unless both natural and human environments are considered, agricultural research will not result in relevant agricultural technology. The income from cropping alone in small and marginal farms is hardly sufficient to sustain the farmers' family. With the decline in farm size (0.15 ha/ person) due to explosion of population, it would be increasingly difficult to produce enough food for the family at 21st centuary. Therefore, the farmer, to be assured of a regular income for a decent living (above the poverty line), a judicious mix of any one or more of these enterprises with agronomic crops should complement the farm income and help in recycling the farm residue/waste. The selection of enterprises must be based on the cardinal principle of minimizing the competition and maximizing the complementarily between the enterprises (Annadurai *et al.*, 1994).

17.2.1 The merits of Integrated Farming Systems

1. Productivity is increased
2. Profitability is enhanced
3. Balanced food production is achieved
4. Helps in sustaining the productivity of soil through crop residue recycled as organics
5. Prevents pollution and conserve the environment
6. Availability of money round the year is made possible
7. Solves fodder crisis, energy crisis, fuel and timber crisis
8. Increases employment opportunity for the family labour
9. Provides opportunity for agri based industries
10. Increases input use efficiency
11. Improves standard of living
12. Decreases encroachment towards forestry for timber and fuel wood.

Integrated Farming system, a component of Farming System Research introduces a change in the farming techniques for maximum production in the cropping pattern and takes care of optimal utilization of resources. A judicious mix of agricultural enterprises like dairy, poultry, piggery, fishery, sericulture, etc. suited to the given agroclimatic conditions and socio-economic status of the farmer, would bring prosperity in the farming.

17.2.2. Basic phases of IFS

1. System description, diagnosis and problem specification
2. On station design
3. On station solution testing followed by on farm solution verification and feed back
4. Extension of research results, training and policy formulation for extension support.

17.2.3. IFS Setting

Prior to fixation of enterprises with crop activity as base, the following norms are to be assessed.

a. Survey the area and assess its major potentiality viz., lowland/ dryland/gardenland

b. Identify major cropping pattern of the region

c. Assess the predominant enterprises of the tract

d. Marketing for the produce should be examined

e. Financial background of the farmer should also be ascertain

Considering the facts indicated suitable model should be fixed to the locality / region.

17.3. INTEGRATION OF SUB-SYSTEMS

The various enterprises that could be included in the farming system are crops, dairy, poultry, goat rearing, fishery, sericulture, agroforestry, horticulture, mushroom cultivation etc. Thus, it deals with the whole farm approach to minimize risk and increase the production and profit with better utilization of wastes and residues. It may be possible to reach the same level of yield with proportionately less input in the integrated farming and the yield would be more sustainable because the waste of one enterprise becomes the input of another, leaving almost no waste to pollute the environment or to degrade the resource base. To put this concept into practice efficiently, it is necessary to study linkages and complementarities of different enterprises in various farming system. The knowledge of linkages and complementarities will help to develop farming system (integrated farming) in which the waste of one enterprise is more efficiently used as input in another within the system.

Generally, there are 3 different farming situations, viz., rainfed, irrigated and wetland farming systems. The improved cropping systems, which form the major component of farming system were best verified and evaluated in farmers holdings at various locations in different agroclimatic conditions. Depending on the ecosystem, the enterprise mix will vary.

Wetland system : e.g. Rice-based cropping system + fish culture +duck / poultry

Garden land system : e.g. Garden land crops with forage + livestock/ poultry (crops with fodder, mulberry + dairy + silkworm rearing)

Dry land system : e.g. Grain / fodder crops + sheep/goat rearing + agroforestry (tree crops/ agri-silvi / agri.horti.)

Arable farming results in narrow profit or less margin to the farmers, warranting development and adoption of integrated farming systems with various combinations of components. Integration of milch cow with cropping resulted in higher net returns through maintaining high yield of crop and animal than cropping alone (Arora, 1980; Siddeswaran and Palaniappan, 1990, and Chinnusamy and Purushothaman, 1992). Increased net returns due to the introduction of poultry-cum-dairy based farming systems over cropping alone was reported by Sirohi *et al.,* (1980) and Govindan (1988). Higher net return could be obtained with duck rearing in the integrated farming systems (Sharma *et al.,* 1985). Higher net income with rice + poultry + fish + mushroom culture under low land rice conditions (Rangasamy *et al.,* 1990 and Govindan *et al.,* 1990), dairy + biogas + sylviculture under garden land conditions (Rangasamy *et al.,* 1990 and Siddeswaran and Palaniappan, 1990) and cropping + goat rearing under rainfed situations (Rangaswamy *et al.,* 1990) were reported. Rice + fish + azolla integrated farming system was found to be the best for maximum net and per day returns over conventional cropping under lowland rice fields in Tamil Nadu, India (Balusamy and Shanmugasundaram, 1994 and Balasubramanian, 1994). Integrated farming systems involving dairy cattle became more relevant under existing ecological and socio- economic conditions in the tropics and would provide an increased quantity of manure for cropping (Swaminathan, 1975). Vikraman Nair *et al.,* (1979) and Sundaresan (1979) found that there was little loss of nutrients from the field as those fodder crops removed were returned to the soil in the form of cattle manure in integrated farming systems. Also by recycling the wastes of dairy cow and poultry birds, the value of nutrients added to the soil could be increased Mohamed Ali *et al.* (1984).

17.4 COMPONENTS TO BE INTEGRATED IN WET LAND ECOSYSTEM

- Cropping
- Fishery
- Livestock-Poultry, Diary, Duck
- Pigeon
- Tree crops, Apiculture, Piggery, Mushroom cultivation, Rabbit, Japanese quail to some extent in the possible areas

17.4.1 Integrating Aquaculture

Natural resources available to a farm family may include water resources, such as streams, ponds and flood prone land. The husbandry of plant and animal organisms that live in water is known as aquaculture. Integrating this form of husbandry into a farming system intensifies the use of natural resources in a sustainable manner through species diversification and nutrient recycling. Integrating fish, animals, trees, vegetable and field crops within a farm is one way of maximizing productivity per unit of land. By-products from one form of resource serve as inputs for other forms, and agricultural wastes can be used to make marginal patches of land more productive.

Fish ponds can be created in wetlands and on homestead land where a nearby stream or spring permits. Fruit trees and vegetables planted on dikes beside the pond can be watered with pond water, which can also be used to water livestock. Manure, crop residues, weeds, tree leaves, rotten fruit and vegetables fertilize the pond. Other crop by-products, such as maize and rice bran can also be fed to fish. Fish convert plant and animal waste into high-quality protein and enrich pond mud, which can then be used to revitalize cropland. Small holders in densely populated areas of Asia have already developed various integrated agriculture-aquaculture systems.

It is not uncommon to find several tanks and ponds amidst the paddy fields, constructed mainly for storing water for irrigating the agricultural crops. However, they are not fully utilized for raising fish in them for generating additional income. Such water resources can be profitably utilized by adopting the scientific aquacultural technology by making use of the byproducts of agriculture such as oilcakes, rice and wheat bran and those of the animal husbandry such as the excreta of the cattle and pigs and the droppings of ducks and poultry. In return, the aquaculture system offers the fish offal and fish meal to poultry and the organically rich bottom soil for manuring the agricultural land. The grass carp which subsists exclusively on various aquatic weeds such as *Hydrilla, Najaa, Ceratophyllum, Lemna, Wolffia, Azolla, Spirodella,* etc. can be fed with the supplementary feed of chopped green cattle fodders such as berseem, lucerne, napier grass, maize leaves, banana leaves etc. Such fodder grasses can be grown on the embankments of the fish ponds. The carps (catla, roghu, mirgal, common carp, grass carp, and silver carp) included under the polyculture system attain the marketable size of 1 kg in an year and a fish yield ranging from 4 to 10 t / ha / yr is achieved under different agro climatic conditions of the country. A financial return of 58 per cent on variable cost has been obtained under this polyculture system of freshwater aquaculture.

Aquaculture has the same objectives and practices of agriculture. It aims to increase by way of all possible means, the production of food far above the level of natural wild production. Those who take to fish culture may range from farms set up by large-scale industrial enterprise to ponds of small family subsistence units. The aquatic animals possess a number of advantages for use in husbandry. The body density of fish and other aquatic swimming animals is nearly the same as that of water in which they inhabit and this enables them to have savings in the energy used for locomotion. Fishes to build up the body utilize this energy. In addition to this, all the aquatic animals are cold blooded and expend no energy in thermo-regulation. The water body is a three-dimensional growing space. Many of the highest aquaculture yields have been achieved through polyculture of fish inhabiting regions of the pond. To understand more about the fish culture, the details regarding composite fish culture, preparation of ponds, how to remove the predatory fishes from the ponds, how to fertilize the ponds, how to stock the fingerlings and supplementary feeding to fish are given in details for the benefit of readers, which are as follows.

17.4.1.1 Composite Fish Culture

Fish culture in inland waters as an economic and paying proposition or trade has been carried on to a great extent in Tamil Nadu, Andhra Pradesh and Karnataka. A sudden fillip to pond fish culture was given with the innovation of polyculture or composite fish culture technique in 1970's. The technology developed under the composite fish culture programme has put India on the threshold of aquapulsion. Though this system of farming with different species of carp together was practiced in Bengal even earlier, the availability of the Chinese carps (silver carp and grass carp) and common carp for farming from mid-seventies changed the potentiality of pond fish production. The principle involved in this culture is fuller utilization of fishpond in a body of water. Three exotic fishes viz. grass carp, silver carp and common carp along with the three indigenous varieties viz. catla, roghu and mirgal are grown in the same pond.

The feeding habits of these fishes are follows:

Catla : Surface (microscopic animal or zooplankton) feeder,

Silvercarp : Surface (microscopic plants or phytoplankton) feeder.

Roghu : Column feeder.

Grass carp : Aquatic vegetation feeder

Common carp : Omnivorous bottom feeder.

17.4.1.2 Preparation of Ponds

Before fish seeds (fingerlings) are stocked in the ponds, it is essential to clear them of aquatic vegetation, predatory and weed fishes, and excessive silt, if present at the pond bottom.

1. If the vegetation in the bottom is scanty and scattered, it can be cleared by manual labour. If the vegetation is dense, manual labour may not be economical or satisfactory and herbicides may have to be used to clear the weeds. The herbicides include sodium salt formulation of 2, 4-D for water hyacinth, lilies, sedges etc. and gramaxone for water lettuce. Grass carp can effectively control some varieties of weeds such as duckweeds, Hydrilla, Najaa, Ceratophyllum etc., if adequate number of these are introduced in the pond.

2. Clearing a pond of predatory fishes such as murrels and the fresh water shark and weed fishes before stocking the same with carp fingerlings will ensure the safety of the stocked fish. These unwanted fish can be eradicated either by dewatering or by poisoning the pond. This can be done most economically during the summer months when the subsequent entry of predators and other undesirable fishes and the escape of the fish stocked, by screening the inlet and outlet with fine mesh wire netting. Pond embankments should be strengthened and raised sufficiently high to prevent possible entry to such fishes from adjacent waters during the rains. Mahua oil cake is used in clearing pond of unwanted fishes. The toxicity of mahua oil cake may remain for about 15 days in the treated pond. Stocking, therefore, should be done after ascertaining complete detoxification of water.

3. In order to supplement the natural fertility of the pond, it is necessary to apply fertilizers. A combination of both organic and inorganic fertilizers is considered more effective than either of them alone. Prior to application of fertilizers, it may be desirable to apply lime in a single dose depending on the soil condition.

Mahua oil cake	-	200 kg/hectare
Cow dung	-	10000 kg/hectare
Super phosphate	-	250 kg/hectare
Quick lime	-	500 kg/hectare

4. Ponds may be stocked with selected species of fish in particular combinations.

With Exotic carps		**Local carps alone**	
Catla	2000	Catla	3000
Roghu	2000	Roghu	3000
Mirgal	1500	Mirgal	2000
Common carp	1500	Common carp	2000
Silver carp	2000		
Grass carp	1000		
	10000/ha		10000/ha

5. To achieve high production of fish it is necessary to provide supplementary feed to the fish in the ponds. Feeding of fish should be done daily with a mixture of powdered oil cake mixed with an equal weight of sieved rice bran.

17.4.2. Integrated Fish Farming

This system recycles the organic wastes such as weeds and crop, by products from fields, wastes from the farm itself. In this system, substances that may otherwise be considered as waste is utilized to increase fish production. Pond bottom humus is used as fertilizer for crops grown adjacent to ponds or directly on their dykes. Live stock manures (Pig, duck, goose, chicken) are used as pond fertilizers for plankton production. Agricultural by products from processing, i.e., wheat and rice, coconut or groundnut cake is also put into the ponds as fish feed. Recently anaerobic digesters are being used in integrated fish farms to produce methane gas as an energy source from agricultural wastes and manures. The nitrogen rich slurry residues are then used to fertilizer ponds. Even with the great variety of pond inputs, the limited quantity of locally available nutrients is the central factor restricting pond area expansion where integrated fish farming is presently practiced. There are various systems of integrated fish farming with rice and the most common are given below.

1. Paddy cum fish culture
2. Paddy-fish-azolla farming
3. Paddy-Fish-vegetable farming
4. Paddy and Fish culture (pond system)
5. paddy-poultry-fish culture

17.4.2.1 Paddy-Cum-Fish Culture

This farming system of growing fish in the paddy fields is commonly practiced in the coastal areas in the States of Kerala and West Bengal. However, the paddy-cum-fish culture in the freshwater paddy fields is rare. This production system has great scope in the States of West Bengal, Assam, Bihar, Uttar Pradesh, Orissa and Andhra Pradesh where only a single crop using the deep water paddy varieties is cultivated. Investigations carried out by CIFRI have indicated that the freshwater fishes can be cultivated in the specifically designed paddy fields with trenches or pools constructed within the paddy plots so that two crops of paddy (tall variety of rice in kharif and high yielding dwarf variety of rice in Rabi) and a single crop of fish, covering both the seasons can be raised. The trenches can be classified as under:-

1. Perimeter type
2. Central pond type
3. Lateral trench type

The depth in the trench may vary between 1 m and 1.5 m. Composite fish culture can be undertaken in these trenches with fingerlings of catla, roghu, mirgal and common carp. Deep-water paddy varieties are usually selected for cultivation during kharif season. Enough care must be taken to avoid the spraying of pesticides during the cultivation. In order to overcome this problem of pesticides affecting the fish, integrated pest management may be adopted. Pesticides like neem-based pesticides, carbonates and selective organophosphates may be used in the low doses so that the fishes are not affected. When rice is harvested during November-December, the water recedes from the rice field and the fishes take shelter in the trenches. With the onset of the rabi season, the rice field is once again prepared and the seedlings of the high yielding variety of paddy such as Ratna, Jaya, Pusa etc., are transplanted. Before the application of the pesticides, an earthen ridge is erected along the periphery of the paddy plot to prevent the drainage of the pesticides into the trenches with the fishes. In this system of integrated farming, a fish yield of 700-1000 kg/ha/year is expected, giving a financial return of 20 per cent over monocropping of rice.

17.4.2.2. Rice-Fish-Azolla Farming

A field study was conducted at low lying wetlands of Lower Bhavani Project (LBP) Area, Agricultural Research Station, Bhavanisagar, Tamil Nadu with a view of evaluating the feasibility and economics of rice-fish-azolla farming. Experiment was conducted with two rice crop seasons (kharif and rabi) planted with rice varieties ADT.36 and ADT.38 with a total duration of eight months with one fish crop harvest. *Azolla microphylla* was applied at the rate of 2.0 t/ha as fish feed cum N fixer five days after transplanting. *Calatropis*, a wasteland weed, was incorporated (12.5 t/ha) as green leaf manure, 10 days before transplanting. Fertilizers were applied at the recommended level (100:50:50 kg NPK/ha). Irrigation was scheduled to maintain 5 cm depth of water throughout the crop growth period. Need based plant protection measures were taken with 5 per cent neem seed kernel extract. Fingerlings of catla, roghu and mirgal were stocked in equal ratio with a total of 3000 fingerlings/ha, 15 days after transplanting. Banana pesudostem with cowdung (1:1) mixture and rice bran were fed at the rate of 5 per cent of the fish body weight. Field trenches were proved with 1.0 m depth and 1.0 m width occupying 10 per cent of the rice area to shelter the fish (Shanmugasundaram and Balusamy,1993).

The results revealed that in low-lying wetlands the most profitable farming system was rice-rice-azolla-calotropis + fish farming. There was a slight decline in rice yield under rice-rice-fish farming due to the presence of fish trenches occupying 10per cent of the area. However, the yield reduction was made to good when azolla and calotropis were introduced. The soil fertility status indicated that in general, there was an increase in the available nutrient contents after completion of cycle. Even after heavy depletion, through crop removal, the nutrient status could be replenished by the addition of organic residues. This might be due to decomposition of azolla, fish manure and fish feed in the system.

17.4.2.3 Rice-Fish-Vegetable Farming

For effective restoration and recycling of organic wastes an experiment was started to integrate rice-vegetable and fish culture in wetlands at Agricultural Research Station, Bhavanisagar, Tamil Nadu. The objective is to maximize the return per unit area and also efficient recycling of crop residues along with fish culture. The experiments consisted of rice alone, rice-rice +

fish, rice-rice-vegetable + fish farming. The experiment was continued for two rice seasons with a total duration of eight months. The study area was 0.25 ha/system. Field trenches (1 m deep x 1 m wide) were provided for fish shelter. Catla, roghu and mirgal were stocked 15 days after transplanting at an equal ratio of 3000 fingerlings/ha. Vegetable crops viz., bitter gourd, ribbed gourd, beans etc. were raised along the raised bund of fish trenches. The results indicated that rice-rice-vegetable + fish culture was more remunerative and it was possible to get two rice crops, one vegetable crop and one fish crop in eight months in rice fields.

17.4.2.4 Rice and Fish Culture (Pond System)

By integrating the fish culture with rice, increased returns could be obtained. Further, the nutrients leached out from rice fields could be effectively utilized for fish culture. With a view to study the feasibility of rice-cum-fish culture under low-lying wetlands, an experiment was conducted at Agricultural Research Station, Bhavanisagar, Tamil Nadu. In one acre holding, conventional cropping system rice-rice-black gram practiced by the farmers, was followed. In another one acre plot, integrated farming system (rice-rice-maize + fish culture) was practiced with an area of 0.85 acre for crop husbandry and 0.15 acre was allotted for fish culture. Fish pond size of 600 m^2 was dug to a depth of 2.0 m. The fish fingerlings of catla, mirgal and common carp were stocked in the fishpond to a density of 3000 No./ha. The excess water in the paddy fields was drained once in four days from 10 days after transplanting of rice. After draining, irrigation was given up to 5 cm depth each time. Whenever fertilizer was applied as top dressing, the subsequent drainage was done six days after fertilizer application. The same practice was followed for plant protection chemical application. Supplementary feeding were given @ 5 per cent of body weight of fishes with rice bran and available farm wastes.

The comparative economics showed that rice-rice-maize+fish culture recorded maximum net return and it increased over by Rs. 6,383 and Rs. 5,795 / ha over conventional cropping of rice-rice–black gram during 1991-1992 and 1992-93 respectively. Drained water increased the yield of fishes and this might be due to increased nutrients available in the drained water from the pond as revealed by increased plankton population.

17.4.2.5. Paddy- Poultry- Fish Culture

The details regarding this system are given in section 17.5 of this chapter.

17.4.3. Integrated Crop-Livestock-Fish Farming

Integrating crops, livestock and fish in smallholder farming systems has many ecological and economic advantages. Such systems are conducive to nature conservation as they promote habitat stability and diversity for wildlife living on the farms and in adjacent areas. As these integrated systems optimize the use of on-farm and adjacent resources, they encourage habitat conservation rather than destruction. Such systems are productive and profitable because they utilize waste as inputs in other enterprises within the farm, and because fish is a highly nutritious and valuable traditional food and it is called as "poorman's meat". They use microenvironments within a farm system, which add to farm productivity and security. The following bioresource flow models, prepared by farmers in Vietnam and Malawi, illustrate how farm enterprises are integrated to exploit different land and water resources and to substitute for external inputs. Making these models helps farmers transform their farming systems by clarifying how existing enterprises can be better integrated and how new enterprises can be better fitted into the farming system.

In Vietnam, the bio-resource flow models show how water from the trenches is used to irrigate vegetables, grown in a mulch of rice straw, on dikes made form trench mud. Immediately after trenching, chicken and cattle manure is put into the rice-field trenches to promote phytoplankton blooms for the fish and shrimp to feed on. Although the shrimps' diet is primarily natural, they are also fed during their first two months with farm-grown by-products, such as germinated rice grain, cassava flour, rice bran, coconut and groundout oilcake, and trash fish from the irrigation canals. Mango and eucalyptus branches are also put in the trenches to keep out cattle and poachers, and to provide an undistributed habitat, which shrimps need. Such integration reduces the need for external inputs. As the shrimps and fish eat rice weeds, weeding expenses can be reduced by 30per cent with no detrimental effect on rice production, because animal manure and fish faces fertilize the paddy.

In Malawi, bio-resource flow models show how fish ponds are integrated with other farm enterprises. The pond water is fertilized with rotten fruits of guava, papaya and avocado, and manure from cattle, sheep and goats. Leaves of leucaena, pumpkin, wild vegetables and cowpea are also put into the ponds to feed the fish, and coarse maize bran is used as a fish food when available and affordable. Apart from fish, other outputs from the ponds are fertile sediments for the vegetable gardens and water for irrigation. The main purchased input on these farms – maize bran – can be supplemented and partly substituted with cocoa yam leaves and napier grass.

17.4.4. Ducks

Ducks can also be integrated with rice based integrated farming systems to get additional return and manure for rice crop. They are considered as biological weeders/animals for weed control especially nut grass. In general, ducks are raised for eggs and meat purposes in India. At some places they are kept for ornamental purposes in parks, sanctuaries etc. There are various breeds of ducks in India and some of them are good producers. Indian runner, both white and fawn, and white and khaki Campbell are good breed for egg production. They weigh between 2 and 2.5 kg. The duckling can be reared in the rearing sheds after hatching, up to 4 weeks. During this time, they should be properly fed. The ration should contain adequate protein and energy (minimum of 2,500 kcal/kg) for rapid growth. For table purpose ducks, the birds can be fed on pelletised feeds, containing 17 per cent protein up to 4 weeks and 15 per cent up to marketing weight. They can be raised on dry or wet mashes. It is better to feed green feeds, which are good quality like berseem, Lucerne etc. The duck mash can be compounded with normal feed ingredients used for poultry. The mach can be fed as dry mash or wet mesh. Refuse from the kitchen can also be utilized by the ducks. However, stale bread infected with fungus, moulds etc., should not be fed.

Plate 48. Duck rearing in wetland ecosystem

17.4.5 Pigeons

Like ducks, pigeons can also be included in rice based integrated farming systems to get additional return and manure for rice crop. An area adjustment to motor room/store room/ rest room can be allotted for pigeon rearing. Rearing pigeon for meat is an accepted industry in Europe and U.S.A. Young pigeon reared for meat is called squabs. These squabs have not started to fly, not eaten anything, other than pigeon milk. They are ready to fly at 26 – 30 days. At that time, they weigh about 500 g and are ready for market. Popular breed used is Red Carneair, White King. Also crosses of these and crosses with Mondain is suitable for squab rearing. There are many other breeds like King breed, Red, Blue and Dark king. All these are based on colour of bird. White king originates from USA. The adult weighs about 750 – 800 g. The Red carneair originates from France, it is not very prolilfic breeder like white king and the adult is slightly smaller, weighing about 650 – 740 g. There are many other breeds like Mondain, which has two varieties such as French and Swiss. These originate from USA. Many housing pigeons are also used for squab rearing like Hungarian, Maltese, and Runt pigeon. But squabs produced out of these tend to be small legs, which are too long or do not tend to be prolific breeder.

a. Housing

Housing should be comfortably dry and built along east- west as long axis. The building constructed with flight areas or intensive production can be practiced. Breeding a pair of pigeon will need 0.4 sq.mt. If they are raised under intensive system, they may require 50 per cent more of area. Each pen holds about 25 – 30 pairs. A comfortable shed should have 4m height at back and 2.5 m at front. The litter material of wood shavings or groundnut shells is kept for height of 7 cm. If shed is built on clay soil, then concrete floor is needed. In the flight area, perches are provided.

b. Nest

Double nests are necessary. When squabs are not old enough, they leave the nest and males continue to feed the squab until they are old enough to market. Nest bore for a pair of pigeon should be 40 cm height and 60 cm width. This width is portioned so that each nest will have 30 cm width with a depth of 40 cm and nest bores are fitted with 20cm flat form, so that birds will have easy access to the nest. A small board can be fixed in front of nest to prevent nesting material from falling out. Straw or wood shavings are used as nesting material. These nesting materials are kept at one corner and birds may take required nesting material from the corner.

c. Breeding

It is always better to select breeder from proness stock. Progeny testing and record keeping is very important. A pair of pigeon should produce about 12 marketable squabs in a year with average live weight of 500 – 700 g. Birds can be mated from the age of 6 months. Some females will have breeding life up to 10 years, while males can be used successfully up to 5 years of age. It is difficult to identify sex. Normally female is smaller in size having finer head. It tends to waddle rather to walk and holds the tail slightly higher up. While handled and checked the pelvic bore is kept further apart. Male is bigger in size, more aggressive and makes low voice. During the breeding season, male tends to struck around the female. Breeding pairs are allowed to select their own males or forced mated. Seeing the male streaking around, the female identifies natural mating pairs. These pairs are either wing banded or leg banded to prevent inbreeding. Forced mating can be accomplished by placing desired males and females together in enclosed nests bore or pen with food and waste for about 2 weeks. To avoid fighting and breaking of eggs, place only mated birds in pen. These mated breeder should not be disturbed, normally they remain together for life. But in case of mates, which have died, they are removed and it is necessary go for forced mating. Old are females to younger males and it mates, to higher yielding female.

d. Hatching and rearing

The incubation period is 17 days. Normally first squab is hatched either of 18-19th day, both parents start preparing nest and take house to sit on eggs. The female birds start laying again where squab is 2 weeks of age; once this occurs the load of feeding lies sold on the male. While the birds are fed, should not be disturbed. If a squab dies in the first couple of weeks, it is possible to put another one along with pair, if it is of same age and size. This is done to make female to lay sooner. If the parents die, the squabs can be hand fed. Squabs, which are meant for breeding, are kept with parents for 6 weeks so that they learn to eat and drink. Normal tendency is to select big birds. But this should not be done.

e. Nutrition

Pigeons can be fed with small whole grains like maize and rice to some extent.

Requirements are

Crude protein	:	13.5 per cent
Carbohydrate	:	6.5 per cent

Crude fibre	:	3.5 per cent
Fat	:	3.0 per cent

The type of grains given to the birds must be locally available and it should be cheap. In non-breeding season, protein level may be reduced. Normally during summer and autumn the birds lead to molt and the protein level is reduced, as it is uneconomical to feed at higher protein level. Normally the birds are fed twice a day. Self-feeders can be less of a problem but the birds tend to waste feed. Cafeteria type of feeding is done sometimes but once again, there is wastage of grains. 25 breeding pairs eat about 3 kg of feed daily. One breeding pair eats about 45 kg of feed and 4 kg of grit in a year.

A 500 g of squab is produced with 3 kg of feed. Along with this, the birds have to be given vitamins (Vit. A and B2, B3) to improve hatchability, fertility and mineral content. Watering is required to remove external parasites and also for drinking. One small basin of water is enough for drinking and bathing. The basin has to be cleaned and fresh water is filled up at least twice a day. Each breeding pair of birds should be provided with 180 liters of water per year (Balagopal,1994).

17.4.6 Dairy Farming

Animals can perform numerous functions in small holder systems. They provide products, such as meat, milk, eggs, wool and hides. They serve socio-cultural functions, as wealth for ceremonial feasts, and as gifts or loans, which strengthen social bounds. Under LEISA conditions, integration of livestock into the farming system is particularly important for:

- increasing subsistence security by diversifying the food-generating activities of the farm family; and,
- transferring nutrients and energy between animals and crops via manure and forage from cropped areas via use of draught animals.

In most of the rice growing areas in India, the farmers are maintaining few work bullocks for ploughing and transporting, milch cows to get adequate milk for their family needs and manure for rice crops. So, dairy farming is one of the economically viable enterprises that could provide constant income throughout the year to farmers. The success of dairy farming depends solely on the availability of inputs like feed and fodder and better marketing facilities to milk. If these essential items are made available, there would be continuous generation of income. An area adjacent to threshing floor or backyard area in all village houses can be allotted for dairying. To get maximum profit from dairying, it is advisable to rear Jersey crossbred cows (F1 crosses), since they come up very well in most of the climatic conditions, consume less feed and fodder, give more milk with high fat content and possess comparatively better disease resistance. However, Holstein- Friesian crosses could be reared for higher milk yield in places of cooler climate as they lack heat tolerance. Buffaloe breeds like Murrah crosses and even non-descript types could be reared for milk production. Buffaloes can digest more percentage of roughage than cows and thrive well on dry fodder. Buffaloes contribute nearly 60per cent of milk to the national milk grid. Higher profit could be obtained from buffaloes (Katherkhan, 1994).

Of several types of input costs, feed and fodder play a significant role in the economy of dairying. Of the total expenditure, nearly 65 per cent goes towards feeding of cattle. To make dairying economically feasible, more quantity of farm grown green fodder should be made available. It has many advantages like high yield and regular breeding. Low yielding cows (6-8 litres/day) could be maintained exclusively on green fodder (45 kg), both leguminous and non-leguminous, and a maximum of around 1.5 kg of wheat bran, 30 g of mineral mixture and 40 g of common salt would be sufficient to provide the required nutrients. Inclusion of processed and enriched straw with urea (2 per cent) and molasses (10 per cent) and mineral mixture (1 per cent) would further improve the quality of dry fodder by 3 times and help to reduce the feed cost considerably. In most of the villages in India, the farmers are feeding diary with rice straw and grasses, cut from the rice field bunds. Besides feeding with balanced ration optimum and comfortable housing facilities, good breeding management and adequate health care should be made available all the time. Very close observation and constant attention in the day to day management of dairy cattle are very important to make the enterprise economically successful. In dairy, taking into consideration of costs of several inputs and getting price of milk, it is found that farmer could easily get a net profit a nearly one rupee per litre of milk production, if family labour is engaged. If fodder is grown in the farm itself, the net profit could be further increased by 15 per cent. It is also observed from field studies that farm labourers and small and marginal farmers are maintaining cows and buffaloes on economical lines than do large farmers. It is mainly due to the maintenance of one or two cows or buffaloes and also providing excellent managerial inputs. Hence mass production of milk should be encouraged to come from masses for equidistribution of income. Calf rearing is another important area to be concentrated as it needs less managerial inputs and also provides considerable income.

17.5 CASE STUDIES

Experience through case studies and experiments conducted at Tamil Nadu Agricultural University, Coimbatore, revealed that by integration of complementary enterprises like livestock, fishery, agro-forestry etc., with rice or wet land crops, additional income and generation of additional employment potential could be achieved.

17.5.1 Model –I (Wetland farming system, Tamil Nadu,India)

A model was developed for 0.40 ha (1.00 acre) considering small and marginal farmers of Tamil Nadu State (Rangasamy *et al.*, 1988 and Ranagasamy, 1992). While programming for the cropping system, due consideration has been given to the crops under cultivation in the locality, the crop rotation that is followed, and water requirement for the fishpond. The details of each system integrated are furnished hereunder:

Components

1. Cropping II. Fisheries III Poultry

Allocation of land

I. Cropping : 0.36 ha

II. Fish pond : 0.04 ha

III. Poultry shed is placed over the pond:

Total : 0.40 ha

I. Cropping programme

Rice (IR 50)	-	Rice (IR 20)	-	Maize (CO.1)	0.16 ha
(May-Aug.)	-	(Aug. Dec)	-	(Jan. – Apr.)	
Rice (IR 50)	-	Rice (IR 20)	-	Groundnut (CO.2)	0.10 ha
Maize (CO.1)	-	Rice (IR 20)	-	Gingelly (TMV3)	0.10 ha

II. Fisheries

The number of fingerlings used for the proposed programme has been arrived at based on the recommended stocking density of 7500 numbers per ha of ponded water. Accordingly, 300 numbers at the ratio noted against each of the following varieties were released in the pond.

Silver carp/Catla (Surface feeder) : 120 (4)

Rohu (Column feeder) : 60 (2)

Mirgal/Common carp (Bottom feeder)	:	90 (3)
Grass carp (Grass feeder)	:	30 (1)

III. Poultry

Five hundred layer chicks are sufficient enough to void required quantum to excreta to meet the feed requirement of 7500 polyculture fingerlings in a hectare of ponded water. Accordingly twenty papkok chicks were used in the study.

Control: The normal crop rotation that is being followed in this tract under low land condition is also included in the programme to compare with income under integrated farming system. The details of which is furnished below.

Rice (IR 50) -	Rice (IR 20) -	Green gram (CO.3)	: 0.20 ha
(May – Aug)	(Aug.-Dec.)	(Jan.-Mar.)	
Rice (IR 50) -	Rice (IR 20) -	Sunnhemp	: 0.20 ha
	Total		: 0.40 ha

Plate 49. Cropping – Fishery – Poultry model at TNAU, Coimbatore, Tamil Nadu

In rice cultivation, even with best management practice we are getting about 5 to 7 per cent ill filled or partly filled grains. This waste product of rice when it is hulled becomes a substitute of 35 per cent of poultry feed. The poultry feed cost can be reduced by about 57.5 per cent by substituting the ill filled/partly filled rice grain, maize grain, oil cakes of crop component, the Lucerne grown around the fish pond and fish waste anticipated through disposal of fish. The reduction in feed cost by way of effective recycling the products of crop component leads to reduction in the cost of production per egg. In commercial poultry farm, cost of production per egg is around 42 to 44 paise. By way of reducing the feed cost, the cost of production per egg is reduced to 19 paise, thus leading to fingerlings which solely depend on the poultry excreta grows at the rate of one kilogram per year yielding about 250 to 300 kg/year, which on marketing will be a bonus.

Plate 50. Cropping-Fishery-Poultry model at Kumulur, Trichirappalli, Tamil Nadu

The growth of fingerlings was solely due to the dropping of the birds kept over the fishpond. The poultry droppings contain protein 30 per cent and fat 2.5 per cent (Table 46). The poultry droppings were found to enhance the nutritive value when compared to periodical fertilization and supplemental feedings of fish with groundnut cake and rice bran. Mortality of birds was almost nil till the end of the program. This may be due to the cool environment created by the fishpond. By converting the groundnut and seasmum seed into its by product viz., oil and cake, the cake can be made free of cost to the poultry feed component and thus can reduce the cost to the poultry feed. The poultry feed cost can be reduced by about 57.5 per cent by substituting the ill filled or partly filled grain, maize grain of the farm, oil cakes of the crop component and the alfalfa grown around the fish pond (Jayanthi *et al.*, 1994).

Table 46. Nutritive value of poultry manure (dry matter basis)

Name of the constituent	Percentage
Protein (crude)	30.00
Fat	2.50
Crude Fibre	13.00
Nitrogen free extract	30.50
Total ash	24.00
Minerals	
Calcium	4.60
Phosphorus	3.00
Sodium	0.40
Potassium	1.37
Copper (ppm)	61
Zinc (ppm)	325
Manganese (ppm)	291

The results of the integrated farming system program was compared with conventional crop activity. The economics worked out for the system as a whole reveal that the mean net income under integrated farming system (IFS) and conventional system were Rs. 34067 / ha /yr and Rs. 17586 / ha /yr respectively. The net profit was Rs. 93.33 /ha /day and it was 93.7 per cent higher than conventional system. The employment opportunity in IFS was enhanced to the tune of 102 per cent in man-days over the conventional system. The income obtained from the integrated farming system program was compared with integrated farming system 59.20 per cent was from cropping, 8.30 per cent from poultry, 7.50 per cent from fisheries and 25.00 per cent from mushroom. i.e. The additional employment generated through farming system over the conventional system was 627 man days / ha / yr.

17.5.2 Integrated Farming System for Coastal Ecosystem

Componenets

i. Crop :

Early duration - Rohini, Cauvery, SSRC 91216 culture

Medium duration - IR20, TRY –1.

Long duration - Pankaj, Jaganath

Flood resistant - IR 43, ADT 40, TRY –1.

ii. Livestock systems : Cows, poultry, goat, fish, prawn and duck can be integrated with cropping in coastal areas rice based farming systems.

Prawn : Water canals in homesteads or coconut groves connected directly or indirectly with back water having free tidal water movements can be converted into productive prawn farms. From an area of 400m^2 of water canal, 12-16 kg of prawn (Indian white variety) can be produced in 90 days. It can fetch Rs. 50-60 per kg. The main expenditure will be only Rs. 2000/- for providing sluice gates to regulate the flow of water. Prawn culture can be integrated with duckery, poultry, agriculture, horticulture or piggery.

Fishery: In fresh water areas, fishes of Indian major carps viz., rohu, catla and mirgal are grown in perimeter trenches in rice plots.

Cows, poultry, goat and duck rearing can be integrated depending upon the situation as given in Model I. The goat breeds like Jamunapari or local breeds are highly suited for this region. But, among the animal components, fisheries and prawn farming are highly economical and mostly preferred by the farmers in the coastal regions.

iii. Tree Crops: Major tree crops suited for coastal region are coconut, cashew nut and jackfruit. The forest species are *Casuarinas* sp. and *Acacia* sp., *Eucalptus teriticornis, Pongamia pinnata, Azadiracta indica* and *Anacardium* sp.

iv. Farm pond : Major farm pond technology which involves digging of a pond in 20% of the area in coastal lowland area and spreading the excavated earth in remaining 80% of the field is gaining popularity among the farmers in coastal region. In the field so developed, the farmers are able to get a samba crop of high yield rice variety in monsoon season and vegetable in winter and summer months.

17.5.3 IFS in Command Area

In command areas wet land, cropping integrated with fishculture, poultry, mushroom cultivation, agroforestry, pigeon rearing and biogas unit.

i. Crops– Crops suitable for submerged conditions (Rice, sugarcane etc.)

ii. Fish culture-as given in model I.

iii. Poultry- as given in model I.

iv. pigeon rearing- as given in model I.

Vermi-composting : Employing suitable earth worms for Tamil Nadu conditions like

(i) Lampito mauritii,

(ii) Perionyx excavatus,

(iii) Drawida wilscii

(iv) Eisenia foetida

(v) Endrilus engeniae

Usually 2 x 1 x 1m size pits are used from which 1 ton of compost can be produced with in 30 – 60 days. A layer of 15-20 cm thick of loamy soil is spread above a thin layer of broken bricks (5 cm). About 2500 to 300 worms/ pit of 2 x 1 x 1m size can be introduced.

17.5.4 Farming System Models for the Tropics

Development of ecologically stable and environmentally sound rice based integrated farming system models for different eco-region to replace or improve the traditional farming systems would maximize the productivity and profitability of the farming systems. Integrated farming system models for different agro ecological regions are given Table 47.

Table 47. Integrated farming system models for different agro – ecological regions

S.No	Farming system model	Region	Suggested by
1.	Composite fish culture	India	Jayaseelan *et al.* (1988)
2.	Pig cum fish farming	India	Jhingran and Sharma (1980)
	--	India	Paulraj (1984)
3.	Poultry cum fish farming	India	Sharma *et al.* (1985)
		,,	Kumar and Singh (1984)
		,,	Govindan (1988)
		Indonesia	Djajadiredja *et al.* (1980)
		Malaysia	Tan and Khoo (1980)
		-	Delmando (1980)
4.	Cropping cum Mushroom Farming	India	Rao (1984)
		,,	Loganathan (1989)
		,,	Rangasamy and Jayanthi (1994)
		,,	Vangriensven (1987)
5.	Rice+Pig+Fish+duck+azolla farming	Philippines	Govina (1990)
6.	Rice+fish+poultry farming	Tamil Nadu (India)	Govindan *et al.* (1990)

7.	Rice+fish+poultry+mushroom farming	,,	Rangasamy (1992)
8.	Rice + fish+azolla farming	,,	Shanmugasundaram *et al.* (1992)
9.	Rice + fish+dairy farming	,,	Veerabadran (1994)
10.	Rice+fish+shrimp+vegetable	Vietnam	Lightfoot and Minnick (1990).

Farming systems research came into prominence as a part of agricultural research in the past decades with a main stimulus of growing realization that most farmers are economically rational (Simmonds, 1986). The meaning of FSR varies differently to different authors and to varied agro-ecological situations, but the objectives of the farming systems, in general, are conversing to the development of suitable location specific farm technologies to raise and sustain the total farm productivity in terms of food, fodder, fuel and to meet the felt needs of the farmers within the circle of this agro-eco-socio-political favorites and constraints (Gomez, 1988 and Norman and Collinson, 1985). FSR has been applied to a variety of activities and viewed the farm in a holistic manner and considered interactions in the system (Byerlec *et al.,* 1982). The FSR approach is directed at small farmers in marginal areas who operate complex systems characterized by important interactions between enterprises (Gilbert *et al.,* 1980; Maxwell, 1984; Simmonds, 1985 and Collinson, 1988). FSR was considered to be very important in providing scientific approach for problem identification and technology development aimed at improving agricultural production systems and human welfare through increased but sustainable agricultural productivity (Plucknett *et al.*, 1987 and Spencer, 1991). John (1991) viewed FSR as on concept open to multiple interpretations existing for the numerous local varieties of farming systems activities. FSR is a research method designed to understand the farmers' priorities, strategies and resource allocation decisions, with a view to identify and develop technologies useful for each farm environment of varied agro-climatic conditions (Kanwar *et al.,* 1992 and Pandey *et al.,* 1992).

The analysis of results available in India and elsewhere reviewed have clearly brought out that specific farming system models are to be developed with combinations of different components for varied agricultural domains with similar agroclimatic zones. Integrated pig-fish-duck-azolla farming system was proposed by Govina (1990) for La Union of Philippines. Higher productivity with better returns of components such as pig, duck, poultry and fish were expressed by Tan and Khoo (1980) by integration in Malaysia. Vangriensven (1987) reported that the significant role of modern mushroom cultivation as one of the component of specialized farming in Netherlands.

However, earlier Balasubramanian (1988) identified sheep and buffalo, poultry and swine as dominant components in the rice regions of Philippines. Similarly, rice – fish integration for small farm holding in Malawi in South Africa was recommended by Noble and Rshidi (1990) in the lowland agro-eco systems. Whereas, Thanh Hung (1992) suggested fresh water prawn and marine shrimp culture for integration to coastal rice farming systems in Vietnam. Lightfoot *et al.* (1992) clearly demonstrated the importance of soil fertility and enhanced residual recycling by integration of rice-fish system in Philippines. Sharma *et al.* (1991) has identified different livestock based farming system models in low, mid and high hill zones of Himachal Pradesh. The possibility and success of duck and fish farming (Sharma *et al.,* 1985), pig-cum-fish farming (Paulraj, 1984), poultry-cum- fish farming (Kumar and Singh, 1984) and poultry-cum-pig-cum-fish farming (Delmando, 1980) were demonstrated in lowland conditions under varied agroclimatic zones with certain specificity. Singh (1992) suggested a model integrating fish-cattle-duck-poultry with horticultural crops under lowland of Northern India.

Introduction of an integrated rice+fish+vegetable farming system for deep-water rice growing regions of West Bengal was successful to optimize the rice production and to sustain the ecosystem (Roy *et al.,* 1990). However, while studying the feasibility of adopting aquaculture to increase the resource productivity in the existing Bangladesh farming systems, Ahmed and Rab (1992) suggested rice-cow-poultry-fish culture farming systems for small farm holdings. Tiwari (1993) while describing several crop-livestock farming systems in India, highlighted crop-livestock-agro forestry, flooded rice-poultry-fishery farming and rice-poultry-fishery-livestock system for resource rich farms and labour intensive farming systems like poultry, piggery, fishery earthworm breeding and mushroom cultivation for small areas of high population, as dominant components in the farming systems.

CHAPTER 18

RICE UNDER ORGANIC FARMING

Indian agriculture poised a faster growth during the post green revolution era between sixties and late eighties, and later there is a sign of either stagnation or decline of productivity in many parts of the country. Expansion of assured irrigated area through major irrigation systems, strong research and development base, large network of extension services and quantum jump in inputs subsidies all have contributed to increase in food especially cereal production in selected pockets. Fine cereals like rice and wheat have been at the command all through the green revolution phases especially on the food baskets of India. The spread of green revolution packages in other potential areas at later stages has resulted in cultivation of cereals like rice. Since rice is a tropical but water guzzling crop, it has been taken up for cultivation on the command areas across the country.

Traditionally in India, rice has been grown as subsistence crop till the green revolution techniques were initiated to increase the production and productivity. Farm produced and homegrown inputs were largely used with less energy intensity for domestic consumption and later the intensity of land has been increased to raise the production and productivity to feed the growing population. To achieve this, huge investment was made on major irrigation projects; new high yielding crop varieties were developed; input delivery systems were strengthened and factory made subsidized external inputs were introduced. All these were possible as the state was keen in developing agriculture sector mainly cereals to avoid starvation. Expectedly, the results were positive from the potential areas where resources were abundant and it continued till late eighties and later cereal productivity especially rice was found to either stagnated or declined marginally (GOK, 1994 and Singh, 1997). It is postulated that the liberalization programs that were initiated during early nineties would aggravate the situation further as the state's role in the form of public investment will be minimized to a great length. The available data clearly indicate that from 1990-91 the share of public investment in Indian agriculture has been declined from 29 per cent to less than 20 per cent (GOI, 2001) in 1999-2000. In this scenario, it is perceived that the cost of cultivation is shooting up and the farmers are finding that agriculture is no more viable proposition and in fact, a large number of farmers committed suicides in a south Indian state of Karnataka (Deshpande, 2002) and also in Andhra Pradesh. Perhaps shooting up of price of factory made external inputs and the government slow withdrawal of investment as well as market intervention and more significantly, shifting of subsistence farming (mainly with homegrown inputs) to commercial farming (largely with purchased inputs) would have also contributed for the present crisis. In other words, the local indigenous farm techniques have been wiped out and replaced by the modern techniques that would have resulted unviable and unsustainable farm enterprise.

Rice as a major cereal crop, grown across the country is also facing the same problem and challenge like any other crop and over 90 per cent is produced and consumed in the Asian continent. Countries like China, India, Philippines, Korea and Pakistan grow a lion's share of global rice production. Though the rice yield levels differ across these countries, suitable agro-environmental conditions favor the farmers to grow this crop. Rice is predominantly grown in the states of Punjab, Haryana, West Bengal, Tamil Nadu and Andhra Pradesh in India. Nevertheless, rice cultivation becomes less remunerative, environmentally unsustainable, and hence alternative crops have been propagated and this option will be less subscribed as a large chunk of Asian population's staple food is rice. It is in this context that alternative farm techniques and strategies for growing rice ought to be found in the larger interest. Owing to the merits of organic cultivation as compared to modern practice, such principle is attracted across the world. Much state supported agencies; NGOs and individuals started experiments on organic methods of production in the recent past. Many European countries provide financial assistance to farmers for organic cultivation in the context of health and environmental implications of conventional farming systems. The recently held 14th Organic World Congress in Canada emphasized the need for promoting ecologically sustainable, economically viable and socially adaptable farming system.

The increase in demand for organic products is mainly due to health implications and scare triggered by bovine encephalopathy (BSE) in livestock, dioxin in animal feed and genetically modified products. Thus scared public in the Europe and America ready to pay premium for organically grown products mainly come from developing countries and even within the developing countries, urban rich consumers prefer to consume organically grown products. This clearly reveals the fact that market is growing at faster pace for organic commodities in the globe and higher premium is mainly due to cost escalation in the organic farming systems (Hazell Peter, 2001).

18.1 CHALLENGES

According to researchers, different farming systems do have implications on the productivity, sustainability, production orientation and ecological situations (Werf and Narayan, 1989 and Pingali, Hossain and Gerpacio, 1997). While modern commercial (conventional) farming system is unsustainable and market oriented the organic (sustainable) farming do have merits like productivity orientation, ecological sustainability and farm complexity.

Traditional rice cultivation involves mainly direct sowing, interculture, weeding, rainfed irrigation and harvest. In such system, rice plants withstood extreme drought, provided more fodder and yield was moderate for subsistence. Research study in the Philippines indicated that in rainfed low land farms, dry seeding of rice was the promising technology for improving rice productivity (IRRI, 1992). Subsequently, the introduction of semi and dwarf varieties of high yielding varieties of rice, resulted in prone and susceptible to pests and diseases susceptibility required more water, demanded more factory made external inputs (especially agro-chemicals) and yielded less fodder. The propensity to use abundant water and more chemicals increased year after year. In some pockets of rice regions in India, indiscriminate and non-judicial use of chemicals was observed that has also led to health hazards on the labour (Mencher, 1994). Nonetheless, such varieties yielded more, which was essential to feed the growing population. Simultaneously, adverse ecological impacts like water logging, salinity and chemical contamination were reported (Pingali, 2001 and Rajendran, 2002). In addition to this, comprehensive studies on organic farming especially rice cultivation are scanty in India. Thus the relevance of organic agriculture and research has been realized all over the world.

18.2 ECOLOGICAL IMPLICATIONS

Application of cattle urine mixed with neem and adathoda (botanical) decoction keeps away the pests and diseases in the rice fields. A couple of organic rice growers confessed that application of neem oil cake along with organic manure as basal fertility agent prevented insect pests and diseases attack; enhanced plant growth and increased yields. Due to non-availability of sufficient organic manure, a few farmers purchased municipal waste and applied the same after due processing. While for transplanted fields, water was applied on alternate days, directly sown (broadcast) field, this was done weekly twice. It is to be remembered that while water scarcity is noted everywhere across the globe, to produce one kilo of rice the farmers in general reported to use an alarming quantum of about 15000 liters (Rajendran, 1992) under flood irrigation system for rice transplanted fields. Such huge water use resulted not only in excessive weed growth but also salinisation and water logging erupted in the long run. In fact, interculture was done with an intention to split root system (for more shoots and clumps) of the rice plants and to reduce the weed growth and instead only the former was realized, weed growth was found to be more. This had resulted in using more human labour for weeding operation and the wage for transplantation upset this cost as revealed by the organic rice growers. Besides, black gram, white gram, pigeon pea and cow pea were raised on the field bunds both on rice seedlings transplanted and direct sown fields and often, this kept the pest and disease well under control and also served as habitation for natural predators like spiders.

18.3 ECONOMIC ASPECTS

Organic rice growers relied more on homegrown items like crop residues, compost and green manure, which brought down cash component of expenditure for inputs. Only a major item that incurred more input cost was farm power sources like human labour, draught animals and machineries. Obviously, rice requires more farm power for operations like ploughing, leveling, transplanting, irrigation, weeding, harvesting and threshing. Of course, between successive operations, rice growers get adequate leisure. A cross section of the organic rice growers reported that rice is fairly remunerative crop as compared with other cereals like ragi, jowar and bajra. On an average rice yields were reported around five tones/ha for transplanted rice and little less than under direct sowing. Since manual weeding was done and homegrown inputs were applied, the input costs were considerably reduced. Perhaps, yield levels under conventional type of farming would have been more but the yield per rupee invested could have been more under organic farming. On an average Rs 20000 was generated as net income from growing rice from one hectare of land and farmers reaped twice as income over investment. Nevertheless, economics of production for different crops shows that horticulture crops were more remunerative as compared with cereals including rice under organic agriculture. It is a clear fact that rice is less productive crop as compared with other horticultural crops. This could be mainly due to accounting of different cost and nature of crop. For instance, perennial crops like coconut and sapota were well established, there may not involve much cost unlike rice. Additionally, the market price for cereals is not commensurate with the increase in input cost. But still the organic farmers preferred to grow rice as one of the crops on their plots mainly for domestic consumption. Notably, another study (ICSIM, 1990) reported that mean yield levels from organic farms in the states of Karnataka and Tamil Nadu were comparable that of conventional farms. Here crops like rice, ragi and jowar registered more yields, groundnut and sesame recorded low yield from organic farms than modern (conventional) farms. Employment generation was found more under organic rice cultivation as weeding was exclusively done with manual labour. Human labour also did some operations like catching rats during pollination and grain formation stages and a substantial amount of human labour was used for weeding operation in the direct sown rice as compared with transplanted rice. Moreover, each and every operation and supervision was done with utmost care and organic growers were concerned with ideology than mere economic returns.

Organic rice cultivation is associated more towards farm oriented self sustained system. Strong local resource endownment, sustained commitment and ideology have influenced the organic farmers to go in for raising rice under organic system in combination with traditional expertise. Arguably, adoption of specific farm practice based on rich local experience helped the growers to sustain the rice production. Followed by this, the local farm based resources like livestock and farm forestry are also found to influence the rice cultivation. Though rice cultivation appears to be less economical as compared with other crops, there is a scope for minimizing the economic cost and environmental loss under organic farming system as compared to conventional farming in the long run. Besides these, environmental balance is maintained, where crops, trees, animal and man can live more harmoniously. Major issues like standardization of necessary inputs like vermicompost and biofertilizers need to be addressed by keeping environment and economic dimensions of agricultural sustainability in the long run. Similarly, marketing for certified as organically grown products is a tiresome process and costly which affects the Indian growers who are interested to export the commodities. This needs to be simplified while keeping the quality of products and human health in view.

18.4 ORGANIC FARMING- NEED

Agricultural production is often affected by natural conditions. Not only cold weather, drought, typhoons, heavy rains and other meteorological phenomena, but soil, water and life in and around cultivated land have great effects on the growth and yield of crops. Moreover, air pollution, water contamination and other adverse effects on the environment caused by man made activities are affecting agriculture seriously.

18.4.1. Agriculture *vs.* Environment

In the process of farmers growing and harvesting crops, the crops change the environment themselves. For example, crops absorb CO_2 in the air by photosynthesis and help purify polluting substances too. On the negative side, activities for cultivating crops are also affecting the environment. The improper use of pesticides and chemical fertilizers is worsening the quality of river water and ground water. As these examples suggest, agriculture brings both favourable and unfavourable consequences to mankind and to the conservation of natural environment.

18.4.1.1. Use of fertilizers

Several chemical substances are let out from agriculture, but nitrogen and phosphorus contained in fertilizers and agrochemicals pose serious problems. The importance of chemical fertilizers to crop yields in India was established as early as 1937. However, fertilizer is no longer the hero of green revolution. The effective methods for reducing the discharge of nitrogen from cultivated land are to apply fertilizers in an amount and at timing suited for crop growth and to introduce slow-release fertilizers. A correct prescription is to apply the right nutrient at the right time i.e. phosphate as basal and the first split of nitrogen after 3 weeks of crop growth and potash along with the second split of nitrogen prior to flowering as NK mixture. In paddy fields, chemical fertilizers partly flow into rivers together with the total drained water after soil puddling for smooth transplanting. There is no risk of groundwater polluted by nitrogen as a result of infiltrated irrigation water. This is a characteristic of paddy field soil; the reduced layer beneath the soil surface reduces nitrate nitrogen into nitrogen gas. The fertilizer recommended must exclude contributions from the basic available nutrients of the soil and contribution from the soil microflora or that used as bacterial inoculants. The industry's view that they are only supplementary is a grave mistake. Option of biofertilizers for paddy is a best remedy for soil fertilizer pollution. Rice yield has been reported to be increased by 11.4% and 24.5% by the use of blue green algae and azolla respectively.

18.4.1.2 Agrochemicals

The frequent and large quantity application of insecticides in recent years has caused a number of problems: increasing resistance of insects to insecticides, insecticide residues in crops, danger of acute and chronic toxicity to man and animals, unexpected outbreak of pests due to reduced natural enemies, appearance of potential insect pests and soil water pollution. In this situation, it is now strongly hoped that dependence on insecticides will be lowered as much as possible and that those methods giving less adverse effects on the environment will be introduced. The use of synthetic sex pheromones and parasitic or predacious natural enemies are environmental friendly means of pest control. Aphids, plant hoppers and green house white flies have the habit of being attracted by yellow objects, while thrips is attracted by white objects. The idea of integrated pest management (IPM) in rice is for making the most of the mechanism of natural control and using many technologies in a harmonious way to hold down insect pests below the level that they will cause economic damage or Economic Injury Level (EIL). Antagonistic microorganisms such as bacteria and actinomycetes are generally used to control soil-borne diseases through inoculating them to seed or plant roots. This inoculation method contributes to decrease the number of pathogen or to increase crop activities. Rhizosphere bacteria are often used to study the control of soil-borne diseases.

18.4.1.3. Green house effect

Agriculture emits green house gases including methane, nitrous oxide and carbon dioxide. In particular, it is estimated that roughly half of anthropogenic methane emissions come from farming activities. About 30% of anthropogenic methane emissions originate from paddy fields and 20% from animals. Paddy fields are considered as an important source for the observed changes in atmospheric methane concentration (Udayasoorian, 1999). The intergovernmental panel on climate change estimated the global emission rate from paddy fields to be 20 to 150 Tg yr^{-1}, averaging 60 Tg yr^{-1}. However, the measurements in paddy fields in various locations of the world show that there are large temporal variations of methane flux and that the flux differs markedly with climate, characteristics of soil and paddy, application of organic matter and mineral fertilizer and agricultural practices.

18.4.1.3.1 Methane emission: Indian Scenario

We have to consider the paddy cultivation area, distribution and various soil types, to estimate methane emission budget from India. Methane budget over India from paddy cultivation would require amongst other parameters, consideration of various soil characteristics, which include texture, pH and soil, mean monthly temperature. In addition to soil characteristics, the mode of cultivation in India, viz., rainfed lowland (40% which includes coastal regions) and other (60% which are mainly irrigated) are also considered. Because of sea, the coastal regions may be rich in salinity, which is known to inhibit methane emissions. These areas are further subdivided into acidic and alkaline categories. The alkaline category under lowland cultivation includes coastal saline, saline and non-saline regions. Methane budget has been estimated by taking the highest emission rates as well as the average emission rates. These consideration lead to methane budget estimates from the Indian paddy fields to be between 3 and 9 Tg yr^{-1}. Indian contribution is only 6% of the projected global methane budget due to paddy cultivation.

Based on the data generated from rice field measurements in Europe and the United States of America, and then extrapolated to Indian conditions, methane emission from Indian rice fields contributed 37.8 Tg yr^{-1} (Environmental Protection Agency). However, actual field measurements made during 1989-1990 in selected rice-growing areas of India showed that methane emissions accounted to 3 Tg yr^{-1}. Of the 42 million hectares under rice cultivation in India, only 16.4 million-hectare are irrigated. Of the remaining area under rainfed conditions, 5.9 million hectare is upland and 19.7 million hectare is lowland. In rainfed low land areas, the rice fields are not always inundated. Thus, in rainfed lowland areas, the methane efflux over a rice-cropping season may be less pronounced than in continuously submerged rice fields. Even under continuously submerged conditions, methane emission from alluvial soil (organic carbon 0.76%) is considerably lower.

a. Methanogenesis

Methane formation in soil is biological energy transformation process mediated by obligate anaerobes at a soil redox potential lower than –140 to –160 mV. Before methanogensis occurs, NO_3, Mn_4^+, Fe_3^+ and SO_4 in the soil must be reduced. All of these reductions are microbiological and energy consuming. Therefore, they affect methane formation to some extent. If a soil contains significant quantities of these oxidants, methane formation is restricted, for the organic matter or the decomposable organic carbon would be first used by the other reducers than by the methanogenic bacteria. The amount of soil organic matter and the decomposable carbon are generally considered critical controlling factors of methane production. The decomposable carbon content in soil is considered to be more significantly related to methane production than the total organic matter content because the latter provides instant energy to microorganisms. Generally the organic matter content of cultivated soils is considered as an index of soil fertility. A cultivated soil with a relatively high amount of organic matter indicates a high level of microorganisms activity. Therefore, a positive relationship between methane emission and the soil organic carbon content is often expected.

Methane produced in the rice fields by methanogenic bacteria may be released into the atmosphere by three different pathways viz., molecular diffusion at the water-air interfaces, release of gas bubbles and plant-mediated transport. The most important one of the three is the flow of methane through the aerenchyma system in rice bodies. 95% of the total methane flux released to the atmosphere throughout the growing season was through the rice plant gas transport system. Methane may be transported through the aerenchyma of vascular plants either by molecular diffusion or convection through flow driven by humidity-induced diffusion; the dissolved methane in soil water surrounding the roots diffuses into the surface water of the roots, and on to the cell wall water of the roots cortex, driven by the concentration gradient between the soil water surrounding the roots and the intercellular spaces in the roots. Methane is then gasified in the root cortex and transported to the shoot via the intercellular space and aerenchyma. Eventually, methane is released mainly through the micropores in the leaf sheaths at the lower leaf position rather than the stomata in the leaf blades.

b. Influence of fertilizer application

Urea and ammonium sulphate account for 80-90% of total nitrogen fertilizer applied to paddy fields. High ammonium sulphate and potassium nitrate usage reduced methane emissions, while urea application recorded highest emission rate (Udayasoorian, 1999). The influence of fertilization with urea and ammonium shlphate on methane emission is heavily dependent on the mode of fertilizer application. In general, methane emission rate were much higher in the fields with surface application than in raked and incorporated applications.

18.5 ORGANIC FARMING PRACTICES FOR RICE

Organic manures have many advantages, which chemical fertilizers lack. The reasons for using organic manures are partly ***agronomic*** - a more productive soil and better crops; partly ***economic*** - making use of the cheap resources in the place of expensive commercial fertilizers and partly ***environmental.*** Organic manures are a more complete plant food and they provide almost all essential nutrients and trace elements while chemical fertilizers supply only specific nutrients, and thus a soil receiving only chemical fertilizers over a long period is likely to have an unbalanced soil nutrient content. The disadvantages are that some organic materials may contain toxic or pathogenic substances that are dangerous to crop or human health. The most serious problem is heavy metals often present in municipal wastes and sewage sludge; it is difficult to standardise commercial organic manures since the quality of raw materials is so variable; organic materials release their nutrients slowly, as the material decomposes and it meets the needs of crops for a steady nutrient supply over a growing season. But, many farmers feel that chemical fertilisers are more effective as the effect is more dramatic and rapid. The main disadvantage of organic manures is that they are bulky and difficult to handle and transport; some organic manure are messy and with unpleasant odour for farmers to handle (Palaniappan and Annadurai, 2001). If society wishes to reduce environmental problems by recycling organic materials, the government must either subsidise their use or offer some incentive to the farmer.

A. Concepts

Mere substitution of organic inputs in the place of inorganic inputs for animal raising and agriculture/horticulture does not form organic farming. Understanding the resource capabilities of the farm and using the same in a conservative manner avoiding over exploitation of land, underground water etc. According to Lampkin (1990), the principle elements to be considered in practicing organic farming are: (1) creating a healthy soil, (2) making nutrient and energy flows in the soil ecosystem (3) keeping the biological life in the cycle and (4) providing sustainable yield.

B. Definition

Organic farming is a productive system which largely avoids or excludes the use of synthetically compounded fertilizers, pesticides, growth regulators and livestock feed additives and to maximum extend possible organic farming systems relays upon crop rotations, crop residues, animal manures, legumes, green manures, off-farm organic wastes, mechanical cultivation, mineral bearing rocks and aspects of biological pest control to maintain soil productivity and tilth, to supply plant nutrients and control insects, weeds and other pests (Palaniappan and Annadurai, 1999) or Organic farming is not merely use of non-chemical fertilizers and pesticides but it is a holistic approach integrating soil, plant, water, soil microflora and fauna, human, animals and the environment. Prime purpose of organic farming is to build of natural resources such as soil and water for future use. The following are the objectives of organic farming.

1. To develop a sustainable agriculture system for guaranteed adequate food production in the foreseeable future.
2. To develop self-sufficient agriculture system which would rely as much as possible upon resources from within its own resources.
3. To develop an alternative strategy over chemical farming which would be a guideline for working of biological processes in natural eco-systems.

C. Need

Because of indiscriminate use of chemical fertilizers for decades, the organic matter content of soils has come down to less than 1 per cent. In addition, the use of pesticides led to pest resurgence and difficult-to-control weeds species. The residues of the chemicals cause concern over the safety of food and sustainable production. The addition of chemical fertilizer like nitrogen in plant caused an infant disease like methanoglobinaemia. Hence, the expectation that organic farming by reverting to the use of manures, green manures, urban waste, rural wastes, etc. can bring sustainability to agriculture with eco-friendliness. Hence, it becomes imperative for the researchers and planners to develop an alternative viable strategy to supplement the chemical farming.

D. Types of organic farming

1. **Pure organic farming**: It includes use of organic manures, and biopesticides with complete avoidance of inorganic chemicals and pesticides.
2. **Integrated Farming:** It involves integrated nutrient management and Integrated Pest Management. It is now widely felt that most efficient crop production system and probably in the long run, the most economical one, consists of a combination of organic and chemical fertilizers, combining the best aspects of both types of nutrient supply.
3. **Integrated Farming Systems:** In this type, local resources are effectively recycled by involving other components such as poultry, fishpond, mushroom, goat rearing etc. apart from crop components. It is a low input organic farming like LEISA (Low External Input Sustainable Agriculture).

The details on rice based integrated farming system are discussed in the subsequent chapters.

18.5.1 Rice response to Organic manures

The major sources of organic plant nutrients in India are farm yard manure, rural and urban compost, sewage sludge, pressmud, green manures, crop residues, forest litter, industrial waste and by-products. The number of biofertilizers such as blue green algae (BGA) and azolla are used extensively to meet the nitrogen demand of the crop. Small quantities of powdered neem cake are also used. These organic nitrogen supplements unlike the fertilizer nitrogen do not suffer any loss in the fields. Phosphorous-solubilising and mobilising organisms such as phosphobacterium and vesicular arbuscular mycorrhizae (VAM) are quite helpful in meeting the phosphorus demand of the crop. Potassium for the crops can be supplied by using potassium rich organic amendments such as burnt rice, rice straw compost using *Tricoderma harzianum* and composted coir pith.

It is the farm wastes and animal manures that supported our crop production system for ages. The rice crop response to application of 12.6 t of FYM or compost per hectare varied from about 100 kg ha^{-1} to 216 kg in Maharashtra and Orissa. The results of long term experiments (10 years) in a sandy loam laterite soils of Bhubaneswar showed that FYM application to supply 45 kg N with and without ammonium sulphate gave the higher grain yield of 2667 kg ha^{-1}. Combined application of 25 t FYM + 20 t press mud per hectare was the best for increasing yield of rice. In rabi season, application of rock phosphate 50 kg P_2O_5 ha^{-1}+ 10 t of FYM registered highest grain and straw yield. Application of press mud to rice crop at 10 t ha^{-1} has increased the yield. The details of the nutrient content of different organic manures and its application are given in the previous chapters (refer chapter 8).

Biogas slurry applied at the rate of 15 t ha^{-1} recorded higher grain yield of rice. Application of poultry manure at 12.5 t ha^{-1} has recorded good response. Green manure application to rice crop before transplanting was a traditional crop management technique in all rice growing regions of the country. Appreciable yield increase in lowland rice with green manuring has been reported from almost all rice growing countries of the world. However, the magnitude of response vary. Azolla, may also be raised either in nursery or as green manure before transplanting of rice and also as a dual crop with rice under lowland condition where the existing temperature conducive for azolla cultivation. Azospirillum is yet another most important associative bacterium found in the rhizosphere of the various grass species capable of fixing atmospheric nitrogen to improve the productivity. Application of non-edible, cheap sources of oil cakes are in vogue among rice growing farmers. Application of FYM + neem cake at 50% each as recommended nutrients

(120:38:38 kharif and 150:50:50 NPK for rabi) has recorded 36.0 8.0 and 11.0 percent increased yield over conventional farming (Kenchaiah, 1992).

Other inputs like application of earthworm castings (vermicompost) and algae especially on the irrigated fields were also noted among the selected organic rice growers. Earthworms proliferate by themselves in the absence of chemicals in the soil. Unfortunately, it was revealed that realizing the fast growing market (demand) for earthworm castings, the fake businessmen entered the field and exploited the farmers. Also in some places, in the state of Maharastra and Karnataka, it was reported that fertile black soil was adulterated with earthworm castings for quick and easy money and farmers were duped. This warrants immediate attention of the policy makers to streamline the market mechanism like licensing and standardization. Scientific studies from the University of Agricultural Sciences in Bangalore, clearly demonstrated that earthworm castings contain five times more nitrogen, seven times more phosphate and 11 times more of potash than the given topsoil. Fragoso and Lavelle (1995) found that litter decomposition by earthworms' help improved the soil nutrients and organic matter. Rigorous scientific screening and standardization is lacking in this crucial aspect. For instance, in India, there are more than 200 varieties of worms and so far only three varieties like *Eudrilus eugeniae* and *Perionyx excavatus* for tropical conditions and *Eisenia fetida* for temperate conditions have been found suitable than others. Unless proper screening and labeling is done, it would resort to black marketing. With regard to earthworms, fields should not be ploughed more deeply and intensively instead light ploughing or rotating top soil with spade created a favorable atmosphere to faster proliferation. Absence of chemical particles (synthetic fertilizers and pesticides) helped the worms to multiply themselves in the soil thereby the excreta of worms enriches soil fertility and plant growth.

Similarly, there is a growing demand for bio fertilizers like algae especially among the rice growers. Commonly used bio fertilizers to fix/ solubilize atmospheric and naturally occurring plant nutrients are blue green algae (BGA), Azolla, Rhizobium, Azospirillum and Azotobater. Such algae are found not only cheaper but also easy to grow by the farmers in their fields with little technical expertise. There is a growing awareness on this not only among organic rice producers and from other farmers as well in states like Tamil Nadu and Karnataka, where the ecological and economical merits of bio fertilizers have been highlighted on the vantage points by the state machineries. Despite these, it was found that only 10 per cent of total demand of bio fertilizers was met by the state departments and there is a potential for private sector to exploit the business on a large scale as it is economical, technically feasible and ecologically sustainable. Perhaps the financial institutions may come forward to start this as agri-business in rural areas and already a couple of large fertilizer companies ventured in this prosperous area for production and marketing. Standardization of production and marketing of bio fertilizers needs to be done at micro level with the active help of agricultural scientists.

Five promising microorganisms like *Nostoc* sp., *Aulosira* sp., *Anabaena* sp., *Oscillatoria* sp. and *Westiellopsis* sp., were identified as efficient nitrogen fixing agents under All India Coordinated Research Project on Bio fertilizers. This has to be viewed in the interest of long term ecological benefits as continuous application of algae for three years farmers need not use synthetic fertilizers because micro organisms will get well established in the soil and be able to generate nutrients, required by rice crop. Despite these, scientific research studies conducted at Tamil Nadu Rice Research Institute at Aduthurai revealed that rice productivity was 20 per cent higher when treated with BGA, Azospirillum and chemical fertilizers than when treated with BGA and Azospirillum only or synthetic fertilizers alone. Organic rice growers on the other hand, argued that continuous application of eco-friendly bio fertilizers enabled the farmers to get more yields. The organic rice growers grew leguminous crops like kolingi and sunnhemp during off farm summer season on the fields. Similarly, some farmers penned sheep flock at nights for manure and they reported that sheep urine and manure enriched the soil fertility. This had helped to increase the soil fertility and in turn yield levels.

Another notable observation is that fodder availability from dwarf and semi dwarf rice varieties was less as compared with traditional ones. But as of now even the organic rice growers do not possess traditional seed varieties of any crop and high yielding crops completely occupied the prime position especially on the assured irrigated regions. Local crop varieties were used only in remote dry land regions. In the selected organic farms, rice varieties developed from the International Rice Research Institute, Philippines have been crossed with local varieties by the SAU's and ICAR stations and only such varieties were grown on the irrigated fields. Thus low availability of fodder from modern high yielding varieties, resorted in fodder shortage for livestock sector and hence purchased feed items was fed to domestic animals. The main benefit accrued from continuous use of organic source of nutrients is the residual effect and enhanced biological activity in the soil. Manurial experiments conducted at IARI, New Delhi and experiments under AICARP and AICRIP reported higher cumulative beneficial effect of organic manure application on long-term basis. While reviewing the green manure research in India, it was observed that residual effect of green manure applied to rice crop was negligible. The reason may be the quantum of manure application was minimum. Whatever the quantity of rice stubbles available in the field may be incorporated for beneficial effect on the succeeding rice crop.

18.5.2 Pest and disease management

Several traditional technologies are being utilized in remote rice growing regions by tribal people to check pest and disease problems and to produce quality crop. In recent years, several bio control agents have been identified for the management of pests and diseases. Also, mycoherbicides are available for the control of weeds. Neem seed kernel extract, neem oil as well as neem cake found to check most of the rice insect pests such as stem borer, leaf folder, whorl maggot, earhead bug, green leaf hopper, when NSKE at 5% and neem oil at 3% and neem cake at 150 kg ha^{-1} was used. Integrated pest management practices excluding synthetic toxic chemicals will suit organic farming. Studies with neem leaf extract on the mycelial growth of rice sheath rot pathogen inhibited the growth. Further details are given in the subsequent chapters. To check weed growth, insect and disease prevalence, the following techniques maybe combined and used for good result.

a. Use of preventive methods such as weed seed free certified seeds and insect and disease free embryo rich crop seeds for good germination and establishment.

b. Use of light, pheromone, sticky traps.

c. Clean cultivation, summer ploughing adopting good water management practices, application of neem cake.

d. Feeding the crops with organic inputs, adopting proper crop rotation, using resistant varieties and selecting weed smothering tall growing crops/varieties.

e. Wherever possible freely using mechanical methods and cultural methods.

f. In rainfed areas, growing intercrops will minimize weed infestation and other pest incidence

g. Wherever required green manure crop raising, recycling crop residues, on farm and household wastes may be resorted.

18.5.2.1 Botanicals for Rice Diseases

Extensive and indiscriminate use of plant protection chemicals had led to serious social and environmental problems. With the increase in awareness on the toxic hazards of chemicals to crops, consumers and environment, the importance of indigenous plant products in plant disease control has been recognized. Surveys and screening of higher plants for anti-microbial properties carried out in different countries have demonstrated the occurrence of anti-microbial substances (Tripathi *et al.,* 1978). These fungi toxicants or anti-microbial substances from higher plants may be exploited to manage plant diseases, especially rice diseases to achieve ecofriendly methods of disease management.

A. Fungal diseases

Different parts of the plants have been found to posses antifungal activity against fungal pathogens of rice (Table 48). The antifungal activity was found to be due to the proteinaceous compounds in *Mentha arvensis,* which was effective against *D. oryza* (Unnikrishna pillai, 1988). Eguchi *et al.*, (1978) reported that 2, hydroxy 1-4 naptha quinone present in leaf extract of *Lawsonia alba* was responsible for the antifungal activity against *D. oryzae* and *P. oryzae.*

Table 48. Efficacy of different plant parts against fungal pathogens of rice

S.N.	Plant parts	Name of the Botanical	Pathogen inhibited	References
1	Leaf	Azadirachta indica	*Pyricularia oryzae*	Krishnamohan *et al.* (1988)
		Prosophis juliflora	*P. oryzae*	Mariappan *et al.* (1995)
		Mentha piperita & M. arvensis	*Drechslera oryzae*	Alice (1984) Unnikrishna Pillai (1988)
		Lawsonia inermis	*D. oryzae*	Devanathan *et al.* (1995) & Sreeramakumar (1990)
		Eucalyptus globulus, Calotrophis gigantea	*D. oryzae*	Devanathan *et al.* (1995)
		Vitex negundo	*Sorocladium oryzae*	Narasimhan *et al.* (1994)
2	Flower	*Achyranthus aspera*	*D. oryzae*	Chakravarthy & Pariya (1997)
3	Pollen	*Lycopersicum esculentum*	*D. oryzae*	Pandy *et al.* (1983)
4	Seed	*A. indica*	*S. oryzae*	Narasimhan *et al.* (1994)
5	Bulb	*Allium sativum*	*D. oryzae*	Alice (1984)
	Oil	*A. Indica*	*S. oryzae*	Narasimhan *et al.* (1994)
6		*M. arvensis*	*D. oryzae*	Unnikrishna pillai (1988)
7	Cake	*A. indica*	*S. oryzae*	Narasimhan *et al.* (1994)

B. Seed borne diseases

Many plant extracts were found to be effective in reducing seed borne disease on rice. Soaking the infected rice seeds in aqueous extracts of *Allium sativum, Mentha pipertia* and *M. arvensis* for 24 hours significantly reduced the seed borne infection of *D. oryzae* (Allce and Venkata Rao, 1981; Unnikrishna pillai, 1988). The leaf extracts from *Adathoda vesica, Cocus nucifera* and *Lawsonia inermis* recorded maximum percentage of seed germination and minimum incidence of *D. oryzae* (Devanathan *et al.,* 1995). Leaf extracts from *Adjantum candatum* and *Polypodium multilineatum* protected the rice seed completely from seed-borne fungi (Kanauja, 1974). When rice seeds carrying *D. oryzae* were immersed for 24 hours in 20 per cent water extract of *M. arvensis* or *A. sativum*, the germination increased by four times, seed infection was reduced by 86 per cent, root length was doubled and shoot length increased by 50 per cent (Jeyarajan, 1988). Extracts of *Coleus aromaticus* and *Acacia arabica* when used as seed treatment significantly reduced the seed-borne infection of *D. oryzae* and increased seed germination (Ravichander, 1987).

C. Soil – borne diseases

The effectiveness of plant extracts/products against soil borne rice diseases have been reported by some workers. Among the plant products, neem products like neem cake, neem seed kernel extract (NSKE) were found effective against many soil borne diseases of rice. Intensity of sheath blight (*R. solani*) incidence was significantly reduced in plots amended with non-edible seed cakes including neem (Rajan, 1980).

D. Air borne diseases

Application of plant extracts has shown significant reduction in the incidence of many foliar diseases of rice with corresponding increase in yield.

a. Brown spot

The extracts from *Lawsonia inermis, Azadirachta indica, Eucalyptus globules* and *Calotrophis gigantea* reduced the spore germination and mycelial growth of *D. oryzae*. In field experiments, they recorded the least disease incidence in addition to increasing the yield (Devanathan *et al.,* 1995). The leaf extracts from *Ocimum tenuiflorum* and *Catharanthus reseus* and seed extracts from *Tribulus terrestris* were also shown to effectively reduce the brown spot incidence (Selvaraj and Narayanasamy, 1994).

b. Sheath rot

Foliar spray of neem oil and neem seed kernel extract (2.0%) will reduce the sheath rot incidence by 10.7 and 8.6% respectively as against 14.7% in control. Mariappan *et al.* (1988) reported that application of neem oil (1.0%) at 17 DAS in the nursery followed by six sprayings at weekly intervals in the main field or application of carbofuran in the nursery (2.5 kg/ha) followed by neem oil spray in the main field as above reduced the sheath rot disease by 4.4 and 7.6% respectively as against 19.3% in control. Narasimhan *et al.* (1994) reported that foliar application of neem oil (3.0%) at 30 and 40 DAT showed minimum sheath rot incidence while application of neem cake (150 kg/ha) combined with foliar spray of NSKE (5.0%) was the next effective treatment. New botanical formulations developed at TNAU were also evaluated. Spraying neem oil formulation viz., Neem oil 60EC (A) neem oil 60 EC (C) and Neem oil + Pungam oil 60 EC (C) twice starting from panicle emergence and 15 days later significantly reduced the sheath rot incidence with significant increase in grain yield (Narasimhan *et al.*,1997). Foliar spray with NSKE (5.0%) twice at panicle emergence and at 15 days later or basal application of neem cake (NC) (150 kg/ha) + NSKE (5.0%) spray significantly reduced the sheath rot incidence.

c. Blast

Leaf extracts from onion (*Allium cepa*), garlic (*A. sativum*) and neem (*A. indica*) were effective in reducing the mycelia growth of *P. oryzea*. The efficacy of aqueous leaf extracts from *Abutilon indicum, A. indica, Prosophis juliflora* and *Zizyphus jijuba* in controlling the rice blast disease have been proved under field condition (Mariappan *et al.,* 1995).

E. Bacterial diseases

The extracts from different plant parts were found to be effective against the pathogen causing bacterial leaf blight (BLB) on rice. Selvam (1995) reported the effectiveness of leaf extracts from *A. odaratissimus, Curcuma longa, Vitex negundo* and *A. Indica* against *Xanthomonas oryzae* pv. *oryzae*.

F. Viral diseases

Neem and their products were reported to be most effective in reducing rice viral diseases. The efficacy of neem seed bitter (NSB) and notchi leaf extract against brown plant hopper *Nilaparvatha lugens*, vector of grassy stunt and ragged stunt virus diseases on rice was reported (Venugopal and Doraisamy, 1994). The feeding activity of *N. lugens* females was significantly reduced on plants sprayed with notchi leaf extract (NLE-10%) followed by

NLE (5.0%) and need oil (1.0%). Many plant extracts were screened against GLH vector and their transmission of RTD. Neem oil (1.0%) and custard apple oil (0.5%) as foliar spray were effective in reducing the survival of *N. virescens* and its transmission of RTD (Mariappan *et al.*, 1988). Illupai and pinnal oil also reduced the survival of GLH and RTD incidence (Mariappan *et al.*, 1987).

Non edible oil extracted from seeds of pungam (*Pongamia pinnata*), mahua (*Madhuca longifolia var latifolia*) and punnal (*Calophilum inophyilum*) were as effective as neem (*A. indica*) in reducing the transmission of RTD by GLH (*N. Virescens*) (Mariappan *et al.*, 1988). The leaf extracts from notchi (*Vitex negundo*), croton (*Croton sparsiflorus*), Bilwa (*Aegle marmelos*) and Ocimum (*Ocimum sanctum*) sprayed on rice seedlings will reduce the survival rate of GLH and RTD transmission. The reduced GLH survival and RTD incidence may be due to acquiring the virus from phloem was very much reduced with concomitant increase in xylem feeding (Saxene and Bancodion, 1988). Pre inoculation spray with leaf extracts of coconut and sorghum reduced the RTD transmission by *N. virescens* by 5 days and helped in increasing the incubation period.

18.5.2.2 Biological control of rice insects

A. Biopesticides

It is hearting to observe the growing awareness among the farmers and policy makers about ecologically sustainable methods of pest management. More and more farmers are coming to realize the short term benefits and long term positive effects of the use of bioagents and other ecologically safe methods to tackle pests. 'Biopesticides' are certain types of pesticides derived from such natural materials as animals, plants, bacteria, and certain minerals. These include for example; fungi such as *Beauveria* sp., bacteria such as *Bacillus* sp., neem extract and pheromones (Table 49) Similarly Canola oil and baking soda have pesticide applications and are considered as biopesticides. The use of these materials is widespread with applications and is considered as biopesticides. The use of these materials is widespread with applications to foliage, turf, soil, or other environments of the target insect pests. In a much simpler way, we can say that these are pest management tools that are based on beneficial microorganisms (bacteria, viruses, fungi and protozoa), beneficial nematodes or other safe, biologically based active ingredients. Benefits of biopesticides include effective control of insects, plant diseases and weeds, as well as human and an important role in providing pest management tools in areas where pesticide resistance, niche markets and environmental concerns limit the use of chemical pesticide products.

Biopesticides in general, have a narrow target range and a very specific mode of action and are slow acting. They have relatively critical applications times and suppress, rather than eliminate, a pest population. They have limited field persistence and a short shelf life and are safer to humans and the environment than conventional pesticide. These pesticides have no residue problems.

Table 49. Biocontrol agents

TYPE OF CONTROL		EXAMPLES
Insect Control	Bacteria	*Bacillus thuringiensis, B. sphaericus, B. popilliae, Serratia entomophila*
	Viruses	Nuclear polyhedrosis viruses (NPV), Granulosis Viruses (GV), non-occluded baculoviruses
	Fungi	*Beauveria, Metarhizium, Entomophaga, Zoopthora, Paecilomyces, Normuraea*
	Protozoa	*Nosema, Thelohania, Vairimorpha*
	Entomopathogenic nematodes	*Steinernema, Heterorhabditis, Romanomermis*
	Others	*Pheromones, parasitoids, predators, microbial By-products*
Weed Control	Fungi	*Alternaria, Colletotrichum*
	Bacteria	*Xanthomonas*
Plant Disease Control	Nematode trapping fungi	*Arthrobotrys*

Advantages

Nearly, 25 million cases of acute occupational pesticide poisoning in developing countries are being reported each year as reported by WHO during 1990. Pesticides cause about 14% of all known occupational injuries. Obsolete pesticides are being stored in developing countries – 20,000 tones in Africa alone. Pesticide residues in agricultural commodities are being the issue of major concern besides their harmful effect upon human life, wild life and other flora and fauna. Equally worrying thing is about development of resistance in pest to pesticides. The only solution of all these is use of 'Biopesticide' that can reduce pesticide risks, as-

- Biopesticides are best alternatives to conventional pesticides and usually inherently less toxic than conventional pesticides.
- Biopesticides generally affect only the target pest and closely related organisms, in contrast to broad spectrum, conventional pesticides that may affect organisms as different as birds, insects, and mammals.
- Biopesticides often are effective in very small quantities and often decompose quickly, thereby resulting in lower exposures and largely avoiding the pollution problems caused by conventional pesticides.
- When used as a fundamental component of Integrated Pest Management (IPM) programs, biopesticides can greatly decrease the use of conventional pesticides, while crop yields remain high.
- Amenable to small scale local production in developing countries and products available in small, niche markets that are typically unaddressed by large agrochemical companies.

Potential of Biopesticides

The efficacy of many of the biopesticide can equal that of conventional chemical pesticides. However, the mode of action will be time from exposure to morbidity and death of the target insect may be 2-10 days. Understanding the fundamental differences in the mode of action of biopesticide vs. traditional pesticides is important, since the use patterns of a biopesticide may be different from traditional pesticides to control a particular pest species.

Types of Biopesticides

Biopesticides are available into three major classes:

I. Microbial pesticides

These pesticides consist of a microorganism (e.g., a bacterium, fungus, virus or protozoan) as the active ingredient. These pesticides can control many different kinds of pests, although each separate active ingredient is relatively specific for its target pest. For example, there are fungi that control certain weeds, and other fungi that kill specific insects. They suppress pest by –

- Producing a toxin specific to the pest
- Causing a disease
- Preventing establishment of other microorganisms through competition or other modes of action.

An example of a most widely used microbial pesticide is subspecies and strains of *Bacillus thuringiensis*, or "Bt". It is naturally occurring soil bacterium that is toxic to the larvae of several species of insects but not toxic to untargeted organisms. Bt can be applied to plant foliage or incorporated into the genetic material crops and as discovered, it is toxic to the caterpillars (larvae) of moths and butterflies. These also can be used in controlling mosquitoes and black flies, Several strains of Bt have been developed and now strains are available that control fly larvae. While some Bt's control moth larvae found on plants, other Bt's are specific for larvae of flies and mosquitoes. The target insect species are determined by whether the particular Bt produces a protein that can bind to larvae to starve.

II. Plant – Incorporated – Protectant (PIPs)

PIPs are pesticide substances that plants produce from genetic material that has been added to the plant. For example, scientists can take the gene for the Bt pesticide protein, and introduce the gene into the plant's instead of the Bt bacterium, manufactures the substance that destroys the pest.

III. Biochemical pesticides

These pesticides are naturally occurring substances that control pests by non toxic mechanisms. Conventional pesticides, by contrast, are generally synthetic materials that directly kill or inactivate the pest. Biochemical pesticides include substances such as insect sex pheromones that interfere with mating as well as various scented plant extracts that attract insect pests to traps. Manmade pheromones are used to disrupt insect mating by creating confusion during the search for mates, or can be used to attract male insects to traps. Pheromones are often used to detect or monitor insect populations, or in some cases, to control them.

Farmers in their traditional wisdom have identified and used a variety of plant products and extracts for pest control, especially in storage. As many as 2121 plant species are reported to possess pest management properties, 1005 species of plants exhibiting insecticide properties and 384 with antifeedant properties, 297 with repellant properties, 27 with attractant properties and 31 with growth inhabiting properties have been identified. The most commonly used plants are neem (*Azadirachta indica*), pongam

(*Pongamia glabra*) and mahua (*Madhuca indica*). About 2-5% neem or mahua seed kernel extract had been found effective against rice cutworm, rice green leafhopper, and several species of aphids and mites. The efficacy of vegetable oils in preventing infestation of stored product pests such as bruchids, rice and maize weevils is well known. Root extracts of *Tagetes* or *Asparagus* as nematicide and *Chenopodium* and *Bougainvillea* as antivirus have also been reported.

Points to be considered

It is important to be careful when using any pesticide, even organic or natural or biopesticide. Even if this product is considered to be organic in origin, it is still a pesticide. Just because a product is thought to be organic, or natural, does not mean that it is not toxic, than many synthetic chemical pesticides have specific modes of action, just as do synthetic pesticides. While some organic pesticides may be nontoxic or are only slightly toxic to people, they may be very toxic to other animals. For instance, the organic pesticide *Ryania* is very toxic to fish. Also, some organic pesticides may be toxic to beneficial insects, such as honeybees, if they are combined with other materials, such as combining pyrethrins with rotenone. The use of an Integrated Pest Management Program (IPM) is important to insure success. To use biopesticide effectively, however, users need to know a great deal about managing pests.

To determine the relative toxicity of any pesticide to humans, check the signal word given on the pesticide label. Least toxic products carry the signal word **CAUTION** on their label. Products with the signal word **WARNING** on the label are more toxic. The most toxic pesticides have the signal word **DANGER** on their labels. Signal words are not an indication of the potential for environmental harm. All pesticides, according to law, can only be used exactly according to label directions. The user must be sure about the label of any pesticide, when and where the label says, they can be used and must follow the instructions exactly as they are written.

Biological pest control is an acceptable and necessary part of modern agriculture. If today these nonchemical methods or Biopesticide are not as popular as chemical ones, it is basically due to lack of understanding of the host-pest-environmental relationships. Hence, there is an acute need for basic and fundamental research, which can provide backup technology for formulating ecological pest management programs for various pests. The necessity is being strongly felt in the third world/developing countries as many of these non-chemical methods are simple, easy to follow and economical, with least fear of environmental pollution.

B. Natural insects

The spiders, *Lucosa* and *tetragnaths* spp are found to be efficient predators of brown plant hopper (NPH). Spiders help in effective management of several crop pests, especially in rice fields that are not sprayed with high doses of pesticides and devour pests. Different types of spiders and its preys are given in Table 49a.

Table 49a. Spiders and its Preys

	SPIDER	PREYS ON
1	Wolf spider *Lycosa pserdonnulata*	BPH, Green leaf hopper (GLH), white backed plant hopper (WPH) case worm, leaf folder and whorl maggot
2	Four legged spider *Tetragnatha* spp.	BPH, WPH, Zigzag leaf hopper (ZLH) and rice white leaf hopper (WLH)
3	Sac spider *Clubiona japonicola*	BPH, WPH, ZLH, WLH
4	Orb spider *Araneus inustus*	leaf folder, BPH, WPH, and whorl maggot
5	Dwarf spider *Callitrichia formosana*	BPH, WPH, ZLH, Whorl maggot
6	Lynx spider *Xyopes jaranus*	BPH, GLH, WPH, caseworm, leaf folder and whorl maggot
7	Orb spider *Argiope catenulata*	LH, BPH, WPH, ZLH, caseworm, stem borer and whorl maggot
8	jumping spider *Satticids spp*	GLH, BPH, ZLH, WPH, Cutworm and army worm, leaf hopper, stem borer moths

Frogs and toads are also important predators of rice pests. The mirid predator, *Cyrtorhinus lividipennis* is effective predatory bug feeding on BPH. The release of 100 mirid bugs or 50-75 eggs/sq.m. at 10 days interval has been found to be very effective in controlling BPH. Of the several natural enemies of rice stem borers, the egg parasitoids *Telenomous* and *Tetrastchius* are promising, with a high degree of natural parasitisation of egg masses under field conditions. The inundate release of Trichogramma at 50 thousand per ha. per week is also found effective. Another indigenous parasitoid, *phentygaster oryzae*, suppresses rice gall midge under field conditions.

18.5.2.3 Botanics for rice storage pests

Grains in storage need plant protection from various fungal and bacterial invasions besides a score of insect pests and rodents. Chemical treatment, though effective, cannot be used because of its inherent harmful effects and environmental hazards. In such a situation, plant derived protectants come handy. The plant products naturally occur near farm sites, and their extracts can be obtained without incurring much expense. They are safe and do not have any residual toxicity. They can be used selectively without disturbing the population of beneficial insects, parasites and predators. Further, there is a little chance of target organisms and insects developing resistance to these plant derived compounds.

1. Farmers in Punjab and Tamil Nadu use leaves of neem *(Azadirachta indica)* to protect grains. Neem extract can be applied over hessian sacks used to store food grains and seeds. The leaves are spread in thick layers on floors of bins and along walls and crushed fruits are stuck on the walls. The leaves of *Trigonella foemumgreekum* (fenu greek) are used to protect the rice grains during rainy seasons. *Artemisia absinthium* is used to save the rice grains from the attack of the augoumois grain moth.

2. Powdered neem leaves and rhizomes of turmeric have been found to repel the rice weevils. *Acorus calamus, Acorus indica, Valariana officinalis, Xanthoxylym armatum, and Artimisia maritime* are other plant species that possess excellent repellent properties against the red flour beetle *(Tribolium castaneum).*

3. Rice grains treated with turmeric powder, mustard oil, and common salt remain free from any insect damage. Several medicinal plants such as *Acorus calamus, Saussrea lappa* and *Skimmia laureola* are powerful insect deterrents. Sandal wood and 'salanin' extracted from neem seeds also protect stored grains.

4. Neem seed powder at one or two per cent mixed with wheat can prevent the incidence of the rice weevil and Khapra beetle, for over 260 days. Similarly, rice, paddy and pulses can also be protected using the neem seed powder for over eight months.

5. Ground Black pepper can eliminate the rice weevils.

6. Garlic oil too has been found to wipe off the Khapra beetle.

7. Oil extracted from the rhizomes of *Acorus calamus* offer sound protection to cereals in storage. Wheat mixed with pounded rhizomes of *Acorus calamus, A. sativum, A. indica, Carrom capitum* and *Xanthosylem armatum* can be free of pest invasion in the bins.

8. Plant materials derived from *cassia abscus, Ipomea hederaceae, Mentha arvenis, Mentha sylvestris* and *Ponganum harmale* have also been found to posses insecticidal property against the rice weevil.

9. Powdered stem wood of *R. speciosa* effectively controls important lepidopteron larval pests at 3-16 kg alkaloid equivalent per acre which makes it one of the most important botanical product insecticide due to the presence of active principles ryanodine and 9, 21- didehydroryanodine.

18.5.2.4 Seed treatment with materials of plant origin for insect control

Materials of plant origin have also been used for treating seeds from time immemorial for the control of insects.

a. Oil

Oil extracted from the seeds like maize and groundnut is used for treating seeds against insect attack. The oil changes the surface tension of protoplasmic contents of the developing embryo and regulates the egg density on the oil treated seed, there by blocking the embryonic development. The following are some examples:

- Non-edible oils like neem, castor, karanja, Undi etc. shows good repellent action against pulse beetle.
- Oils of clove, cedar wood and karanja at 25-100 ppm and acetone extracts of pepper provides protection up to 30-60 days to wheat seeds against *Sitophilus oryzae and Rhizopertha dominica.*
- Mixing kernel oil (0.1 per cent), Kernel powder (1 per cent), Cake (5 per cent), leaves (1 per cent), and leaves of neem (0.5 per cent) and Bebulgum (1 per cent) with wheat seeds provides protection against *Corcyra cephalonica.*

b. Active principles

Some of the examples of the active principles in plant products used in pest management of seed/grain are given below:

i. Antifeedants

Azardirachtin	---	*Azardirachta indica*

ii. Insecticides

Thijic acid	---	*Thuja plicata*
Pipercide	---	*Piper nigrum*
Oleoresin amide	---	*Chrysanthemum cinerifolium*
Limoneme	---	*Citrus aurantiora*

Eugenol	---	*Pogostemon heynearvs*
1-liphenyl thio Urea	---	*Allium cepa*

iii. Repellent

Turmorone	---	*Curcuma longa*
Capric acid		
plargonic acid	---	*Gossypium hirsutum*

18.5.2.5. *Traps*

Various kinds of trap can be constructed to catch insects or rodents. The most common is the light trap, set up to catch night flying insects. The highest catches are obtained by the ultraviolet lamps, but electric and kerosene lamps can also be used. A wooden frame work is encored firmly in the ground, a light source is mounted on the frame and a shallow bowl of water is placed immediately underneath. A Kerosene lantern can be suspended from a simple bamboo tripod. A Couple of spoonfuls of oil can be added to the water so that, when the moths fall in, after being attracted by the light, the oil sticks to their wings and they cannot fly again. The lamp can be set up at different heights above the ground. If the light intensity is varied, a greater variety of insects or more insects will be attracted. The trap must be sturdily built so that it cannot be blown down or knocked over the animals. Species, which can be caught with light traps, include, American bollworm *(Heliothis armigera)*, army worms *(Spodoptera* sp.*)*, brown rice plant hopper *(Nilaparvata lugens)*, cutworms *(Agrotis* sp.*)*, green rice leaf hopper *(Nephotetix nigropictus)*, rice gall midge *(Orseolia oryzae)* and tomato hornworm *(Manduca quinquemacula)*. The optimum timing for placing the traps depends on the life cycle of the insect and the stage of the crop. The best time is soon after the adult moths have emerged but before they have laid their eggs. The opposite of attracting pests is to repel them, and repellents can be used to keep animal pests away from crops and economic trees.

18.5.3 Organic certification

Organic farming practices stipulate certain guidelines/procedures to be followed in raising crops/animals. Certification of produces as organic requires periodical inspection of the field and records quality and analysis of produces. In India, there are a few certifying agencies who does regular inspection and certification. Primarily certification is required for export and to sell the produce as organically grown. Organic fields should be always away from fields grown under chemical agriculture. To convert inorganic agriculture to organic agriculture, about three-season or three years period is required to declare the produce fully organic. If there is any reduction in yield due to the use of organic inputs, a premium price at 25% to 30% over the production cost/ruling market price is permissible for all certified products. Tamil Nadu Agricultural University, Coimbatore is an identified certifying agency for organic produce, in addition to other agencies in India. The application packs and details on certification may be obtained from the Chairman, Certification committee, Centre for Soil and Crop management studies, Tamil Nadu Agricultural University, Coimbatore-641 003, Tamil Nadu, India.

18.6. CASE STUDIES

18.6.1 Organic Rice Production in Taiwan

The production of organic rice is up to 6,000 tons every year in Taiwan. The cultivation of organic rice system includes selection of local rice variety, increasing soil fertility, raising seedlings, land preparation, transplanting, fertilizers application technology, weeds control, pest control, water management, harvesting, and processing. Non-polluted paddy fields are recommended to be planted with the high quality rice variety or resistance to pests, such as Tai-keng 2, which is resistant to blast and high rice quality. For increasing soil fertility, farmers are requested to plant green manure on the organic farm. There are many organic fertilizers, such as diary manure, bone meal, and compound organic fertilizer, etc,. The lower input of organic fertilizer was considered in organic rice production. It would take much time and cost to apply organic fertilizers with labour. We can use auto-spread machine, which designed, by Taichung DAIS to reduce the cost of input for organic rice. To control weeds in organic rice cultivation, there are several methods including flooding deeply at early growth stage, heating weed seeds by applying residues of pulses, and raising duck in paddy field, etc. To pest control, residues of bitter tea to control *Pomacea canaliculata* (Lamarck), the oil of bitter chinaberry to control blast, and *Bacillus thuringiensis* to control stem borer, etc, were applied in organic rice. The average grain yield is about 4500~5500 kg/ha. Government before 2002 emphasized the verification of organic rice. The non-government system of verification had been developed since 1999. They will take the verification work of organic rice in the future.

Management of organic rice cultivation

The management of organic rice cultivation can be discussed with selection of locality and variety, increasing soil fertility, raising seedlings, land preparation, transplanting, fertilizers application technology, weeds control, pest control, water management, harvest and processing and the details are given below.

(1) Selection of locality

There should be no pollution such as factories and stock farms around the farm. Then, we should consider soil fertility. For example, alluvial soil is much better than red soil. The soil texture of clay loam and sandy loam are the bests. The selected land should be good for ventilation, plenty in water source and enough sunshine. To avoid the pesticide pollution from adjacent ordinary farm, the organic farm should be in grouping and separated from other farms.

(2) Selection of variety

The variety should be not only with good quality suited for consumers, but also with the characteristics of disease and insect-resistant, anti-lodging and have some special yield potentiality. So far, some popular varieties used are Taikeng 2, 4, 5, 9, 10, 16, Koashing 139, Taichung sen 10 & Tai sen 2, etc.

(3) Increasing soil fertility

Soil fertility is important for organic rice production. Soil diagnosis can be the reference of using organic fertilizer. Due to lower organic content contained in Taiwan soil, farmers are requested to plant green manure on the organic farm during the fallow period to provide nutrition and speed up crop growth. Planting green manure can be effective to regulate soil pH, improve soil aggregate structure, increase organic content, conserve soil moisture content, increase absorption capacity of nutrients, etc. To increase soil fertility and avoid loss of nitrogen from leaching out with rain in winter, the green manures, such as rape, vetch, Egyptian clover, *Sesbania sesben* or soybean are grown (Table 50).

Table 50. List of green manures planted in Taiwan

Green manurc		Fresh weight (t/ha)
Legume	Sesbania	25-30
	Vetch	25-30
	Egyptian Clover	20-30
	Sunnhemp	20-30
	Lupine	25-30
	Chinese milk vetch	10-34
Non-legume	Rape	20-35
	Buckwheat	50-60

(4) Raising seedlings

To ensure seedling quality, the seeds should be selected from the field. For rice organic cultivation, it is more suitable to use media with sawdust compost or bark compost. There is no obvious difference between seedling media and soil. The rice seedling can grow well in both materials. The well composted sawdust, soil plus sawdust (1:4) and peat can replace the soil as growing media.

(5) Soil preparation

The soil preparation of organic rice farmland should be done two or three times. For the first time, planted green manure should be ploughed under drying condition or with shallow irrigation for rough preparation. This has to be completed in 15-20 days before the second time of soil preparation. The second soil preparation can be completed with re-ploughing. The third soil preparation includes leveling, usually proceeding with depth of watering. To ensure seedling survival rate, easy for irrigation and drainage, the soil should be crushed and leveled precisely which may be useful for effective weeds control by flooding. The first soil preparation should include compost as basal fertilizer and mixed with rice straw or residues of previous crop. Irrigation and drainage should be practiced to keep the soil moisture content for about 60-70% field capacity, it can speed up the organic material decomposition.

(6) Transplanting

Because the soil is in muddy condition after land preparation, it is not suitable for transplanting immediately till the soil gets condensed. We have to wait for 2 days for transplanting. The rice may not be transplanted too deep, the depth in 2-3 cm under ground would be better. It would get better ventilation for rice if transplanted in north and south direction. Row distance in 30 cm width, and hill spacing in 15-18 cm apart would be better. Too close of spacing would make the rice plant weak and affect the grains maturity. Besides, it may also reduce the resistance to diseases. Too crowed and bad ventilation may induce pests. This is not beneficial to rice yield and quality, because of resulting lodging and plant pest and diseases.

(7) Technique for using fertilizer

The fertilizer for organic rice includes green manures and organic compost. Green manure turning over can reduce the application rate of organic compost. The amount of application depends on green manure variety, application rate, decomposition and mineralization rates. When we multiply the quantity of green manure incorporated with its nutrient content, we can get the amount of nutrient supplied and from that we can calculate the quantity of organic fertilizer to be applied. The following Table 51 gives the important organic manure used in Taiwan with their nutrient content.

Table 51. N content, ratios of C/N, N/P and N/K in 10 common organic materials of compost at Taiwan

Material	N (%)	C/N	N/P	N/K
Chicken manure	1.1	8.3	2.3	1.8
Hog manure	3.6	8.1	1.9	4.0
Diary manure	2.2	15.4	3.1	7.1
Sawdust	1.4	35.0	1.1	1.3
Rice hull	0.6	76.7	10.0	0.8
Rice straw	1.2	33.5	8.6	0.5
Rice bran	3.2	11.4	1.7	2.0
Bone meal	12.4	3.3	2.5	20.7
Peanut hull	1.7	18.2	18.9	2.8

a. Making compost

The first consideration for the use of organic materials is nearby low cost, balanced nutrition and easy to pile up, turn over and decomposition. In making compost, we must consider the soil texture and fertility, selection of proper organic materials. Compost-broadcasting machine is used for compost application. The process of composting is a series of microbial activity.

b. The fertilizer rate

The proper organic fertilizer should be used in organic cultivation according to soil texture and fertility. Therefore, before planting, one must analyze the soil sample collected randomly to know how much organic matter and the main elements (such as N, P, K) in soil. This is the reference of managing soil using organic materials. To keep balanced nutrition, one has to use alternately different fertilizer and soil inspection. The chicken manure is most equalized between three major components. The nitrogen content in vegetable meal is high, so it can be used for topdressing. Anything like agricultural and husbandry wastes can be made for the use after decomposing thoroughly. For example, diary manure, hog manure, chicken manure, mushroom waste, rice bran compost, bark compost, soybean meal and vegetable seed meal.

c. Method of fertilizer application

It would take much time and cost to apply with labour. We can use auto-spread machine, which designed by Taichung DAIS to reduce production cost of rice.

d. Time of fertilizer application

The use of fertilizer can be distinguished as basal fertilizer and top dressing. The basal fertilizer is used in soil preparation, and should be ploughed deeply to soil layer. It is the best time to use 15-25 days before transplanting. Top dressing is used during rice growth period. The time to apply should be 7-10 days earlier than conventional cultivation, because the organic fertilizer release nutrients slowly than chemical one. Top dressing is used to speed up tillering, mid growth, and full of grain. It is not expected for organic cultivation, if the tiller number is too much in earlier stage of growth and luxury in the mid period. Too crowded rice plant would propagate pests, so the top dressing should be applied especially during panicle initiation stage. As for usual rice cultivation, it is the proper time to apply fertilizer when the formation of panicle with 0.2 cm. Because organic fertilizer needs the time for mineralization to release nutrition, we suggest the time for fertilization for first and second crop is 8-10 and 6-8 days respectively before panicle formation.

(8) Weeds control

Rice cultivated in organic way uses no chemical herbicides. There are many weeds in Taiwan paddy field, including *Scirpus juncoides, Scripus maritimus, Alternanthera sessiles,* Umbrella plant, False Pinpernel, *Ammannia baccifera, Echinochloa crusgalli* (L.) etc. Most of them propagate with seeds and some are with agamogenesis. The methods of controlling weeds in Taiwan are as follows:

a. Land preparation precisely and flooding deeply.

b. Mulching with rice hull.

c. Crop rotation with legume, in order to reduce weeds and its density.

d. Weeding with hands and physical methods.

e. Inhibiting the germination of weeds by heat with residues of pulses.

f. Cultivating azolla for covering.

g. Using weeding machine to rid of weeds.

h. Biological control: let ducks eat new-borne weeds in tillering stage of rice. The ducks walk 3-4 km/day on the field, making the water turbid in which weed seed cannot sprout or the seedling cannot grow up. Thus it can suppress the weeds' growth.

The first weeding in organic rice should be done 12-15 days before transplanting, and complete rough plough, fine plough and rake leveling at the same time. And then, last deep irrigation (5-10 cm) for 3 days before transplanting, is given. Do the same things for the second time thus the weeds can be ridded effectively. After transplanting, as the rice seedling is stand up, mulching with 4 t/ha rice hull and filling with water slowly, and then it can rid most of weeds. Accompanying with riding weeds with manpower, the weeds can be controlled more effectively. Applying undecomposed vegetable organic fertilizer over the soil, the organic materials would be decomposed by the soil microbes. The oxygen in the soil under flooding is not enough for weed seeds to sprout. Besides, feeding ducks in the paddy field or propagate with Azolla may stop weed's germination and growth.

(9) Pests control

The common diseases and pests in Taiwan are brown plant hopper, rice leaf folder, rice plant weevil, rice stem borer, rice blast, brown spot, bacterial blight, and sheath blight. The ways of organic cultivation to prevent diseases and pests are as follows:

a. Cultivating disease-resistant and pest-resistant variety

The experts of rice breeding cultivate good quality variety with disease-resistant and pest-resistant e.g.,Taikeng 2 is resistant to Rice blast, Bacterial leaf blight and Brown plant hopper, etc.

b. Biological control

The insects can be prevented and reduced by predators such as living things like parasite insects and the microbes. In organic farm, there is no pesticide application therefore many beneficial animals, such as spider, frog, and others may be allowed for pest control. In order to reduce the density of harmful insects, one has to create ecological environment that benefits natural enemy. For example, the spiders and insects capturing harmful insects always rest at the base of rice plant. After harvest, most of them will be in the soil after ploughing. Before next soil preparation, a few of them can survive. Therefore, the better methods are to leave weeds between fields, soil preparation in rough way for the first time, irrigating water to let more natural enemy flee. Riding weeds at the second soil preparation, remained weeds 2-3 cm high as to protect more natural enemy.

c. Field management

There is a close relationship between the blight and field management. After soil preparation, the Sheath blight floats on water with fungus or stick to something, floating to the corner of wind direction. Picking up and drying may reduce the Sheath blight quantity and its effectiveness. The Rice stem borer destroys the first crop, the quantity of insects that can live through the winter are much related with the destroyed degree of first crop rice. There are two third of larva of stem borer that can live through the winter and one third in the stubble. After harvest of second crop, piling up rice straw beside the farm or covering the surface of soil for horticulture crop, then there would be more adult insects. Therefore, rice straw in the second crop has to be cut in small piece with harvest machine for drying, burying stubbles into soil by tractor. Followed by flooding to kill larva and pupa that live through the winter to reduce the source of insects, which may harmful to the first crop rice. Fertilization affects the occurrence of blight very much. Too much nitrogen makes the rice excess growth with dark and flourishing which is suitable for harmful insects to lay eggs. Application of slag and potash fertilizer can increase the hardness of rice's tissue and reduce the damage of pests, especially to the rice blast, brown spot and bacterial leaf blight. Bacterial leaf blight often occurs after wind blowing as bacteria infect the wound caused by friction. When bacterial leaf blight occurs, don't get into the farm before the dews dry up in the morning as to avoid infection to human.

d. Escape of disease

Transplanting in proper time can flee the diseases. For example, transplanting first crop at the end of January, may not suffer to Rice blast occurring around on the first ten days of April, while the leaves of rice would be old matured and won't be infected. Because of no chemical in organic cultivation, it may keep away from pests damage due to the rice plant become strong which increase the resistance.

(10) Water management

For organic cultivation, we have to change the tradition that the rice should be submerged in water during the growth period. The rice plant only needs much water during the stages of establishment, panicle formation and heading. Other times, we can use rotational irrigation because of the less water demand of rice and requirement of soil aeration. Water management of paddy soil includes shallow flooding, deeply flooding, rotational irrigation, flow continuous, drainage and drying for inhibiting tiller growing. Ten days after transplanting of first crop, and seven days after transplanting of second crop are the rice establishment stage. But in the area of rice weevil infestation, shallow irrigation is needed. If we use rice hull to control weeds, we should drain the field immediately, for keeping rice husk drying, and then practice irrigating. The speed of irrigation should be slow. This can avoid flushing rice hull and increasing the effects of weed control. Hereafter, irrigation with shallow water for 5-7 days each time. During the highest tillering stage, the field should be drained for inhibiting tiller growth and balancing the redox. As for others, we use irrigation by turn, exception at panicle formation and heading stages. During two growth stages of rice, the technique of irrigation is much related with rice yield and quality. In terms of variety, the indica is more resistant to drought than the Japanica. For example, Taichung sen is more resistant to drought than any other japonicas. The leaf of this variety will be curved when in long time shortage of water. Taikeng 9 can't withstand against drought. Therefore, we have to use different irrigation and drainage management to different variety in the process of organic cultivation.

The technique of rice water management is different in various growth stages. After transplanting, the rice needs to be flooded with 4-6 cm depth of water for 25-30 days and 20-25 days in the first crop and second crop respectively. The reasons are; the temperature in the first crop is low after transplanting, providing enough water and nutrition for rice growth and Flooding water can cause reduced form of soil.

Thus results in less oxygen in the soil to get enough energy to start metabolism of life. This can stop weed seeds sprout. However, during those periods, if the rice grows slowly, we should drain for 1-2 days to get enough oxygen and change ventilation to stop rice root from rot. Till maximum tillering stage, the rice field needs long drainage for soil aeration and inhibiting tiller growing and balancing the redox during 25-30 days and 20-25 days after the first and second crop respectively until the field is split up as wide of finger. The duration of soil aeration can be lasted for more than 1 week. However, as the rice leaves are curved, flood the water immediately.

It is important to practice irrigation and drainage in organic rice cultivation. The reasons are:

1. Flooding too long time makes the soil in reduced form thus it makes soil contain toxic substances accumulation.
2. Too long drainage makes upper soil water content insufficient when aeration and drying the soil, which results in rice root growth deeply. It can use the soil nutrient in deep layer sufficiently, besides it can prevent rice plant from drought damage during the late period.

3. It can make water and nutrient insufficient during draining, and terminate growth immediately to inhibit rice plant excessive growth and ineffective tiller. And it can make plant strong to prevent plant lodging.

There are three stages namely need flooding at 7 cm depth during rice growth (active tillering, panicle formation and heading).

(11) Harvest and processing

It must be careful to clean harvest machine when harvest organic rice. The proper period of harvest time is at the grains getting golden yellow colour stage, and just a few yellow-green grains at the base of panicles. If the rice be harvested too early prior to maturity, the immature grains increase; if harvested too late, the grains may over mature thus apt to cause broken rice and increase right-linkage starch content as to affect quality of rice tasty and smelling. So harvest rice at the proper period. The grain drying should be according to three temperature changing stages, the first stage is drying rice at 55°C until the water content of rice lower to 20%, then change to 50°C and keep the same temperature until its water content down to 16%, the last stage is keeping temperature at 45-47°C until its water content lower to 14-15%, then turn off the power and let the rice in oven machine until it reach to room temperature. Under this condition, the husk rice, the smelling quality and the percentage of complete grain can be increased.

18.6.2 Organic Rice Production in California, USA

The best examples of organic rice production in California are the farms of Bundberg brothers. Totally 100 acres are under organic rice in which a 2-year rotation that alternates rice with purple vetch (*Vicia benghalensis*) and fallow is practised. Dry unsprouted rice seed is drilled directly into the soil until appropriate moisture is available for sprouting, at which point the field is rapidly and briefly irrigated. After the rice plants have become fully established, the fields are flooded until they are drained in preparation for harvest. Purple vetch is planted in the fall following rice harvest and again in the fall of fallow year. The vetch supplied about 120-130 kg of N/ha. In the spring of fallow year, vetch is chopped and disked along with largely decomposed rice straw leaving mulch on the field. The field is shallow tilled with an implement to control weeds. Seeding into mulch followed by intermittent flooding in the early stages of rice growth and development also breaks the life cycles of various weeds, pests and pathogens and eliminating the use of pesticides. It has the disadvantage of significantly lowering the yield but the profits obtained average about US $ 65/acre more than in conventional farms due to premium price they obtain for organic rice.

18.6.3 Organic Farming in India

The organic farming and sustainable agricultural movements have also percolated in India in recent years and many persons and voluntary organizations are advocating less use of chemical fertilizers and more use of organic sources of nutrients to substitute chemical fertilizers. There is, however, a difference in the sustainability concept of developed and developing countries. The developed countries are more concerned with maintenance of environmental quality and ecological balance and would be satisfied with status quo in productivity with their mode of sustainable agriculture. In developing countries like India, however the sustainable agricultural system aim at increasing productivity along with proper care of environment and ecology. From the mid 80's onwards, experiments in organic farming have been conducted in various parts of India inspired by the Japanese agriculturist, Masanobu Fukuoka. India is taking to organic concepts and systems with fresh determination. Farmers are getting convinced that this is the natural way to get out of the stranglehold of chemicals. The Friend's Rural Centre at Rasuila in Hosahangabad District of Madhya Pradesh was one of the first to try out organic farming methods. At Bolpur in West Bengal, the Tagore Society of Rural Development under the guidance of Dr. Mrinal Kanthi Mitra also tested organic farming methods with encouraging results. These efforts resulted in fairly high yields of wheat, barley, lentils and chickpea. Similarly the Mohanpur experiment in natural farming conducted by the Joint venture of ITTI, Delhi and Gandhi Peace Foundation gave encouraging results. Intensive agriculture has polluted the environment, caused soil erosion, depleted non-renewable resources and damaged soil fertility. The use of high yielding varieties in monoculture has resulted in pest epidemics. Solutions to these problems, according to the Centre for Science and Environments' Second Environment report (1984-85), is lead back to traditional farming systems and increased use of organic manure and multiple cropping with legumes.

In Dharwad district of Karnataka, about 50 farmers practising organic farming have joined together to form a club. Village to village programme to spread the efficiency and economic gains of this method is finding favourable response. Such experiments are also being tried in Uttar Pradesh, Tamil Nadu and Andhra Pradesh. From the growing awareness of these practices, it is obvious that organic farming is seen by many as a way to excessive use of toxic chemicals in agriculture and food production. Organic farming is native to this country. Whoever tries to write a history of organic farming will have to refer India and China. Organic farming sustained agriculture of these countries. The revival of organic concept appears to stem from an incident that occurred in India about 70 years ago (Gunjal, 1991). Sir Albert Howard, an English Soil Microbiologist, developed the indoor process for the manufacture of compost from vegetables and animal wastes.

Organic farming is basically a simple idea, beginning with soil, compost, natural cycles that need to return garbage, sludge and wastes back to the land and personal health benefits that result from eating quality, nutritious food (Oldeman *et al.*, 1990). In India, the Agriculture-Man and Ecology programme in Pondicherry conducted studies to compare ecological and conventional farms with reference to sustainability indicators, especially soil fertility management, diversity and integration of cropping and livestock activities (Sivasubramanian and Dejong, 1990). Seven farms were selected from Tamil Nadu, one each from Pondicherry and Karnataka. The farmers are less than 2 ha in size and less than 20 per cent is irrigated. The farms were selected for similar agro geological and socio-economic situation. Crop grown were rice, groundnut, ragi and sorghum. For both the cereal crops, the yield was higher with organic farming and it was slightly better than the state average yield (Van der werf and Jager, 1991).

A. *Basmati rice under organic farming*

Basmati rice has its pride of place among special rices and it fetches a premium price in the international market. Basmati has been mainly a crop of the North. In south India, the warm humid conditions that prevail during grain ripening stage are not favourable for getting higher head rice recovery. Mr. R.C. Kannan, an enterprising farmer in Tamil Nadu raised Basmati rice under organic farming during the dry season (May-September). Only organics were applied to the fields and nimbicidine was used for pest management. The crop gave an average yield of 25 bags per acre (4938 kg/ha) and straw worth Rs. 625/-. The farmer spent Rs. 3745/- for raising the crop since he purchased the seeds from outside. The cost of seed would come down if he uses his own seed. In spite of the high initial expenditure, he could get a profit of Rs. 16,329/- acre as could be seen from Table 52 .

Table 52. Economics of organic farming basmati rice

Particulars*	Rs. P.
Seed cost (30kg @ Rs.30/-per kg)	900.00
Nursery - 3 ploughings	100.00
Labour charges for applying 4 bundles of Kolingi leaves to nursery	15.00
Hand levelling	10.00

Main field preparation		
Rotavator(1 ploughing)		150.00
Tiller (2 ploughings)		100.00
Puddler (1 round)		50.00
Land levelling	110.00	410.00
Labour for shifting green manure to the fields		200.00
1 load farm yard manure		100.00
Cost of plucking seedlings -		500.00
bundles @ Rs.15/-per 100 bundles		75.00
Transplanting charges 25 women @ Rs.12/- per woman		300.00
Weeding twice		300.00
Irrigation		300.00
Plant protection - 3 bottles of Nimbicidine		135.00
Harvest		800.00
	Total	3745.00
Income from whole rice 900 kg @ Rs. 20/kg		18,000.00
Income from broken rice 180 kg		1,440.00
Income from straw		625.00
Total income		20,065.00
Expenses per acre		3,745.00
Net Profit per acre		16,320.00

*Note: Unit cost at the time of report

B. Other studies

After examining the environment and economic dimensions of rice cultivation from six farms in four south Indian states - Karnataka (2), Kerala (1), Pondicherry (1) and Tamil Nadu (2), the following informations are observed. Basically, rice is grown on the farms for domestic consumption and some farmers gave a portion of rice products to their friends and relatives and however, other crops like fruits and vegetables were grown for both domestic and marketing. The resource endowments were highly diversified and found more among organic growers. Such farmers maintained a good number of livestock (for dung, milk and draught power), agro-forestry (for green manure) and adapted mixed crops including horticulture for sustainable income. These enabled the organic growers to avail and use the home grown resources as inputs in growing crops including rice. The rice growers collected seeds from their fields, dried, processed and stored them in eco-friendly containers like bamboo bins, wooden grainage and paddy straw bin. Inside the containers, some farmers kept leaves of tulsi, lotus, neem and notchi plants, others put dry chillies to check pests like white moth, which was found as common infested on the seed. While a couple of farmers preferred to sow the seeds, some went in for raising nurseries and used for transplantation. It was observed that local crop varieties were preferred for cultivation, though such varieties extinct from such localities and the organic growers stored some seeds as treasure of troves. For raising seedlings, green leaves like pongamia and neem (sometimes paddy husk) were applied as compost, which enabled the soil to be porus and this prevented wastage of seedlings and increased shooting capacity of clumps. Row method of transplanting paddy was practiced, which helped the farmers to remove the weeds, control rats and sufficient aeration was found and all these resulted in increasing the yield levels significantly (Vander werf and Jager, 1991).

CHAPTER 19

RICE CULTIVATION UNDER PROBLEM SOILS

The problem soils are soils, which need special management practices to overcome the soil related constraints in addition to nutrient deficiencies in order to obtain optimum yield. There are different kinds of problems soils in India.

19.1 SALINE, SODIC AND SALINE- SODIC SOILS

a. Saline soils

In saline soils, the EC will be more than 4 dS/m, pH will be around 8.5 and exchangeable sodium will be less than 15%. The main problem associated with this is poor water intake by the plant in the soil. To alleviate this problem, the excess soluble salts should be removed by surface scrapping of the soil or flushing the soil with good quality irrigation water. Salt tolerant varieties like CO-43, TRY I, TRY 2 etc., can be grown in the soil. Abundant application green manures, organic manures, application of Zn SO_4 can be used for getting maximum benefit.

b. Sodic and saline sodic soils

i. Sodic soils

The main problem associated with the sodic soil is excess Na (ESP more than 15) and higher pH (more than 8.5). Under this situation, the clay will be much dispersed leading an ill drained condition with poor internal drainage and poor N use efficiency due to higher volatilization and denitrification of applied fertilizer N and non availability of Zinc. All these problems put together both directly and indirectly affect the rice crop growth.

ii. Saline- sodic soils

In saline sodic soils, the pH will be less than 8.5 but the other two parameters viz., EC (more than 4 dS/m) and ESP (more than 15 per cent) will be higher than the normal soil. Such soils will have both the problem of poor water intake and other nutritional problems. These soils should be managed according to the intensity of the above two problems. In the saline sodic soils as well as sodic soil, the excess sodium needs to be removed by accessing the gypsum requirement and application of fairly powdered gypsum. The sodium has to be exchanged with Ca from the gypsum and the exchanged sodium has to be removed through the internal drainage by dividing the field in to small plots with drainage channels and impounding a thin film of water. The collected drainage water should be carefully disposed in the main drainage channels. In saline sodic soils, after removal of exchangeable Na, the soluble salts must be leached out by using good quality water. To increase the fertility status, the FYM/green manure (dhaincha) and other organics have to be incorporated. Chemical fertilizers particularly P & K can be given as basal dose at the time of last ploughing. To provide sufficient Zn, 37.5 kg $ZnSO_4$ / ha can be applied just on the surface without incorporation in the soil while taking up transplanting. About 25 per cent N over and above the recommended dose needs to be given in split application by considering the duration of varieties. The varieties like ADT 36, ADT 43, TKM 9 ETC., of short duration varieties and medium duration varieties like CO 43, TRY 1 and 2 could be grown in these problem soils.

19.2. EFFECT OF SALINITY AND SODICITY ON THE RICE PLANT

Rice is generally reported as a moderately salt-tolerant crop, but no rice variety can withstand high salinity throughout its growth cycle, and no rice is grown as a dry land crop in salt-affected land. Soil solutions high in sodium chloride with electrical conductivity values of 6-10 dSm^{-1} are associated with 50% decrease in yield, but sodium sulphide is considered somewhat less toxic. Such values have only minor significance in the field. First, the salt tolerance of rice varies throughout its growing cycle (Pearson and Ayres, 1960). The plant is tolerant during germination but young seedlings are sensitive until the age of at least 4 weeks. Damage to the rice plant during transplanting increases its sensitivity to salinity. An increase in salt tolerance occurs during tillering, but the plant again becomes sensitive during flowering. Sensitivity again diminishes during the maturation period. Second, varieties differ considerably in susceptibility to salinity (Castro and Sabado, 1977). Farmers growing rice in salt-affected soils have, without intent, naturally selected varieties with salt tolerance. Finally, at given salinity level, rice plants are more salt sensitive at higher light intensity and lower relative humidity and can withstand a certain level of salinity better during the dry season.

High sodicity caused by sodium bicarbonate and sodium carbonate affects rice. Various factors are involved in depressing yields. The high pH (more than 8.5) of sodic soils induces zinc deficiency. A high bicarbonate level may cause calcium deficiency by lowering the level of dissolved calcium, which is precipitated as calcium carbonate. Potassium deficiency and excess of reducing substances may be other growth-inhibiting factors (IRRI, 1972). Iron deficiency in intermittently flooded sodic rice lands has been reported. Sodic soils are invariably high in exchangeable sodium. Although rice tolerates such high levels better than most of other crops, it appears that a sharp decrease in yield is caused by an exchangeable sodium percentage (ESP) of 15 or more. However, it is difficult to isolate the effect of the high sodium adsorption from that of high pH. A management problem for rice on sodic soils is that applied ammonium or urea nitrogen is easily lost by ammonia volatilization.

19.3 CAUSES OF SALINITY AND SODICITY IN RICE SOILS

In general, soil salinity is caused by the presence or by intrusion of seawater, or by surface evaporation of soil water of an initially low salt content. Whereas intrusion of seawater may cause harmful salinity even in high-rainfall areas, evaporative concentration of salts is important only where evaporation exceeds precipitation. Both mechanisms can act simultaneously so that high salinity is often caused by intrusion of moderately saline water, the evaporation of which concentrates the salts.

Harmful sodicity is related principally to the presence of sodium carbonate and sodium bicarbonate. Accumulation of those salts in the soil and soil water occurs when groundwater with a high content of bicarbonate ions evaporates. Initially, calcium and magnesium carbonates precipitate upon evaporation, immobilizing part of the bicarbonate ions. But if excess bicarbonate is present (so-called residual alkalinity), high pH values will develop, even at initially low concentrations of sodium bicarbonate. In strongly sodic soils, sodium carbonate and bicarbonate become prevalent salts, and sodium will replace calcium and magnesium in the clay complex (sodic soils). This leads to the formation of sodium clays, which are strongly dispersed and, hence, highly impermeable to water. Highly alkaline conditions due to residual alkalinity in the rice-growing zone are limited to semiarid and arid areas. Chances for residual alkalinity in groundwater and surface water are especially high where granite rocks dominate in the catchment areas. In marshy areas alkalization may result from microbial reduction of sulfate. In the process, organic matter is oxidized to bicarbonate while sulfate is transformed to gaseous hydrogen sulfide (Janitzky and Whititig, 1964).

19.4 TYPES OF SALINE AND SODIC (ALKALINE) LANDS

Depending on their source and occurrence in the landscape, forms of soil salinity and alkalinity may be classified as:

- ***Marine salinity,*** derived directly from seawater intrusion either by surface flow or by seepage;
- ***Interflow salinity or sodicity*** derived from lateral influx of drainage water mineralized by weathering of salt-bearing crystalline or sedimentary rocks situated above the level of the salt- growing lands;
- ***Groundwater salinity or sodicity,*** typically in depressions where groundwater occurs at shallow depth and salts have accumulated near the soil surface because of capillary rise from the water table; and
- ***Surface water salinity or sodicity*** due to evaporation of soil moisture brought in by river floods or irrigation and which, because of low permeability of the soil or rapid evaporation, has no chance to drain vertically.

There are, of course, transitions between these forms of salinity. Salinity occurring in rice growing areas is caused mainly by chlorides and bicarbonates of sodium sulphide, but sodium sulphide is also found locally in harmful concentrations. High concentrations of the soluble salts of calcium and magnesium are not important in relation to rice because they are considered positively sodium and occur mainly in the most arid zones where rice is rarely grown.

This is the right time to know about different forms of soil salinity and alkalinity in detail. Because no one in this world can forget the impact of Tsunami waves on the coastal belts of Tamil nadu, Andhra Pradesh, Orissa, West Bengal, Kerala and Andaman Nicobar Islands during December, 2004. For the benefit of the readers, the informations are given in detail here under.

19.4.1 Marine Salinity

Marine salinity is confined to low areas bordering seacoasts. Some saline land is found along every coast, but the width of the affected area varies greatly. Marine salinity is limited to land that is below the high tide level and that is not protected by man. There are important differences in degree of salinization, depending on climate, tidal regime, and regime of streams entering the coastal plains. Generally speaking, salt accumulation and severeness of salinity increase in the drier climates and diminish strongly in an equatorial climate without a pronounced dry season. Extremely saline deltas are found in arid climates. e.g. east coastal regions of Tamil nadu, India. There, the intruding seawater evaporates, leaving behind salt crusts and groundwater with a salt concentration higher than that of seawater. On the other extreme, salinity in the rainy perhumid zones, e.g. East Sumatra and the coastal area of Kalimantan, Indonesia, is confined mainly to the frequent inundation of tidal flats that are normally occupied by mangroves. Little rice is grown on land influenced by either of the two extreme conditions of marine salinity.

In the arid zones, good quality water to reclaim the land for rice is often not available; the lower Nile delta is an exception. In the perhumid zones, the tidal land offers a score of difficulties for reclamation, such as poor accessibility and poor soil conditions (potential acid sulfate soils) (Moormann and Pons, 1974). Most rice land influenced by marine salinity is found in zones with a monsoon type of rainfall. Such land is widespread in the lower part of the major and major deltas of Asia, where due to increasing population pressure, rice cultivation extends to areas well below the levels of high tides. In such areas, it is not possible to completely control salinity by the traditional method of ponding fresh water. In a few areas such land has been desalinized by empoldering and artificial drainage as in the Red River delta in northern Vietnam.

Many coastal rice- growing zones in tropical Asia, however, are not or at least not completely, protected against periodic saltwater intrusion. In this respect, the coastal areas periodically subjected to floods caused by cyclones or typhoons suffer most unless they are protected by high dikes. Such floods affect much rice-growing land around the gulf of Bengal in India and Bangladesh in the wet season. The impact of Tsunami waves on the coastal areas of Tamil Nadu and Andhra Pradesh that occurred recently (December, 2004) is clearly indicated the seawater effect on agricultural lands. Crop damage may vary from slight (if the flood strikes when the plants are in salt-tolerant growth stage, i.e. seedlings, ripening) to severe (if the flood coincides with flowering). Protection against this type of unpredictable, generally catastrophic salinity cannot be achieved by improved reties. Major engineering measures such as dike building and empoldering are required for sustained improvement. Rice growing in such areas is hazardous, and the benefits of improved local water management and improved salt-tolerant varieties can be undone by one tidal wave.

The tidal regime influences the degree of marine salinity. Seawater penetrates deeper inland where the tidal amplitude can, however, be helpful in reclamation. The land, protected by dikes, is drained during low tide, and no salt water enters the protected area during high tide. Reclamation in such coastal areas is often pushed much further seaward than it is along coasts with small tidal amplitudes. Often the mean tide level as well as the spring tide levels fluctuates seasonally. In most of the coastal areas in Thailand and in the Mekong delta spring tides are highest in the dry season, leading to strong seasonal salinization. When sufficient fresh water is available during the wet season, the salts are washed away and a good rice crop can be grown. In contrast, high tide levels in the Philippines coincide with the onset of the wet Southeast monsoon (May-July) and farmers are often obliged to delay planting for several months. If irrigation water is available during the dry season, however, rice can be grown on the same fields without any risk of salinisation. The influence of rivers and streams entering coastal areas varies from place to place, which means the relation between salinity and river regime must be studied locally. Man's interference with river flows in areas with potential marine salinity has had both positive and negative effects. While the tapping of fresh river water is a helpful factor in reclamation of saline coastal lands, a decrease in discharge of coastal rivers due to water use lead to salinization of other areas.

19.4.2. Salinity and sodicity by interflow

Salinity by interflow can be expected in areas with salt bearing crystalline of sedimentary rocks at relatively shallow depths in the higher parts of the landscape. Seasonally high rainfall promotes the transfer of salts from the weathering zone to spring levels in valleys and along foot slopes. Drainage water from weathering zones of felsic rocks such as granites often has alkalizing properties and may result in highly alkaline soils near spring levels of interflow water. Interflow salinity is quite common in monsoonal areas with a pronounced dry season. Rice-growing areas influenced by this type of salinity or alkalinity is found in the Deccan plateau of India, and in the northeast region of Thailand (Sinanuwong and Takaya, 1974a,b). Smaller areas are found in most dry zones of countries where lowland rice is grown. The relations between salinity and climate are clear on the rice-growing plains along the Vietnamese coast from Phan Rang to Hue. Interflow salinity and alkalinity are common on the dry plains of Phan Rang and Phan Thiet, but diminish sharply toward the higher rainfall zones to the north. They disappear entirely on the high rainfall plains, such as those of Danang and Hue. Where salt-bearing rocks are rare or absent, salinity by interflow is generally unimportant.

In areas with interflow salinity, the lower parts of toposequences tend to be saline at the surface or at shallow depth. Salt moves down from salt-bearing rocks by interflow and appears at the surface in the lower parts of the landscape where salt accumulates in the upper horizons because of evapotranspiration. Suppression of salinity by ponding good quality rain and irrigation water in the bottomlands of valley leads to a distribution of salt-affected lands in rice-growing areas. Many examples of these patterns are observed in Asia. The land-use pattern in the valleys around Hyderabad, India, is remarkable in this aspect. In northeastern Thailand, the higher parts of the landscape are mainly in shifting cultivation while in a score of places, people gather salt form the saline zone above the paddy fields. The transition between salt-affected zones in lower slopes and nonsaline paddy fields below is remarkably sharp. Often, the outside of the paddy bund bordering a saline zone may show salt efflorescence at the surface, while the inside of the bund is not or is only slightly saline. Saline-alkali spots tend to be confined to the areas near the spring levels; saline but non-alkaline soils also occur toward the lower areas, as e.g. in the dry zone of Sri Lanka. When the catchment areas are completely cleared, interflow increases and the water table in depressions may increase concurrently. This can lead to a local worsening of lower-slope salinity.

19.4.3 Groundwater salinity and sodicity

Groundwater salinity and sodicity are associated with groundwater at shallow depths. An excess of evapotranspiration over rainfall causes a net upward movement of water through the soil by capillary action and leads to surface efflorescence of salts during dry periods. This mechanism plays an important role in areas affected by seawater and in areas with interflow salinity, but it is most typical in inland areas with large depressions occupied by alluvial and lacustrine plains and by river terraces of different ages. The high water table may be due to continued influx of surface or groundwater from elsewhere. Irrigation schemes without proper drainage often lead to this type of salinization. For instance, in the southern Indus valley in Pakistan (Sind province), large areas became strongly saline after the introduction of extensive irrigation, which led to a rise of the water table from a depth of 20 to 30 m to 1 2 m within 20 years. The groundwater itself need not be saline, although it often is, because long-term evaporation of continuously recharged water of a low salt content inevitably leads to strong salinity in arid climates. Often, the most saline soils are concentrated in the lower parts of the landscape, e.g. around intermittent lakes in back swamps river channels. Land affected by groundwater salinity is often most profitably used for rice growing, and paddy cultivation on salt-affected land is rapidly increasing in many parts of Pakistan and India.

19.4.4 Surface water salinity and sodicity

Alluvial depressions in areas with semiarid and dry monsoonal climates collect and retain part of the surface runoff from higher areas. If the permeability of the soils in the depression is slow and little percolation takes place, surface salinization occurs by evaporation of soil moisture and surface water even though such areas have a dry, nonsaline subsoil. As in the case of groundwater salinity, micro topography plays an important role in salinization, with the lowest spots most strongly affected. Most of the highly alkaline soils near Lahore, Pakistan, where rice grown on a large scale, are in this category. The high levels of exchangeable sodium in such soils cause strong dispersion of the clay and thus low permeable, which prevents internal drainage and increases the chance for further accumulation of soluble salts.

19.5 VARIABILITY AND DYNAMICS OF SALINITY AND SODICITY IN RICE FIELDS

Due to the high mobility of water- soluble salts, salinity is extremely variable in time and place. Salt on or near the surface may be removed by vertical and lateral drainage at the onset of the wet season or salinity may be diluted by fresh water. In low areas, salt may be brought in by seepage or surface runoff, and salt may concentrate on slightly higher, nonflooded spots by evaporative concentration. As noted, coastal salinity often depends on largely unpredictable factors such as cyclones (e.g. around the Gulf of Bengal) or, less frequently, on the flow cycle of rivers that provide fresh water during the rice-growing season. Sudden seawater infringement may destroy a rice crop completely or may do relatively little damage, depending on the stage of crop development. When rice is planted at different dates in a salt-affected area, a single short-duration seawater infringement may produce a checker-board effect of badly and slightly affected fields.

In paddy fields affected by salinity, the amount of rainfall and its distribution will alter the salt injury pattern from year to year, or even within a season. The major production constraint in such lands is crop insecurity. In a year with copious, well-distributed rain, injury may be minimal but in other years, crops may suffer when there is not enough rainwater or irrigation water to suppress salinity in the root zone during the critical stages of plant growth. The quality of irrigation water, especially when provided by smaller reservoirs, is also variable and salt contents tend to be higher in unusually dry years. In most cases there is a clear microvariability of salinity and degree of injury of the rice crop. Microrelief variations are one cause. In partially flooded fields, slightly higher spots are often distinctly more saline, because of evaporation from the soil and capillary rise of saline groundwater when the high spots are temporarily dry. Typically, in northeast Thailand and similar salt-affected areas, termite hills in paddy fields have a fringe of saline soils along their base where no rice will grow. Man-made irregularities such as somewhat higher strips along bunds may show similar fringe salinity. Micro variation in internal soil conditions such as miniscule differences in hydraulic conductivity of the soil profiles may also be involved in the typical spotty salinization. Furthermore, saltwater subsurface flow may follow preferential lines or aquifers, as in sandy lenses in a more clayey make agronomic studies, including varietal testing for salt tolerance, extremely difficult. Although alkalinity also is often spotty and shows lateral variability closely related to microtopography, it is less variable in time than salinity. The movement of the readily soluble sodium bicarbonate and sodium carbonate is strongly hampered by the slow permeability of those dispersed clayey soils. Furthermore, the exchangeable sodium in sodic soils is not mobile.

19.6 RECLAMATION AND MANAGEMENT OF SALINE AND SODIC SOILS

The major reason that rice is grown on saline land, where virtually no other crop will grow, is the aquatic nature of the crop. Rice will grow in land that is highly saline when dry, provided the soil can be flooded with low-salinity water. Floodwater dissolves and dilutes the salt in the surface soil layers and drainage, and percolation induced by ponding remove part of the salt from the rice plant's shallow root zone. In most of the areas in Asia where saline land was reclaimed by ponding, no artificial drainage was provided and the salt was simply leached down to a level where it was no longer dangerous to the rice crop. Some salt may leave the affected area by natural drainage, through interflow, and by creeks and rivers, but a large portion remains in the subsoil. In tidal areas, the natural surface drainage during the wet season can remove salt from the surface soil, but unless dikes and flap gates are constructed, salinity returns in the dry season. For example., in areas adjacent to the sea on the Mekong delta of Vietnam, surface salt accumulates in the dry season. Water ponded in the paddies during the early rainy season is commonly drained to evacuate the salt and rice is transplanted only after later rains full the paddies. In several marine estuaries of West African rivers, saline mangrove land is similarly used (Moormann and Pons, 1974). The only land reclamation measure is the cutting of the mangrove. The crop is planted during the flood season after the fresh floodwater has diluted and washed out enough salt to allow rice to grow. When the seasonal floods subside, seawater reintrudes and the land reverts to a saline condition. Use of such unprotected saline land is frequently risky, especially if the river floods subside early.

Protection of such tidal land by dikes will decrease the risks of salinity damage, but there are disadvantages as well. Premature reclamation of mangrove marshes by diking in India and Bangladesh caused considerable land subsidence leading to increased seepage of saline water and greater damage from cyclonal floods (Nagaraja, 1964). Rice cultivation is also advantageous for highly alkaline/sodic soils. Because of their poor permeability for water and the relatively low mobility of exchangeable sodium, the alkaline/sodic soils generally do not improve under paddy rice cultivation as quickly as saline soils. But in highly alkaline/sodic soils, flooding has a different beneficial effect. Chemical reduction of the surface soil after flooding increase the concentration of carbon dioxide, which leads to a drop in pH and an increase in dissolved calcium by dissolution of the calcium carbonate that is usually present in such soils. The decrease in pH (down to values between 7 and 7.5 if sufficient organic matter is present) is favourable for the availability of phosphorus and many micronutrients, whereas the increase in dissolved calcium favours the removal of exchangeable sodium. The presence of rice plants seems to promote the mobilization of calcium and sodium beyond that in upland flooded soil (CSSRI, 1974). Sodic soils often amended with gypsum or pyrite in north India and Pakistan, which help speed soil improvement. Nonamended paddies, however, will improve to the same level of productivity with more time.

Gypsum is the most economic and effective amendment for reclamation and crop production in sodic soils. The quantity of gypsum depends on the nature of soil, its stage of deterioration and crops to be grown. Soils with pH lower than 9.0 (1:2 soil: water suspension) seldom need gypsum for growing rice and barley. But if the pH is above 10.0, no crop is likely to grow unless gypsum is applied. For growing shallow rooted crop such as rice and wheat or barley in sequence, gypsum should be applied @ 12-15 t ha^{-1} depending upon the sodic status and soil texture. The agriculture grade gypsum (available in packed gunny bags) having at least 70% purity should be powered and mixed thoroughly in top 8-10 cm by shallow cultivation followed by planking. After this, the field should be irrigated and kept ponded for at least 15 days, so that applied gypsum may react with the soil. Farmyard manure @ 25 t ha^{-1} should be applied along with gypsum for better yields if it is available. The gypsum requirement of soil is generally determined in laboratory. However, for general use, Table 53 serves as a guide to work out the quantity of gypsum needed.

Table 53. Gypsum requirement (GR) of different soils (t ha^{-1})

Soil pH	Gypsum requirement (t/ha)		
	Light	Medium	Heavy
9.2	1.7	2.5	3.4
9.4	3.4	5.0	6.8
9.6	5.0	7.5	10.0
9.8	6.8	10.0	14.6
10.0	8.5	12.5	15.0
10.2 & above	10.0	15.0	15.0

Total elimination of sodicity is not always desirable for lands where rice is grown in rotation with winter wheat, such as in large tracts of northern India and Pakistan. That would drastically increase the permeability of those often silty and sandy soils and lead to greater losses of irrigation water when the lands is used for rice. Although this excessive permeability would deteriorate the physical condition of the soil for the upland crop in the rotation. In irrigation projects in saline areas, the tendency is to practice multiple cropping with wet-season rice followed by one or more intermediate or dry-season dry land crops. For economical use of irrigation water, this approach seems logical. But the dynamics of salinity are definitely against such an approach. When dry land crops are grown in the dry season with intermittent irrigation, a moist zone forms between the subsoil saline layers and the surface. Evapotranspiration becomes important and salinity develops rapidly in the surface soil layers. The speed of salinization varies. Salinization has occurred within one dry season in medium-textured soils at Kalasin, northeast Thailand, and at Richard Toll in the Senegal River basin. Many irrigation projects in the drier rice-growing areas suffer from the same lack of insight into the dynamics of salinity. Salinization of nonsaline rice fields due to injudicious engineering is found in an increasing number of irrigation projects in areas with interflow and groundwater salinity. The storage of water in reservoirs and its transport through unlined irrigation canals above the level of the paddy fields in the low-lying areas generally lead to rising water tables. This frequently affects rice lands below and adjacent to canals and reservoirs but it can affect the salinity of large areas if no appropriate drainage is foreseen. When the upward movement of saline groundwater in such cases exceeds the downward movement of good quality rain or irrigation water, paddy lands will be affected by salt. Such small-salinized areas are typical in Northeast Thailand adjacent to and below small reservoirs constructed since the 1950' s in salt-affected lands. This kind of localized salinity was also found below reservoirs on the Deccan plateau (Hyderabad area, India) and, to a lesser extent, in the dry zone of Sri Lanka.

19.6.1 Agrotechniques for Sodic soils

1. Nursery: Raising the nursery in sodic soils, the seedling should be raised in a normal soil since rice seedling are very sensitive to sodic condition at 1–2 leaf stage. A seed rate of 50–60 kg may be followed to raise the seedling for transplanting of one ha of sodic land as compared to 30–40 kg ha^{-1} required for transplanting under normal soils. The higher seed rate for sodic soil is to maintain higher plant population as safeguard against mortality and poor tillering in the initial reclamatory stage.

2. Varieties - Depending upon the location, suitable varieties may be grown as given below. For e.g.

i. P-2-221, IR-8, IR-8-69, Ratna, Palman 579 and Kalarata- Recommended for sodic soils of Indo Gangetic plains.

ii. CSR–4, Bhurarati, Dasal, Jhona. 349, Kalarata and IR-50- Recommended for heavy black sodic soils of western Madhya Pradesh.

iii. Damodar, Gitu, Dasal, Recommended for highly sodicity conditions where amendments are not added.

iv. CO-43, Try-1 and TRY-2- recommended for Tamil Nadu

3. Field preparation - The physical conditions of sodic soils are often poor. So, the sodic soils should not be tilled when too wet, as it spoils the soil structure further. At proper soil moisture content, 3 to 4 shallow cross ploughings should be done to pulverize the soil. After last ploughing, the field should be leveled with leveler. Puddling in standing water, an important operation in normal soils for successful rice cultivation, is not beneficial in sodic lands during the initial years of reclamation (2 to 3 years after gypsum application) as these soils are poorly permeable and water remains standing in the field for several days and generally weeds are also a serious problem. In first year, puddling is distinctly a harmful practice in these soils.

4. Seedling age and time of transplanting - Since young seedlings are relatively sensitive to sodic conditions, older seedlings are therefore recommended for transplanting in sodic soils. Transplanting should be done, when the seedlings are 5-6 weeks old, as compared to 3-4 weeks old in the case of normal soil. For better yields, more than 40 days old seedling should be used under condition of delayed transplanting.

5. Plant population - To maintain the desired plant population in sodic soils, it is necessary to transplant more seedlings per hill in a closer spacing, as safeguard against seedling mortality and reduced tillering ability. In the initial years of reclamation, 4-5 seedlings per hill should be transplanted, as compared to 2-3 seedling per hill in normal soils. Spacing may be followed for different varieties based on the concerned state government recommendation.

6. Nutrient management - Low organic matter content (<0.1%) and high pH impart inherent poor fertility to sodic soils. Poor air water relationship and high pH of many sodic soils further influence the transformation and availability of applied fertilizer nutrients.

a. Nitrogen - High pH of sodic soils favours greater volatilization losses of N added through ammonical fertilizers or urea or the decomposition of organic matter. Among the different nitrogen sources, the losses of N are higher in the case of urea followed by calcium ammonium nitrate and ammonium sulphate. During early reclamation years, ammonium sulphate proved superior to calcium ammonium nitrate and urea as source of nitrogen for rice but due to its non-availability and also higher unit nitrogen cost, urea may be used profitably in sodic soils. Best yields are obtained when N doses are raised by about 25% in first year in comparison to normal soils. Semi dwarf rice varieties require higher dose of nitrogen i.e. 150 kg N ha^{-1}. For efficient management of N, its application in 3-4 splits is advocated to reduce nitrogen losses. Part of N can also be applied as foliar spray (Gupra and Abrol, 1990). Slow release nitrogenous fertilizers such as sulphur coated urea (SCU) and urea super granules (USG) have been found to be more effective than urea in sodic soils (Rao and Ghai, 1987).

b. Phosphorus - Phosphorus deficiency is not a major constraint/ problem in these soils. Though rice is an efficient P absorber, yet due to shallow root system, it cannot utilize P from the lower layers and is affected by the low fertility of the surface soil layer. Therefore to obtain satisfactory rice yield, sufficient quantity of phosphorus should be applied at the time of last ploughing/transplanting or as per the recommendation.

c. Micronutrients - Zinc is the most important input which affects the rice yield in sodic soils. In general, application of 25–30 kg Zinc sulphate ha^{-1} as basal is recommended.

7. Water management - A poor moisture transmission characteristic of sodic soil helps in maintaining submerged moisture regime with ease as compared to normal soil. Due to this, the total irrigation requirements are also considerably reduced in comparison to normal soils. The crop should not suffer due to lack of water at establishment, vegetative, panicle initiation, milk and grain development stage.

8. Weeding - Weeds are not a serious problem in sodic soils for the first 2-3 years of reclamation. After 2-3 years of cropping, weeds appear in the rice field, which can be controlled by hand weeding. Application of Butachlor @ 1.5 kg a.i. ha^{-1} at 4-5 days after transplanting in a standing water is recommended in controlling weeds.

9. Cropping Pattern - Choice of cropping pattern in sodic soils is very important to enhance the process of reclamation. It is usually ideal to initiate reclamation by leaching in the month of April followed by raising Dhaincha as green manure crop. Rice (kharif)-berseem/wheat/barely (Rabi) is recommended for sodic soils condition of northern India as given in Table 54. It is recommended that as far as possible, sodic lands should not be left fallow for any considerable length of time.

Table 54. Crop rotation for Sodic/ Sodic Soils

Season	1st year	2nd year	3rd year
Summer	Leaching / Dhaincha (G.M)	Dhaincha (G.M.)	Dhaincha (G.M.)/Fallow/ Fodder Cowpea / Fodder Maize
Kharif	Rice / Dhaincha	Rice	Rice
Rabi	Barley / Wheat	Barley / wheat / Raya / Berseem / Sugarbeet	wheat / Raya / Berseem / Snji / Sugarbeet

10. Amendment - Gypsum is used as an good amendment for reclamation and crop production in sodic soils. Details are given in the section 19.6.

11. Distillery effluent- The distillery effluent can be considered as a liquid manure and controlled application of the treated effluent can increase the productivity of crops there are a no of distillery units in India. For the production of every one litre of alcohol, nearly 12-14 litres of effluent is discharged. Every distillery unit is generating 5-10 lakh litres of raw effluent everyday. Some units have installed biomethanization plants for producing biogas from the raw effluent and discharging the treated effluent with 90% reduction in BOD and 63% reduction in COD. These values are found to exceed the limit prescribed by PCB. Experiments conducted on sodic soils revealed that in the areas where effluent in available, it should be diluted 50 or more than 50 time with Irrigation water along with application of organic amendment neem leaves @ 6.25 t/ha or FYM at 12.5 t/ha to increase the rice yield, without much detrimental effects on soil health (Annadurai *et al.*, 1999).

Plate 51. Raising Daincha for reclamation of sodic soil.

19.6.2 Agro techniques for Saline soils

1. Rice varieties for saline soils- Selection of salt tolerant rice variety is the first right step to achieve best yield and to get best results from the applied inputs. Some of the improved rice varieties identified for different states are given in the Table 55.

Table 55. State wise recommended salt tolerant rice varieties

States	Rice Varieties
Tamil Nadu	CO-43, TRY-1, TRY-2
Gujarat	Jaya, IR-8
Maharashtra	Panvel (resistant to blast and blight), Kalarata 1-24, Bhura rata 1-10, SR 3-9, MK47- 22, Damodar (CSR-1).
Goa	Vytilla, Mohan (CSR-4), AU-1, MCM-2

Karnataka	Vytilla, Balikagga, Karikagga, Arya, Kl-21, 19,68; MR – 261, SR 26 – B, Getu (CSR-3).
Kerala	CSR –4, C – 23-2-1
Orissa	SR26-B, Patnai 23, Luni, Raspanjari, Valuki, Rajmal, FR 43b (more suitable for flood prone areas) Nonabokra, Pakhira, Dahikera, Kartik Sola, Cuttack Dhani, Pokkali, Getu (CSR-3).
West Bengal	Pankaj, Mahsuri, SR 26B, Kalamota (sel) CSR –1, CSR-2, CSR-4, CSR-6.

Source : Biswas and Bandyopadhyay (1987).

2. Preparatory tillage - Before the onset of monsoon, a thorough initial deep ploughing (30-45 cm) is most essential because of extremely bad physical condition of these soils. It can be achieved by a tractor drawn sub-soiler, which avoids turning up of the subsoil successively. The field should be given 2-3 harrowing followed by leveling to have a fine tilth.

3. Avoiding summer fallow - As far as possible, summer fallowing should be avoided. This could be accomplished by growing a second crop, following the kharif rice. If sufficient water is available, a second rice crop can be grown which not only gives additional production but results in the increased yield of subsequent kharif rice due to leaching of salts from the root zone. In the case of non-availability of sufficient water the crops, which require less, like chilli, cotton and barley could be grown.

4. Division of field into sub plots- Field should be divided into 10-15 subplots per hectare. Subplots should be connected to central drain between them to serve as an outlet for the surplus water. This is needed to pond the irrigation water or rainwater for leaching of salts from the soil surface.

5. Pre-sowing irrigation - A heavy pre-sowing irrigation to leach down the accumulated salts from the root zone is very essential for early establishment of transplanted rice seedlings. Wherever available, the pre sowing irrigation should be given with good quality canal water.

6. Time of transplanting - Non – availability of good quality irrigation water is a common problem in saline soil areas. Thereby, transplanting time is dictated by the onset of monsoon for kharif rice. Transplanting of rice should be avoided until salts from the upper 8-10 cm have leached out. Normally kharif rice transplanting is done after receiving 600 mm of rainwater. Accordingly time of transplanting may vary between mid July to mid August. It has been observed that mid July is the most suitable time for coastal belt of West Bengal, while mid to late June is most suited for coastal zone of Gujarat. Due to salinity hazard, mortality of young seedlings and poor tillering are common in these soils. Higher seed rate and spacing are advisable to counter these adverse effect. Therefore, to have adequate plant population, the number of seedlings per hill should be increased to compensate any loss of seedling mortality. In coastal saline soils, 3-4 seedlings per hill are required in kharif to compensate the yield loss due to seedling mortality. Number of seedlings of rice should be increased during rabi season.

7. Nutrient management

a. Nitrogen - Mineralization of N decreases with increasing salinity. Besides the effect of excess salts, chloride, which is the dominant anion in coastal saline soil, suppress the nitrification process when present in excess concentration. High soil salinity also decrease microbial activity of nitrifying and atmospheric N fixing organisms in soil – rendering N fertilizer use efficiency only 20-25%. Nitrogen deficiency is wide spread in saline soils and nitrogen requirement of rice in saline soil is higher than in normal soils.

The nitrogen can be supplied through the following ways,

i. Green manuring with dhaincha

ii. Use of bulky organic manures

iii. Placement of N fertilizer below the soil surface

iv. Split application: For best result, 65-75% of total N may be applied at active tillering stage and the rest at panicle initiation stage.

b. Phosphorus and Potassium - There is not much response to P and K application in this soil. But whenever P is needed, it is to be applied preferably in the form of single super phosphate.

19.6.3 Rice based cropping systems for saline soils

The selection of crops and cropping sequence to be followed in saline soils is of paramount importance. Since crops vary in their tolerance to salinity and are either too sensitive or semi tolerant to tolerant to a given level of salinity (Mass and Hoffman, 1977). As discussed earlier, the underlying principles is not to allow the soil being fallowed in dry season, which will enhance the accumulation of salts on soil surface. Taking a second crop following kharif rice still lower the salinity level at root zone and will permit better crop production in next kharif season. The second crop may be rice if sufficient water is available. Otherwise crops like Chilli, cotton, barley, etc. can be taken which are salt-tolerant as well as need less water. Table 56 presents some of the prominent cropping systems found to be efficient and profitable for different parts of coastal saline soils in India.

Table 56. State-wise cropping systems for coastal saline soils of India

States	Seasons			Situation	Remarks
	Summer	Kharif	Rabi		
Gujarat		Rice	Wheat	Irrigated	Wheat may be taken
		Cotton	Wheat	Irrigated	With late rains
Orissa		Rice	Ragi / horse gram / cotton barely	Limited irrigation or rainfed	—
		Rice	Wheat/ cow pear green gram / groundnut	Irrigated having no possibility of submergence	The second crop may be taken on residual soil moisture
Karnataka		Rice	Cow pea / ragi / radish	Rainfed	-do-
Andhra Pradesh		Rice	Rice/black gram / green gram / sunnhemp / sugar / jowar / ragi / pillipasara fodder	Rainfed	-do-
		Rice / jowar / bajra / redgram / gingelly etc.	Ragi / maize/ chilli/tur/tobacco etc.		
Tamil Nadu		Rice	Rice	Irrigated	—
Maharashtra		Rice + vegetable (bitter gourd, cucurbits, bhendi, guar, pumpkin etc. on bunds	—	Rainfed	—
	Green gram	Rice	Maize for grain	Laetrile soil of costal	Max. grain yield and net return
	Fallow	Rice	Ground nut	Areas	Second max. grain yield and net return
Pondicherry	—	Rice	Oil seeds / pulses / sugarcane /	-	—

			cotton / fruits and vegetables		
Goa	Vegetables	Rice	Vegetables	Limited Irrigation	—
West bengal	Rice	Rice	Rice	Irrigated	The best for obtaining higher yield and reducing salinity
	Mesta	Rice	Wheat / barley	Limited irrigation	Good economic viability
	Dhaincha (green manure)	Rice	Rice	—	Most suited for obtaining higher yield and improving soil characteristics
	-	Rice + fresh water fish (carp and *Macrobranchium rosenbergel*	Vegetable on bunds + brackish water fish (*P. monodon L. panasia)*	Coastal saline soils of sunderban with no available good water	Most judicious and profitable land use.

Source : Biswas and Bandyopadhyay (1987)

19.7 OTHER PROBLEM AREAS FOR RICE CULTIVATION

A. Acid Soils

Soils low in pH and responding to application of lime may be considered as acid soil; for the purpose of characterization and classification, has considered soils having pH less than 5.5 in 1:1 water extract as acid soils. (Muthy *et al.* 1976). Soils with low pH are observed in red, lateritic, sandy, sandy loam and silt clay loam soils. Acid soils are characterized by

i. Shallow to medium deep, coarse to medium textured

ii. Porous and permeable, low water receptivity

iii. Crusting of surface soil due to impact of raindrops.

iv. Dominated by oxide of Fe & Al (more then 95% of the total chemical content).

v. Dominant clay minerals like kaolinite and illite

vi. Low CEC (3-15 C mol kg^{-1} soil)

vii. Higher fixation capacity for P & K

viii. Devoid of bases and usually acidic in reaction.

ix. Poor N status

x. Low organic matter content

xi. Low availability of P to crop plants due to P in Fe – P & Al-P bound forms

xii. Moderate to good storage of non exchangeable K, whose availability depends on the mechanism of release in relation to the K bearing minerals feldspar and mica and the presence of clay minerals of hydrous mica group.

The major problems due to acid soils are:

i. Poor N status due to low organic matter content.

ii. Higher rate of P & K fixation.

iii. High leaching losses of nutrients

iv. High percolation rate (demands more irrigation water for rice cultivation)

v. Crust formation on the surface (obstructs emergence of seedlings after heavy showers in direct seeded rice)

vi. Toxicity due to Fe (in some cases, due to Al and Mn) and

vii. Susceptibility to erosion due to natural high slopes and light texture.

Agro techniques for Acid soils

1. Soil Compaction: To increase the production of rice in these soils by way of reducing deep percolation, artificial soil compaction may be introduced. Soil compaction with the help of a roller (800 kg) by 4-6 passes will reduce the percolation loss.

2. Lime application : To correct the soil acidity, lime or other basic materials like calcium silicate are to be applied. This will raise pH and decrease P fixation and prevent the release of nutrients like Fe, Al, and Mn becoming toxic. Industrial wastes like basic slag and blast furnace slag from the steel industry and lime sludge, press mud, cakes from paper mills are economical. Addition of amendment to the extent of ¼ lime requirement as estimated by woodruffs Buffer method has been found to be most economical and beneficial. (Panda and Nada, 1985). Rice plants prefer a pH range of 6 to 7. To maintain this pH level for different acid soils, the amount of lime to be applied can by guided by the following Table 57 given by ICAR. In India 47% of total acid soils have the pH ranging between 5.6 – 6.5 which may not require liming for rice since submergence itself can take care of that. But rest 53% of acid soils having the pH below 5.5 need attention for their amelioration through liming.

Table 57. Lime requirements

pH	Average quantity (t ha^{-1}) of powdered lime stone			
	Sandy	Sandy loam	Loam	Silt & clay loam
4	2.8	5.6	7.3	10.0
5	73	44	5	6.2
6	0.8	1.1	1.7	2.2

3. Organic materials: Addition of organic materials such as paddy or wheat straw, FYM, water hyacinth compost, green manure, azolla etc. along with fertilizer enhance the growth and yield of rice on these soils. Straw should be added 4 weeks before transplanting of rice for effective decomposition. Green manuring *in situ* with sunnhemp and dhanicha will save 45-60 kg N ha^{-1} to rice crop in acid – lateritic soil. Green leaf manuring can also be practiced by incorporating lopping of legume shrubs like neem, pungam, *Glyricidia maculata, Leucaena leucocephala*, etc., 10-15 days before planting.

4. Inorganic nutrients: The recommendation given by the state government may be followed for the application of inorganic N, P and K fertilizers. Due to higher fixation of P in these soils, application of powdered rock phosphate, (Panda and Nanda, 1985; Patnaik and Bhadra-Chalam, 1985) and partially acidulated rock phosphate (Shinde, *et al.*, 1978), are reported to be superior to water-soluble sources of P. These are also considered to be useful for adding Ca and Mg in addition to P and for reducing the rate of P fixation. These are to be applied 2 to 3 weeks before flooding and planting of rice.

5. Bronzing - Bronzing of rice plant is related to low dry matter production as a result of which there is a build up of high concentration of Fe in the tissues causing toxicity. Rather being simply due to high level of Fe, bronzing is often related to insufficient supply of P, K, Zn and sometimes Ca & Mg.

Bronzing of rice covers three disorders in different places

i. Iron toxicity (soil pH 4.5 – 5.0)

ii. Management toxicity (bottom land & Bahal between two slopes)

iii. Injury due to hydrogen sulphide (swampy low land soil).

Measure to overcome bronzing of rice include

i. Liming

ii. Drainage provision

iii. Cultivation of tolerant varieties

iv. Application of calcium, phosphate and potash to soil.

v. Foliar application of urea

vi. Transplanting of rice about 3 weeks after flooding since water soluble and exchangeable iron concentration increases with flooding & reaches a peak in about 3 weeks, after which it decreases.

B. Acid Sulphate Soils

Acid sulphate soils are those soils that have, some where within 50 cm depth, pH below 3.5 (Entisols) or 4 (Inceptisols) that is directly or indirectly caused by sulphuric acid formed by oxidation of pyrite (cubic FeS_2) or rarely due to other reduced sulphur compounds. Potential acid sulphate soils are poorly drained and highly pyretic with a nearly neutral or slightly acidic reaction in the field. They become acid sulphate soils if pyrite oxidizes after drainage (Van Breemen and Pons, 1978).

Agro techniques for Acid sulphate soils

1. Amelioration -Amelioration to be followed to improve acid sulphate soils are as follows;

i. Flushing the soils to make it free of salts.

ii. Application of lime or other basic carriers to correct the soil acidity and related problems.

iii. Application of oxidized agents like NO_3 or MnO_2 to keep the system in a fairly oxidized condition since liming would favour microbial activity due to rise in pH and the presence of high organic matters in these soils might bring in acute reduced conditions to release more iron into the soil solution (Patnaik and Bhadrachalam, 1985).

2. Soil management - The acid sulphate soils are difficult to manage because drainage of these soils results in more oxidation causing further soil degradation. These soils are therefore to be managed in water logged conditions for low land rice. The hydromorphic, topographic and hydrologic settings of these soils are normally suitable for growing lowland rice. Moreover, acid sulphate soils are generally well supplied with plant nutrients, partly because of relatively high contents of 2:1 type clay minerals and organic matter. Application of lime to correct acidity has a short term ameliorating effect only. The retention of calcium sulphate (formed as a result of liming) is also likely to produce free H_2SO_4 causing a decrease in pH. So frequent top dressing of lime is required during one crop of rice to get over this difficulty. Lime may be substituted by basic carries like carbonate and sulphate which are easily leached out and have an ameliorating effect comparable to that of lime and the regeneration problem after drying the magnesium treated soils is slower (Patnaik and Bhadrachalam, 1985).

3. Nutrient management – In addition to liming, Nitrogen @ 80-100 kg ha^{-1} should be applied through urea in three splits (50% basal + 25% each at 30 DAT and Pl stage or 50% at 15 DAT + 25% each at 30 DAT and Pl stage) to harvest the higher yield. Application of 60-80 kg P_2O_5 / ha is also recommended as basal. Availability of Zn is generally low because of inherent low Zn content or as a result of liming. To correct Zn deficiency, either soil or spray application is needed.

4. Varieties - Some of the rice cultivars found to perform well, are RD-15, SR-26B (Bandyopadhyay, 1988) RD-19, Khao Dawk Mali (Bandyopadhyay, 1986) and Vyttila-3 (George *et al.*, 1987).

C. Sandy soils

Sandy soils are containing predominant amounts of sand resulting in higher percolation rates and nutrient losses. Compacting the soil with 400 kg stone roller or oil drum with stones inside eight times at proper moisture level (moisture level at friable condition of soil which is approximately 13 per cent) once in three years could reduce the percolation losses.

D. Red soils/ Hard pan soils

Hand pan occurs in red soil areas due to the movement of clay and iron hydroxides and settling at shallow depth, preventing the root proliferation. These soils can be reclaimed by chiseling the soils with chisel plough at 0.5 m interval first in one direction and then in the direction perpendicular to the previous one, once in three years. Application of FYM or composted coir pith at 12.5 t ha^{-1} could bring additional yields of about 30 per cent over control.

E. Surface crusted soils

The soils having weakly aggregated soil structure are easily broken by the impact of raindrops resulting the formation of clay crust at the soil surface. The clay pan prevents the emerging seedling and arrests the free exchange of gases between the soil and atmosphere. The surface crust can be easily broken by harrowing or cultivator ploughing and its formation can be prevented by improving the aggregate stability by the application of lime at 2 t ha^{-1} or FYM at 12.5 t ha^{-1} or any other organic amendment.

F. Heavy textured clay soils

The clay soils are containing major amounts of clay resulting in the poor permeability and nutrient fixation. Such soils can be reclaimed by the addition of river sand at 100 t ha^{-1} or managed by deep ploughing the field with mould board or disc plough during summer to enhance the infiltration and percolation.

G. Distillery effluent

The disposal of effluent from sugar factories is creating environmental problems. Recent experience with direct discharge of distillery effluent to sodic soil to improve the productivity of the sodic soil is the new information. Distillery spent wash is acidic (pH 3.8-4.2) with considerable quantity of magnesium. Organic compounds mainly the humic related melanoidins improve the biocatalytic potential of the treated soil. Hence, 2 lakhs litres of distillery spent wash can be added to an acre of sodic soil in summer months. Natural oxidation is induced for a period of six weeks with intermittent ploughing once in a month. In the second month (after 45-60 days), fresh water may be irrigated and drained. Such a treatment reduces the pH and exchangeable sodium percentage and increases the productivity of the sodic soil.

Dilution with water

Distillery effluent contains Nitrogen 12200 ppm, Phosphate 1500 ppm, Potash 120000 ppm, calcium 1800 ppm and iron 300 ppm. Since the effluent has the electrical conductivity, 50 times diluted effluent can be irrigated to rice for getting higher yield (Annadurai *et al.,* 1999). However, the effluent irrigation should not be adopted in lands having drinking water sources.

One time application

Distillery effluent contains both macro and micronutrients. Because if its high salt content, it can be used for one time application to fallow lands, about 20 to 40 t/ha of distillery effluent can be sprayed uniformly on the fallow land. It should not be allowed for complete drying over a period of 20 to 30 days. The effluent applied field has to be thoroughly ploughed two times for the oxidation and mineralization of organic matter. Then, the crops can be cultivated in the effluent applied fields by conventional methods.

H. Irrigation of pulp and paper mill effluents to rice

Pulp & paper effluents contain lot of dissolved solids stabilized organic matter. The properly treated matter can safely be used for irrigation with appropriate amendments *viz.*, press mud @5 t ha^{-1} or fortified press mud @ 2.5 t ha^{-1} or daincha as *in situ* green manure. Though there were perceptible changes in soil pH, EC, available NPK, exchangeable cations, exchangeable sodium percent and sodium absorption ratio, there is no detrimental effect due to sodium either on soil or plants grown in sandy loam soil with good drainage facilities, the treated effluent can be used for irrigation in these soils also. The rice crop varieties like IR 20, TRY 1 and CO43 can be raised alongwith recommended doses of amendments (press mud @ 5 t ha^{-1} or fortified press mud @2.5 t ha^{-1} or daincha as *in situ* green manure at 6.25 t ha^{-1}).

I. Treated tannery effluent irrigation to rice

Leather industry is one of the major foreign exchange earners in India. Though the export of leather and its products earn more than Rs. 7000 cr per year besides providing direct employment to more then one million people, it is also one of the major sources of pollution. It is estimated that approximately 17,000 t of hides and skins are daily converted into leather. This process uses about 35-40 litres of water per kg of leather processed. This implied approximately 680 million litres of waste water are being generated per day. The effluent and sludge disposed from these industries diverted into rivers and on the land, has led to extensive degradation of productive land since the tannery effluent typically consists of high concentration of salt (sodium, chloride and sulphate etc.) and chromium (Cr) and these pollute surface, subsurface land and water resources as well. The tannery effluent varied considerably in pH (6.17-8.17) EC (10.4-23.0 dSm^{-1}), Na (2042-24765 mg L^{-1}) and CR (620-26200 μg L^{-1}) concentration. The tannery sludge contained higher Cr (1179-16158 mg kg^{-1}). The salt and Cr rich tannery wastes disposal have a great impact in polluting the environment and the continuous disposal onto the land makes the soil more alkaline. The rice varieties like TRY1, CO-43, Paiyur-1 and ASD-16 can be grown in these areas.

J. Compaction technology for fluffy paddy soils

As already discussed in the previous chapters, they are characterized by low bulk density of the topsoil resulting in the sinking of farm animals and labourers as well as poor anchorage to paddy seedlings. For such soils, passing of 400 kg. stone roller or oil drum with stones inside eight times at proper moisture level (moisture level at friable condition of soil which is approximately 13 per cent) once in three years to prevent the sinking of draught animals and labourers during farm operations.

CHAPTER 20

SITE SPECIFIC DELTA MANAGEMENT SYSTEM

20.1 CAUVERY DELTA ZONE (CDZ)

The Cauvery river basin is a historical one and forms one of the important river basins in Tamil Nadu, which is also considered as the granary of Tamil Nadu. The Cauvery Delta Zone (CDZ) lies in the eastern part of Tamil Nadu between 10°00' – 11°30' N latitude and between 78°15' – 79°45' E longitude. It is bound by the Bay of Bengal on the East and Palk in the South.

The CDZ has a total geographic land area of 1.45 million ha which is equivalent to 11.13 per cent of the state area. Cauvery is the river transversing the delta. The east flowing water of this river is checked at Kattalai Bed regulator and again at Upper Anaicut, which are 136 and 184 kms away from Mettur dam respectively. Grand Anaicut built by the ancient King Karikal Chola is the principal irrigation diversion where the water gets trifurcated into Cauvery, Vennar and Grand Anaicut Canal for three irrigation commands besides Coleron acts as full drain of excess water during flood times and incidentally serves for irrigation to a very little extent. Water gets distributed in Cauvery Delta mainly through Cauvery, Vennar and Grand Anaicut Canal systems, before reaching the sandy coasts of Bay of Bengal covering the land mass of Thanjavur, Thiruvarur and Nagapattinam Districts, and parts of Trichy, Karur, Perambalur, Pudukkottai and Cuddalore districts in Tamil Nadu.

20.1.1 Soils of CDZ

Soils of CDZ are mostly transported in origin by water currents of Cauvery. The Zone is largely flat, alluvial terrain gently sloping towards east and most of the area in CDZ is around 20-25 m above mean sea level whereas in some areas, the New Delta accounts slight elevation ranging from 50–120 m above MSL. The CDZ is one of the seven agro-climatic zones of Tamil Nadu. Out of 22 major soil series in Thanjavur District, the soils at old delta are classified into Kalathur, Adanur, Padugai and Alangudi series and the same at new delta are classified into Pattukkottai and Madukkur series. The subgroup classification (USDA) is Udorthentic chromusterts (Vertisol), Vertic ustropepts (Inceptisol) and Typic ustifluvents (Entisol) in old delta and Typic haplustalf in new delta. Soils are low in soil organic C and total N, of near neutral pH and had high levels of extractable P and NH_4 extractable cations.

20.1.2 Ayacut Area

For the total water required for the entire Cauvery Delta Zone, Mettur dam alone contributes 71.8% of water, while lower Bhavanisagar and Amaravathi with 0.25 and 0.03% respectively. The total area benefit from Mettur is the highest followed by lower Bhavani and the least from Amaravathi. Depending on Mettur reservoir, 9 districts including Karaikal are benefited. There are 9 river systems existing. Among them, Cauvery delta system flows over 58.2% of the total area. Among the districts, major area is in Thanjavur district (27.9%), followed by Thiruvarur (24%) and Nagapattinam (21.6%) in utilizing the water from Mettur dam.

20.1.3 Ayacut Area in Erstwhile Thanjavur District

All three rivers in Cauvery, Vennar and CMP (Grand Anaicut) have its Ayacut area in Thanjavur, Thiruvarur and Nagapattinam districts. Among them, Cauvery and Vennar are the major contributors compared to CMP (GA). Thiruvarur and Nagapattinam are the districts getting more or less equal benefit from Cauvery and Vennar compared to Thanjavur. Whereas, Thanjavur district get the major benefits from CMP.

20.1.4 Irrigation Systems of Cauvery Delta

The Delta can be broadly divided into three regions based on the irrigation systems.

1. Cauvery system comprising of Kumbakonam, Mayiladuthurai and Sirkazhi and parts of Papanasam and Thanjavur. The soil type is alluvial clayey texture and the groundwater potential is good.

2. Vennar system comprising of Nagapattinam, Thiruthuraipoondi and Nannilam. The soil is heavy clay with poor ground water potential. The texture is sandy in the coastal areas and water logging is the common problem.
3. Grand Anaicut canal system (GA) comprising of Orathanad, Pattukkottai parts of Thanjavur, Mannargudi. The soil is sandy loam with good drainage.

The data on irrigation availability from the Cauvery command depends mainly on the date of release and the onset of monsoon rains. The time of onset, intensity and frequency especially in the catchment area decides the water storage in the reservoir. The rainfall probability during the early months viz., June and July are low. With the available storage and uncertain precipitation and high evapotranspiration demand, not much option is left out to go in for rainfed cultivation other than to go for sesamum or some green manure crops or multi purpose crops like cowpea, black gram etc. Moreover, considering the delay in the release of dam water and labour availability, some portion of the delayed kuruvai need to be direct sown to compensate the time loss and shift the peak labour demand.

20.1.5 Land Utilization-CDZ

The Land utilization pattern is given in Table 58 and the source of irrigation and area irrigated are given in Table 59.

Table 58. Land utilization (ha)

Area	Thanjavur	Thiruvarur	Nagapattinam
Total Geographical area	339657	209709	271533
Forests area	3426	2452	4633
Barren and Uncultivable lands	2201	113	33419
Land put to non agricultural uses	77122	36939	47206
Cultivable wastes	10647	2982	4585
Permanent pastures and other Grazing lands	1746	768	989
Miscellaneous tree corps and groves not included in the net area sown	6288	2219	10333
Current fallow	8396	6875	5608
Other fallow lands	26656	4901	13036
Net area sown	203175	152460	151724
Area sown more than once	64675	92520	103189
Total cropped area	2,67,850	2,44,980	2,54,963

Table 59. Irrigated area and Irrigation source

S. No.	Source	Net area (ha)		
		Thanjavur	Thiruvarur	Nagapattinam
1	Canal	158353	142399	128328
2.	Tank	7913	—	—
3.	Tube wells	7116	4872	—
4.	Ordinary wells	829	730	—

20.1.6. Rainfall Pattern

Cauvery Delta Zone having *humid tropical monsoon* type of climate, the CDZ has the average annual rainfall of 1093 mm. Out of which, North East Monsoon alone contributes 52.5 per cent followed by South West Monsoon with 30.5 per cent. Hot weather seasons account 11.4 per cent in the total while the winter season has the least of 5.6 per cent only. The summer shower that one rain wetting could be there in each of the summer months of April and May which ranges from 20.8 to 35.2 mm. There exists a linear relationship between the intensity of rainfall received and that of nearness to the sea as well as East longitude. However, it is observed to be insufficient for long duration summer crops. Maximum temperature is as high as 40.0°C with mean minimum temperature 18.40°C. The rainfall distribution plays a vital role in deciding the cropping pattern.

The following are season wise average rainfall

- Winter - 49 mm
- Summer - 97 mm
- South West Monsoon - 299 mm
- North East Monsoon - 648 mm

Out of the normal rainfall received during the South West Monsoon, June and July months contribute 40 to 50 mm each, and August and September contribute 100 to 125 mm each. The temperature and wind velocity are usually high resulting in higher level of evaporation. Bright sunshine prevailing during this period is highly favourable for rice production and yield is always higher in kuruvai season as compared to thaladi/samba rice crop.

20.2 CROPPING PATTERN

20.2.1 Main cropping seasons

S. No.	Name of the season	Period
1.	Kuruvai	June to September
2.	Samba	August to January
3.	Thaladi	September to February

The *Kuruvai* and *Thaladi* crops are sown in the same field making these lands as double cropped wet land while Samba grown lands are known as single cropped wet lands. In addition to the above, summer crops are raised to a small extent during March – May where ground water facilities are available.

Cauvery delta said to be Granary of the Tamil Nadu. Rice is the dominant crop in the delta. The nature of soil, flatness of the land, availability of water from June to January is conducive for the cultivation of paddy. In double crop lands, first crop Kuruvai stands from May 15 with the help of ground water potential and synchronizes with South West Monsoon (SWM) period. During the season, short duration paddy varieties (105 days) are being cultivated. Farmers are growing paddy in **kuruvai** to the extent possible in view of its high productivity due to plenty of sunlight and less damage from pests and diseases. After the harvest of Kuruvai crop, thaladi is planted with medium duration paddy varieties (135 days). This season synchronizes with North East Monsoon (NEM) period, which starts from 2nd week of September. In a single crop area normally a long duration (150 days) and medium duration paddy crop (135 days) are raised and this is called as **samba** and late samba respectively, which stretches out in both monsoons from August to January.

Out of a total area of 1,15,800 ha under **kuruvai**, Thanjavur has more area (45,000 ha) while Thiruvarur and Nagapattinam districts have 35,840 and 35,500 ha respectively. After the harvest of Kuruvai, Thaladi crop will be raised in most of the kuruvai paddy crop area. Major paddy is under single crop Samba (323800 ha). All three districts are having more or less same area under samba crop (1,03,000 to 1,12,000 ha), In total, the normal area under paddy in Cauvery Delta Zone is about 5,41,500 ha (5.4 lakh hectares). Due to poor rainfall and non-availability of water from Mettur, the area under paddy in Cauvery Delta Zone is in descending order. Among the three districts, Nagapattinam is the most affected district as it is in the tail end area of Cauvery river. As against the normal area of 30,180 ha under Kuruvai, during 2002-03, paddy is raised only in 17,402 ha, 98,746 ha in samba as against normal area of 1,03,000 ha and 17,067 ha in Thaladi as against normal area of 26,820 ha. In Thanjavur and Thiruvarur districts the short fall under area coverage is almost similar.

20.2.2 Existing cropping pattern

The important cropping systems are rice - rice - fallow, rice - rice - pulses, rice - rice - sesame and rice - rice - cotton. The climate is tropical. There are three major rice growing seasons namely Kuruvai, Samba and Thaladi as indicated above. The rice fallow summer crops viz., pulses, cotton, sesame etc., are grown under no tillage condition utilizing the residual moisture and nutrients in the old delta whereas the same are grown as summer irrigated crop with splash irrigation in the new delta. Groundnut and maize are raised in the light clayey soil of new delta with single crop of rice. In some pockets of old delta, vegetables like brinjal, cucumber etc., are grown routinely after one rice crop.

Rice-Rice-Rice fallow crops (Pulses / Cotton / Gingelly) are the normal cropping pattern being followed in Cauvery Delta Zone. Paddy, Groundnut, Sugarcane and Pulses are the major crops while maize, red gram, sesame, vegetables and banana are the minor crops. Uncertainty in the timely release and insufficient quantity of irrigation water coupled with depleting ground water affect the Kuruvai paddy in Cauvery Delta Zone in the past five years (2001-2005). This makes an imperative change of the existing cropping pattern in CDZ during Kuruvai season.

In the Cauvery New Delta Zone, under normal years, farmers go in for *kuruvai*. otherwise only a single crop of samba rice is cultivated under delayed release of water.

20.2.2.1 Double cropped wet lands

These lands are characterized by wet lands with dual source of irrigation viz., Canal water and filter point or tube wells.

a. Kuruvai rice

Kuruvai rice is cultivated from June to September in the Delta region. The nursery is grown from the last week of May and spans up to June end. In the pre-monsoon period, with the available ground water, the nursery and initial vegetative phase of the crop is raised. Tillering stage onwards, the crop is raised with canal water. Preferably a short duration rice cultivar with 105 – 110 days duration is chosen. The season relatively has the advantage of good sunlight and reduced incidence of pests and diseases. The productivity levels are more in this season compared to *thaladi*. The major problem encountered in this season is the coincidence of harvest with the monsoon rains. Hence, early crop escapes and if delayed, *kuruvai* rice is subjected to flood and lodging at maturity phase. This renders the straw moist and the quality of the straw becomes poor due to secondary fungal infestation.

Normally, planting starts early from the western parts of the delta and as it progresses east, the planting is delayed. The tail end crops usually suffer from excessive rains and lack of proper drainage. The slope is also very gentle and water gradient is less. Varieties like IR 50, ADT 43 are cultivated in the early *kuruvai* seasons. As the season advances, ADT 43 and ADT 36 are preferred. Under situations of excessive flooding at the maturity phase, the regular cultivars germinate in the field itself prior to harvest and the farmers incur total loss. To overcome the situation, ADT 37 was released which has the dormancy and withstands such aberrant weather to some extent. But ADT 37 seeds are short bold type and hence the consumer preference is less. The major portion of the *kuruvai* rice production comes as marketable surplus.

b. Thaladi rice

Following the *kuruvai* rice, the second crop in the delta is invariably rice. The farmers are left with no other alternative other than rice as the season coincides with NE monsoon season and coupled with canal water supply. Drainage is a big problem and no other crop survives except rice. Normally a medium duration fine grain variety with 135 days duration is grown from September second fortnight to January. The popular cultivars are ADT 38, IR 20, Co 43, Co 47, White Ponni and ASD 19. The crop is mainly grown for personal consumption and the excess alone comes as marketable surplus. The field at the time of harvest is dry and they harvest the crop and leave it in the field for over night as "*Arikatchal*" and then thresh the next day to reduce the moisture content in the grain and straw and to increase the threshing efficiency. The straw is also harnessed and utilized for cattle consumption. In kuruvai, where the major portion of the straw is left in the field and ploughed *insitu* prior to thaladi planting. Whereas, in the thaladi season, the crop is cut closer to the ground and straw is conserved for cattle use. While incorporating kuruvai stubbles for raising thaladi crop, sufficient quantity of super phosphate or rock phosphate should be applied to hasten the decomposition process.

c. Late thaladi

If *thaladi* sowing is delayed under late *thaladi* conditions farmers go in for ADT 39 of 125 days duration, which withstands the winter climate also. Under delayed conditions beyond November, ADT 36 is planted with risk.

20.2.2.2 Single cropped wetlands

a. Normal samba

In the mid and tail end areas having no filter points or secondary irrigation sources, farmers depend only on canal water. Even under normal release periods, the water reaches late due to the proximity or topographical conditions. Such farmers go in for a single crop of rice preferably a long duration rice with 150 days duration such as ADT 44 or CR 1009 with higher yield potential.

b. Late samba

If samba sowing is delayed due to delayed release of canal water or delayed onset of monsoon, ADT 38 or ADT 39 is grown to compensate the time delay and it is called as late samba.

20.2.3 Major crops grown

In Cauvery Delta, rice is the principal crop. Copious supply of irrigation water in the canals of Cauvery for over eight months a year provides the suitable environment for rice in the delta and this is in vogue from the early times. In the last two decades, growing double crop of rice is in the ascend. A short duration kuruvai crop during June to September followed by a medium duration thaladi rice between September and January is catching more area. However, samba lands with either medium or long duration rice between September and January continue in most parts of the delta region for obvious reasons. Planting of kuruvai rice in the delta largely depends on the date of release of water from Mettur reservoir. Subsequent to the release of water, the first reach fields are prepared and planted earlier than the tail ends.

Water is generally flushed from the reservoir at the first instance to facilitate faster reach of irrigation water even to the tail ends. In spite of this, the tail ends take more time to get soaked because of the soil type and prolonged drought during summer. Dates of kuruvai planting vary between the extremes in the delta. It is not uncommon, sometimes the kuruvai crop is skipped for want of irrigation water in time during first season. Kuruvai planting is normally expected to start on 12th June every year. Now-a-days, it gets extended even upto September depending upon the storage position in the reservoir. Subsequential delay in planting of kuruvai in the double crop wetlands results in insufficiency of water for the thaladi crop, particularly at its maturity phase. Planting of samba rice in the delta is more uniform because of availability of plentiful water through the canal source as well as continuous downpour during North East monsoon. Labour shortage poses problem during this period because of continuous rains and overlapping of kuruvai harvest and thaladi/samba sowing and planting.

Table 60. Area under different crops grown in CDZ (2003)

S. No	Crops	Thanjavur		Thiruvarur		Nagapattinam		Total	
		Area (ha)	(%)	Area (ha)	(%)	Area (ha)	(%)	Area (ha)	(%)
1.	Kuruvai rice	33137	12.37	21082	8.61	26849	10.53	81068	10.56
2.	Thaladi rice	33137	12.37	21082	8.61	26849	10.53	81068	10.56
3.	Samba rice	116371	43.45	119925	48.95	108375	42.51	344671	44.89
4.	Summer rice	6265	2.34	1513	0.618	1547	0.607	9325	1.21
5.	Maize	202	0.09	2	0.001	20	0.008	224	0.029
6.	Rice fallow pulses	14694	5.48	64694	26.41	74073	29.05	153461	19.99
7.	Summer pulses	1544	0.57	594	0.242	—	—	2138	0.278
8.	Sugarcane	11653	4.35	2660	1.09	3837	1.50	18150	2.36
9.	Banana	3475	1.30	385	0.157	592	0.232	4452	0.58
10.	Vegetables (incl.chillis)	778	0.29	213	0.087	434	0.170	1425	0.186
11	Cotton	1086	0.41	1204	0.491	1284	0.504	3574	0.465
12	Groundnut	8450	3.15	1922	0.785	3412	1.34	13784	1.78
13	Gingelly	7278	2.72	4108	1.68	517	0.203	11903	1.55
14	Soybean	799	0.3	37	0.02	129	0.051	965	0.126

Even before the harvest of thaladi / samba rice, blackgram or greengram seeds are broadcasted as relay crop, when the soil moisture is optimum. Final irrigation is extended to the crop even after its maturity in order to sow these pulses under optimum field moisture condition. Pulses are cultivated under no tillage condition utilizing the residual moisture and nutrients and by the dew set during January and February months. Cotton is an another rice-fallow crop, dibbled in the harvested rice fields in waxy condition. This crop requires assured water supply through lift irrigation sources throughout its growth period of about 4½ months. Gingelly is sown during April, grown as a rainfed crop in the delta. In light clay loamy soils, blackgram is grown during summer season with splash irrigation. Groundnut and maize are grown after single crop of rice in light clayey soils. During summer months, vegetables like brinjal, bhendi, tomato, cluster beans, gourds, greens and water melons are grown with partial irrigation through temporarily dugout 'urakkinaru'.

Mostly vegetables are grown in small patches under hired water. Flowers like rose, arali, jasmine, crossandra and chrysanthemum are grown in low lying padugai thidals adjacent to irrigation belts. Banana and sugarcane are cultivated in more area in the recent past. Thus, amidst the predominance of rice growing in the delta, the main crops that are altered in the sequence include blackgram, greengram, cotton, gingelly, groundnut, maize, vegetables and greens. The quantum of rainfall differs between years and also between seasons within a year. This pattern of rainfall coupled with the release of canal water largely decides the cropping pattern in the Delta for a particular year. The crops grown in the CDZ are given in the following Table 60.

20.3 ALTERNATE CROPPING

20.3.1 Reasons for Alternate Cropping

1. Delayed supply of water through canals or late release of water from Mettur reservoir and Monsoon failure
2. Delayed planting of kuruvai causes the crop damaged in the North-East monsoon rain at its harvest stage.
3. Delayed planting makes postponement of thaladi planting, which suffers for want of water at its later stages.
4. Problem in marketing the kuruvai paddy due to its quality.
5. Discouraging kuruvai, which consumes more water as compared to other dry crops.

20.3.2 Suggested alternate cropping systems

20.3.2.1 Old Delta areas

Planning will be based on the following situations taking into condition of the water requirement of different cropping systems as given in the Table 61.

1. Normal water supply in Cauvery
2. No water supply from Cauvery

i. Normal water supply in Cauvery

The irrigation water availability is normal or near normal and even if the time of water is delayed by 1 or 2 fortnights, a normal cropping pattern of kuruvai and thaladi followed by rice fallow crop like pulses and cotton, vegetable etc. can be adopted in Cauvery and Vennar system. Normal cropping pattern is successful only when there is adequate South West monsoon rain in the catchment area of Cauvery and good storage position in the Mettur dam.

Table 61. Water requirement of different cropping systems

Cropping system	Field duration (days)	No. of irrigation	Water requirement (mm)
Rice-Rice-Cotton	347	49	2438
Rice-Rice-Black gram / Green gram	312	45	2371
Rice-Rice-Sesame/Black gram	332	44	2331
Rice-Rice-Groundnut	287	42	2251
Black gram-Rice-Cotton	302	31	1495
Blackgram-Rice-Blackgram/Sesame	297	29	1429
Blackgram-Rice-Sesame/Blackgram	317	28	1389
Maize-Rice-Cotton	337	38	1772
Maize-Rice-Black gram / Green gram	332	36	1729
Maize-Rice-Sesame/Black gram	352	35	1669

ii. No water supply from Cauvery

If adequate canal water is not available for raising kuruvai in time, then planting should be done solely, depending on the rainfall received during the season. About 50 mm of rainfall is normally received from May and taking advantage of this rain, summer ploughing has to be done on a larger scale in the delta. This operation not only conserves moisture and checks weed growth but also makes dry sowing easy. During June-July, an average rainfall of about 90-100 mm can be expected. Using this rain, short duration pulses like green gram and black gram, and millets like bajra and finger millets can be raised as dry crops in clay soil of Cauvery and Vennar system. Sesame can be grown in loamy and light textured soil. The coastal tract characterized by the sandy loam soils can be used for growing vegetables like melon, cucumber, gourds and onion. This soils of Cauvery alluvium rich in K and lime, are best suited for growing vegetables like bhendi, tomato, cluster bean and brinjal.

In August, another 90-100 mm rain can be expected and this will be sufficient to support the above mentioned dry crops. If the storage position improved in Mettur reservoir late in the season, one or two wettings can be given to these crops, which would push up yield considerably. But on no account, water should be let in the canal for rice cultivation, after these dry crops are raised except for occasional wettings alone. In the Cauvery basin, 1 lakh acres of rice could be irrigated by existing filter points. Nowadays, there is a drop in water table due to over exploitation of underground water in these areas. In this assured irrigated area, the normal practice of kuruvai planting can be adopted in the June and harvested in September well before the onset of monsoon season. In the delta, except the filter point irrigated area, a single rainfed crop of rice (samba) can be raised in August-December. Using the August rain (90-100 mm) fields can be prepared after the harvest of dry crops, rice can be dry seeded using a high seed rate (100 kg/ha). In heavy soils, furrows at an interval of 2-3 m can be formed to drain excess water. The initial rain in August-September will be sufficient for germination and establishment. After the on set of North East monsoon, this fields can be converted into wet land and water can be impounded. In the rainfed samba areas of Cauvery and Vennar basin, short duration pulses can be relay sown in December by utilizing the residual moisture and nutrients. In the filter point irrigated areas, cotton, sesame and vegetables like tomato, bhendi, cluster bean, brinjal and maize may be raised. Alternately, one rice fallow pulses after thaladi, or samba and another summer pulse in April-June with irrigation can be raised. Depending upon the soil type, the following cropping pattern can be followed in old delta areas.

a. Clay soil

Black gram/Green gram-Rice-Pulses/Cotton

Cumbu-Rice-Pulses/Cotton

Finger millet-Rice-Pulses/Cotton

Green manure-Rice-Pulses/Cotton

b. Light textured soil

Sesamum-Rice-Pulses/Gingelly

Pulses-Rice-Pulses/Gingelly

Vegetables-Rice-Pulses/Gingelly

c. Coastal sandy loam soils

Melon-Rice-Pulses/Gingelly

Cucumber-Rice-Pulses/Gingelly

Gourds-Rice-Pulses/Gingelly

Onion-Rice-Pulses/Gingelly

d. Coastal Alluvium soils

Bhendi-Rice-Pulses/Gingelly

Cluster bean-Rice-Pulses/Gingelly

Amaranthus-Rice-Pulses/Gingelly

20.3.2.2 New Delta areas

Planting in new delta areas will be based on the following situations, taking into consideration of the water requirement of crops and cropping systems as followed in old delta regions.

i. Under Assured ground water availability

1. Blackgram – rice – rice fallow pulse / gingelly
2. Soybean – rice – rice fallow pulse / gingelly
3. Maize – rice – rice fallow pulse / gingelly

ii. Under uncertain ground water availability

1. Blackgram – rice – rice fallow pulse / gingelly

Suggestions on water management

- Conversion of high level wet lands into horticultural plantation crops.
- Provision of sprinkler and drip irrigation under existing pump sets so that suitable crops can be cultivated with minimum water in May and June.
- Desilting and deepening of existing ponds, tanks to recharge the groundwater.

20.3.2.3 Farmer's practice

Bottlenecks for introducing crops other than rice during kuruvai as expressed by the farmers. In spite of the following constraints, the farmers are ready to accept Black gram – rice – rice fallow pulse / gingelly.

a. Groundnut

- Increased seed cost.
- High cost of labour for harvesting and stripping of pods.
- High price fluctuation.
- Low revenue.
- Poor filling of pods.

b. Maize

- Non availability of seeds locally.
- High water requirement – summer showers alone is not sufficient.

c. Soybean

- High seed cost.
- Low price for the produce.
- Poor filling of pods during the season, in spite of heavy pod set.
- Keeping quality of the seeds is poor, hence seeds have to be purchased every time.

d. Vegetables

- Price fluctuation.
- Marketing problem.
- Pilferages will be more as the farms are aloof and farm houses / village is away.
- Lack of cold storage facilities.

e. Black gram / gingelly

- Excess moisture or rainfall affect the crop growth and yield.
- Yellow mosaic virus infestation during the season in blackgram.
- Maintenance of optimum population in gingelly is a problem.
- Price fluctuation.

20.3.2.4 Research findings

Research works were carried out on the alternate cropping system for the Cauvery New Delta Zone at Soil and Water Management Research Institute, Thanjavur. Among the various systems studied, Rice ADT 43 + *Sesbania rostrata* direct sown 10 : 1 – Rice ADT 39 – Redgram VBN 1 recorded the highest rice grain equivalent yield of 17,150 kg/ha, net return of Rs. 72268/ha/yr and C : B ratio 3.74 and was closely followed by Rice ADT 43 – Rice ADT 39 – Redgram VBN 1 with rice grain equivalent yield of 16587/ ha/ yr, net return of Rs.68872/ha/yr and C : B ratio 3.61 (Table 62).

It is also confirmed that the crops like blackgram, groundnut, soybean, and maize could be successfully raised as alternate crops in the place of rice during kuruvai under contingent situation.

Table 62. Details of different cropping systems tried at SWMRI, Thanjavur, Tamil Nadu

Treatment	Kharif	Rabi	Summer
C_1	Rice ADT 43	Rice ADT 39	Redgram VBN 1
C_2	Rice ADT 43 + *S.r.*D.S 10:1	Rice ADT 39	Redgram VBN 1
C_3	Rice ADT 43 + *S.r.*T.p. 10:1	Rice ADT 39	Gingelly TMV 6
C_4	*Sesbania rostrata* (seed)	Rice ADT 39	Gingelly TMV 6
C_5	Blackgram ADT-5	Rice ADT 39	Blackgram ADT 5
C_6	Groundnut VRI-2	Rice ADT 39	Blackgram ADT 5
C_7	Soybean CO-1	Rice ADT 39	Sunnhemp (seed)
C_8	Sunnhemp (GM)	Rice ADT 39	Cotton MCU 7
C_9	Maize (Ganga-5)	Rice ADT 39	Soybean CO1

S.r. D.S---- Sesbania rostrata Direct sown

Different crops like pulses, sesame, sunflower, maize and millets like sorghum and cumbu were raised as alternate crops during kuruvai and compared with rice at Tamil Nadu Rice Research Institute, Aduthurai. All the crops performed well under both irrigated and rainfed condition. The green gram in particular gave higher yield in both conditions, maize in irrigated and sorghum in rainfed condition (Table 63).

Table 63. Results on alternative cropping for kuruvai at TRRI, Aduthurai, Tamil Nadu

Crop	Variety	Yield (kg/ha)		B/C ratio	
		Irrigated	Rainfed	Irrigated	Rainfed
Blackgram	CO 5	740	610	2.28	2.36
Green gram	KM 2	1100	875	3.00	2.90
Redgram	APK 1	850	725	2.32	2.41
Cowpea	CO 7	925	830	2.52	2.76
Sesame	TMV 4	750	610	2.04	2.03
Sunflower	CO 4	825	0	2.48	0
Maize	COH 3	3500	0	3.50	0
Sorghum	CO 28	2510	1810	2.28	2.01
Cumbu	COH 8	2220	0	2.62	0
Rice	ADT 43	5040	0	1.88	0

The systems are so tailored to give a combination of cereal, pulse and oil seed crop to cater the carbohydrate, protein and fat dietary requirements of the delta. The crops suggested for kharif are taken with most care to withstand moisture stress. The sequence is so arranged that, under scarce water situation, one may choose sesame for kharif. Under more moisture by slight showers, black gram will be more remunerative. If moisture is adequate initially but uncertain in distribution, green manure can be resorted to as they can be ploughed *in situ* if insufficient rains or can be raised for seed purpose. Similarly, cowpea also can be used for green manure cum vegetable cum grain option depending upon the rainfall. Groundnut as a rainfed crop can be grown if initial moisture is sufficient. Maize can be cultivated in certain pockets where the soil water holding capacity is good.

20.4 CONTINGENT CROPPING

20.4.1 Kuruvai season

i. Under assured irrigation (Bore wells and filter points)

Old delta : Maize, Pulses, and Vegetables

New Delta : Maize, Black gram, Sugarcane

ii. Rainfed crops suggested when there is no water supply from Cauvery during Kuruvai

During SWM season (June to September) about 240 mm rainfall will be received. The following rainfed short duration crops may be grown wherever the rainfall is sufficiently uniform and well distributed. When the rainfall is limited, green manure crops can be grown (or) summer ploughing can be taken up for direct seeding during August – September.

Old delta

Clay soils : Black gram, Cumbu, Finger millet (Ragi), Green manure

Light textured soils : Sesame, Pulses and Vegetables

Coastal sandy loam : Melon, Cucumber, Onion and Gourds

New Delta : Sesame, Black gram, Groundnut and Green manure

iii. Normal water supply from Cauvery (Areas with supplemental irrigation)

Old delta

Heavy clay soil : Rice

Well drained soils : Maize, Pulses and Vegetables

New crops introduction

Spices and horticultural crops are being cultivated in the coconut garden as inter crops and in some places, cole crops also perform better with remunerative yield. Hence, Cole crops like Cauliflower and Beet root may be tried in the new delta having well drained soils. Similarly, grapes can also be tried in Vallam areas, Thanjavur, Tamil Nadu.

20.4.2 Samba and Thaladi seasons

In the entire Cauvery Delta Zone, except in irrigated areas a single crop of rice can be grown during August to December as semi dry rice. During the August rains, rice varieties such as CR 1009, ADT 44 can be grown as dry seeded paddy. If rain delays to September, medium duration rice varieties like ADT 38, ADT 39, ADT 46, CO 43 and improved White Ponni may be sown in summer ploughed dry soils. If rains received only during October, with the onset of monsoon, direct seeding of short duration paddy varieties such as ADT 36, ADT 39, ADT 42, IR 36, IR 64 and MDU 5 can be raised.

20.4.3 Technologies for direct seeded rice in CDZ

The following technologies should be followed for direct seeded rice.

i. High seed rate : 100 kg/ha (Less seed rate can be used for profuse tillering varieties like CR 1009)

ii. Seed hardening with 1% KCl

iii. Basal application of FYM besides P & K, if not soil tested

iv. Herbicide application: Thiobencarb 2.5 lit/ha (or) Pendimethalin 3.0 lit/ha on 8 days after sowing (if adequate moisture is available in the soil) or on the day of receiving rainfall.

- After the onset of monsoon, the fields can be converted into wet lands
- In case of late receipt of canal water up to October/November, medium duration varieties and short duration varieties may be planted during the last week of October and first week of November respectively
- Due to subsequent good storage position after June and unexpected water release of Cauvery there after, wet seeding with long duration varieties during August – September and with medium and short duration varieties during September–October can be done.

20.4.4 Technologies for wet seeded rice

i. Perfect land leveling

ii. Correct seed rate (75 kg/ha)

iii. Sowing with modified drum seeder to maintain plant population

iv. Proper weed control using pre emergence herbicides

However, large scale adoption of these contingent crop plans by the farmers under the exigency of paucity of water can be implemented through appropriate integrated approaches.

20.5 POPULARIZING ALTERNATE CROPPING IN CDZ

An imperative change of the existing cropping patterns in the Cauvery Delta is need of the day in view of the uncertainty over the timely release and insufficient quantity of irrigation water from Mettur dam by introduction of crops like pulses, maize and vegetables in the place of rice during Kuruvai. The crops like maize, pulses and vegetables can be grown as alternate crops during kuruvai in the filter point irrigated areas of Cauvery old delta except the places where the soils are ill drained in nature with heavy clay. The same crops could be raised in the canal irrigated area even during the normal years of water supply. The crops like pulses, bajra, ragi and sunnhemp can be grown as rainfed crop if the canal water is not timely released. Similarly the crops like pulses, sesame, vegetables can be cultivated during kuruvai in the loamy and light textured soils. The coastal tract characterized by sandy loam is can be used suited for growing vegetables like melon, cucurbits, gourds and onion. The soils of Cauvery alluvium rich in K and lime, are more suited for raising crops like bhendi, cluster bean and brinjal. In the bore well irrigated areas of new delta, crops like pulses, maize can be grown as alternate crops in the place of rice during Kuruvai. The crops like gingelly, black gram, cowpea and groundnut can be grown as rainfed crop if water is not available in the canal and rice can be grown during thaladi followed by rice fallow crops during summer. Since there is good demand for medicinal plants now-a-days, suitable medicinal plants can be grown as rainfed as well as irrigated crops to fetch higher income. The alternative cropping pattern could successfully be introduced among the farmers of Cauvery Delta (both new and old) by conducting awareness campaign and OFT as well as OFD through village adoptions.

CHAPTER 21

NEW METHODS OF RICE CULTIVATION

Rice being a semi-aquatic plant, transplanting method of establishment is more favourable than direct seedling methods. Transplanting of rice seedlings is most widely practiced and in South East Asian countries, seedling broadcast planting method has been developed for better management and high yield in rice cultivation. Method of establishment is one cultural practice, which influences the performance of rice through its effect on growth and development. Transplanting of rice seedlings is done in two ways (1) random planting and (2) straight row / line planting.

21.1 NORMAL METHODS OF PLANTING

A. Random planting

In random planting, the rice seedlings are transplanted at random in the field without any definite planting geometry. As such, rice is transplanted randomly by women, who transplant much less number of hills/unit area resulting in low yields. Even with fertilizer application is no way comparable to improved practice of line planting, which ensures even plant stand and easiness in intercultural operations. In villages, cycle tyres are used to count the number of plants per unit area.

Plate 52. Dividing rice field for manual random transplanting

Plate 53. Spreading rice seedling bundles in main field.

B. Line Planting

In line planting, the seedlings are planted in straight rows with definite spacing (using ropes) within the rows and plants. Proper plant spacing is one of the important factors to obtain a good yield in rice, particularly transplanted condition. Row method of planting facilitates uniform crop growth, permitting more efficient use of inter row space for weeding operation and other field management operations such as spraying and fertilizer top dressing. Based on several experiments across the agro-climatic zones of Tamil Nadu, it is found that line planting enhanced the productive tillers, yield attributes and grain yield of both tall and semi-dwarf cultivars compared to random planting. A plant spacing of 15 cm x 10 cm, 20 cm x 10 cm and 20 cm x 15 cm is recommended for short, medium and long duration rice varieties, respectively for general cultivation in Tamil Nadu.

Plate 54. Rice-Line planting

Plate 55. Rice- line planting – A field view

21.2 NEW METHODS

21.2.1 Seedling throwing

Transplanting method of rice cultivation has been highly accepted and adopted for many decades. However, this method is labour intensive and time consuming. During peak cropping season, the transplanting operation gets delayed for want of adequate women labourers. Such a delay leads to planting of aged seedlings late in the season and the crop did not get right seasonal condition for expressing its yield potential. Rice being a semi-aquatic plant, gets well established under well puddled soil situation. A shallow planting (3 cm) of rice seedling is sufficient for better rooting and tiller production. Since, rice establishes well under varied methods of planting, a less labour intensive method would be helpful in planting crop in the right season. In the field experiments, they have compared throwing of rice seedlings (seedlings uprooted with about 5 g of soil adhering to their roots and dropped on land while standing with about 5 g of soil downward on to a well puddled soil) with conventional planting. The broadcast planting resulted in considerable saving in labour costs and gave paddy yields similar to conventional transplanting. The broadcast seedlings anchored themselves and established erect in 7-10 days. Though the number of hills/ m^2 was low in broadcast planted crop compared to transplanted crop, increased number of panicles as a result of enhanced tillering was recorded in broadcast planted crop (Table 64).

Table 64. Rice yield and economics of seedling throwing/broadcasting technology

Variety	Var. Ptb 20		Ptb 20		Jaya	
	45		30		22	
	BCS	TPS	BCS	TPS	BCS	TPS
No. of hills/m^2	24	36	24	36	40	60
No. of panicles/m^2	328	260	293	277	456	408
Grain yield (kg/ha)	2178	2092	2205	2100	3297	2902
Economics: (Rs./ha)						
Cost of transplanting	200	700	200	700	200	700
Cost of produce	9882	9413	9998	9630	1257	11161
Profit derived by the technology	469	—	368	—	1366	—
Net profit (savings in labour cost + additional yield)	969	—	868	—	1866	—

BCS – Broadcasting seedlings; TPS – Transplanting seedlings

Source : Kandasamy *et al.* (1996).

In an another study conducted at TNAU, Coimbatore, the new method of rice planting by throwing seedling gave an yield of 6.11 t ha^{-1} during kharif season, which was comparable with the yield of line planted crop (6.21 t ha^{-1}) and higher than random planted crop (5.87 t ha^{-1}). However during rabi season, being cool and monsoon season, the incessant rainfall during early establishment stage affected the establishment of thrown seedlings and resulted in low yield (4.97 t ha^{-1}) compared to line planting (5.65 t ha^{-1}) but gave at par yield with random planted crop. On an average, the woman days required for throwing seedlings over one hectare was 34, which accounted a saving of 56.4% and 30.6% of labour required for line planting and random planting respectively. Hence broadcasting of seedling not only resulted in considerable saving in labour but also increased the income compared to transplanting (Kandasamy *et al.* 1996).

21.2.2. Kharhan seedling method

Here, thirty days old conventional seedlings were uprooted from the first nursery and transplanted very closely (10 cm x 10 cm) in the second nursery using 4-5 seedlings per hill and the seedlings will be allowed to grow in the second nursery up to 75 days. Seventy – five days old kharhan seedlings were superior to conventional seedlings. Kharhan seedlings had thicker culm and better growth and there by, more food reserves in comparison to conventional seedlings. Kharhan seedlings are also less susceptible to damage by insect pests. The crops raised with Kharhan seedlings were taller than the conventional seedlings and hence it escaped submergence. Kharhan seedlings of 75 days old were superior to conventional seedlings (Table 65).

Table 65. Effect of type seedlings on rice yield

	Type of seedlings	Rice grain yield (t/ha)
1.	45 days old conventional seedlings	1.44
2.	60 days old conventional seedlings	1.77
3.	75 days old conventional seedlings	1.82
4.	60 -75 days old kharhan seedlings (30 days in I nursery + 30 days in II nursery)	2.37
5.	60 days old kharhan seedlings (30 days in I nursery + 30 days in II nursery)	2.48

21.2.3 System of Rice Intensification (SRI)

SRI involves the use of certain management practices, which together provide better growing conditions for rice plants, particularly in the root zone, than those plants grown under traditional practices. SRI was developed in Madagascar in the early 1980s by Father Henri de Lauhanie, a Jesuit priest. In 1990, Association Tefy Saina (ATS) was formed as a Malagasy NGO to promote SRI. It has since been tested in China, India, Indonesia, Philippines, Sri Lanka and Bangladesh with positive results.

Most rice farmers plant fairly mature seedlings (20-30 days old), in clumps, fairly close together with standing water maintained in the field for as much of the season as possible. These practices seem to reduce the risk of crop failure. It seems logical that more mature plants should survive better, that planting in clumps will ensure, that some seedlings should result in more yield; and that planting in standing water means the plants will never lack water and weeds will have little opportunity to grow.

21.2.3.1 Six practices in SRI

1. Young seedlings

Seedlings are transplanted early. Rice seedlings are transplanted when only the first two leaves have emerged from the initial tiller or stalk, usually when they are between 8 and 15 days old. Seedlings should be grown in a nursery in which the soil is kept moist but not flooded. The seed sac should be kept attached to the infant root, because it is an important energy source for the young seedlings. The young seedlings should be planted so carefully that the root tip is not left pointing upward. The effect of young seedling (14 days old) over normal seedling (21 days old) was studied on growth and yield of rice and the yield of rice in various treatments was furnished in Table 66.

2. Single seedling

Seedlings are planted singly rather than clumps of two or three or more. This means that individual plants have room to spread and to send down roots. They do not compete as much with other rice plants for space, light and nutrients in the soil. The effect of number of seedlings/hill was studied on growth and yield of rice and the yield of rice was furnished in Table 66.

3. Wider spacing

Seedlings are planted in a square pattern with plenty of space between them in all directions. Usually, they are spaced at least 22.5 cm x 22.5 cm. It helps for vigorous root growth and more tillering. The square pattern facilitates *in situ* incorporation of weeds by Cono weeder. Roking marker for squere planting of rice seedlings at 25 x 25 cm^2 spacing in a 2 M width bed area, is also used for new women labourers (Plate 62 and 63).

Table 66. Effect of spacing and age of seedlings on the yield of rice

Number of seedling with spacing	Young seedling (kg/ha)		Normal seedling (kg/ha)	
	Kuruvai 2003	Late Thaladi 2003	Kuruvai 2003	Late Thaladi 2003
One seedling at 20.0 x 20.0 cm	6951	5377	6329	4859
Two seedling at 20.0 x 20.0 cm	6553	4868	5142	4572
Three seedling at 20.0 x 20.0 cm	6355	4402	5802	3905
One seedling at 22.5 x 22.5 cm	7444	6138	7714	5160
Two seedling at 22.5 x 22.5 cm	7745	5019	7384	4710
Three seedling at 22.5 x 22.5 cm	7641	4538	7120	4230
One seedling at 25.0 x 25.0 cm	7149	4978	6725	4142
Two seedling at 25.0 x 25.0 cm	6951	4286	6329	4450
Three seedling at 25.0 x 25.0 cm	6752	4093	6329	3567

4. Lesser seed rate

SRI method requires much lower seed rate (5-8 kg/ha) than traditional methods (75-100 kg/ha).

Plate 56. SRI – Nursery preparation

Plate 57. SRI – Nursery Management

Plate 58. SRI – On farm trail near Aduthurai

Plate 59. SRI – Nursery

Plate 60. SRI – Using LCC

Plate 61. SRI – Using Conoweeder

5. Moist but unflooded soil conditions

Rice has traditionally been grown in water under submerged condition. In SRI method, soil is kept moist but not flooded during the vegetative period, ensuring that more oxygen is available in the soil for the roots, occasionally the soil should be allowed to dry to a point of cracking except in saline soils. This will allow oxygen to enter the soil and also induce the roots to grow and search for water. In SRI method, unflooded condition is only maintained during vegetative period and from flowering to harvest, 1-3 cm of water is kept in the field as is done in the traditional method.

6. Cono weeding

Weeding can be done by cono weeder. The cost of conoweeder will be around Rs. 300 to 1000/- depending upon the type and materials used. First weeding should be done on fourteen days after transplanting and this should be continued up to 40 days at 7-10 days interval. At least two or three weedings are recommended. This practice of churning of soil seems to improve soil structure and increases the aeration of the soil. Thus the incorporation of weed biomass into the soil results in enrichment of CO_2 near to root zone, increases the biological activities, increases soil microbes population and activities, results in better nutrient availability in soil and uptake by plants. If conoweeder is not available, rotary weeder is recommended.

Plate 62. Rolling marker for square planting

Plate 63. Rolling marker for square planting - in operation.

21.3 RESEARCH FINDINGS

The System of Rice Intensification (SRI) was evaluated through various trials and demonstrations at TRRI, Aduthurai; SWMRI, Thanjavur and AEC &RI,Kumulur, Trichy, Tamil Nadu.

a. On Farm Trial at Thanjavur, Tamil Nadu

On Farm Trials were conducted during Kuruvai 2002 in Arupathy and Kathiramangalam villages of Nagapattinam district. In both villages, trails were laid out with two treatments namely SRI and conventional methods of rice cultivation. Enhanced grain yields of 7650 and 5680 kg/ha were harvested at Arupathy and Kathiramangalam villages respectively (Table 67) representing an increased grain yield of 45.7 and 40.3 per cent over that of conventional method of rice cultivation. Similar trials were also conducted at Edamelaiyur village and at TRRI, Aduthurai (On Station Trial), in which the SRI method of cultivation recorded higher yield of 7314 kg/ha which was 31.7 per cent higher than conventional method of cultivation.

Table 67. Results of SRI – Kuruvai 2002-03 (ADT 43) in On Station trial (OST) and On Farm Trial (OFT)

Trial	Productive tiller m^{-2}		Grain yield kg ha^{-1}		Per cent increase over control
	SRI	Conventional	SRI	Conventional	
OST					
TRRI, Aduthurai	545	445	7314	5554	31.7
OFT					
Arupathy	845	429	7650	5250	45.7
K. Mangalam	426	391	5680	4050	40.3
Edamelaiyur – 1	322	—	5730	3816	50.2
Edamelaiyur – 2	459	—	7443	4976	49.6

Source: TRRI Annual report, 2004

b. On Station Trial

The various technologies of SRI in different combinations were evaluated during late Thaladi 2002 and Kuruvai 2003 at TRRI, Aduthurai along with conventional method of rice cultivation. The SRI technologies i.e., use of young seedlings and one seedling per hill, square planting, use of cono weeder and intermittent irrigation produced higher yield of 7061 and 7546 kg/ha (Table 68) during late thaladi and kuruvai seasons, 2003 respectively, representing an enhanced grain yield of 48.81 and 53.2 per cent over that of conventional practices. Increased grain yield under SRI was mainly attributed to more number of lengthy productive tillers with increased number of filled grains per panicle and lesser number of unfilled grains.

Table 68. Evaluation of different components of SRI methods on rice grain yield (ADT 43)

Treatments	Grain yield Late Thaladi 2002 (Kg ha $^{-1}$)	Grain yield Kuruvai 2003 (Kg ha $^{-1}$)
YOSC I	7061	7546
NMRH I	5143	4530
YMRH I	5241	5527
NORH I	5463	4596
NMSH	5586	4794
YORH I	5169	5461
NMSC I	5750	5092
NOSH I	5652	5059
YOSH I	5698	5594
NOSC I	5874	5390
YMSC I	6137	5958
NMRH (F)	4745	4927
YOSC(F)	6428	5759

Y: Young seedling, O : One seedling, S: Square planting, C: Cono weeding
N: Normal seedling, M: Multiple seedling, R: Random planting, H: Hand weeding
I: intermittent flooding, F: Continuous flooding

Besides, experiments on Integration of Site Specific Nutrient Management with ICM (SRI) in Rice" sponsored by IRRI, Philippines were laid out in three places of Cauvery Old Delta. Under this programme,.two on farm trials in Arupathy and Keelamaruthuvakudi village and one OST at TRRI, Aduthurai were laid out during Kuruvai 2003. In OFT and OST trials, SRI method gave higher grain yields compared to that of conventional method of cultivation (Table 69).

Table 69. OST on Integration of SSNM with ICM on rice yield (kg/ha)

Treatments	Grain yield (kg/ha)		
	TRRI, Aduthurai	Aarupathy	Keela Maruthuvakudi
Conventional method - FFP- NPK	5497	4235	5406
Conventional Method - LCC N, FFP - P& K	5974	5023	5999
ICM - LCC N, FFP - P & K	7694	6993	6922

FFP- Farmers Fertilizer Practice

C. OFT at Trichirappalli, Tamil Nadu

The three OFT trails were conducted at Lalgudi block of Trichirappalli Dt. Tamil Nadu. All inputs including conoweeder and all technical guidance were given to them and the results revealed that higher yields were obtained by all the farmers in SRI method. i.e. with a B/C ratio of 2.12 to 3.97 (Annadurai and Kavimani, 2004).

Name of the farmers/ Operation	A. Krishna moorthi, Thirumanamedu, Lalgudi (TK), Tamil Nadu	A. Natarajan, Mission Street, Sangamarajapuram, Anbil, Lalgudi (TK), Tamil Nadu	S. Durairaj, Main Road Thirumanamedu, Lalgudi (TK), Tamil Nadu
Cost of cultivation	Rs.16113	Rs. 16113	Rs. 16113
Gross income	90 kg / 3 cent – 30 kg/cent x 100x2.5	1120 kg / 25 cent	24 kg /cent – 2400 kg/ acre
	1) Paddy yield :7.5t/ha – Rs. 320/60kg. Rs.5.30/kg of grain 7500 kg x Rs. 5.30 – 39750.00	For 1 acre – 11200 kg 11,200 x Rs. 5.30 = 59360.00	For 1 ha 6000 kg x Rs. 5.30 = 31800.00
	2) Straw yield :12.5t/ha – Rs. 240/t 12.5 x 240 – 2880.00 **Rs. 42630.00**	19 t x Rs. 240 = 4560.00 **Rs. 63920.00**	9.5t x Rs.240 = 2280 **Rs. 34080.00**
Net return	42630.00 – 16113 = **Rs. 26517/-**	63290 – 16113 = **Rs. 47807**	34080 – 16113 = **Rs. 17967**
B/C ratio	$\frac{42630}{16113} = 2.65$	$\frac{63920}{16113} = 3.97$	$\frac{34080}{16113} = 2.12$

Plate 64. SRI- Using conoweeder at Kumulur, Tamil Nadu

21.4 AEROBIC RICE

Water in irrigated rice production has been taken for granted for centuries, but the "looming water crisis" may change the way of rice production in the future. Water saving irrigation technologies that were investigated in the early 1970s, such as saturated soil culture and alternate wetting and drying, are receiving renewed attention from researchers. These technologies reduce water input, though mostly at the expense of some yield loss. Farmers in Asia who confront scarcity or high costs of water have already started to adopt these technologies. In China, various forms of AWD and reduced floodwater depths have been developed and massively adopted by farmers (Li, 2001). Surveys in north – central India and in Central Luzon, Philippines show that farmers who operated pumps to irrigate their fields consciously apply some form of AWD to save on energy costs.

21.4.1. Concept

Aerobic rice is a new concept to further decrease water requirements in rice production. it is commercially grown in Brazil and is being pioneered by farmers in northern China. In the heart of the rice – wheat belt in India

(Haryana, Punjab and Uttar Pradesh), innovative farmers have begun to grow rice aerobically under furrow irrigation in raised bed systems. Over the centuries, lowland rice has proven to be a remarkably anaerobic character. The shift from anaerobic to aerobic systems will have major consequences for weed, disease, and insect pest management, nutrient and soil organic mater dynamics, and green house gas emission and sequestration. Weed control is an especially crucial issue in most water – saving irrigation technologies.

A fundamental approach to reduce water inputs in rice is to grow the crop like an irrigated upland crop such as wheat or maize. Instead of trying to reduce water input in lowland fields, the concept of having the field flooded or saturated is abandoned altogether. Upland crops are grown in non-puddled aerobic soil without standing water. Irrigation is applied to bring the soil water content in the root zone up to field capacity after it has reached a certain lower threshold (e.g., halfway between field capacity and wilting point). Since it is not possible to apply irrigation water to the root zone only, some of it is lost by deep percolation and is unavailable for uptake by the crop. Typical field application efficiencies vary from 60-70% using surface irrigation. (e.g., flash or furrow irrigation) to more than 90% using sprinkler or drip irrigation. Evaporation can also be reduced with this technique since there is no continuous standing water layer. De Datta *et al.,* (1973) tried growing rice like an upland crop using furrow irrigation in the 1971 dry season at IRRI. Using the high yielding lowland variety IR20, total water (irrigation plus rainfall) savings were 56% and irrigation water saving was 78% compared with growing the crop under flooded conditions. However, yield decreased from 7.9 to 3.4 t ha^{-1}. Upland cultivars that performed equally well under flooded and dry land irrigation were used, but their yields of around 5.0 t/ha were much lower than those of lowland cultivars.

21.4.2 Research findings

Studies on non-flooded irrigated rice using sprinkler irrigation were conducted in the United States in Texas and Louisiana (McCauley, 1990, Westcott and Vine, 1986). Experiments used commercial rice cultivars under lowland cultivation. Irrigation water requirements were 20-50% less than in flooded rice, depending on soil type, rainfall, and water management. The highest yielding cultivars (producing 7-8 t ha^{-1} under flooded conditions), however, had yield reductions of 20-30% compared with flooded rice. The most drought – resistant cultivars produced the same under both condition, but their yields were much lower (5-6 t ha^{-1}). Under economic conditions prevalent in the US, the adoption of irrigated dry land rice (using existing cultivars) was not economically attractive.

New varieties must be developed, if growing rice like an irrigated upland crop is to be successful. Lowland cultivars have been selected to give high yields under continuously flooded lowland conditions. They generally suffer a yield loss when the soil water content drops below saturation. Upland varieties have been developed to give stable, though low yields in adverse environments where rainfall is low, irrigation is absent, soils are poor or toxic, weed pressure is high, and farmers are too poor to supply high inputs. Therefore, IRRI recently coined the term "aerobic rice" to refer to high yielding rice grown in unpuddled aerobic soil. The aerobic rice, which can be rainfed or irrigated, should be responsive to high inputs and should tolerate flooding. It has to combine characteristics of both upland and high – yielding lowland varieties. Evidence for the feasibility of aerobic rice comes from Brazil and north China. In Brazil, aerobic rice cultivars have come out of a 20-year breeding program to improve upland rice with yields of 5-7 t ha^{-1} under sprinkler irrigation in farmers' fields (Bouman, 2001). These varieties are grown commercially on 2,50,000 ha in the state of Mato Grosso. In north China, aerobic rice cultivars yield up to 6-7.5 t ha^{-1} in farmers' fields using flash irrigation in bunded fields (Wang and Tang, 2000). It is estimated that these cultivars are now being grown on some 1,20,000 ha in the north China plains.

Plate 65. Aerobic rice field with sprinkler irrigation at Kumulur, Tamil Nadu

21.4.3. Areas to be grown under aerobic rice

The aerobic rice can be raised in the areas

- Where rainfall is insufficient to sustain lowland rice production (estimated to require 1,200 – 1,500 mm) but thought to be sufficient for aerobic rice (estimated to require some 800 mm). Here, maize is the dominant crop and rice is an attractive alternative through the benefits of crop diversification. Moreover, the price of rice is supported by the government and farmers perceive this as advantageous.
- In pump – irrigated areas where water has becomes so expensive that lowland rice production was abandoned.
- Where water is scarce during the first part of the growing season (necessitating irrigation), but floods occur in the second part, upland crops such as maize and soybean cannot withstand flooding, but aerobic rice can.

21.4.4 Causes for yield decline and future thrust

In both Brazil and China, however, it has been reported that high initial yields are difficult to sustain and that yields may decline severely after 3-4 years of continuous cropping. Brazil has adopted a rotation system in which aerobic rice is only cropped once every 3-4 years. The causes for the yield decline are not understood, but likely reasons are the buildup of soil borne diseases (such as nematodes) or toxic substances. Water has long been the cheapest herbicide, but this may not be so any more in the near future. Breeders have to respond to the challenge of breeding varieties that perform well under non permanently flooded conditions. The development of aerobic rice varieties is probably the most ambitious challenge of all.

CHAPTER 22

PRECISION FARMING

In irrigated rice systems of South and Southeast Asia, many farmers have achieved tremendous yield gains during the past 30 years, often approaching the economically attainable yields. However, average yields of irrigated rice have to increase from 4.9 t/ha in 1991 to about 8 t/ha in 2025 (Cassman and Pingali, 1995). Therefore, irrigated rice farmers will have the greatest need to fine-tune their farm management to produce the bulk of the future increase in rice supply. At the same time, the positive awareness of environmental aspects of farming needs to be ensured for the farmers, future yield increase will require improved crop care, integrated resource management approaches and more knowledge intensive strategies for the efficient use of all inputs. They will have to move from simple general decisions and recommendations to much more knowledge of intensive, site-specific crop management. Increasing yields and input use efficiency will require new farming concepts that focus on fine-tuning of seed, nutrients, water, pesticides, energy and labour inputs for smaller management units. Precision farming (PF), prescription farming or site-specific crop management are frequently used new terms to describe such emerging technologies. (Robert *et al.,* 1995).

22.1 PRECISION FARMING-OBJECTIVES

- To improve crop performance and environmental quality.
- To manage spatial and temporal variability associated with all aspects of agricultural production.

- To match resource application and agronomic practices

22.2 CONCEPT

Two phrases

1. 'Spatial variable', 'GPS based', 'Prescription', 'Site-specific' or 'Precision',
2. 'Farming': 'Agriculture' or 'Crop production'.

Variations occur in crop or soil properties within a field:

- These variations are noted, and often could be mapped.
- Management actions are taken as a consequence of the spatial variability within the field.

22.3 COMPONENTS

1. Geographical Information Systems (GIS)
2. Global Positioning Systems (GPS)
3. Sensors
4. Variable Rate Technology (VRT)
5. Yield Monitoring (YM)

22.3.1 Geographical Information Systems (GIS) and its uses

- GI is a software that provides data storage, retrieval, and transformation of spatial (field) data.
- It provides data such as soil type, nutrient levels, etc, in layers and assign that information to the particular field location.
- A fully functional GIS can be used to analyze characteristics between layers to develop application maps or other management options.
- Field location is stored by the latitude and longitude of that position
- Several maps can be created showing the variability of nutrient levels, soil type, topography, pest incidence and yield.
- Most GIS software packages also have some form of data smoothing or contouring
- Simulation models for different purpose like the flow of water, crop growth, soil erosion, nutrient and pesticide leaching.

- GIS helps in integrating geographical data on various aspects such as soil, weather and field history along with simulation models.
- GIS support to agriculture is the engineering component, in which the research findings are translated into operational systems for use at farm level.
- GIS can support this engineering activity by providing a good platform for storage of base data, simple modeling, presentation of results, development of a user interface, and in combination with a GPS, controlling the navigation of farm.

22.3.2 Global Positioning Systems (GPS)

The ability to know where you are and where you are going is crucial to so many, to date Global Positioning Systems (GPS) achieved it. To day, Agriculture demands improved productivity and efficiency, cutting costs, saving time and ensuring the entire agricultural enterprise is more efficient and accountable, and it is essential to compete in domestic and global markets. Positioning wireless and information technologies are changing the farmer relationship with land, bringing them quite naturally down to earth. Farmers can now manage every aspect of an agricultural operations to improve over all productivity and efficiency- from planting to harvesting, land leveling and literally work their land by square foot instead of square miles.

- GPS farming systems provide precise guidance for field operations or collect data on tillage applications, planting, weeds, insects and disease infestations, irrigations etc.,
- Increase farm and farm productive area
- Precisely plant seeds at optimal rates at different areas of the field
- Saves time by performing multiple operations in one pass over the field
- Create and maintain detailed records of crops, weeds, insects, diseases, topography for prescription farming and water management.

22.3.3 Sensors

Sensors are being developed to determine crop stress, soil properties, pest incidence, etc. Sensors mounted on a tractor are used to measure soil and crop properties. As the tractor passes over the field, as a scout goes over the field on foot, or as an airplane or satellite photographs the field from the sky. Measurement of plant and soil properties as the tractor or combine travels over the field is a developing area in precision farming. Currently, yield monitors are the primary sensing system that makes measurements "on-the-go". Commercial sensing systems are available that claim to measure soil properties and make application rate changes "on-the-go".

There are also sensors that can be carried by a scout to the field and used to spot-check the health of plants and soil properties. They use light reflectance on the leaf to determine chlorophyll production. Several soil chemistry kits are also available to measure soil pH, nitrogen, phosphorus and potassium in the field without waiting for soil sample results.

22.3.3.1 Remote sensing

PF needs information about mean characteristics of small, relatively homogenous management zones.

- These may be obtained from soil tests for nutrient availability, yield monitors for crop yield, soil samples for organic matter content, information in soil maps, or ground conductivity meters for soil moisture.
- Remote sensing technology is to obtain spatially and temporally variable information for PF.
- Remote sensing imaginary for PF can be obtained either through satellite-based sensors or video digital cameras on board small aircraft.
- In the first approach, the multi-spectral images used for anomaly detection. These anomalies can be in the forms of disease/pest, weed growth, water stress, etc.
- Using the reflectance in the visible part of the spectrum, it is possible to detect diseases and identify weeds from crops.
- The difference between remotely sensed surface temperature and ground-based measurements of air temperature has been established as a method to detect water stress in plants.
- Correlating variation as in spectral response to specific variables such as soil properties or crop yield. Soil physical properties such as soil water, organic matter and soil texture can be correlated to spectral reflectance.
- Vegetation spectral response used to infer other soil conditions.
- Crop yields for many crops like rice, wheat etc. have been found to be highly correlated with spectral vegetation index during maximum vegetative cover.

- The yield map generated from spectral images can be used to farm management units.

22.3.4 Variable Rate Technology

Variable Rate Technology (VRT) includes computer controllers and associated hardware to vary the output of fertilizer, lime and pesticides. Controllers are manufactured by several companies and can generally take an application map, along with the GPS to locate field position, and control the hardware that varies the application rate.

22.3. 5 Yield Monitoring

Yield Monitoring is the most important direct method to assess the field production and how it should be better managed. A yield monitor measures the crop as it is harvested. YM usually measure crop weight, impact forces, or the time an array of light beams is broken. As the yield is measured, data are stored on a computer along with the GPS coordinates at the point where the yield was measured. Mapping software can then create a yield map. The yield map can immediately provide two important pieces of information viz., yield variability, and yield production. Yield variability is illustrated on a map by a change in colour, where each colour represents a range in yield.

22.4 FARM AND SITE-SPECIFIC CROP MANAGEMENT

We consider "Precision farming" or SSCM as a scientific concept that is applicable to farm and fields of all sizes, including those found in Asia. It is proposed to characterize three terms for modern farming operations.

1. Total Farm Management (TFM)
2. Technology Application Domain (TAD)
3. Site Specific Crop Management (SSCM)

22.4.1. Total Farm Management

This TFM is an information-based agricultural management system to provide an optimum of the balance between profitability, food production and sustainability within a **single farm**. This definition emphasizes optimization of the management of the whole farm, including various cropping systems and, if applicable, other types of agricultural production. At this level, a farmer has to make decision about the allocation of major inputs and any optimization attempts should also consider relationships with other farms in a neighborhood of varying size. The later includes buy and sell operations, rental of machinery and labour, credits, distribution of seeds, manure and post harvest activities.

22.4.2. Technology application domains (TAD)

These are for a **specific technology**, the **smallest uniform spatial units** that can be treated differently. Within a SSCM approach, the size of TADs can be different for different crop management operations. This definition of TAD emphasizes a **specific technological solution for implementing a certain crop management operation according to site characteristics**.

22.4.3. Site-specific crop management (SSCM)

This is the crop-specific use of all local soil, crop and climatic parameters to make precise application of production inputs to technology application domains (TAD) with different characteristics. The typical SSCM components of a particular crop are varying the depth of soil tillage based on topography, changing varieties or sowing rates according to soil types, adjusting the rate of fertilizer application according to variation in soil test values, varying rates of pesticide spray based on actual crop stress. Therefore, this definition emphasizes management of a **single crop** according to site characteristics.

Therefore, SSCM decisions and operations may include **continuous adjustment,** applications specific to patches within a single field, uniform application to single field, or uniform applications to groups of several fields with similar properties.

22.4.3.1 Practices of SSCM

A typical SSCM application includes the following steps:

- **Knowledge Capture:** Identify and quantify (map) the variability of the key input parameters at the scale needed to make a decision about the specific SSCM operation.
- **Knowledge Interpretation:** Translate the information into decision about management (application) by understanding the sources of variability and its impact on yield.
- **Applications:** Identify the available application technology and the optimal size of the technology application domain (TAD). Apply differential amounts of inputs to TADs.

22.4.3.2 Potential benefits of SSCM

The basis of SSCM is that fields are highly variable (both between and within) and new technologies are available that can characterize that variability and delineate meaningful management zones to optimize supply and demand of nutrients, water, energy and other resources according to their variation in time and space. Where there is less supply (e.g. certain fields or certain locations within a field) application has to be greater than in physical land units with higher supply and vice versa. Hence, the degree of variability, the quality of the input information, the approaches used to translate it into application decisions and the costs associated with differentially managing a field determine the benefits obtained from SSCM.

22.4.3.3 Opportunities for SSCM

22.4.3.3.1 Need

Crop management recommendations over the past four decades in Asia were driven by increasing use of externally provided inputs and the so-called "Package approach" based on blanket recommendations over wide areas (Byerlee, 1996). If a question is raised whether we should and can apply the principles of SSCM to manage rice fields in Asia, We believe yes, because recent research in lowland rice areas has demonstrated that

- There is large variability in stable soil properties, soil nutrient supply, nutrient use efficiency, other production factors, grain yield, and economic performance between rice farms or between single rice fields (Cassman *et al.,* 1996; Dobermann *et al.,* 1997a; Olk *et al.,* 1996; Dobermann and Oberthuer, 1977) and
- Even within very small rice fields, tremendous variation in yields and yield components exists that is caused by micro variation in soil nutrients, land leveling, crop emergence, weeds and other pests (Baki, 1993; Dobermann *et al.,* 1997a; Dobermann *et al.,* 1997b; Dobermann *et al.,* 1994)

This will lead to achieve the significant gains in productivity and input use efficiency by soil and crop management technologies that are much tailored to the specific characteristics of individual farms, rice fields and variation within fields (Cassman *et al.,* 1997; Dobermann *et al.,* 1996). Farmers already integrate a lot of knowledge and often already have a sense of the variability they have in fields and crops.

22.4.3.3.2 Feasibility

While much of the purpose of PF in developed countries is to breakdown variability across large fields to small uniform units, in Asia, much of this division has already taken place due to the typically small farm size. Although India has made much advances in agricultural research, but still the blanket recommendations are very much in vogue for adoption over larger areas. These blanket recommendations are no more useful to enhance productivity gains, which were witnessed between 1960s and 1980s. Now, to enhance growth rate in productivity, Site Specific Crop management (SSCM) practices have to be developed.

By definition, SSCM practiced in developing countries focuses on managing spatial and temporal variability within a single and usually large field. In a broader sense, however, SSCM should be viewed as farming philosophy aiming at replacing prophylactic application of agricultural inputs by demand base application. Because of this, SSCM is relevant to Indian conditions, which have varied type of soil and climate. Conceptually adjusting tillage, sowing, fertilizer or pesticide rates separately for many small fields or farms (< 1 ha to < 10 ha) in an Asian domain are similar to adjustment according to soil variation within a large field (> 10 ha to > 100 ha) found in North America. There is a continuum, however, as there is still variability within these small fields. We feel that first major step is to refine regional recommendations to individual field level recommendations-further advances can be made in the future to within field recommendations. Generally, SSCM in Asia can be built around much less sophisticated technology than implementing SSCM in large fields, where GPS (Global Positioning System), mapping systems, computer technology and VRT (Variable Rate Technology) are minimum requirements.

The practical options for SSCM in irrigated rice are very much determined by the general shifts in these production systems, including:

- Labour shortages- Fewer agricultural field workers
- Water shortages for rice-Less water for agriculture, urban demands and alternate uses increase
- Less land for rice production-As cities grow over production areas and alternate land use increases
- Shifts in rice quality preferences-As economy develop through out the region, quality preferences will become more important in some countries.

Generally land preparation, crop establishment, harvest and post harvest activities are labour-intensive farm operations and farmers will increasingly seek ways to reduce costs associated with them. They will also seek ways to add value to their production by improving the quality of harvest products or improving their use for various by-products.

22.4.3.3.3. Technologies

Although precision farming (PF) or SSCM stated as a technology led development, it is not just synonymous with yield mapping and variable rate technology (VRT) for managing spatial variability within the field. Instead, PF should be viewed as a system approach to crop production in which the goal is to reduce decision uncertainly through better understanding and management of uncontrolled variations. SSCM aims at improving input - output characteristics of soil and crop system as they vary in space and time.

22.4.3.4 SSCM decision on region basis

Even in a SSCM approach, recommendations and decisions at regional scales play a role. At scales such as a district, important agronomic advice can be given to a farmer. Information captured may include

1. Information about most suitable crops and varieties for a given environment
2. General recommendations for soil fertility management based on delineation of soil types.
3. Weather forecasting and real-time variation in weather, and
4. Inputs application, Pest forecasting and seasonal variation in pest populations.

This would help the farmer to make decisions about varieties, seedling age, planting date, fertilizer application, pest management and harvest time. The major tools involved in this are remote sensing (e.g., high resolution radar images), GIS crop models (Mathews *et al.,* 1997; Singh *et al.,* 1991) and pest models (Kropff *et al.,* 1995). Mass media (TV, radio, newspapers), HAM Radio communication and the existing extension systems are the main information providers.

A. Selection of variety

Most rice farmers already select their varieties based on knowledge about adoption to site characteristics and agronomic fitness. In general, choice of the variety for a place as well as season is very important. Sowing the crop at appropriate time in the season will give higher yield. The production and productivity are generally higher in dry season (Kuruvai/ Kar/ Sornavari) due to favorable weather parameters. Bright sunshine hours, favorable temperature regime, optimum moisture, marked response to nutrients application, relatively less pest and disease incidence are some of the factors contributing to higher grain yield in dry season.

During the wet season (Thaladi/Samba/Pishanam), factors like reduced sunshine hours during rainy season, cloudy weather, cool temperature, flat terrain fields, excess moisture, ill drained conditions due to continuous rain, poor response to applied nutrients, incidence of pest and diseases etc., are responsible for lesser yields. Most modern varieties released are resistant to some common pests and they are also screened for adaptation to some soil stresses such as severe P and Zn deficiency or Fe toxicity. Among the various research options to break the yield barrier, exploitation of heterosis and restructuring of rice plant ideotype for better harvest index are important. Hybrid rice is found to be a viable technology and it has been proved successful in China. In Tamil Nadu three hybrids viz., CORH 1 (1994), CORH 2 and ADTRH 1 (1998) were released for cultivation. The ADTRH 1 rice hybrid has got the yield advancement of 10 t/ha over the best high yielding varieties like ADT 36, ADT 43 and ADT 45. Ever since the evolution of dwarf varieties that helped to break the centuries old genetic yield barrier, rice scientists have been looking for still more efficient technology for raising genetic yield potential. There is probably some scope for refining selection of varieties in a SSCM approach with regard to

- Fitness for dry or wet seeding
- Resistance to pests
- Nutrient requirements
- Grain quality and
- Seed health

Though biotechnological approach and transgenic lines may be developed for enriching with vitamin A and iron, It is felt necessary that more variety-specific information should be jointly established by breeders, agronomists, plant protection specialists and post production specialists (e.g. millers) and released through the national seed distribution and extension systems.

B. Land preparation

SSCM options for land preparation vary widely. In many small to medium size farms, the lowest feasible TAD for ploughing with a four-wheel tractor would be a whole field, i.e., ploughing depth would be uniform. Examples of field-specific SSCM decisions include

- Is there need for ploughing and puddling in each rice crop?
- Is there need for occasional deep ploughing?
- Is precision leveling feasible?

In the last years, many farmers have switched to shallow tillage machinery such as hydro tillers. This may lead to formation of shallow plough pans and a reduction of the rooted layer.

C. Crop establishment

Two types of site-specific crop establishment decisions and operations are important: (1) assess whether a (whole) field is suited for a particular crop establishment method, and (2) vary sowing rates or transplanting density according to variation in soil properties and micro relief within a field. Decisions about suitability for a specific crop establishment technology and varying the sowing rate require expert knowledge and basic soil and climatic information. If this information is available, farmers could do field-specific or spot-based variation within a field in either sowing/planting by hand or modified mechanical transplanters or row seeders that allow easy adjustment on the go. For large farms, drill seeders with VRT features are available and can be modified for use in dry-seeded rice. Although many farmers will do their own judgement, a more quantitative approach is warranted, which would probably be under the responsibility of the existing extension service or, if they exist, crop consultants. Access to specialized equipment will likely be through contractors.

D. Precision seed sowing

Determination of appropriate seed sowing rate and depth at which the sowing should be done will improve the uniformity in the plant population hence reduction in the spatial variation of yields.

E. Precision weed Control

The following three components are involved in precision weed management:

(a) Preventing weed effects through adapted crop management

(b) Improvised decision making on weed control

(c) Developing precision control technology

The improvement of weed management systems, on the basis of the above three-component strategy, requires quantitative insight into both crop-weed interactions and the dynamics of weed population in space and time. Some of the recent advances like 'Variable weed control techniques' minimizes the pollution and increases the efficiency of the chemicals and reduces the cost. The other areas concerned with precision farming are irrigation, farm machinery, control systems and spatial database of all crop and field attributes.

F. Precision water management

The water management technology is highly location specific, and as such the various decisions for irrigation scheduling vary with soil characteristics, crops to be grown, climatic conditions, quality of water and social and other factors. In arid and semi-arid regions owing to water shortages and multivarious competing demands for the limited supply of available water, there is increasing need to irrigate as efficiently as possible. This situation emphasizes the need for using scientifically sound technology for deciding when and how much to irrigate. Most of the conflicting reports concerning the effects of irrigation on plant growth resulted from irrigation scheduling not always based on sound knowledge of soil-plant-atmospheric water relations. There is an utter lack of literacy about real value of scarce water resource among farmers. There is also a mismatch between water delivery and crops needs. These deficiencies result in over or under-irrigation and sub-optimal production efficiency and degradation of natural resources. A major emphasis is placed to describe interactive effects of soil, plant, atmospheric factor and input resources like water, fertilizers on plant water status, crop water use, and possible alterations and adjustments in various management strategies to ensure high input efficiency of various resources. Crop water use is primarily influenced by evaporative demand of atmosphere, but plant and soil factors also play their important role, not so much so in altering water use, but to a great extent in improving water use efficiency. In this aspect, the controlled irrigation depth, optimum dose of fertilizers, suitable crops/their cultivars and crop sequences under limited supply of water, proper agronomic practices for efficient resource management under problematic soils and water conditions are important factors to increase the proportion of water passing through the crop in transpiration, and minimizing evaporative and other losses of water from soil system. Water and fertilizers improve not only internal plant water status but also each other's efficiency. Recently, the crop coefficients and irrigation management through simulation modeling using the concept of water management response indicators have been developed for Indian conditions. However, there is an urgent need to develop optimum crop plans and irrigation management in canal command areas by adopting suitable combinations of water, fertilizers and other inputs to ensure not only high use efficiency of these resources but also high economic returns on sustainable basis to the farming community from agriculture. A multidisciplinary and farmers' participatory approach for efficient irrigation management is required to achieve desired results from irrigated agriculture.

Site-specific water management is closely linked with the technologies available for soil tillage (e.g., puddling requirements). In particular, precise leveling is the most important factor in efficient irrigation management (Hill *et al.,* 1991), but it would also reduce variability in weed growth, soil properties and rice growth caused by heterogeneous water flow patterns within fields (Dobermann *et al.,* 1997a). Various kinds of reduced irrigation, including reduced water depth, periodical flush irrigation, or sprinkler irrigation are options, but in most of them a single field would be treated homogeneously. Bypass flow in cracking clays may cause huge unproductive water losses during land soaking, i.e., during the initial irrigation flush to achieve water saturation (Tuong *et al.,*1996). Dry shallow tillage soon after harvesting reduces soil drying and cracking during the fallow period and the water need for the subsequent rice crop. In SSCM, soils and fields where this may be beneficial need to be identified. The same relates to decisions about need and type of drainage. The primary focus is on capture of suitable field- or farm specific information to make a decision about the most appropriate water management technology. At this stage, we do not know who would be the best choice for this.

In Tamil Nadu, the area under rice is about 2 million ha and 70 per cent of which is irrigated. The dominant system of water management in rice cultivation is irrigation up to 5 cm depth one day after disappearance of previously pounded water. However, many farmers are unable to follow this recommended water management practice due to difficulties in water flow control and uncertain water availability. Typically water input in rice fields varies between 1000 and 3000 mm depending on environmental conditions and the length of the growing period. It is estimated that the water supply demand gap for irrigated crops will be about 21 billion m^3 in 2025. Water shortage in Tamil Nadu has already resulted in reduction of the irrigated rice area and in a shift towards less water demanding cropping activities. To sustain present food sufficiency and to meet future food requirements, India has to realize an annual growth rate of at least 3 per cent in productivity of rice. Considering the future food requirements, new methods of rice cultivation must be identified to increase water and crop productivity. The practices of system of rice intensification (SRI) developed at Madagascar has been reported to increase grain yields of rice and to save water based on the synergistic effects of adopting various cultivation practices simultaneously, i.e., the use of young and single seedling per hill, wider spacing, square planting, cono weeding, limited irrigation and frequent loosening of the top soil to stimulate aerobic soil conditions, as discussed in detail in the chapter 21-New methods of rice cultivation.

G. Precision Pest management

Field-specific decisions may be based on both qualitative and quantitative information. There are three major agronomic options for site-specific management of weeds, insects, diseases, and other pests like rodents, snails, nematodes etc.,

- Control via combination with other SSCM measures
- Homogeneous, prophylactic control
- Variable rate control based on observation or sensing (spray only in needy places)

Hand-held sprayers dominate for applying pesticides, but there are improvements in their design allowing more accurate and variable adjustment of rates during field operation. The heavy machinery (or even aircraft) used for variable rate application of pesticides in regions such as North America is only of interest for few large farms of Type III.

Ecologically sound site-specific pest management includes measures such as selection of resistant varieties, pest specific bio control agents, reduction in the amount of pesticides used, substitution of less hazardous chemicals for the more hazardous ones, and use of pesticides or non-chemical control measures based on knowledge of pest pressure. Following an integrated pest management (IPM) concept, application of insecticides could be restricted to spots with high infestation only, *i.e.*, the farmer would use a TAD of much smaller size to manage within-field variation with simple means. With the farm yields in tropical Asia expected to rise to 7-8 t/ ha and more, we will see an increasing need for disease control using fungicides, cultural practices, or improved host plant resistance (Heong *et al.,* 1995). Presumably, most of these are prophylactic measures applied to whole fields with less scope for managing variability within a field. Unlike fertilizer management, much of the information required can be captured by observation so that site-specific pest management would mainly be the responsibility of a farmer or farm manager. On this situations, development of location specific insect forewarning system and pest forecast modules assumes significance. Training to improve farmer's knowledge about capturing the information and translating it into application decisions plays a vital role.

Precision leveling allows precise water management as one important measure for field-specific weed control, particularly in direct-seeded rice (Williams *et al.,* 1990). Weeds, particularly annual species tend to occur in patches (Baki, 1993) so that spot applications or variable rates of herbicides are feasible even when farmers use hand-held sprayers. Identification of suitable post-emergence herbicides will be required for this approach.

H. Precision Post-production - Harvest

SSCM can be interpreted in various ways after harvest. Firstly, yield maps such as those presently being developed automatically by yield monitors (such as in the USA etc.) can be used as guides to refine management both of fields and within a field. They can be used to identify management effects (e.g., variety, fertilizer, rotation, etc.) on yields, and so management can be refined. While yield monitors are not expected to be common, good record keeping of yields on a field- or parcel-specific basis could be used to provide such management guidance in small farms.

The second aspect relates to maximizing returns during the post-production system. While farmers can see their yield, much of the subsequent effects of their harvest and losses are unseen. For example, variety, harvest date and post-production management (e.g., drying, storage) can have large effects on head rice yields and quality (e.g., discolouration). These are factors that are essentially hidden from the farmers, who will likely lose interest once the grain leaves their farm gate. However, as premiums for quality (taste and head rice yields) become important, they will drive for improvement in the post-production systems, and add incentives for farmers (if credit systems allow) to maximize the added value of their crops. A major consideration in this is the method of harvest and handling. The major SSCM aspect of this is therefore variety selection and timely (i.e., optimum moisture content) harvest. Although options for byproduct use (e.g., hulls, straw) are generally known, they will require system (total farm) management. For example, straw removal (e.g., integrated animal systems, mushroom production) will require fertilizer substitution and straw enrichment for animals. Farmers will need greater information on variety effects on quality and the effects of harvest time (grain moisture content) and byproduct use options. This will require training to improve farmer's knowledge and incentives for the multiple players in the post-production chain. Mechanized harvest will likely be done by contractors.

I. HAM- Wireless communications

Use of internet, intranet and wireless communication in villages for knowledge revolution among farmers through high order transfer of technology (TOT). The web camera addition saves time travel etc., in getting the quick remedial measures from the research station or scientist for distant places.

J. Precision in nutrient management

Precision in nutrient management aims at optimization of nutrient resources and increasing the nutrient use efficiency through various means like soil testing, balanced application of primary, secondary and micro-nutrients, proper placement of fertilizers, water management, pest management, maintenance of optimum plant density, conjunctive use of bio-fertilizers and crop residues etc. The most accurate nutrient management is obtained in PF, also known as site-specific management. The PF technology allows the farmers to fine tune the application of fertilizers according to the needs and conditions within the field. This has become possible with the use of computer for management of large amount of data. Using the GPS, soil sample can be taken from within small areas and be accurately identified in a GIS. Fertilizer applicator or drill can use the GPS in conduction with soil test data stored in a GIS to apply needed nutrients to individual areas using variable-rate application equipment. Planting machinery can be equipped to change seeding rates in response to soil parameters in GIS, and irrigation scheduling can be linked to the GIS to apply water in response to soil infiltration rates or water-holding capacity. These systems are expensive and need further development, but they provide valuable tools that allow farmers to manage land, nutrient inputs and crop productivity more efficiently.

22.4.3.5 Nutrient management in Asia

In irrigated rice systems of south and southeast Asia, farms and individual fields are small. Only small equipment can be used. In the early stages of the Green Revolution, response to N, and later also P, was nearly universal. Blanket approaches have dominated rice nutrient management in the past. Fertilizer N recovery efficiencies average only 30 to 40%, unbalanced plant nutrition occurs, and yield growth has stagnated since the mid-1980s in some of the most important regions (Dobermann and Cassman 2002). Revitalizing yield growth, increasing profitability, and sustaining soil fertility are key targets for farmers in these environments. On-farm studies demonstrated that major uncertainties were caused by variability occurring among small fields in the supply of nutrients from indigenous sources and, consequently, the response to fertilizers (Angus *et al.,* 1990; Cassman *et al.,* 1996; Dobermann *et al.,* 2003a). Soil tests are of less use for predicting this variability because they fail to account for the dynamics of nutrient supply under submerged conditions and do not measure nutrient inputs from other indigenous sources such as irrigation or N_2 fixation (Dobermann *et al.,* 2003b). Site-specific management in these environments has focused on managing nutrients at the scale of a single, small field, including in-season N management decisions. One line of research has focused on corrective, in-season N management using tools such as a chlorophyll meter (Peg *et al.,* 1996). Typically, this involves no or a small application of N before planting and assessing leaf N status ("greenness") at intervals of 7 to 10 days beginning from mid-tillering stage of rice. Each time, a decision on whether to apply N is made, which requires local calibration to empirically define thresholds and suitable N amounts. A simple leaf colour chart (LCC) can replace the chlorophyll meter in this approach (Balasubramaniam *et al.,* 1999; Furuya, 1987; Singh *et al.,* 2002; Yang *et al.,* 2003). Evaluation of these methods has generally shown that the same rice yield can be achieved with about 20 to 30% less N application, whereas increase in yield is less common or relatively small. The gains in NUE are usually large.

Some of these uncertainties were addressed in a broader concept for site-specific nutrient management (SSNM), which integrated information collected at different scales for making field-specific decisions on N, P, and K management (Dobermann *et al.,* 2004). Key components are measurement of grain yield in nutrient omission plots to obtain field-specific estimates of the indigenous supply of N, P, and K (Dobermann *et al.,* 2003), a decision support system for predicting total NPK requirements and N splitting patterns before planting, and in-season upward or downward adjustments of pre-determined N doses at critical growth stages, based on chlorophyll meter or LCC readings. Variations in crop nutrient requirements due to regional differences in yield potential, location-specific crop management practices, and the overall nutrient input-output balances are taken into account in addition to spatial variation in indigenous nutrient supply. From 1997 to 2000, this strategy was evaluated in permanent on-farm experiments at 179 sites in eight irrigated rice domains of Asia (Dobermann *et al.,* 2002). On average, grain yield increased by 11% and the N fertilizer rate decreased by 4% as compared to the baseline farmers' fertilizer practice. Average profit increased by US $ 46 / ha per crop cycle (12%) and the recovery efficiency of fertilizer N increased from 30% to 40%. Benefits of SSNM varied widely among sites and were strongly affected by the overall crop management quality. Average yield increases ranged from 0.1 to 0.6 t/ ha among the eight rice domains studied. SSNM was most profitable at sites in China, southern India, and the

Philippines, with average profit increases ranging from US$57-82 /ha per crop in those domains. The performance of SSNM improved over time due to a gradual learning process and re-capitalization of investments in P and K applications made in the first year. Extra costs and other socio-economic issues may complicate the adoption potential of SSNM, but the gains in profitability appear to be attractive enough for farmers. More recent data show remarkable consistency of increases in yield, profit and NUE.

22.4.3.5.1 Site Specific Nutrient Management (SSNM) for irrigated rice

A. What is SSNM?

Site Specific Nutrient Management (SSNM) is an approach for feeding rice with nutrients as and when needed. The SSNM approach ensures application of fertilizers for location and season – specific conditions. With the SSNM approach, fertilizers are applied

- at right time
- in optimum amount
- in the optimum ratio and
- with the appropriate method
- to meet the deficit between rice demand for nutrients and the supply of nutrients from soil and organic inputs.

B. Need

The SSNM approach is needed for proper management of fertilizers in rice because,

- soil fertility varies from field to field, the demand for nutrient inputs therefore varies among fields.
- Demand for nutrient inputs varies from season to season, year to year and at different crop growth stages within a season

Such variations in soil fertility and demand for nutrients are not effectively handled with blanket fertilizers recommendations, consisting of fixed rates and timing of fertilizers for large areas. The SSNM approach provides farmers with guidelines for dynamically adjusting fertilizer application to local conditions.

C. Benefits

- Increase profit through optimal fertilizer use

- Sustain soil fertility through balanced application of nutrients (fertilizers).
- Reduce pest and disease incidence through balanced nutrient use.
- Protect soil, water and environment from contaminations by avoiding excess fertilizer use.

D. Components

The following tools and techniques are promoted for proper management of fertilizers.

1. Leaf colour chart (LCC) to determine the correct time and amount of N fertilizer application.
2. Omission plot technique to calibrate the P and K fertilizer rate.

1. LCC?

Using LCC, farmers can determine the right time of application of N fertilizer by matching the leaf colour with the colour strip of LCC. Farmers can decide on N fertilizer application, when leaf colour measurement falls below the critical value.

Guidelines

1. Start leaf colour measurement by LCC in transplanted rice at about 14 days after transplanting (DAT) and 21 days after seeding (DAS) in wet (direct) seeded rice.
2. Measure the colour of the top most youngest fully expanded leaf in ten randomly selected healthy hills as follows;
   - Select the top most fully expanded leaf in the selected hill
   - Place the middle of the selected leaf on the LCC colour strip
   - Do not detach the leaf from the plant
   - Place the leaf under the shade of your body to prevent falling of direct sunlight on LCC for accurate reading
   - Match the leaf colour with the LCC colour strip and record the number, matching the colour strip. The recorded number is the LCC reading for the leaf measured.
   - If the colour match falls between two-colour strips, take the mean of two colour strip numbers as LCC reading. For Example, if the leaf colour matches with LCC shade numbers 2 and 3, then the LCC reading will be taken as 2.5.

- Measure the leaf colour from each of the 10 selected hills in the same way and record the LCC readings.
- Repeat leaf colour measurement at 7-day intervals up to the flowering stage in respect of short duration varieties and 10 days interval up to flowering in respect of medium and long duration varieties. The last LCC measurement is at about flower initiation.
- Whenever more than 5 out of 10 LCC readings fall below the threshold LCC value (Threshold LCC value is 4 for varieties other than Improved White Ponni and hybrids, and 3 for Ponni and direct seeded rice), apply 35 kg N/ha in kuruvai and 30 kg N/ha in Thaladi seasons.

Plate 66. SSNM field view

2. Establishing and managing omission plots

The objective of P and K omission plots is to develop on site – specific P and K fertilizer recommendation. The difference in grain yield between a fully fertilized plot and a P or K omission plot illustrates the deficit between the crop demand for P or K and indigenous supply of P and K, which must be met by fertilizers. The field layout for omission plot is given in Fig. 14.

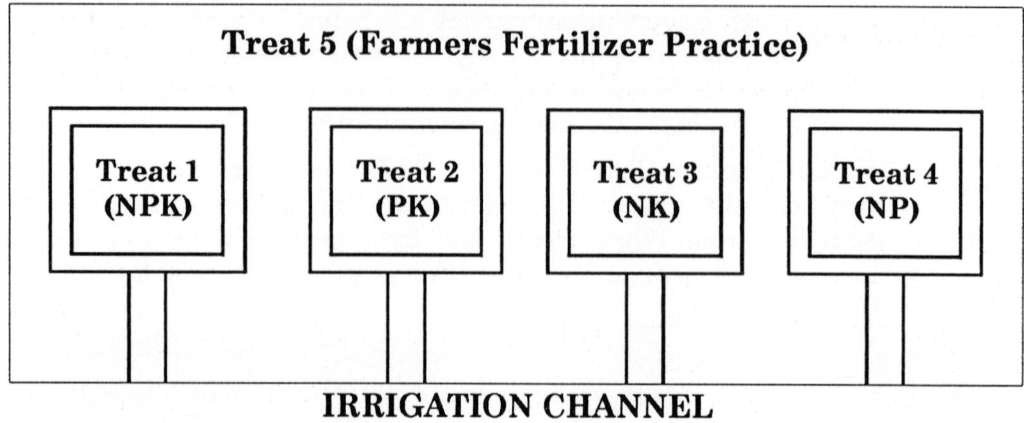

Plot size for each of treatments 1, 2, 3 & 4 = 5 m x 5 m (25 m^2)
Plot size for treatment 5 (remaining area of the field)

Fig. 14. Field layout of omission plot trial.

- Establish omission plots both in wet season as well as dry season.
- Measure four plots of 5m x 5m size each within a farmer's field. Do not place the plots in a corner of the field. Separate each plot by about 3 m distance from the next plot to minimize risk of fertilizer movement between the plots.
- Construct bunds of sufficient height surrounding each plot to prevent water flow between the plots and the adjacent field. Maintain the bunds nicely throughout the crop season to avoid fertilizers and water contamination from one plot to another or from outside the plot. Provide guard channel all around the plot. It is essential to prevent movement inward and outward of fertilizer with floodwater and irrigation water.
- All the crop management practices including pest management should be under optimal conditions and uniform in the treatment plots.
- Apply fertilizers to the plots strictly as per the treatment with the involvement of the farmers.

Plate 67. SSNM – Omission plot – Field lay out

Treatment details

T_1 : NPK : Recommended N, P, K, application without limitation.

T_2 : PK : Require P and K without limitation (ON) – N omission plot

T_3 : NK : Required N and K without limitation (OP)- P omission plot

T_4 : NP : Require N and P without limitation(OK)- K omission plot

T_5 : FFP : Farmers Fertilizer Practice in the remaining area

Determination of grain yield

At crop maturity, harvest the sample from all the five treatment plots in the following way :

- Harvest all the hills of a central $5m^2$ area of each treatment plot separately and properly label each harvested sample.
- Thresh and clean the harvested grain. Record the grain yield at 14% moisture level.
- Convert the sample weight to yield in t ha^{-1}
- Determine the P and K response

  P response = Yield in NPK plot – Yield in P omission plot

  K response = Yield in NPK plot – Yield in K omission plot

- From the P and K response, P & K requirement of the field will be calculated by the researchers based on the ready reckonar table as suggested by Fairthurst and Witt (2002) (Tables 70 and 71).

Table 70. Recommended P_2O_5 rates according to target yield and P-limited yield

Target yield (t/ha) →	4	5	6	7	8
Yield in 0-P plots (t/ha)			Fertilizer P_2O_5 rate (kg /ha)		
3	20	40	60		
4	15	25	40	60	
5	0	20	30	40	60
6	0	0	25	35	45
7	0	0	0	30	40
8	0	0	0	0	35

Table 71. Recommended K_2O rates at medium straw input*

Target yield (t/ha) →	4	5	6	7	8
Yield in 0-K plots (t/ha)			Fertilizer K_2 rate (kg /ha)		
3	30	60	90		
4	0	35	65	95	
5	0	20	50	80	110
6	0	0	35	65	95
7	0	0	0	50	80
8	0	0	0	0	65

*2-3 t/ha straw input

Source : Fairhurst & Witt, 2002. Rice: A practical guide to nutrient management

E. Advantages of SSNM

The SSNM strategy aims to achieve sustainable, large and economic yield through proper nutrient and crop management by

1. Following plant need based N management strategies using the Leaf Colour Chart (LCC) and avoiding over dose of N coupled with pest and disease incidence.
2. Using nutrient omission plots to determine soil indigenous nutrient supply (particularly for P and K)

3. Providing the crop with a balanced supply of nutrients (N,P,K and micro nutrients)
4. Replacing nutrients (Particularly P and K) removed with grain and straw to avoid depleting soil nutrient reserves.

F. Economics

Adoption of SSNM always recorded higher grain yield viz., averaging 500 to 650 kg/ha in Cauvery Delta of Tamil Nadu than in Farmers Fertilizer Practice (FFP). Therefore, the farmers can get Rs. 2250 to Rs. 3000/- per hectare as profit, besides a reduction in N fertilizer dose and avoiding excess cost on pesticide spray in SSNM plots.

Thus, SSCM has to include land preparation, crop establishment, water management, pest management, and nutrient management plus all their interactions as the basic components. Such SSCM will help to increase production and profitability while protecting the environment. The new system will, however, come with new demands in terms of management and extension expertise. In countries where SSCM has evolved, the technology is advancing fast, but the theoretical work has lagged behind. Development of sound procedures for information collection and accurate decision-making is insufficient and the long-term biophysical and economic benefits of SSCM remain to be demonstrated. Further, many of the input response functions still have to be verified under a range of conditions. Nevertheless, we believe that the concept of SSCM is right and that many of the initial difficulties can be overcome as the technologies developed. In China, SSCM may provide the basis of sustaining high rice yields in the coastal regions as well as achieving substantial yield increases in the central and western parts of the country where yields are currently lower. We believe SSCM can lead to substantial improvements in input use efficiency (i.e., product/unit input). One important opportunity will be to design integrated farms (e.g., rice-animal) where rice by-product use (e.g., straw) is examined in terms of the optimal flows of nutrients, water, energy, and labour at the whole farm level. SSCM aims at integrating the knowledge generated by various scientific disciplines into more complex but site-specific guidelines for action at the farm level. The principles of "precision farming" or SSCM are applicable to farms and fields of any size, including those found in Asia, but the specific technological solutions differ from case to case.

CHAPTER 23

RICE BREEDING AND BIOTECHNOLGY

23.1 BREEDING APPROACHES

23.1.1. Cytogenetics

The cultivated Asian rice *Oryza sativa* and the African rice *O. glaberrima* have the chromosome number 2n = 24. On the other hand, the wild *species O. latifolia, O. minuta, O alta, O. eichingeri* and *O. coarctata* have 2n = 48 chromosomes and are tetraploids. Hybrids between *O. sativa* and the wild tetraploid species *O. latifolia* and *O. minuta* have been sterile. So also, the crosses between the two cultivated species of *O. sativa* and *O. glaberrima are pollen* - sterile. Natural mutants in rice with 4n = 48 have been recorded. Triploid plants (3n = 36) also occur in the fields due to natural crossing of tetraploid and diploid plants. Haploids have also been recorded very rarely. Induced autotertraploids in rice have been produced in many countries by research workers. But due to the large size of the grains, which are unattractive, the reduced number of spikelets, partial sterility and fewer tillers, these autotertraploids are of no practical value. Twelve linkage groups of various characters in rice have been studied in detail by many scientific workers in different countries.

Characters in rice genetics

Tall and Dwarf : In crosses between tall and dwarf varieties, the segregation has been as 3:1 for tall to short. A segregation of 1:3 has also been reported in a similar cross.

Hairy and Non-hairy vegetative parts: Ratios of 3:1 and 15:1 for hairiness and non – hairiness have been obtained in F_2.

Tillering : When varieties with high and low tillering are crossed, the segregation in F_2 indicated the possible involvement of polygenes.

Lodging : The non – lodging habit has been reported to be controlled by a single recessive gene.

Floating growth habit : The crosses between normal and floating rice in deep water conditions have given 3:1 ratios indicating duplicate recessive genes are responsible.

Colour of vegetative parts : The colour of internodes, leaf sheaths and leaf blades segregate into 3:1, 9:7 and 27:37 ratios for purple vs. non – purple. A ratio of 15:1 indicating duplicate genes for colour in leaf sheath has also been reported.

Exsertion of the Inflorescence: The panicle exsertion and non – exsertion has been found to be controlled by three gene pairs with exsertion being dominant possibly by one major gene and two modifiers.

Lax v. Dense Panicle : In crosses between lax panicle and dense panicle showing complementary gene action has been obtained.

Awned and Awnless : An awned condition of the lemmas has been noted as the dominant character. Different ratios like 3:1, 15:1, 9:6:1, 12:3:1 and have been obtained by various workers. Inhibitory factors like 13:3 have also been noted.

Hairy of Smooth Glumes : The smoothness of the glumes is recessive to pubescence with the segregation ratio of 3:1

Grain characters: Generally, short grains are dominant over long grains and the grain length is governed by multiple genes. The starching nature of the grain is dominant over the non-starching or glutinous condition. The glutinous or waxy nature of the grain is due to a single recessive gene. Xenia has been noted in glutinous and non-glutinous crosses. The chalky and normal endosperm is controlled by two complimentary factors and F_2 ratio of 9:7 for resistance to susceptibility.

23.1.2 Conventional Breeding Methods

23.1.2.1. Varietal collection

There are eight thousand botanical varieties of rice in existence in the world and more than four thousand *indica* varieties have been identified in India. The collection and maintenance of as many varieties of rice as possible from different sources is the primary step for crop improvement. These are the initial materials for selection of suitable parents for economic and genetic studies. Direct introduction of rice varieties from one country to another has not been very encouraging. Introduction of varieties from one state into the other has been successful in many cases in India.

23.1.2.2. Mass Selection

Mass selection consisting of selecting phenotypically resembling, desirable individual plants and breeding them enmass, irrespective of their genotypic values, is now almost obsolete. A variety developed by this method will be pure only for physical features, which are easily seen and differ in quantitative or yield attributes. The improvement attained by this method is rather temporary and this procedure has to be repeated again. Hence, this method is no longer adopted for crop improvement in rice.

23.1.2.3. Pure line selection

Johnannson's pure line concept has been utilized in rice crop with the single plant as the basis for selection. The natural crossing in the cultivated rice in India is low and varies from 0.1 per cent to 4.0 per cent; due to the high self – fertilizing nature of the crop, inbreeding and maintenance of pure lines are easily achieved.

23.1.2.4. Hybridization and Selection

In the recent years, hybridization has played a very important role in the crop improvement of rice. Yield increase have been obtained by hybridization and so also quality improvement of the grain size, texture and colour. Phenomenal success has been obtained in the incorporation of disease resistance, resistance to lodging, drought and salinity by hybridization.

23.1.2.5. Inter Varietal hybridization

The easy crossability and the high fertility of the inter varietal hybrids in rice have afforded much scope for exploiting fully this procedure for combining desirable traits available in certain varieties. Among the cultivated species of rice, three geographical races, namely. *Oryza sativa* var. *japonica* and var. *javanica* have been recognized. These races are distinguished by their differences in morphological and physiological traits as well as by their photoperiod and fertilizer responses.

Indica – Japonica hybridization to combine the high fertilizer response, nonlodging and nonshattering grain characters of the *japonicas,* with hardiness and tropical adaptability of the *indicas,* has given successful results. Phenomenal success has been achieved by improving the plant type of tall varieties through hybridization of Taiwanese dwarf *indicas* with tall *indica* varieties.

23.1.2.6. Inter specific hybridization

While agronomically useful genes like yield and quality are found in cultivated plants, those constituting hardiness and resistance have to be sought in the wild species through interspecific crosses.

23.1.2.7. Ploidy Breeding

Ploidy breeding in rice has been attempted mainly to produce gigas character and to over come sterility in interracial and interspecific crosses. However, ploidy breeding in rice has not yielded much success in the isolation of economically useful traits.

23.1.2.8. Mutation breeding

Numerous instances of the occurrence of both spontaneous as well as induced mutations affecting almost every part of the rice plant and many of its physiological characters have been altered by X-ray and chemicals like Colchicine were the mutagens employed in the earlier years and almost all the induced mutations were recessive and uneconomical. Recent attempts with gamma irradiation have shown much scope for getting economic mutants artificially. Certain spontaneous mutations that were recorded in rice have been found economically useful. The first improved strain from Madras, namely, GBS 24 is reported to be a probable mutant isolated from a variety Konamani of Godavari district.

23.1.2.9. Heterosis

The exploitation of hybrid vigour in rice crop had not been widely taken up by rice breeders because of the self-pollination breeding mechanism of the crop and the practical difficulty in producing hybrid seeds on a mass scale. However, hybrid vigour has been successfully exploited in recent yeas in China. The identification of a few male sterile lines and respective maintainers has paved the way for exploiting the hybrid vigour. V. 20A, V. 41A, Zhen- Shan 97A and Er-jiu- man 1A are four cyto-sterile lines extensively using in this hybrid rice programmes. Research work is in progress in different countries to develop more cyto- sterile lines based on their agroclimatic conditions.

23.1.2.10. Breeding Techniques

For artificial crossing, a panicle from preferably a main tiller is selected. Panicles, which are on the second or third day of blooming, are the best for the purpose. The spikelets at the top of the panicle, which may be already fertilized, and the younger spikelets that are found at the base of the panicle are removed. After the removal of unwanted spikelets in the panicle is enclosed with a brown paper bag. As the temperature inside the bag increases the spikelets begin to open much earlier than under natural conditions. At this stage, the anthers are not fully mature to result in dehiscence since the opening of the glumes has been artificially hastened with the increase temperature. Artificial flower opening can be hastened by the 'wet cloth method'. By this method, the panicle is held tight with a wet cloth for about 10 minutes between 8 a.m. and 10 a.m. i.e., before the normal time of flower opening. Emasculation by clipping method is widely used; hot water or hot air methods are not popular. Suction emasculation by vacuum system is used for mass emasculation. The anthers are easily removed by means of fine forceps and emasculation is thus effected. If by any chance, the anthers show any dehiscence, the spikelets that group has to be discarded. The stigmas, which also protrude with the artificially hastened anthesis, are ready to receive the pollen. After dusting them with the pollen collected from the male parent, the whole panicle is kept enclosed in a muslin cloth bag to prevent access of any foreign pollen.

Plate 68. Rice trial nursery

Plate 69. Rice trial nursery after irrigation.

23.1.2.11. Tissue Culture

Haberlandt as early as 1902, saw the possibilities of totipotency in plant cells, which can be exploited, through the cultured cells. Earlier studies have mainly dealt with the physiological aspects of hormonal regulation of growth and morphogenesis. In recent years economic exploitation of tissue culture is becoming increasingly important in a variety of crops including rice. Genetic variation of plants is brought about mainly through mutational and wide hybridization techniques. The tissue culture techniques however, have been exploited in recent years in creating the variability with greater efficiency through mutation, recombination, selection, gene transfer, protoplast fusion, organelle and DNA uptake. Callus initiation through tissue culture in plant explants like root, shoot base, internodes, apical meristem, endosperm, embryo and anthers have been used in cereals. In rice, Japanese workers have worked with tissue explants like rice embryo and endosperm. Another culture in rice has been taken up recently in Japan and other countries including India. Anther culture is done mainly in F_1 generation. This is to produce homozygotes in second generation itself and thereby reduce the time taken for selection in conventional breeding methods. A few anther-cultured varieties have been developed and released for cultivation in Japan and China.

23.1.3 Bio Technological Approaches

23.1.3.1. Aromatic Rice

Modern plant breeding approaches have greatly helped in the development of improved, semi dwarf, high yielding aromatic rice varieties like Pusa Basmati 1, Kasturi, Haryana Basmati 1 etc. These varieties are gaining popularity gradually. Undue emphasis on the breeding of extra long slender export quality aromatic rice has neglected the improvement of other non basmati aromatic rice which are very popular in their native areas of cultivation. Right now the need is to realize the importance of our indigenous non-basmati aromatic rice and to undertake the noble task of their conservation and improvement. The leading aromatic fine quality rice in the world trade, popularly known as basmati rice is used for the special preparations like Biryani and Pulao. They are the most preferred and therefore, command a very high premium price in both domestic and international markets. In India, basmati rice is traditionally grown in the districts of Karnal, Ambala and Kurukshetra of Haryana, Kapurthala, Gurdaspur, Amritsar, and Pathankat in Punjab, Dehradun, Saharanpur, Bijnore. Muzaffarnagar, Rampur, Bareilly, Udham Singh Nagar and Pillibhit in Uttar Pradesh and Uthranchel. Basmati rice is also grown in Delhi, Rajasthan and Jammu and Kashmir. From a total area of about 7,00,000 hectares under basmati rice in the country, nearly 6,00,000 tones of milled rice are produced annually (DRR bulletin on basmati rice, 1994).

The initial approach in the improvement of aromatic rice was through single plant selection. Hybridization was used only after the development of semi-dwarf high yielding varieties. However, most of the programmes employed followed by pedigree selection for handling direct crosses and segregating generations. Three high yielding aromatic rice varieties released viz. Pusa Basmati 1, Kasturi and Haryana Basmati 1 were all developed by pedigree selection. Though these high yielding varieties have become popular among the farmers, they are not able to replace the traditional basmati rice varieties like basmati 370 or Karnal Local etc. To develop high yielding export quality aromatic rice, Plants with short stature, sturdy stem, more number of productive tillers, long slender grains, high head rice recovery, acceptable cooking quality, high volume expansion, soft texture, intermediate amylase content and resistance to major disease and pests are to be selected. This requires rare combination of genes, which can be achieved only through the screening of larger number of segregating populations. The quality analysis must be taken up from the F_3 grains. There is a greater scope to improve selection for semi dwarf plant height, better yield and early maturing plant type. Since the expression of aroma gene is depending on lower temperature at the time of flowering, (Asoka *et al.*, 1985), early maturing types can fit by late sowing.

Hybrid breeding seems to be promising in developing heterotic basmati varieties. Efforts are on to develop good CMS, maintainer and restorer lines. Experimental basmati hybrids have shown 20-25% yield superiorities over Karnal Local and Pusa Basmati–1. Genetic engineering and biotechnology hold good potential for plant breeding as it promises to expedite the time taken to produce crop varieties with desirable traits. With the use of molecular techniques it should be possible to hasten the transfer of desirable genes among varieties and to introgress noble genes from related wild species. Polygenic traits which were previously very difficult to analyze using the traditional plant breeding methods, can be easily tagged using molecular markets. Techniques which are particularly promising in assisting selection for desirable traits involve the use of molecular markers such as Random Amplified Polymorphic DNAs (RAPDs), Restriction Fragment Length Polymorphisms (RFLPs), Microsatelites and PCR based DNA markers such as Sequence Characterized Amplified Regions (SCARs), Sequence Tagged Sites (STSs), Amplified Fragment Length Polymorphisms (AFLPs) using F_2 and back cross populations, near isogenic lines, double haploid and recombination inbred lines.

Molecular techniques like RELP, AFLP, RAPD, PCR etc., and antisense RNA could be of great help to quality breeding programme. Recent successes in mapping genes for aroma (Ahn *et al.*, 1992) and Kernel length (Ahn *et al.*, 1993); molecular characterization of waxy gene and characterization or microsatilite sequence closely linked to waxy gene (Bilght *et al.*, 1995) have opened up the possibilities to accelerate the speed of aromatic rice improvement. Gene for aroma and elongation could be isolated from indigenous rice and transferred to high yielding, semi dwarf, and long slender grain background. Wide hybridization coupled with embryo rescue could be useful to evolve semi dwarf, long slender grain background. Wide hybridization coupled with embryo rescue could be used to transfer useful genes for disease and pest resistance from wild rice species to basmati background.

23.1.3.1 Research priorities

Most of the quality traits are quantitatively inherited and need continuous efforts to combine. Quality traits have also shown negative correlation with yield (Anon, 1997). Among the aromatic rice, genetic variability for different traits is also low with respect to resistance against biotic stresses. Some kind of intermitting in the segregating generations might be helpful in isolating the desirable types. Lot of genetic variability is available in the aromatic rice for quality traits. Some of the areas, which need an immediate attention to researchers, are as follows.

a. Preservation and exploitation of genetic variability

Collection, characterization, evaluation and maintenance of indigenous aromatic rice germplasm from being lost to other countries. Development of a database on all the indigenous aromatic rice germplasm is the need of the day.

b. Characterization of factors affecting aroma formation

All the factors which are responsible for aroma (environmental, cultural, soil or nutritional) must be characterized to produce better quality aromatic rice and to identify other non traditional areas for cultivation of aromatic rice e.g. basmati rice could be tested in non basmati aromatic rice areas like Siddharthnagar (native of Kala Namak).

C. Chemical characterization of aromatic compounds

There is a need to :

1. Identify aroma components and their distribution pattern in different aromatic rice cultivating areas.
2. Explore their biosynthetic pathways.
3. Develop rapid quantitative methods for aroma estimation and
4. Understand the factor affecting aroma trait.
5. Such studies would help to
6. address the problems of loss of aroma
7. identify non traditional areas of cultivation of aromatic rice
8. improve these cultivars with respect to aroma and other quality traits.

D. Non basmati aromatic rice

Available genetic diversity among non basmati aromatic rice remained unexploited. There is a number of small and medium grain aromatic rice, which can surpass basmati types with respect to several quality traits. Many of these can be cultivated in the areas where basmati cannot be grown. Small grain cultivar Tilakchandan is much superior to basmati types with respect to aroma, grain elongation on cooking test and digestibility, and it performs well under waterlogged conditions. Efforts must be directed to identify and promote such varieties for cultivation under different agroclimatic conditions and explore the international market for their export. Many of these are much popular than basmati in their native areas of cultivation. Some of the potential aromatic cultivars are Kala namak. Tilackchandan, Chinnar, Sakkarchini, Dhanla, Badshabhog (Utter Pradesh), Ambemohar (Maharastra). Champaran Basmati, Kesar, Sonachur (Bihar), Badshabhog (Bihar & West Bengal), Bindil (Water logged conditions of Utter Pradesh) and Vishnubhogs & Dubraj (Madhya Pradesh).

E. Development of heterotic basmati hybrids

Development of heterotic hybrids is also under way at New Delhi, Kapurthala and Pantnagar centres. Two basmati hybrids viz., Pusa 3A/PRR 69 and Pusa 3A/PRR 71 were found to be promising with 30-40% higher yield than Pusa Basmati and Kasturi.

F. Development of disease / insect-pest resistant varieties

Aromatic rice, both traditional as well as improved ones are highly susceptible to different biotic stresses. In case of severe infestation the crop damage is very high. This discourages the farmers to grow aromatic rice. Donors have been identified for biotic stresses but they are limited in number and poor combiners. Hence, there is an urgent need to identify the suitable donors and incorporate resistance against the biotic stresses.

G. Development of production and post harvest technologies

Proper agronomic and post harvest practices should be developed and standardized. Timely sowing and transplanting is critical for ensuring high yield and better quality. Package of practices for different agro-climatic conditions should be developed and optimized. Aromatic rice is relatively more susceptible to different storage pests. Unscientific storage leads to both quantitative and qualitative loss, therefore, proper drying, milling and storage technologies should be developed.

23.1.3.2 Basmati rice

The word basmati is derived from two Sanskrit words : VAS (aroma)and MAYUR (Ingrained from the origin); while combining, the later is changed to mati. In Hindi, Vas is pronounced as Bas, thus, the word "Basmati" is derived. Hence the presence of aroma of fragrance is essential for Basmati. However, all aromatic rices are not basmati. Basmati rice should be long slender, transparent with aromatic gains and should posses high volume expansion, linear kernel elongation, non sticky nature and soft texture on cooking (Table 72). During green revolution period, development of high yielding varieties (HYVs) of rice was concentrated to produce more quantity of food in order to cater the needs of ever increasing population and hence area under basmati rice decreased considerably. India after having attained the self sufficiency in rice production by seventies, basmati rice got in for export importance for their smell, non sticky and soft texture of cooked kernels. Besides, our policy makers realized its importance due to its potential. In mid sixties at IARI, New Delhi, a systematic basmati varietal improvement programmes was started by Dr. M.S. Swaminathan. Later on this programme was extended to DRR, Hyderabad and CRRI, Cuttack.

Table 72. Component quality characters and their minimum acceptable range of quality as Basmati

S1.No.	Component character	Minimum acceptable range
1	Aroma	Fine appealing smell at the time of cooking and in the cooked rice
2	Appearance and texture of cooked rice	Well separable, non sticky, no hard core after cooking.
3	Kernel dimension	
	Length (L)	6.5 mm or more
	Breadth (B)	2.0 mm or less
	L/B ratio	3.5 or more
4	Milling recovery	40% and above
5	Alkali spreading value at 27-30° C in 1.7% KOH	3-4
6	Water uptake	250 ml and above
7	Volume expansion	3.5 times and above
8	Kernel length after cooling	10 mm and above
9	Elongation ratio	1.5 and above

A. Varietal status of basmati and other aromatic rice

Almost every region in India has its own popular aromatic rice variety from centuries (Table 73). It may be basmati of north or Mologozhukulu of south of Prasadbhog of east or Ambemohar of west India. The salient features/ quality characteristics of some of the basmati rice varieties are given in Table 74. Small/medium grained aromatic rice is classified as non-basmati aromatic rice. Although, the grain length is now considered as one of the most important characters for export purpose the aroma is still rated as one of the most essential traits followed by taste and elongation after cooking. Though the Indian basmati is still rated as the best in the world markets and fetches high price (US$ 800-1000/tonne), it is likely to face stiff competition in near future from Pakistan, Thailand, USA and Australia. Most of the major rice growing countries are now emphasizing on the development of high yielding, semi dwarf, long slender grain aromatic rice varieties.

Table 73. Popular aromatic rice varieties

State	Name of variety
Andhra Pradesh	HR 22, HR 47, HR 59
Bihar	Badshabhog, Ramtuisi, Mohanbhog,NP 49, NP 14
	Basmati 23, Madhumalti
Himachal Pradesh	Ranbir Basmati
Jammu & Kashmir	Geeriga Senne, Gandhasall
Karnataka	Kallmoonchh, Chinoor, Jauphool, Dubraj, Banspatri, Kallikamod, Tulsiamrit, Kadamphool.
Madhya Pradesh	Ambemohar 102, 157 & 159, Chinoor, Jirasel,
Maharastra	Tuijapuri 1, Pankhali 203
Orissa	T 412, Badshabhog
Punjab & Haryana	Karnal Local, Basmati 370, Puse Basmati –1, Kasturi
Rajasthan	Basmati kota, Kalabadal, Nawabikalam, Kamod
Tamil Nadu	Jeeraga Samba
Uttar Pradesh	Basmati 370, Kasturi, Karnal Local, Kalanamak, Bhataphool, Pusa Basmati 1, Tericot

Table 74. Quality characteristics of some popular basmati rice varieties

Name of variety	Kernel			Milling (%)	Head rice recovery (%)
	Length (mm)		Elongation ratio		
	Uncooked	Cooked			
Pusa Basmati 1	7.20	13.91	1.93	70	55
Kasturi	6.94	12.37	1.78	69	61
Haryana Basmati –1	7.06	12.43	1.76	72	58
Basmati – 370	6.84	13.09	1.91	70	58
Karnal Local	7.13	16.25	2.28	68	40

B. Germplasm collection and evaluation

Although many of the short and medium grain aromatic rice are superior to basmati types in many respect, they are cultivated by farmers in small areas primarily for their personal consumption e.g. Kala Namak variety in eastern Uttar Pradesh. Due to stiff competition from other high yielding varieties and lack of priority in conservation and improvement programmes, most of the aromatic rice varieties and land races have gone out of cultivation. Most of traditional aromatic rice varieties are tall and photosensitive with average to strong aroma and 100 to 200% kernel elongation on cooking but have long growth duration and are poor in yield. Cultivation of these land races is confined to the localized regions. These are not well characterized and documented for different morphological and quality traits. Recently about 1000 aromatic rice germplasm from India and neighbouring countries were collected under the ICAR project entitled "Improvement of Basmati rice for increased Productivity and Export" and evaluated for yield and biotic stresses at Kapurthala, Kaul Pantnagar, R.S. Pura, Ludhiana and IRRI. After deleting the duplicates and purifying the mixtures, about 420 lines were selected which are being maintained at DRR, Hyderabad. Some germplasm lines are also being maintained at the aforesaid centres. In addition a large number of aromatic rice germplasm is also being maintained at CRRI, Cuttack and IRRI, Philippines. At IRRI, the aromatic rice germplasm collections were analyzed for isozyme variations. Maximum variation was observed for phosphoglucomerases and aminopeptidases. Most of the basmati types showed closeness to *Japonicas* but available variation indicated scope for further sub grouping among basmati types (Singh *et al.*, 1990). Green quality analysis of a set of 357 aromatic rice showed a wide range of variation for various traits cultivars. Under the basmati improvement programme, the available germplasm is exchanged among the research centres and screening is done for major diseases and pests.

C . Achievements

The varietal improvement of basmati rice was started in 1920's at Kala Shah Kaku, now in Pakisthan and at Nagina in U.P. All the earlier attempts were made to develop high yielding varieties through pureline selection from available land races. Cultivar Basmati 370 selected by Sardar Mohammed Khan in 1933 in Punjab is most classical example. A number of other aromatic rice varieties like Dehradun Basmati (T33), Hansraj, NB-10, N-12 etc. were developed in Utter Pradesh. In sixties, efforts were made to develop semi dwarf aromatic rice varieties by introducing dwarfing gene through hybridization followed by selection at Kaul, Kapurthala, Pantnagar and IARI, New Delhi under All India Coordinated Rice Improvement Project (AICRIP). A number of semi dwarf aromatic rice varieties were developed but most of these could not be compared well with Basmati 370 or T3 with the exception of aroma. They do not posses many of the export qualities. In the year 1989 world's first high yielding semi dwarf basmati rice variety Pusa Basmati 1 was released for cultivation. (Pusa 150/karnal Local). It's yield potential is 4.5 t/ha with 25-30% higher yield than other traditional basmati varieties. It matures in 135 days and possesses all essential qualities of basmati rice. Another aromatic rice variety, Kasturi, was also released in the year 1989 which combines higher yield potential, resistance to blast and other quality traits in semi dwarf back ground, but have mild aroma. Kasturi which is a derivative of the cross basmati 370/CR 88-17-1-5. It matures in 125 days and yields about 4.5 t/ha, which is more than a tone that of Basmati 370.

ICAR launched a basmati research network in the year 1990 entitled "Improvement of Basmati rice for increased productivity and export" at Kaul, Kapurthala, Ludhiana, IARI New Delhi, CRRI Cuttack, Pantnagar and DRR Hyderabad with the main objective of incorporation of genes for resistance to diseases and pests *viz.,* blast, BLB, WBPH and stem borer, development of production and post harvest packages and study of exploitable genetic variability among aromatic rice cultivars. Among the cultures *viz.* IET 10367, IET 10650, IET 11341, IET 11348, and IET 12019 were identified as promising. IET 10367 was released as Haryana Basmati 1. IET 11348, an early selection of basmati 370, was released as Ranvir Basmati in Jammu and Kashmir for general cultivation. Purified HBC 19 (Karnal Local) was released as Taroari Basmati in 1992 in Haryana state. Taroari Basmati is considered as the best quality export rice, primarily because of its extra long fine grains. Another variety, Basmati 385 developed in Pakistan was introduced in India through IRRI.

In 1994, ICAR started another project entitled "Genetic enhancement of quality rice for higher productivity and export" to improve the aromatic rice. Through this network, two cultivars *viz.* IET 13548 and IET 13846 were identified for On-farm trials (ICAR, 1996). Development of requisite lines for basmati hybrids is an important objective of the UNDP/FAP funded project on "Development and use of Hybrid Rice Technology". A CMS line Pusa 3A is developed and about 15-20 restorers with basmati traits have been identified. The experimental hybrids yielded 15-20% higher than the existing basmati varieties (Anon, 1997).

23.2. NON CONVENTIONAL BREEDING APPROACHES (ON RICE GRAIN QUALITY IMPROVEMENT)

23.2.1 Golden Rice

Golden rice refers to the genetically engineered rice that makes and stores vitamin A precursor in seed. It is called Golden rice because the grains of such rice appear pale yellow/golden in colour due to the accumulation of significant levels of Beta Carotene in the endosperm. Golden rice has been developed by the concerted efforts of the research team headed by Dr. Imgo Potrykus at the Swiss Federal Institute, Zurich and the German Scientist Peter Beyer. Dr. Potrykus transferred two genes called *Phytoene synthase* and *Lycopene cyclase* (from daffodil plant) and another gene called *Phytoene desaturase* (from a soil bacterium called *Erwinia carotivora*) through genetic transformation into the *Japonica* rice variety Taipei 309. The presence of these three genes ensures that a metabolite called geranyl pyrophosphate (which normally accumulates in large quantities in rice endosperm) is being converted to Beta-carotene. Beta-carotene is a precursor for the synthesis of an essential vitamin called Vitamin A. The deficiency of Vitamin A in human beings results in many disorders like impaired vision, night blindness, impairment of immune system etc. It is hoped that when Golden rice is consumed by the masses, there may be a drastic reduction in the number of people suffering from Vitamin A deficiency.

The Golden rice technology has been gifted by its developers to the developing countries for the benefit of its poor people suffering from Vitamin A deficiency. In India, we need to develop local varieties with vitamin A genes using this technology. Hence a programme for transfer of these genes into the background of Indian rice cultivars has been initiated by the Department of Biotechnology, Govt. of India under Indo-Swiss cooperation. Four research groups viz. Directorate of Rice Research, Hyderabad, Indian Agricultural Research Institute, New Delhi , University of Delhi, South campus, New Delhi and Tamil Nadu Agricultural University, Coimbatore are engaged in this effort of transferring the Golden rice traits into Indian rice cultivars. It is expected that Indian version of Golden rice may be available for cultivation by farmers in the next 4-5 years.

23.2.2 *Bt* Rice

Bt rice refers to the genetically engineered variety of rice that carries and expresses gene (s) from the ubiquitous bacterium *Bacillus thuringienesis*. These genes encode the proteins called crystal proteins or delta endotoxin, which are toxic to certain groups of insect pests. These toxins are least toxic to other insects and are safe for higher vertebrates like fish and mammals. Other *Bt* crops like cotton, corn and potato are a great commercial success and are now being grown on more than 50 million ha in over 12 countries. Though there are several *Bt* genes available, like *Cry* 1A, *Cry* 2, *Cry* 5 etc. Current *Bt* rice being produced and evaluated in India *Cry* 1Ab or *Cry* 1 Ac genes. The toxin produced which is effective in controlling lepidopteran pests like stem borers and leaf folders. Since there are no rice varieties with resistance to stem borer and chemical control is expensive and environmentally disruptive, *Bt* rice, it is hoped, will be readily accepted by rice farmers. However, there is a threat of insect developing resistance against *Bt* toxin, if *Bt* rice is not properly deployed.

In India, *Bt* rice have been developed by many ICAR and SAUs, Indian Agricultural Research Institute, New Delhi, Punjab Agricultural University, Ludhiana and Bose Institute, Calcutta. In addition, many *Bt* rice have been developed at the International Rice Research Institute, Philippines and are currently undergoing evolution under biosafety conditions in India. It is proved that *Bt* rice is safe for human consumption. The *Bt* rice developed both in India and abroad after successful evaluation and validation may enter the Indian markets in another 3-4 years only if the issues related to the Intellectual Property Rights of the *Bt* genes and techniques used in genetic transformation are satisfactorily sorted out.

23.3. PROSPECTS OF BIOTECHNOLOGY FOR IMPROVING SALT TOLERANCE IN RICE

Continuous increase in area under salt stress poses a great threat. About 25% of world cultivated land suffers from excessive salinity. Out of 30 m. ha in South and South East Asia under saline soil, only 16 m. ha are climatically, physiologically and hydrologically suited for rice cultivation.

Most of the saline soils in South East Asia are coastal soils, while in South Asia, they are Inland saline. A total of 10 m. ha of agricultural land alone is succumbed to salinity stress in India of which, 7.5 m.ha is under coastal salinity. Since the cost of reclamation of saline soil under vast area is very high, breeding salt tolerant cultivars seems to be the only viable alternative. The agro techniques for saline and sodic soils are given in the chapter 17. Conventional plant breeding so far, not made any major break through in releasing salt tolerant genotypes. Because of the lack of salt tolerant genes in cultivated rice varieties, transfer of genes for salt tolerance from wild relatives and traditional cultivars is being tried and it is not successful due to crossability barriers.

Screening of rice varieties from available germplasm for salt tolerance has been well documented (Mandal, 1990-91; Mishra and Singh, 1990-91; Yan and Tan, Wells *et al.,* 1991; Ahmed and Gupta, 1991 and Mandal *et al.,* 1993). About 90,000 accessions have been evaluated at IRRI (Balakrishna, 1994) and the donors for tolerance have always been the traditional cultivars (Pokkali, Nonabokra, Getu, Damodar, etc.). *Porteresia coarctata* T, a salt tolerant weedy halophytic wild relative of *Oryza sativa* can be used as a sourse for introgression of salt resistance gene(s) into high yielding cultivated rice varieties. But these gene pools remain unused so far, owing to their sexual incompatibility with *O.sativa.* Biotechnology has been identified as a potential tool to over come their barrier in addition to its application in short –circuiting the duration of plant breeding and creating new genetic variations through tissue culture. Transgenic plants engineered for salt tolerance through recombinant – DNA technique can be a direct approach. Besides, use of molecular markers for selection of salt tolerant plants is imperative, as there is no internationally standardized screening method conceptually fit for reliable screening and evaluation.

23.3.1 In vitro culture and somaclonal variation

In vitro culture in rice has received considerable attention during the last decade and several methods for regenerating plant via organogenesis or somatic embryogensis from different explant tissue have been reported (Yoshida *et al.,* 1983; Maherchandani, 1991; Chowdhury *et al.,* 1992; Binh *et al.,* 1992 and Tal, 1993). Induction of somaclonal variation has emerged as a potential tool for elucidation of various concepts of somatic genetics such as, function, expression and regulation of genes, as well as in the applied areas including its exploitation for crop improvement through broadening the spectrum of genetic variation and production of secondary metabolites of commercial interest. Since it arises through relatively mild conditions of *in vitro* growth, the genetic and chemical stresses applied on tissue cultures can lead to selection of tolerant cell lines from which the improved plants can be regenerated.

Yano *et al.* (1982) cultured immature embryos in the nutrient medium containing 17.5-67.5% sea water and obtained calli showing various degrees of salt tolerance. After four subcultures, they obtained salt tolerant plants from 17.5-37.5% sea water medium. Similarly, Viji *et al.* (1996) reported two salt tolerant lines, which had higher germination percentage, shoot length, root length, total dry matter production compared to IR 50 (control under salt stress). Proline accumulation, total protein content and enzyme activities e.g., acid phosphatase, peroxides and amylases were also significantly higher in the salt tolerant lines.

Under organogenesis, somatic embryogenesis, which is a single cell event, has its own merit specifically for breeding abiotic stress resistance (Salinity). This technique eliminates the absence of constraints of differentiation and diplontic and haplontic competition; the recovery variation is tremendously improved. The somatic embryo-derived regenerants are the true mutants (devoid of chimaerism) and serve as valuable test materials for reliable screening. Cells of a two year old suspension culture of rice grown under 1.5% NaCl stress for three months were reported to give rise the plants through embryogenesis, but higher salt stress caused physiological injuries and prevented the adapted embryogenic cells from differentiation beyond germinating stage (Binh *et al.,* 1993). It indicated that at least a lower level of salt tolerance could appear in rice even at a very early stage of ontogenesis. Production of large population of somatic embryos is essential for any fruitful screening against abiotic stresses. A number of efficient protocols are available for *in vitro* induction of somatic embryos. With the fast progress in automation in plant tissue culture, the bioreaction technology that use liquid media can also be employed for mass production of somatic embryos. This reduces the cost towards labour and media (Sharma Navin, 1992).

Vajrabhaya (1991) selected salt tolerant cell lines of *indica* rice by screening embryogenic calli on media containing 1.0 and 2.0% NaCl. In the third (R_2) generation of selecting plants, there was 94.0% survival as compared to 2.0% in the control under salt stress in the R_3 generation. Plants regenerated from tolerant calli were tolerant when compared to those derived from sensitive calli (Li, 1990) which envisaged association of cellular salt tolerance with whole plant tolerance. But contrasting reports do also exist. This is because for some instance, extensive somoclonal variation among regenerants that accompanies the selection process may be deterrent (due to their deleterious pleiotropic effects) to the successful recovery of heritable salt tolerance. However, somoclonal variation for slat tolerance has its genetic basis and is shown to be heritable in cell lines, regenerants and their progenies (Yamada, 1983; Narayanan and Rangasamy, 1990).

23.3.2 Anther culture

Anther culture technique encompasses creation of usable genetic variability and increases the selection efficiency to obtain quick homozygosity (an one stroke process) from even F_1. It provides fertile double generation haploid progeny with fixed desired character combination that might otherwise disappear in the course of an extended series of segregating generations. The characters, those expressing predominantly additive genetic effects, their values in double haploid lines could reach the heterotic level of the hybrid (Ba bong and Swaminathan, 1995). This technique has been used effectively to produce several commercial varieties of rice and wheat in China. Recently, in India, an anther culture rice variety 'Parag 401 (ACR 401) evolved from a cross between Prabhavati and Basmati 370 was released for irrigated ecosystem in Maharashtra (Patil *et al.,* 1997). Anther culture is in progress at CRRI, Cuttack; IARI New Delhi and CSSRI, Karnal. But not a single variety has been evolved so far for incorporating salt tolerance through this technique.

In vitro haploid induction from anther culture was attempted previously. In rice replacing sucrose with maltose and adding silver nitrate to the calus induction medium substantially increases the anther culture response of indica varieties (Lentini *et al.,* 1995). Besides, other conditions such as, optimization of media recipe, cold treatment, proper stage of anther selection, to name a few have been standardized (Hoan, 1992; Raina, 1992a; Sree rangasamy *et al.,* 1992; Gosal *et al.,* 1993; Zhenhua *et al.,* 1993; Mandal and Bandyopadhyay, 1996). However, the yield of double haploids per anther from *indica* varieties still about 20-fold lesser than that from japonica varieties. Anther culture response is reported by Miah *et al.,* 1985; Quimio and Zapata, 1990 and to be controlled by a few genes (oilgogenic). Lenthi *et al.* (1997) have been able to transfer anther culture response of *japonica / Indica* crosses, Draz *et al.* (1991) have shown that crosses having japonica as female parent could produce high callus induction efficiency. Reddy (1992b) employed tissue culture techniques for production of haploid rice plants from anthers and selected rice lines tolerant to salt stress. For selection of cell lines for salt tolerance, callus selection media were salinized with 1,2,3% NaCl and 25, 50 and 75% sea water. Plantlets were regenerated and TC_2 plants, when subjected to 13.5 ds NaCl/m. showed better tolerance than the control. There was differential genotypic response of 25 *indica* cultivars for callus initiation and green plant regeneration. Frequency of plant regeneration was the higher in Rasi (47.5%).

F_1 anthers from a crosses of a salt resistant (8-10dS/m) line, IR 13146-13-3-3 with BR 4608-R_1-R_2 and BR 4909 - R_1-R_2 (medium duration tall parents) were cultured following appropriate procedures (Mlah and Pathan 1993). Fifteen homozygous double haploid plants were identified from the regenerated plantlets. Seeds of such DH lines were grown under normal and salt stress (8-10dS/m) condition. Some of these DH lines emerged as salt resistant / tolerant. Recently, anther culture has been attempted to raise DH lines (following colchiploidy) from crosses of CSR 21, 22, and 23 with the salt tolerance donors CSR 10 and CSR 11 at CSSRI, Karnal. In India, the selection of such DH lines against salt stress is under way. DH lines produced through anther culture are more amenable to tag genes of interest by RFLP and PCR mapping (Lentil *et al.,* 1993).

Anther response is reported to be poor for in vitro culture in wide crosses (Sree rangasamy *et al.,* 1992) of *Oryza latifolia* (as male parent) with *O.sativa* varieties such as IR 50, CO43, IR64, IR 58025A and IR 62829. In certain combination such as crosses between *O. sativa* and *O. malampuzhaesls*, most of the pollens of R_1 hybrids were highly sterile as revealed from iodine – potassium iodide sustainability. Similarly, Jena (1994) was successful to obtain a hybrid between *O. sativa* var. IR 36 and *Porteresla coarctata*, but the hybrid was completely male sterile. However, anther culture at present, in wide crosses is not bright and warrants successful recovery of male fertile distant hybrids through elaborate hybridization followed by embryo rescue. Further, optimization of the protocol favouring good anther culture response is a prerequisite. Available literature do not reveal any report on anther culture of *O. sativa*/, *P. coarctata* cross combination.

23.3.3 Wide hybridization

Wide hybridization (intergeneric or inter specific) is sought for instances when there is lack of gene(s) for a trait in the available germplasm. Some progress has been made by using this approach to increase salt tolerance of cultivated wheat. However, in most cases such distant crosses are sexually incompatible and R_1 hybrids are seldom recovered or are highly sterile. This problem can be circumvented by embryo rescue and somatic hybridization.

23.3.4 Embryo rescue

Hybrid embryos are often formed but fall to develop further. The embryos rescued from degeneration by allowing them to grow on an artificial medium. Plants regenerated from such embryos carry a combination of genes of two species or genera. Mohapatra *et al.*, (1992) attempted wide crosses of *O. sativa* / *O. malampuzhaensis* and revealed that endosperm degeneration starts 7 – 14 days after pollination leads to embryo abortion. Therefore, 7 – 10 days old embryos are rescued and cultured in hormone – free ¼ strength (Murashige and Skoog, 1962) medium and incubated in dark. Crossing of a high yielding variety of *O. sativa* with a pure line of *Porteresia coarctata* T. to raise F1 following embryo rescue and back crossing of F1 with the *O. sativa* variety for five to six generations while retaining the gene(s) for salt resistance in the back cross derivatives with the common *O. sativa* genetic background, is a sure approach of improving salt tolerance. Embryo rescue is essential for at least first few back cross generations. Crosses between these two genera were unsuccessful until Jena (1994) reported a hybrid between IR 36 and *P. coarctata*. This triploid hybrid was much smaller than either parent and was completely male sterile. Sarkar *et al.,* (1993) observed that *P. coarctata* accepts *O. sativa* pollen readily, but not vice versa. Seraj *et al.,* (1996) were successful to raise triploid and tetraploid hybrids from *P. coarcta* (4x) / *O. sativa* (2x) and tetraploid *O. sativa* (2x)/*P. coarctata* (4x) respectively. Embryos were rescued 10-12 days after pollination and cultured on 1/4 MS medium with 0.45% agarose. The morphology of the triploid hybrids was close to *P.coarctata.* Whereas, tetraploid hybrids grew vigorously and were taller than tetraploid *O. sativa* varieties. Spikelets and flowering pattern of tetraploid hybrids resembled *P. coarctata* and other traits including leaf texture and shape were intermediate. Backcrossing is being carried out, followed by embryo rescue to introgress salt tolerance into *O.sativa* varieties. Still, optimization of tissue culture variables for better response of immature F_1 embryos (to be resized) is needed for fruitful employment of this technique especially for the above intergeneric cross combination.

23.3.5. Somatic hybridization

Fusing protoplasts of different origins to produce novel somatic hybrids provides new opportunities for study of somatic cell genetics and genetic manipulation of crop plants. This technique paves the way for transfer or combining useful character without detailed genetic or molecular knowledge of the genes underlying such characters (Jones and Karp, 1985). Introduction of desirable agronomic characters and also disease resistance from sexually incompatible wild species into cultivated potato germplasm has been successful by this technique. In Brassica, both at intergeneric (*B. juncea* + *Morlcandla arvensis*) and interspecific (*B. juncea* + *B. splnescense*) level, somatic hybrids have been successfully raised at Biotechnology Centre, New

Delhi for harnessing genes imparting biotic and abiotic stresses including salinity (Chopra and Sharma, 1991).

Extensive progress has been made in the understanding of properties and behaviour of protoplasts (Fowke and Gamborg, 1980 and Vasil, 1984). Protoplast isolation follows routine procedures (Potrykus and Shillito, 1985). Several workers have attempted and became successful in protoplast fusion through polyethylene glycol (PEG) mediation (Douglas *et al.,* 1981) and electrofusion (Bates 1989; Tempelaar and Jones 1985). Protoplast isolation from leaves of *P. coarctata*, its fusion with protoplast from cell suspension cultures of *O. sativa* and recovery of heterokaryons following adoption of advance selection procedures have been successfully accomplished in the Cocking's laboratory at Nottingham (Cocking, 1988). Use of metabolic inhibitors and more recently the nuclear and cytoplasmic DNA analysis has facilitated rapid selection and verification of somatic hybrids. For successful protoplast fusion, it is essential that both the fusion union (protoplasts) of different species or genera should have better response for protoplast isolation, its culture, capacity to reform cell walls, proliferate and regenerate plants normally diminishes over time. This causes difficulty in regeneration, because by the time, the selection and testing (hybridity) stages are completed, the selected cells may fail to regenerate. Hence, a rapid and reliable screening procedure for testing of hybridity is imperative.

23.3.6 Recombinant – DNA technology

Recent advances in molecular biology paves the way for transfer of any foreign gene into an organism, irrespective of its phylogenetic status. These genes have stable expression in the recipient and are inherited in the progeny both under laboratory and field condition. Since salt tolerance is assumed to be polygenically controlled, this technique would be of any utility if there is favourable mutation of a gene encoding the crucial enzyme involved in the biosynthetic pathway for production of osmoprotective metabolites leading stress tolerance. There is a great biological diversity for adaptation to high salinity stress ranging from the extreme halophytic bacteria and algae to mangrove vegetation of estuaries and coastal belt. *Dunallella viridis*-a green algae can thrive in the saline Dead Sea, whose salinity is eight times that of normal sea water (Rana, 1986). Salt tolerant genes have so far not been identified and characterized in detail in any crop plants. However, a great array of micro organisms harbouring stress tolerant gene(s) is available and these can serve as potential source of genetic variability. Of approximately 150 metabolites tested in bacteria, only the betaine series possess potent biological activity in promoting growth under strongly inhibitory levels of osmotic stress. The most active molecules include glycine-betaine and its precursors, proline- betaine. Free – proline is also active but under lower levels of stress. Betaine accumulation in salinised plants (e.g., spinach) is strictly proportional to external salinity (Coughian and wynjones, 1980). It is highly localized in cytoplasm, more possibly in chloroplasts, while salts accumulate in the vacuole (Andrew *et al.,* 1986). Betaine is a major osmoticum not only in some halophytic higher plants (Guy *et al.,* 1984), but also in halophytic bacteria (Galinski and Truper, 1982) and extraneous supply of betaine protects bacteria to thrive well in salt stress medium. It prevents damage of cells from cell dehydration by balancing the osmotic strength of the cytoplasm with that of the stress environment.

The first osmotic tolerance gene (*OSM*) identified is a proline overproducing mutation in *Salmonella typhlmurium.* The excess production of the amino acid (Proline) is accompanied by allosteric inhibition of the first enzyme of the proline biosynthetic pathway (r-glutamyl kinase) by proline and its analogues (Csonka and Baich, 1983). Several mutations for proline overproduction have been identified in the structural gene (*pro B*) for r-glutamyl kinase. The mutant allele pro B 74, which gave the most pronounced osmotic tolerance, is a single base pair change causing a 100 fold loss of sensitivity of r-glutamyl kinase to feed back inhibition by proline. Similarly, osmotic mutation of a gene (*bet / osm*), which is the crucial enzyme, involved in the synthesis of glycine – betaine from chollne has been mapped to lac operon in the *E.coll* linkage map. Selection of proline and / or its analogue (s) overproducing mutant cell lines has been reported in *Nicotiana sylvestris* (Dix *et al.,* 1984), *H. vulgare, Solanum tuberosum, Brassica napus* and *Vigna radiata.* The above findings in micro organisms and crop plants lead to the possibility of isolation, cloning, and transfer of genes encoding stress resistance by suitable vector. The success of genetic transformation entails selection of totipotent recipient cells for transformation that are capable of callus formation and regeneration into plantlets, modes of delivery of DNA, identification and isolation of a few markers for selection.

In all genetic transformation experiments, the integration of donor DNA appears to be random and like transposon, it may inactivate some important gene's expression or a whole operon itself due to its position effects. This problem can be mitigated by a highly promising technique, targeted homologous recombination. This involves transfer of a cloned DNA containing a mutated copy of a target gene into the cells, which somehow finds its target in the cell's nucleus and replaces it in its normal chromosome context (Welntroub, 1990). Site specific recombination systems are well known between bacteriophage and bacteria (e.g., *E. coli, Salmonella* spp.). This involves integration of phage DNA into bacterium genome via a pathway utilizing enzymes that recognize specific sites and catalyze recombination events between them (Suzuki David *et al.,* 1989). Similarly, Ganesan(1985) proposed to construct artificial minichromosomes in rice, which could be used to transform the haploid cells in another culture and then be made homozygous diploid cells for regeneration of whole plants. Microinjection and electroporation have widened the scope of genetic manipulation through direct transfer of desired gene(s). The protocol for plant protoplast transformation is given by Potrykus *et al.,* (1985). Recently, Reddy (1992a) employed electroporation through somatic hybridizer involving T 309 (*a japonica* rice cultivar) suspension culture derived protoplasts and plasmid containing GUS and NPT II genes. Transgenic rice plants can also be produced by PEG-mediated plasmid uptake into protoplasts (Li *et al.,* 1995).

A simple, but efficient reproducible transformation system can be adopted for production of transgenic plants in rice. A large number of transformed plants have been regenerated using micro projectile bombardment of even young callus and immature embryos with plasmid DNA containing genes sought for transfer (Wan and Lemaux, 1994). DNA hybridization analysis confirmed integration of the transgene(s) in callus and 84-85% transformation frequency for the gene(s) have been reported. Most transgenic callus lines carried multiple copies of the transgenes and most lines were regenerable. Transgenic tobacco plants with high osmotolerance (drought and salinity) have been produced by Kishore *et al.* (1995). They could over express the enzyme, pyroline-5-carboxylate synthetase that catalyzes the conversion of glutamate to pyroline-5-carboxylate. The latter, then reduced to proline. The transgenic plants produced a high level of the enzyme and synthesized 10-18 fold more proline than the control plants. More recently, expression of a late embryogenesis abundant protein gene, HVA 1, from barley conferred tolerance to water deficit and salt stress in transgenic rice (Xu *et al.,* 1996). Large number of such transformed plants were produced following biolistic – mediated transformation method. The barley HVA 1 gene was regulated by gene promoter and led to more accumulation of HVA 1 protein in both leaves and roots of transgenic rice plants.

23.3.7 Molecular marker assisted selection

For effective employment of plant cell, tissue and organ culture with a view to identify putative salt tolerant genotypes, a reliable selection criterion through *in vitro* culture and at whole plant level under salt stress is very important. The most widely used approach in seeking gene(s) related to stress resistance entails the assumption that novel mRNAs and proteins that appear in response to an environmental stress. The approach has great advantage that it can lead quickly to gene isolation, either via DNA clones (Geriach *et al.,* 1982) or via amino acid sequences of the stress related proteins, which can be used to construct synthetic oligonucleotide probes. Since salt stress tolerance may be largely a property of plasma membranes (Tripathy and Kar, 1995) and tonoplasts, any unique salt stress induced protein may be associated with these structures. These proteins are essentially required for proliferation of cells in saline medium. Adaptation to salt stress elicits a relatively small set of cellular proteins. The most active one is the 'osmotin' – a 26 KD protein that accumulates in vacuoles of salt stressed plants. But, its biological activity is not yet known. Saqlan Naqvl *et al.* (1992) employed SDS-PAGE analysis of roots excised from salt stressed and non-stressed plants of a salt tolerant IIRIs line, IR 10198-66-2 and observed significant induction of at least two proteins both with isoelectric point (PI) of about 7.0 and having 27,000 and 25,500 KD molecular weight in the salt stressed plants. No such proteins were observed in the control gels at the same location.

Changes in stress inducible protein profile pattern has been shown in salt sensitivity 'Randhunipagalu' and salt tolerant 'Getu' by Reddy (1992a). In the salt tolerant variety, he observed 13 new (induced) polypeptides of which, the 26KD polypeptide was significant. Two dimensional analysis of *P. coarctata* seedlings also revealed the induction of a 26 KD polypeptide under salinity stress. Whereas, no such common demarkable (26KD) polypeptide was observed in sensitive cultivar, but instead, a single 15KD polypeptide was induced in shoot. No difference in protein profile pattern in root proteins was revealed through 2-D analysis. Reddy *et al.* (1992) with another set of material also observed salt stress (NaCI) responsive polypeptides (SRPs) with molecules weight of 15 and 26KD polypeptide is the same as the one induced by a variety of other stresses including ABA stress and desiccation caused by air drying at the seedling stage.

If there are no morphological markers, the next approach is tagging genes of interest through RFLP (Restriction Fragment Length Polymorphism) mapping. The technique offers an unique opportunity like restriction enzymes, which can cut DNA at specific sites, can distinguish between plant strains even if they differ by a single base in the restriction enzyme site. The introgression of important genes from wild species into cultivated plants (through back crossing) is the most significant contribution of RFLP technology (Tanksley *et al.,* 1989). This feature is well illustrated in rice in which a RFLP map already exists (McCouch *et al.,* 1988). The RFLP markers that are highly linked to traits of interest can be used to select back cross derivatives. RFLP mapping of QTLs controlling N_2- fixation (Wu *et al.,* 1995), seedling vigour (Redona and Mackill, 1996), agronomic traits including yield (Xiao *et al.,* 1996), resistance to sheath blight (Xhikang *et al.,* 1995) and root growth parameters for drought avoidance (Champoux *et al.,* 1995) in rice has been done.

Invitro tissue culture systems provide the opportunity to observe and select large populations of variants within the limited time and space a definite advantage over screening of large plant populations in the field. It has been further shown that compared to conventional mutation breeding programme, tissue culture may result in mutations at a higher rate. This techniques has facilitated genetic manipulation of both inter-and intragenomic level utilizing the source of genetic variability for salt tolerance from diverse origin. Elucidation and use of special proteins and mRNAs induced under salt stress and more precisely the RFLP markers are quite helpful in determining the genetic status of *in vitro* derived regenerants for salt tolerance. The biotechnological approaches considered in the present context could be a supplement to the conventional plant breeding and hasten the progress of genetic improvement and for salt tolerance, in particular.

APPENDIX - I

COST OF CULTIVATION FOR RICE

Working out cost of cultivation for one ha of transplanted rice: For e.g., Medium duration ADT-39 (120-135 days)

S.No.	Particulars of operation	Farm Pair / Tractor	Labour required		Unit rate* Rs.	Amount*
I.	**NURSERY PREPARATION** (20 cents)		A		B	
1	Ploughing, Puddling and leveling the nursery area (20 cents)	1	2	-	100 72/ea	244.00
2	Cost of FYM 1000 kg and spreading	-	2	-	350 / MT	494.00
3	Trimming, Plastering, Digging corners in the nursery	-	2	1	72/ea,72/ea	216.00
4	Seed material 40 kg	-	-	-	Rs. 12/-kg	480.00
5	Seed treatment with chemicals	-	-	-		79.60
	I. Carbendazim 80 g				Rs. 62 for 100g	
	II. Azospirillum – 3 pkts				Rs. 10/pocket	
6	Basal application of DAP 40 kg	-	-	-	Rs. 9.73/kg	389.20

7	Labour for seed treatment, DAP application, sowing	-	2	-	72/ea	144.00
8	Pest management	-	-	-		
	I. Butachlor (or) Thiobencarb 160 ml (Herbicide) 8th DAS				170/lit	27.20
	II. Monocrotophos 36 WSC 40 ml + labour for pest management	-	1	-	316 /lit	84.64
9	Pulling out of seedlings	-	1	10	72/ea. 72/ea	720.00
II.	**MAIN FIELD PREPARATION- one ha**					
1	Ploughing with tractor	3hrs/ha	-	-	150/hr	450.00
2	Application of FYM @ 12.5t /ha	-	-	-	350/MT	4375.00
3	Labour for spreading FYM	-	4	4	72/ea	576.00
4	Puddling with tractor	4 hrs/ha	-	-	150/hr	600.00
5	Trimming field bunds, digging the	-	10	-	72/ea	720.00
	corners and plastering bunds, seedling dipping in Azospirillum and its cost –5 pockets	-	-	-	10/pkt	50.00
6	Transport & planting of seedlings	-	2	30	72/ea 72/ea	2304.00
III.	**PERTILIZER APPLICATION**					
1	Recommended dose of 150:50:50kg N:P_2O_5;K_2 O/ha					
	Urea 330.00 kg/ha				5.00 /kg	1650.00
	Super 312.50 kg/ha				3.25 /kg	1015.63
	MOP 83.50 kg/ha				4.60 /kg	384.10
2	Labour for applying fertilizer 1- Basal & 3 times- Top dressing-	-	3	3	72/ea	432.00
3	Basal application of Zinc sulphate 25kg/ha	-	-	-	Rs. 18.40/kg	460.00

4	Cost of Gypsum for basal – 500 kg/ha	-	-	-	0.86/kg	430.00
5	Cost for Soil application of Azospirillum 10 pkts/ha	-	-	-	10/pkt	100.00
6	Labour for applying Znso4, Gypsum and Azospirillum	-	1	2	72/ea,	216.00
IV.	**WEED MANAGEMENT**					
1	Pre-emergence herbicide application Butachlor 2.5 l/ha	-	-	-	170/lit	425.00
2	Labour charge for application	-	1	1	72/ea,	144.00
3	Second weeding on 30-35 DAT	-	-	20	72/ea	1440.00
VI.	**IRRIGATION MANAGEMENT**					
1	Labour for irrigation- total cost	-	15	-	72/ea	1080.00
VII.	**PEST AND DISEASE MANAGEMENT**					
1	Spraying Monocrotophos at 100 ml/ha				316/lit	316.00
2	Spraying of Carbendazim at 1000 grams/ha				260/500g	520.00
3	Dusting Sevin 10% at 25kg/ha				10/kg	250.00
4	Labour cost for pesticide application/spraying		3	3	72/ea,	432.00
VIII.	**HARVESTING**					
1	Labour for harvesting, bundling, transporting & threshing		15	45	72/ea,	4320.00
2	Labour cost for winnowing, drying, bagging and staking		5	10	72/ea,	1080.00

***subject to change.** The cost of cultivation of medium duration variety ADT-39 was worked out, considering the prevailing cost of inputs, cost of labour etc., at Lalgudi Block, Trichirappalli, Tamil Nadu.

Abbreviations used: ea-each person, hr-hour, MT-metric tone, lit- litre, pkt- pocket, kg-kilograms, g- gram.

EXPENDITURE : Cost of cultivation = Rs. 26648.37

INCOME

Yield of paddy	=	8 t/ha
Marketing price	=	Rs. 5.50/ka
Yield of paddy straw	=	12 tones/ha
Marketing price	=	Rs. 500/ton.
Gross income	=	Rs. 50,000

ABSTRACT

Field operations		Amount (Rs.)	Percentage over expenditure
I.	Nursery preparation	2878.64	10.80
II.	Main field preparation	9075.00	34.05
III.	Fertilizer application	4687.73	17.59
IV.	Weed management	2009.00	7.54
V.	Irrigation management	1080.00	4.06
VI.	Pest and Disease management	1518.00	5.70
VII.	Harvesting	5400.00	20.26
Total		26648.37	100.00
VIII.	Other charges		
Land tax (wet land)		250.00	
Supervising charges		750.00	
Interest rate of capital		708.75	
Total		28,357.12	

COST - BENEFIT RATIO

1. Net profit = Gross income – gross expenditure

   Gross income = Rs. 50000.00

   Gross expenditure = Rs. 28357.12

   Net Profit = Rs. 21642.88

2. Cost benefit ratio = $\frac{\text{gross income}}{\text{gross expenditure}}$ = 1.76

Return per rupee invested = 1:1.76

GLOSSARY

The glossary given in this part, contain the following list in alphabetic order. The glossary in relation to rice science alone are given in this book. For a more complete glossary, we refer the readers to the books like Agriculture: Facts and Figures (by Subbaian, Annadurai and Palaniappan, 2000-published by Scientific Publishers, Jodhpur, India), Agronomic terminology (by Indian Society of Agronomy, New Delhi), Glossary of terms in crop production (by Ramamoorthy, Annadurai and Subbaian, 2005 published by Scientific Publishers, Jodhpur, India, 420p) and the web site :www.irri.org.

GLOSSARY OF RICE AND RICE BYPRODUCTS

Amazake- As a beverage or ingredient in Japanese cuisine, this thick fermented liquid is made from cooked rice.

Amylopectin-This is a type of starch found in highest amounts in medium, short, jasmine, and waxy rice. Makes rice more tender with a greater tendency to stick or cling together.

Amylose-This type of starch is found primarily in long grain rice, which makes long grain rice cook up fluffy and separate with a lesser tendency to stick or cling together. Raw or uncooked grains with a high content of amylose starch tend to look translucent. Uncooked grains of rice with a low content of amylose starch look opaque.

Arborio Rice--A type of rice, named after a town in northwest Italy, that is one of the most popular varieties used for Italian risotto recipes. Arborio rice is a short-grain variety that is very plump and like other Italian rice varieties, it contains more starch than many other types of rice. The starch is an important component for creating creamy dishes, such as risotto, so it is important not to rinse the rice before cooking or else much of the starch can be washed away. Besides risotto, Arborio rice is used in dishes such as stir-fried rice, sushi, pudding, and recipes where rice is molded into shapes. It is available in well-stocked food stores and Italian markets.

Aromatic Rice-Brown or white rice with a natural aroma and flavor similar to that of roasted nuts or popcorn. Various types grown; cooks dry and separate or moist and tender.

Black and Red--Rare Asian grains with black and red husks and bran coverings; also cultivated in California. This exotic grain is normally added to commercial blends.

Black Japonica-A registered trademark. This type of rice is musky, spicy, and aromatic. The rice was developed from a black short-grain variety of Japanese rice, and a mahogany-colored medium grain type of rice both grown together in the same field. This rice is typically used for rice stuffing or other side dishes.

Boil-in-Bag Rice-This type of rice is usually parboiled and pre-cooked. It is pre-packaged in a colander-style plastic bag. The product is pre-measured for quick, convenient cooking in 10 minutes.

Brewers' rice -The portion of total milled rice consisting of broken pieces which can pass through a 5 1/2 /64-inch sieve, used in the process of brewing beer and other fermented products.

Brown rice -It is the least processed form of rice. It has the outer hull removed, but still retains the bran layers that give it that characteristic tan color and nut-like flavor. The outer layer of the bran gives this rice a chewier texture than white rice.

Calas-The word "Calas" originates from the African word for "rice", and refers to a deep-fried pastry made from rice. This dish is popular in New Orleans, Louisiana.

Calrose-Type of medium grain rice grown in California.

Della Rice - A cross of long grain rice and basmati rice which was developed in the United States. It has an aroma and flavor similar to basmati and cooks dry and separate.

Dolmades – An Arabic term meaning "something stuffed", dolmades is a Greek dish prepared of grape leaves stuffed with rice, meat, lentils, and seasonings. This dish is typically served as an appetizer or as an entrée.

Enriched rice -Blend containing ordinary milled rice and a small percentage of milled rice heavily fortified with thiamin, niacin, and iron phosphate to raise the vitamin and iron content slightly above the level present in brown rice. When the yellow-colored riboflavin is added to the enriching agents, white pigments such as calcium oxide, talc, and titanium dioxide are also included to make the finished product appear white.

Glutinous Rice-White, brown, or black rice characterized by a broad, short grain; which stick together during cooking; mainly used in Asian sweet snacks; also known as Sweet rice, Sticky rice, or Waxy rice.

Head rice -The portion of total milled rice consisting of whole kernels

Kasmati Rice -A long-grain variety of aromatic rice developed in the U.S. that is considered to be similar to Basmati in taste and texture. It is served in Indian and Middle Eastern food dishes.

Jasmati Rice- A hybrid variety of aromatic rice that is a cross between Thailand's jasmine rice and India's basmati rice. It has been developed from a variety of rice known as Della that is related to Bertone rice from Italy, to be grown in the warm climates of the southern U.S. This rice is soft textured with a mild rice flavor.

Organic rice- Organic rice is rice that is certified by an independent body, to have been grown and processed according to set "organic" standards. While there is no single definition for "organic", most definitions share common elements. For example,"organic" as applied to most field crops generally means that: 1. No synthetic or artificial chemical pesticides and fertilizers have been used,2. Soil fertility is maintained through "natural"processes such as growing cover crops and/or the application of composted manure and plant wastes. 3. Crops are rotated in fields to avoid growing the same crop year after year in the same field. 4. Alternate (non-chemical forms) of pest control are used to manage insects, diseases and weeds – e.g., beneficial insects to prey on pests, mulches to suppress weeds, etc.

Parboiled rice - Rough rice which has been subjected to a steam or hot water treatment prior to milling; parboiling increases the percentage of head rice and the vitamin content of milled rice. This procedure gelatinizes the starch in the grain, and results in firmer more separate grains. It also takes a few minutes longer to cook.

Pearl Rice - California short-grain; kernels very plump and almost round; white only. Grains are soft and stick together when cooked.

Polished Rice - Regular milled white rice, often referred to as "white" or polished" rice is the most common form of rice. The outer husk is removed, and the layers of bran are milled until the grain is white.

Popcorn Rice – Grown in Louisiana, this long-grain rice is a cross between Basmati and regular rice. It tastes like buttered popcorn, but the scent is stronger than the flavor.

Precooked rice-It has been milled, cooked, and dehydrated before packaging. Precooked rice, in its dried form, is more porous so that boiling water can penetrate the grains and rehydrate them in a short time.

Red rice- An annual grass, adapted to an aquatic habitat, that reproduces by seed. The first seedling (primary) leaf is green, cylindrical, and lacks a blade. The second leaf has a sheath, blade, ligule, and auricles. Leaves have short, stiff hairs on their upper and lower surfaces. The mature plant has round, hollow, jointed culms, and rather flat, sessile leaf blades. The leaves have short, stiff hairs on their upper and lower leaf surfaces. The plants tiller profusely, becoming brushier than white rice plants. Panicles are loose and open and droop slightly. The grain shatters easily when ripe. The many types include those with short, medium, or long grains; those with straw-colored, red, or black hulls; and those with short or long awns on the spikelet.

Rice bran -Coarse outer bran layers, the embryo (germ), and small bits of the starchy endosperm. It is an ingredient in cereals, mixes, and vitamin concentrates, and non-food grades of bran are used in livestock feeds.

Rice bran oil-It is a high quality cooking oil extracted from rice bran.

Rice flour -It is ground milled or brown rice. It is non-allergenic, making it valuable as a wheat substitute for those allergic to gluten and wheat flour products. It can also be extruded to produce rice pasta, chips and other snacks, as well as breakfast cereals.

Rice hulls -These are the outer covering or husk layer that encloses the rice kernel. Hulls are inedible, but can be burned as a source of fuel for power plants or rice mills, used as a mulch, or a component in abrasives or other products.

Rice polish (white bran) -Inner layers of the bran removed during polishing

Rice Cakes – This is a low-calorie snack made of puffed rice.

Rice Cereals – Are whole grain rice, puffed or crisped, and generally used in breakfast cereal, rice cakes, or even candy bars.

Rice Couscous – Cracked rice, which resembles a cracked wheat product called *couscous.*

Rice Diet – A low fat, low salt diet. For more information, try.

Rice Flour – Rice is ground into flour, which can be used in place of wheat flour for baking. Rice flour is gluten-free and non-allergenic. This type of flour has a wide variety of uses such as baby foods, sauces, gravies and cosmetics.

Rice Milk – Is a dairy substitute, and can be used as a beverage, or a wide variety of recipe applications.

Rice Noodles – Are made from rice flour and typically used in Asian cuisine.

Rice Paper – Is dough made from rice flour and water. Rice paper is used as the wrapper in Vietnamese and Thai spring rolls.

Rice Starch – Is a major component of milled rice. Rice starch can be used as a thickener in making sauces or desserts.

Rice Syrup – Is a mild sweetener, which is easily digestible. Rice syrup may be substituted for other sweeteners when baking.

Rice Vinegar – Made from rice, this vinegar has a gentle, but tart flavor and is milder than other types of vinegar.

Rice Wine – Is a sweet, golden wine made from fermenting freshly steamed glutinous rice.

Rijsttafel – The actual word 'Rijsttafel' is a combination of two Dutch words, which when literally translated means "rice table". This word describes an elaborate Indonesian meal, which consists of rice and accompanying dishes such as meat, seafood, and vegetables.

Rosematta - Left-on bran gives this *parboiled* rice from South India an attractive rose color. Sought after for its firm texture and warm flavor.

Rough Rice – Rice as it comes from the field. Also, known as paddy rice, the rough rice kernels are still encased in an inedible, protective hull, which must be removed. The hulls can be recycled and used in electrical co-generation.

Rice starch -It is the major component of milled rice. From the endosperm of the grain, it makes up 90 to 93% of the milled dry weight. It is used as a thickener in making sauces and desserts, and can be hydrolyzed to make a sweet syrup.

Screenings -The portion of total milled rice consisting of broken pieces about 1/4 to 1/2 the length of a whole kernel

Second heads -The portion of total milled rice consisting of broken kernels at least half as long as a whole one

Starchy endosperm -Bulk of the endosperm within the aleurone layer, consisting largely of starch granules embedded in a proteinaceous matrix. The starchy endosperm also contains sugars, fats, and fibers. The bulk of the starchy endosperm survives the milling and polishing process as the milled white rice.

Saffron – Derived from the dried stigmas of a purple-flowered crocus *(Crocus sativus)* used to color and flavor foods.

Sake – Traditional Japanese wine made of fermented rice.

Short Grain Rice - Brown or white rice kernels averaging about 5 millimeters in length that are much thicker than long-grain varieties. Grains are soft and stick together when cooked. Also, referred to as "*Valencia Rice*".

Sushi – Japanese dish of cooked and seasoned rice served with a variety of other ingredients, which usually includes fish.

Tahdeeg – When cooking rice, sometimes some of the rice sticks to the bottom of the pan, Iranians call this Tahdeeg. It is prized and is eaten in a variety of ways.

Total milled rice (polished rice) -Bulk of the starch endosperm consisting of head rice, second heads, screenings, and brewers' rice.

Wehani - A California-grown hybrid rice with Indian basmati in its ancestry; reddish color; very nutty aroma and flavor.

Whole Grain – Brown Rice is a whole grain - nothing is added, nothing is removed from the kernel except the outer hull.

Wild Pecan - An aromatic long-grain rice grown in Louisiana. It is named for the strong aroma and milder, but distinct nutty taste.

Wild Rice - Is the seed of a plume-topped wild aquatic grass found mainly in the North Central United States and Canada. It is expensive due to short supply, hand gathering, thrashing, and is often mixed with other types of rice or grain.

ALPHEBATICAL LIST OF GENERAL GLOSSARY

A

A line. The male sterile parent involving cytoplasmic or cytoplasmic genetic male sterility developed to produce hybrid seed. It is also known as a cytoplasmic male sterile line.

Abnormal growth. Growth that deviates from the normal type or form due to various environmental factors such as disease, pests, soil condition, humidity, temperature, etc. absolute plating efficiency. The number of individual cells that give rise to colonies relative to the number of cells plated; expressed in percentage.

Absorption. The process by which a substance passes from one system into another, e.g., from the soil solution into a plant's root cell or from the leaf surface into the leaf cells.

Absorption, active. Movement of ions and water into the plant root as a result of metabolic processes by the root, frequently against an activity gradient.

Acaricide. A pesticide that kills mites and ticks.

Accession. (1). An addition (of a variety or strain) to a national or other register of varieties obtained by field collection or exchange. (2). A variety, strain, or population registered at a research center and worth conserving.

Achene. Small, dry fruit with one seed.

Achlorophyllous. Lacking chlorophyll.

Acid equivalent (a.e.). The portion of pesticide that is theoretically converted into acid.

Acid soil. A soil with a pH value of less than 7.0; for practical purposes, soils with pH below 6.6.

Acid sulfate soil. A soil with a pH value of less than 4.0 when air-dried; contains a high amount of sulfate and yellow jarosite mottles. Abnormal growth of plants in this soil is due to (i) toxicity from aluminum, iron, and hydrogen sulfide; and (ii) phosphorus deficiency.

Actinomycetes. (1). Bacteria with branching filaments. (2). A nontaxonomic term applied to a group of organisms with characteristic intermediate between the simple bacteria and the true fungi.

Active ingredient (a.i.). The potent portion of a compound (such as fertilizer, insecticide, fungicide or herbicide applied on the soil or plant) used as basis to estimate the chemical effect. Useful for comparing the strength or toxicity of chemicals.

Active vegetative stage. The growth stage of plants when there is rapid development of leaves, tillers, branches, or stems.

Acuminate. Gradually tapering to a sharp, prolonged point (e.g., rice leaves).

Acute rodenticide. A poison for rodents that acts swiftly and causes death shortly after ingestion. An example is zinc phosphide.

Adapt. To fit or adjust to a particular environment or a set of specific climatic conditions through a change in the plant's growth and development.

Adaptive research. Research conducted to validate, modify, or calibrate a new technology to specific soil, climate, socioeconomic, or environmental characteristics of a given area.

ADP. Adenosine-5'-diphosphate, a complex sugar-phosphorus compound formed as a result of expenditure of energy and the loss of a phosphate group from the energy-rich **ATP** (adenosine triphosphate) compounds.

Adsorption. (1). The taking up of molecules or ions at a surface, including exchangeable cations and anions on soil particles. (2). The sticking of a liquid, or gaseous or dissolved substance to a solid, resulting in higher concentration of the substance.

Adventitious roots. Roots developing from the part of a plant other than roots, which are formed from nodes of the plant. After 10-20 days of growth, all roots of the rice plant are adventitious roots.

Aerenchyma. The lysigenous intercellular spaces in the parenchyma layer forming a system of air passages which ramifies through the leaves, stems, and roots of a rice plant.

Aerial branching at nodes (Br). New shoots which develop at nodes high on the culm following the cutting of panicles.

Aerial roots. Roots that grow above the ground from the nodes.

Aerial tillers. Tillers that grow above the ground level.

Aerobic. Growing only in the presence of molecular oxygen. Having molecular oxygen as part of the environment.

Aerobic digestion. The partial biological decomposition of suspended organic matter in wastewater or sewage under aerated conditions.

Aeroponics. A technique in growing plants wherein the plants derive their nutrients and water from a mist of air and aqueous solution that come in contact with the roots.

Agar. Mucilage derived from a seaweed. It forms a gel with water. It is used to solidify culture media on which microorganisms are grown.

Aged rice. Rice that is kept at least 4 month after harvest expands more on cooking and less sticky than cooked, freshly harvested rice.

Agrarian system. A historically constituted and durable mode of exploitation of the environment; a technical system adapted to the bioclimatic conditions of a given area and which complies with its social conditions and needs at that moment.

Agricultural production system. The whole structured set of plants, animals, and activities selected by a farmer for his production unit to achieve its goals. It is a global system that is finalized by farmer socioeconomic objectives and related management strategy.

Agroecological zones. Geographical mapping units based on climatic conditions and landforms that determine relatively homogeneous crop growing environments.

Agronomic characters. Plant characters related to crop production usually observed during plant growth; e.g., height, maturity, tiller number, panicle size, yield and quality factors.

Agronomy. (1). Science of agriculture that deals with all aspects of field crop production and soil management. (2). An applied ecological science.

Ahu rice. An early rice similar to "Aus"; grown in Assam, India.

Albino. A plant that lacks chlorophyll and turns white; a deficiency in normal pigment.

Aleurone layer. (1). The peripheral layer of endosperm of the grain beneath the seed coat which envelops the endosperm and contains oil and protein. (2). It is a layer of high-protein cells surrounding the storage cells of the endosperm. Its function is to secrete hydrolytic enzymes for digesting food reserves in the endosperm.

Alien gene. A gene transferred to the cultivated species from another related or unrelated species.

Aliquot. A small part or portion of a given amount (such as 5 ml out of 100 ml solution).

Alkali soil. A soil containing sufficient exchangeable sodium to adversely affect crop production. Its sodium adsorption ratio is greater than 15 and pH is > 8.5.

Alkaline soil. A soil with a pH value higher than 7.0; $CaCO_3$ in the upper horizon from a few percent to 95%.

Allelemorph. In Mendelian inheritance, a pair of alternative forms of a gene in which one may be dominant and the other recessive.

Allelism. The relationship between alleles in different parents. When the alleles of two parents belong to the same gene (locus), they are allelic; otherwise, non-allelic.

Allelochemicals. Nonnutritional substances produced by a plant that affect the behavior, growth, health, or physiology of another plant or insect.

Allelopathy. The phenomenon of suppressing the growth of one plant species by another through the release of a toxic substance.

Allogamous. Capable of cross-fertilization or having cross-fertilizing nature.

Allogamy. Cross fertilization; opposite of autogamy.

Alloploid/allopolyploid. A polyploid containing genetically different sets of chromosomes from two or more species, e.g., *O. minuta* with BBCC genomes is an allopolyploid.

Alluvial. Pertaining to alluvium; a clayey, silty, sandy, or gravely material deposited by a stream or other bodies of running water. Alluvial soil- Soil that has been deposited by any form of running water. Alluvium- Mineral material (sand, silt, and clay) deposited by rivers.

Alternate host. An individual or plant other than the main or preferred host upon which a parasite (or its spores, eggs, larvae, etc.) or a disease organism (pathogen) could live.

Aman rice. A term used in Bangladesh and east India for lowland rice grown in the wet season during June to November.

Ambient air. Air that reflects surrounding environmental conditions.

Amensalism. An interaction between two organisms in which one organism is suppressed by toxins produced by the second.

American wild rice. Not a species of rice. It is a wild grass found in the USA with grains similar to those of rice. Its scientific name is *Zizania aquatica*.

Ammonia volatilization. The escape of nitrogen from the soil or floodwater as ammonia gas; causes loss of urea and the ammonium form of nitrogen fertilizer from floodwater or saturated soil surfaces.

Amylase. The enzyme responsible for catalyzing the breakdown of starch into sugars; may be active in one of two forms: a-amylase and b-amylase.

Amylopectin. (1). A component of starch that has a high molecular weight, long-branched chain structures, and does not tend to gel in aqueous solutions. (2). A type of starch molecule composed of long-branched chains of glucose units (a polysaccharide). (3). The major and branched fraction of starch, a glucose polymer with 5,000--19,000 glucose units and mean chain length of 18-22 glucose units. This fraction contributes directly to gel consistency and has a long linear chain fraction in high amylose starch.

Amylose. (1). In cereal endosperm, it is the starch fraction of molecules or residues made up of glucose units, straight short chains (see nonwaxy endosperm). (2). The essentially linear fraction of starch, a polymer of glucose (dextrose) with up to 1,000 glucose units. It gives a blue complex with iodine and contributes directly to cooked rice hardness. It is measured colorimetrically by its blue complex with iodine in acetate buffer.

Amylose content. The value representing the starch fraction of milled rice, or the amount of starch in the grain that determines its eating and cooking quality. Low-amylose rice varieties are moist, sticky, and glossy after cooking. Rice with a high amylose content cooks dry and fluffy.

Anaerobic. The absence of molecular oxygen.

Anaphase. The stage of cell division (mitosis and meiosis) in which the chromosome halves move toward the opposite poles of the spindle.

Anatomy. Study of the internal structure of living beings.

Androgenesis. The development of a haploid individual from a pollen grain or microspore.

Angiosperm. Flowering plant. One of a group of plants whose seeds are enclosed in a mature ovary (fruit).

Annual. Yearly; used for plants, which complete their life cycle (seed to seed production and death), in 1 year or less.

Antagonism. In plant nutrition, the interference of one element with the absorption or utilization of an essential nutrient by plants.

Antagonistic symbiosis. A symbiotic association, which is destructive to one of the symbionts or partners, involved in the association.

Anther. The saclike structure of the male part (stamen) of a flower in which the pollen is formed. Anthers normally have two lobes or cavities that dehisce at anthesis and allow the pollen to disperse.

Anther/microspore culture. In vitro culture of anthers or microspores which allows the production of homozygous diploid (doubled haploid) plants.

Anthesis. The action or period of opening of a flower; the period of pollination, specifically the time when the stigma is ready to receive the dispersed pollen and fertilization takes place.

Anthocyanin. Any of a class of soluble glycoside pigments that are responsible for most of the blue to red colors in leaves, flowers, and other plant parts.

Antibiosis The antagonistic association between two organisms producing detrimental effects on one of them. It could also mean an association between one organism and a metabolic product of another.

Antibiotic. A chemical substance produced by certain microorganisms that retards or inhibits the growth of other microorganisms. Substance that acts to destroy or inhibit the growth of a microbe (e.g., bacteria or fungi).

Antibody. A protein produced in a warm-blooded animal in reaction to an injected foreign antigen and capable of reacting to that antigen.

Antidote. A remedy to counteract the toxic effects of a pesticide (e.g., atropine sulfate for carbamate and phosphate poisoning).

Antigen. Foreign protein and occasionally complex lipids, carbohydrates, and some nucleic acids, which upon injection into a warm-blooded animal, induce the production of antibodies.

Apiculate. Ending in a short-pointed tip.

Apiculus. A small extension of the lemma or palea.

Apomixis. An asexual method of reproduction in which the seed develops without the union of egg and sperm. The resulting embryos have the same genetic constitution as that of the seed parent.

Appendage. Any external outgrowth of a plant that does not have any apparent essential function.

Application rate. The amount of fertilizer, insecticide, or herbicide applied per unit area or volume in experiments or commercial production.

Applied research. Research in which results can be used immediately by the farmer and can be applied to the peculiar problems of a country or a region.

Aquatic plant. A plant that grows and develops in standing water and provides sufficient aeration to the parts under water.

Aridic. A soil moisture regime that limits plant growth during much of the growing season.

Aril. An outer covering of the seed arising from the stalk of the ovule or below the hilum, the pulpy inner pod.

Aroma. A distinct smell or odor.

Artificial manure. Any form of nutrient source that is not of natural origin.

Artificial medium. A substance having an agar base with different chemicals or other constituents developed to replace the normal food or diet in rearing insects or pathogens.

Asexual reproduction. Reproduction that does not involve fertilization or the fusion of sexually dissimilar gametes.

Assimilates. Building blocks of carbohydrates manufactured by the leaves of the rice plant.

Assimilation. (1). The utilization of inorganic and organic substances in cell synthesis. (2). Uptake of CO_2 by a leaf or canopy during the day time or plant foods utilized in building up a protoplasm and cell walls of the plant. (3). Computer modeling of the rice crop growth.

Asymmetrical. Parts not equal or symmetrically arranged.

Auricle. (1). A pair of small ear-like appendages borne at the base of the leaf blade and usually arising at the sides where the ligule and the base of the collar are joined. (2). An ear-shaped appendage, usually occurring at the junction of the leaf sheath and the blade that may not be present in older leaves.

Aus rice. A photoperiod-insensitive, rainfed, drought-prone, lowland, or upland rice, broadcast and transplanted during the early part of the wet season from March to September in Bangladesh and from April to August in east India.

Autogamy. Fertilization arising from self-pollination.

Autopolyploid. A polyploid which possesses duplication of the same set of chromosomes, or more than two sets of chromosomes, coming from same species.

Autosomes. Chromosomes not associated with sex of the bearer.

Autotroph. A cell or organism that manufactures its own food from CO_2 and other small inorganic compounds by the process of photosynthesis.

Auxins. A group of growth regulators or plant hormones (natural or synthetic) that may stimulate cell growth and induce cell elongation or division; often induce adventitious roots, root development, and other growth processes including seed germination.

Available nutrient. An element in the soil, which can be readily absorbed and assimilated by growing, plants.

Available water. Water in a soil that can be readily absorbed by plant roots.

Axil. The upper angle between a leaf or a branch or a stem from which it arises.

Axillary. Arising in an axil.

Azolla. A water fern that fixes nitrogen symbiotically with the blue-green alga Anabaena.

B

B line. The fertile counterpart parent of a cytoplasmic male sterile A line which is used as male parent to maintain the A line. It is also known as a maintainer line.

Backcross. (1). A breeding method in which a desired character such as insect resistance is transferred into an improved variety carrying it as a recurrent parent to reinforce or increase the gene frequency of the character. (2). Fl hybrid which is again crossed to either of its parents.

Backcross breeding. A method for developing a new variety in which a simply inherited trait is transferred through the backcross method.

Backcross nursery. Frequently used in hybrid rice breeding. Breeding nursery where male sterile plants identified among the test crosses (CMS x elite maintainer line) and their backcross progenies are crossed with the male parents with the objective of transferring cytoplasmic male sterility into the nuclear genotype of the elite maintainer line.

Backcross parent/recurrent parent. One of the parents of a hybrid with which the hybrid is repeatedly crossed.

Backcross ratio. The proportion of heterozygotes to recessive homozygotes expected in a backcross.

Bactericide. A chemical compound that kills bacteria.

Bacteriostatic. A chemical or physical agent that prevents multiplication of bacteria without killing them.

Bagging. (1). The act of covering the reproductive parts of a plant to prevent cross pollination or physical damage. This is an important process in cereal breeding. (2). Filling plastic or hemp bags with grain.

Bao. A floating rice mixed crop of aus and aman rice, i.e., early and late varieties sown during the early part of the wet season. A practice frequently used in Assam, India.

Basal application. Broadcasting or applying fertilizer into the soil before seeding or transplanting the crop.

Basal roots. Roots developing from basal nodes and growing into the soil.

Basal tillers. Tillers produced from basal nodes near the soil surface; the term 'basal' is used to differentiate the tillers produced on the upper nodes from those developing at the base.

Bed planting. A method of planting in which seed is planted on beds. Often two or more seed rows are planted on each bed.

Benchmark. A point of reference for measurements as in topographical surveys and land leveling.

Biennial weeds. Weeds that complete their life cycle in two years.

Bioassay. The measurement of the potency of any physical, chemical, and biological components by means of the response it produces in a test organism.

Biochemical oxygen demand (BOD). The quantity of oxygen used in the biochemical oxidation of organic matter in a specified time, at a specified temperature, and under specified conditions.

Biodegradable. Types of material subject to break down by biochemical processes.

Biodiversity. The totality of all the species of plants and animals in an area.

Biological control. The action of natural enemies--parasites, parasitoids, predators, and pathogens--in maintaining another organism's density at a lower level than it would occur in their absence. If control is facilitated by man, it is called applied biological control. If not, it is called natural biological control.

Biological denitrification. Loss of nitrogen from the soil due to microbial activity. Microbial reduction of oxidized inorganic nitrogen compounds serving as nominal electron acceptors in anaerobic respiratory pathways lead to nitrogen losses to the atmosphere as N_2 or N_2O.

Biological factors. Characteristics of the plant that influence the health, vitality, and quality of harvested products.

Biological N_2 fixation. The conversion of inert atmospheric N into usable form by living organisms.

Biology. The study or science of living things.

Biometrics/biometry. The science that deals with statistics to explain biological phenomena.

Biotechnology. (1). Any technique that manipulates living organisms (or parts of organisms) to make or modify products, to improve plants or animals, or to develop microorganisms for specific uses. An applied biological science--e.g., recombinant DNA technology. (2). The various industrial processes that involve the use of biological systems; the collection of microbial and other biochemical processes carried out on an industrial scale. It includes, but is not limited to, the industrial aspects of genetic engineering. Other areas of biotechnology deal with fermentation technology (antibiotics), hybridoma technology (monoclonal antibodies), and agricultural technology (plant and animal transformation).

Biotic. Having a mode of life.

Biotic environment. Insect pests, plant pathogens, weeds, other crop plants, small animals, and human beings make up the biotic environment.

Biotic/abiotic stresses. Limitations imposed on development which occur because of biological (biotic) or physical (abiotic) factors.

Biotypes. (1). Individuals or populations of plants or animals that are morphologically alike but physiologically different. (2). A population of insects capable of surviving on and damaging varieties that are resistant to other populations of the same insect species.

Bisexual. Said of a flower with both stamen and pistil.

Bivalent. A pair of homologous identical chromosomes derived from two parents united in the first mitotic division.

Bivoltine. Having two generations in a year.

Blanket drive. A term used when farmers form a line and move across a field or area to control pests, such as rats.

Blotched white. White spots that give a pie -bald appearance.

Blue-green algae. Algae that contribute to the maintenance of soil fertility especially in rice fields.

BOD. Biological oxygen demand; the amount of oxygen required to stabilize the demands from aerobic action in the decomposition of organic matter.

Bold grain. Grain that is broad in shape.

Boliland rice (*Sierra Leone*). Rice grown in fields that may be flooded for 2-4 months.

Boot. A rapidly growing panicle enveloped by the flag leaf sheath. In tissue culture, this refers to the panicle collected when the distance between the collar of the flag leaf and subtending leaf is about 7 to 8 cm.

Booting. Bulging of the flag leaf sheath due to the growing panicle inside.

Booting stage. The reproductive phase of rice growth and development when the developing panicle causes a swelling of the culm. The swollen area is referred to as the boot.

Border effect. In experiments, the difference in performance of plants planted at the edges of a plot and of those planted at the center of the plot.

Border row. A row of plants around test plots which protect test entries from a border row effect, such as a greater than normal growth due to lack of competition for nutrients and light, lower insect populations than on plants surrounded by other plants, etc.

Boro rice. An irrigated, high-yielding, cold-tolerant, relatively pest-free and photoperiod-insensitive rice cultivated during the winter months in India and Bangladesh.

Bract. A reduced or modified leaf at the based of a flower or flower head. A leaf from the axis of which a flower arises.

Bracteole. A bract borne on a secondary axis, as on a petiole; a reduced leaf.

Bran. The outer layers of a cereal grain, including the pericarp, tegmen, embryo, aleurone layer, and a small portion of the starch endosperm, which are removed in milling.

Branch. A division of the stem or axis of growth.

Breed. A particular type or variety. An artificial mating group derived from a common ancestor for genetic study and domestication.

Breeder seed. Seed of the highest genetic purity produced for maintaining purity of a variety. It is used to produce foundation seed.

Breeding. The art and science of improving plants and animals genetically for the benefit of mankind.

Breeding line. A genetic line bred in a crossing program, before it is named and officially released for commercial cultivation.

Bristle. A short and blunt seta. A stiff hair.

Broadcast. To apply seeds or granules by hand or machine over a surface area. To spread randomly.

Broadcasting. The action of spreading seeds, fertilizer, or pesticides on the surface of the field at random, by hand or by machine.

Broadleaf weeds. Weeds belonging to either subclass Monocotyledonae or Dicotyledonae. They are identified by their fully expanded, broadleaf structure and netted venation.

Beaked hull. The tip of the lemma is curved over the palea.

Broken grain (brokens). Grain that has broken into two or more pieces during milling.

Brood. All individuals that hatch at about one time from eggs laid by one set of parents and which normally mature at about the same time.

Brown plant hopper. An insect that causes burnt-like appearance on rice when in large numbers. Scientific name: *Nilaparvata lugens.*

Brown rice. Rice grain with its hulls removed but not polished.

Brown spot. A disease of rice caused by the fungus *Helminthosporium oryzae*, with leaf symptoms consisting of brown and oval spots with gray or whitish centers. The disease is also observed on the grains. The disease is closely associated with abnormal or poor soil and its occurrence serves as an indicator of poor soil conditions for rice production. It is sometimes called "poor man's disease."

Bud. A short embryonic stem tip bearing leaves or flowers or both.

Buffering. Chemically, the ability to resist change in pH.

Bulb. An underground fleshy offspring of a plant that develops roots and shoots.

Bulk. All the seeds of the same crop species mixed together.

Bulk density, soil. The mass of dry soil per unit volume, normally expressed in megagrams per cubic meter or grams per cubic centimeter.

Bund. An embankment used to control the flow of water; a division between fields.

Bush. A low woody plant with a number of a branches at, or near, ground level.

BVP. Basic vegetative phase; the juvenile growth stage of a plant that is not affected by photoperiod.

By-product. A substance obtained during the manufacture of another substance, e.g. bran is a by-product of milled rice.

C

Cage wheel. Cage-type steel wheel commonly used on hand tractors.

Calcareous soil. Soil that contains sufficient calcium carbonate or calcium-magnesium carbonate to effervesce visibly when treated with cold 0.1N hydrochloric acid.

Calciphytes. Plants that require or tolerate considerable amounts of calcium or, are associated with soils rich in calcium.

Calibration. Adjustment of model parameters to obtain model behavior that corresponds with observed behavior.

Calibration of sprayer. A series of steps done primarily to determine the volume of spray solution and time needed to spray a given area of field considering other factors (e.g., swath of the sprayer, pressure exerted inside the sprayer, or the sprayman's walking speed at an even pace).

Callus. A mass of thin walled undifferentiated cells, developed as a result of wounding or culture on nutrient media. Plant cells which proliferate on a defined medium and lack morphological differentiation.

Calorie. A unit of measure of energy in food.

Calyx. The outer ring of flower made up of separate or joined sepals, usually green and leaflike.

Capsule. (1). A fruit that sheds the seed when dry, consisting of more than one carpel. (2). a plastic structure enveloping a chemical. (3). A relatively thick layer of muco-polysaccharides that surrounds some kinds of bacteria.

Carbohydrate. A chemical compound manufactured and accumulated by plants--e.g., sugar, starch, or cellulose. The name is derived from the fact that the relative proportions of C, H, and O in simple carbohydrates is $(CH_2O)x$.

Carbon cycle. The sequence of transformations whereby carbon dioxide is fixed in living organisms by photosynthesis or by chemosynthesis, liberated by respiration and by the death and decomposition of the fixing organism, used by heterotrophic species, and ultimately returned to its original state.

Carbon-nitrogen ratio. The ratio of the weight of organic carbon to the weight of total nitrogen in soil or organic material.

Carrier(s). (1). Transport proteins that bind specific solutes and undergo conformational change in order to transport the solute across the membrane. (2). A material necessary to hold pesticide and fertilizer elements in a form suitable for application.

Caryopsis. A small one-seeded dry indehiscent fruit (as in grasses) with a thin membranous pericarp adhering so closely to the seed that fruit and seed are incorporated in one body forming a single grain (wheat and barley). In rice, brown rice is the caryopsis.

Catalase. An enzyme that catalyzes the degradation of hydrogen peroxide to water and the oxidation by hydrogen peroxide of alcohol to aldehydes.

Catalyst. A substance which accelerates the rate of a chemical reaction without itself being consumed in the process. Enzymes are catalysts.

Catch crop. A crop planted usually very early in the cropping season. It matures early or earlier than the main crop to produce some food before the main crop matures. Could be a short-duration crop grown in between two croppings of a main crop.

Cell. A unit mass of protoplasm surrounded by a cell membrane and containing one or more units of nuclear material.

Cellulase. An enzyme that breaks down cellulose.

Cellulose. A long-chain complex carbohydrate compound (polysaccharide) with the general formula $(C_6H_{12}O_6)n$. It is the chief substance forming cell walls and woody parts of plants.

Celsius. A temperature scale devised by Anders Celsius in which the freezing point of H_2O or melting point of ice is set at 0° and the boiling point of water is set at 100°. Also called centigrade.

Center of distribution. Center of origin; the hypothesized geographical area from which any organism or species has spread.

CentiMorgan (cM). Unit of recombination. Named after the geneticist Morgan. One cM is equivalent to one percent recombination.

Central shoots. The new growth that emerges from the middle point of the plant.

Cereal. Seeds of flowering plants of the grass family (Graminaceae) cultivated for the food value of their grains, e.g., rice, wheat, sorghum, maize, oat, barley, rye, and millet.

Certified seeds. Seeds used for commercial crop production produced from foundation or registered seeds under the regulation of a legally constituted agency. In hybrid rice, they are first generation seeds produced directly from CMS x restorer lines grown as per certification standards.

Check. (1). A row or plot of a selected variety included in an experiment for comparison with other treatments. (2). A standard for testing and evaluation.

Check row marker. A manually operated (usually pulled) light tool (common in Laguna, Philippines) used to mark the soft rice field surface in perpendicular directions to form a grid; used as guide for transplanting of rice seedlings.

Check row planting. A planting pattern where each hill is aligned in rows and columns and if the row and column are equidistant, each hill has also diagonal alignments.

Chemical name. The systematic name of a chemical compound according to the rules of nomenclature set by the International Union of Pure and Applied Chemistry.

Chemical oxygen demand (COD). A measure of the oxygen-consuming capacity of inorganic and organic matter present in water or wastewater.

Chemodenitrification. Nonbiological processes leading to the production of gaseous forms of nitrogen (molecular nitrogen or an oxide of nitrogen).

Chisel. A narrow tool that is used to break up soil. A very sharp hand tool used to cut and shape wooden objects and tools.

Chlorina. Yellowish green appearance as a result of chlorophyll deficiency. Plants are often smaller than normal.

Chlorofluorocarbon (CFC). Any of several gaseous compounds that are derivatives of methane, contain carbon chlorine and fluorine, and are used especially as aerosol propellants and refrigerants; suspected to be a major cause of ozone depletion.

Chloroplast. A plastid containing chlorophyll; seat of photosynthesis in eukaryotes.

Chlorosis. Yellowing or discoloration of the normally green parts of chlorophyll-bearing plants due to diseases or nutritional/physiological disorders.

Chromosome. Structural unit of the cell nucleus which carries in linear order the genes responsible for the determination and transmission of hereditary characteristics. The term is applicable not only to the nuclear chromosomes but more loosely to the DNA of viruses, bacteria, and the chloroplast and mitochondria of eukaryotes.

Clean tillage (clean culture, clean cultivation). A process of frequent cultivation or ploughing to prevent growth of all vegetation except the particular crop desired during the growing season.

Cleistogamy. Fertilization taking place in an unopened flower.

Clipping. (1). To cut leaves with scissors or shears. (2). To cut off briskly the margins, ends, or a small portion of an object or plant.

Clod. A compact coherent mass of soil, varying in size. Usually produced by breaking up soil by ploughing, especially when performed on clayey or fine-textured soils that are either too wet or too dry. Seldom occurs in sandy soil.

Clone. (1). A group of cells, tissues, or plants which are in principle genetically identical. (2). Progenies resulting from sexual or vegetative propagation. A group of plants vegetatively produced from the same original stock.

Cluster. Things grouped together because of similar characteristics or physical proximity. Closely crowded inflorescence.

Cluster spikelets (Cl). Spikelets that show a clumped arrangement on the primary or secondary panicle branches with two or more spikelets per branch. Bunched arrangement of spikelets in groups of 10-48 is termed "super clustering."

CMS (cytoplasmic male sterile) line. A line whose anthers produce no pollen or only abortive pollens. A genetic factor which induces sterility is present in its cytoplasm. No seed is set on this line by selfing. But its pistil is normal and can produce seeds when pollinated by a restorer or maintainer line.

Co-dominance. The situation in which both alleles at the locus influence the phenotype of the heterozygote resulting in a new phenotypic class.

Coarse-textured soil. Sandy, loamy sand, or sandy loam soil.

Coding. The process by which the sequence of nucleotides within a certain area of RNA determines the sequence of amino acids in the synthesis of a particular protein.

Cohesion. The tendency of a substance to cling together; the mutual attraction among molecules or particles comprising a substance that allows it to cling together as a continuous mass.

Cold tolerance. Ability of a plant to grow and develop normally under cold conditions. Tolerance for low temperature in a plant.

Coleoptile. The cylinder-like, protective membrane of certain species that protects the plumule as it emerges through the soil. The coleoptile is photoperiod-sensitive and stops growth when exposed to light, allowing the plumule to break through and continue growth.

Collar. The joint between the leaf sheath and leaf blade.

Collection. A gathering of different strains, varieties, or species for preservation until evaluated or multiplied. A set of varieties used and maintained by a researcher is called a working collection.

Colony. A distinguishable localized population within a species.

Colorimeter. An instrument for chemical analysis of liquids by comparison of the color of the given liquid with its standard colors.

Combining ability. The ability of a genotype (inbred, pureline, or synthetic/ composite) to transfer its desirable traits to its crosses. General combining ability is the average performance of a strain in a series of crosses; specific combining ability is a deviation in a cross from performance predicted on the basis of general combining ability of the parents.

Combustion. A chemical reaction, especially oxidation, accompanied by the production of light and heat.

Compaction. Increasing soil bulk density and decreasing porosity by application of mechanical forces to the soil.

Compatibility. The characteristic of a substance that allows it to be mixed in a formulation without undesirably altering the characteristics or effects of the individual components.

Compatible. Said of plants with morphologic and genetic traits enabling them to cross-pollinate, bud, or graft easily to produce progenies.

Competition. A rivalry between two or more organisms for a limiting factor in the environment.

Competitive ability. The success of one species or a variety to produce a greater proportion of offspring in the next generation at the expense of another.

Complementary DNA (cDNA). DNA synthesized by reverse transcriptase from an mRNA template.

Complementary RNA (cRNA). Synthetic RNA produced by transcription from a specific DNA single-stranded template.

Complete fertilizers. A classification of fertilizer materials guaranteed to contain three of the major required elements, namely, N, P, and K.

Complete heading. When all panicles of a plant (or a plot of plants) have emerged.

Component technology. The cultural techniques used in the management of a cropping pattern. These include choice of variety, time, and methods of tillage and crop establishment, fertilization, field-level water management, pest management, and harvesting.

Composite variety. A variety consisting of a number of components, such as inbred lines, all types of hybrids, populations, etc. A stage in the process of creating new varieties. To have common characters, such as similar growth period, degrees of resistance to lodging or to a pathogenic agent. The stability of the composite variety is limitless. The term is used whenever the plant involved is not to be extended.

Composite. In plant breeding, said of a population developed by putting together several individuals with distinct genotypes.

Compost. Organic residues or a mixture of organic residues and soil that have been piled, moistened, and allowed to undergo biological decomposition.

Concentration. The quantity of active ingredient expressed as mass per unit volume, or an amount of a substance based on its proportion to the whole or a specified amount and expressed as mass per unit volume of the resulting solution or mixture, e.g., g/l, ppm.

Condensation. A process in which vapor molecules cohere to each other and change to liquid.

Conditional response. Plant's response to quantitative or qualitative reactions.

Conservation. The preservation of materials for future use that can be propagated.

Consumptive use. The water used per unit of time in plant growth by the combined processes of evaporation, transpiration, and retention in the plant.

Contact herbicide. A herbicide that kills only the plant tissue at and very close to the site of application.

Contact insecticide. An insecticide that penetrates the outer covering or cuticle of the insect and kills it.

Contiguous. Touching or near each other.

Continuous submergence or flood irrigation. A method of applying irrigation water in which the soil is submerged from transplanting up to about 2 weeks before harvest.

Coordinated trial. Trials conducted in many places made up of the same treatments organized by a research center or a group of members with regional and national focus.

Corm. The swollen part of the base of the stem that is capable of producing new shoots.

Cotransformation. The simultaneous transformation of an organism with two or more genes, one or more of which is silent (non-selected) while at least one gene confers antibiotic resistance or another identifiable or selectable trait.

Cotyledon. First leaf developed by the embryo in seed plant. In most dicotyledon seeds, they are thickened and are storage sites of reserve food for use by the germinating seedling. Also called "seed-leaf."

Critical nutrient concentration. The nutrient concentration in the plant or its organs at the time the nutrient becomes deficient for growth.

Critical photoperiod. The longest photoperiod at which the plant will flower, or beyond which the plant will not flower.

Crop. Plants on a farm that are managed for economic purposes, producing a physical product for farm use or sale.

Crop damage. Any reduction in quantity or quality of yield that results from injury caused by environmental factors, chemicals, or pests.

Crop development. The sequence of processes and events involved in producing new tissues and organs throughout the crop cycle. Change in growth stage (phenology) and morphogenesis (new organs).

Crop growth rate. The crop's rate of dry matter accumulation.

Crop injury. Visible and measurable symptoms and/or signs caused by physical or biological pathogens, insects, weeds, and other factors.

Crop loss. A reduction in value and/or financial return due to damage; often measured as the difference between actual yield and attainable yield due to the effects of one or more pathogens or pests.

Crop removal. Absorption of soil nutrients by crops.

Crop residue management. The operation and management of crop land to remove stubble, stalks, and other crop residue or maintain them on the surface to prevent wind and water erosion, to conserve water, and to decrease evaporation.

Crop rotation. A planned sequence of growing crops in a regularly recurring succession on the same area of land, as contrasted to continuous culture of one crop or growing different crops in a haphazard order.

Cropping pattern. The spatial and temporal combination of crops on a plot and the management used to produce them.

Cropping system (farming). The crop production activity of a farm. It comprises all components required for the production of a set of crops in a farm and the relationship between the crops and the environment. These components include all necessary physical and biological factors as well as technology, labor, and management.

Cropping system (research). The set of techniques performed on plots which are handled in an identical way. Each cropping system is defined by the kind of crops and their succession order and the itineraries of techniques applied to these several crops, including the choice of varieties for selected crops.

Cross. A hybrid between two genetically dissimilar parents.

Cross cultivation (cross ploughing). The tillage of a field, orchard, etc., in which the field is cultivated in one direction followed by cultivation at right angles to the first.

Cross-fertilization. The fertilization of egg nuclei (ovules) of one plant by the pollen from another plant.

Cross-pollination. The transfer of pollen from the flowers (florets) of one plant to the stigma of another plant. It may or may not lead to fertilization.

Crosswise. Across, from side to side, when ploughing in a field.

Culm. The stem of a grass or sedge. The round smooth-surfaced, ascending axis of the shoot, consisting of hollow internodes joined by solid nodes.

Culm length. Length of the rice stem from the base of the plant to the base of the panicle or neck node.

Cultigen. A plant species known only in cultivation with its origin from domestication; for example, *O. sativa* and *O. glaberrima*.

Cultivar. (1). A variety. (2). A cultivated variety; the international term for variety.

Cultivation. (1). A tillage operation used in preparing land for seeding or transplanting or later for weed control and for loosening the soil. (2). The processes used in growing field crops, vegetables, plants, fruits, trees, flowers, and fish.

Cultural control. The use of agronomic practices such as soil tillage, varying planting time, fertility level, sanitation, water management, crop diversification, crop rotation, and short-duration cultivars to reduce pest populations or stresses.

Cultural practices. Activities or operations that are usually carried out in raising field crops, e.g., land preparation, seed selection, crop establishment, fertilization, etc.

Cumulic. Derived from accumulation. Descriptive of a wetland type where 100 mm of accumulated water will stay for more than 7 days when the soil has been puddled, even without rain or irrigation.

Cuticle. In plants, it is a thin layer over the aerial parts (particularly leaves) It functions to protect the plant from injury and to prevent excessive water loss.

Cutting. A section of a plant capable of developing into a new plant.

Cytogenetics. Branch of biology that deals with the correlated study of genetics and cytology.

Cytokinins. A group of plant growth-regulating substances and hormones (natural or synthetic) that regulate cell division and shoot formation.

Cytology. Branch of biology that deals with the study of structure, function, development, reproduction, and life history of cells.

Cytoplasm. All the protoplasm of a cell except the nucleus.

Cytoplasmic heredity/inheritance. The transmission of characters from parent to offspring through the cytoplasm of the germ cell.

D

Dapog. Method of raising rice seedlings on banana leaves, polythene, or a cement slab. Seedlings are raised in the same way as in the wet bed method. Because the seeds are sown thickly (3 kg seeds per square meter), seedlings look like a mat or carpet and can be rolled for transplanting when they are 10-14 days old.

Day length (photoperiod). Duration of the daylight period including twilight during a given day; the number of hours between sunrise and sunset.

Days to heading. Number of days from seeding to heading.

Dead furrow. A furrow left in the field that is slightly wider than twice the width of a plough bottom, usually occurring at the completion of a field.

Dead heart. Dead rice tiller caused by the attack and gridling of the tiller's base by stem borers.

Debris. Rubbish, vegetable matter, dry weeds, etc.

Decomposition. The process of resolving into constituent parts.

Decumbent. (1). The stem lying on the ground and tending to rise at the end. (2). Lying flat with the apex ascending.

Deep placement. A method of applying fertilizers in which the fertilizers are placed directly in the reduced zone of the paddy soil.

Deepwater rice. A long-duration rice that grows under rainfed, nonflooded conditions for 1-3 months, then is subject to flooding with water depths more than 50 cm for a month or longer. Plants elongate as the flood rises, above the water, in normal flood conditions, but may be submerged in deep flash floods. Maximum water depth varies greatly.

Deficiency. Any inadequacy or shortage of substances essential to growth and development of plants. Lacking in some quality, faculty, or characteristic necessary for completeness.

Defoliator. Any chewing insect that feeds on the leaves of plant, or chemical that removes foliage.

Dehiscent. Splitting spontaneously when ripe.

Dehull, dehusk. To remove the hulls or husk from a grain.

Dehumidify. The removal of moisture from the air.

Delta. A fan-shaped alluvial deposit at a river mouth formed by the deposition of successive layers of sediment from upstream areas.

Delugic. Derived from deluge. Descriptive of a land type where the water levels stay for more than 2 weeks at a depth greater than 30 cm, which is above the normal height of bunds or dikes, during high-rainfall months.

Demonstration plot. A plot that shows the characteristics or evidence for the local people to observe; planted specifically as a demonstration at a research station.

Denitrification. The biochemical reduction and loss of nitrate or nitrite to gaseous nitrogen, either as N_2 or oxides of nitrogen (N_2O, NO, NO_2).

Density. Quantity or number per unit volume or area.

Dentate. Toothed, usually with the sharp teeth pointed outward.

Depth. A measurement from the top to the bottom, e.g. from the surface of the water to the ground or to the bottom of a container.

Descriptors. Terms used to describe or characterize cultivars or to distinguish plants or seeds collected in germplasm programs.

Desiccator. A glass jar, fitted with an airtight cover, containing some desiccating (drying) agent such as calcium chloride at the bottom.

Determinate. Said of a condition when the terminal bud ceases to grow, preventing elongation of the main axis and promoting auxiliary growth.

Detritus. Dead organic matter, mostly from plants and usually in particulate form, including dissolved organic matter, but not inorganic matter and various organisms often associated with it.

Deviation. Departure of an observation from its expected value.

Diapause. A period during which growth or development is suspended.

Dibbling. Planting seeds in holes made by a pointed implement such as bamboo or wooden stick. It is usually practiced along sloping or hilly areas.

Dicotyledon. A plant with two cotyledons or seed leaves, netted leaf venation, and branched tap roots. Floral parts occur in groups of 4 or 5.

Diffuse. Widely spread.

Digestibility. (as applied to organic wastes). The potential degree to which organic matter in waste water or sewage can be broken down into simpler and/or more biologically stable products.

Dihybrid. A hybrid for two different genes. Heterozygous for two pairs of alleles.

Diluent. Any gas, liquid, or solid material used to reduce the concentration of an active ingredient in a formulation.

Dioecious. Having male and female flowers on separate plants of the same species.

Diploid (2n). An organism having two chromosomes of each kind.

Direct seeding. A system of planting rice in which seeds, either pre-germinated or dry, are broadcast or sown directly in the field, with either dry land or wetland preparation; no transplanting process is involved.

Disc floret. One of the central tubular flowers of a composite flower head.

Disciplinary research. Research conducted by specialists of a particular discipline.

Discoloration. The changing of color of grains, leaves, or stems due to disease, weather, water stress, insect damage, or lack of nutrients.

Discrepancy. Disagreement or inconsistency as in data of an experiment

Disease. A harmful deviation from the normal functioning of physiological and biochemical processes caused by plant pathogens (fungi, viruses, bacteria, and parasites).

Disease control. The use of chemicals or resistant varieties to control or eradicate the disease present.

Disease incidence. (1). The frequency of occurrence of a disease; commonly, the proportion of plants affected in a given population. (2). The number of plant units infected expressed as % of the total number of plant units assessed.

Disease resistance. Ability of the plant to survive and grow without symptoms of the disease that has infected it.

Disease severity. The degree of infection of disease in an individual plant; refers to the area of plant tissue affected by the disease expressed as % of the total area assessed.

Dissect. To separate and expose the parts of an organism for examination

Dissemination. The distribution of germplasm, information, etc.

Dissimilation. The decomposition or transformation of an organic substrate to yield energy for use by the organism or organisms.

Dissolve. To cause to pass into solution, such as to dissolve sugar in water.

Diverse. Having various forms or qualities.

Diversity. The condition of being different or having differences.

DNA. Deoxyribonucleic acid. The molecule that is the repository of genetic information in all cellular organisms and many viruses. The information coded in DNA determines the structure and function of the organism.

Dominance. Inter-allelic/intragenic interaction with complete suppression of one allele by another.

Dominant. Form of expression of a gene, in which the phenotype of the dominant form is expressed over the recessive form.

Dominant gene. A gene that is fully expressed in a heterozygote. A dominant gene may partially or entirely suppress the expression of another allelic gene (recessive gene).

Donor. In plant breeding, a variety that serves as a source of characteristic such as insect resistance.

Donor parent. A target parent in the crossing program expected to contribute the desired characteristics, e.g. donor for resistance, quality, etc.

Dormancy. A physical or physiological condition of a viable seed that suspends or prevents germination even in the presence of otherwise favorable germination conditions.

Double cropping. Two crops grown in sequence on the same land in a production cycle.

Double cross. A cross between first generation hybrids of four separate inbred lines.

Double transplanting. Used when seedlings once transplanted are uprooted again some 2-4 weeks after the first transplanting and planted in the final field.

Dough stage. The stage in rice development which occurs during the ripening phase when the milky caryopsis inside the developing grain turns into soft dough and later into hard dough.

Drench. To wet thoroughly by covering with water or another liquid.

Drift. The movement of airborne spray or dust particles outside the intended contact area.

Drilling. Sowing seeds in furrows or holes by hand or machine.

Drought. An insufficient supply of moisture from precipitation or soil for optimum plant growth.

Drought tolerance. A total expression of the plant's ability to stay alive, grow, and ultimately produce grain, with part of its life cycle under water stress.

Dryland. (1). Pertaining to soils rarely or never flooded, or to crops grown in such soils. (2). land that, except for limited periods, does not hold moisture in the rooting zone in excess of that held at field capacity.

Dryland rice. Direct-sown rice grown in a manner similar to wheat, and which depends on rainfall for moisture.

Dry matter partitioning. The distribution of dry matter to various plant organs.

Dry season. The season when no rain is expected.

Dry weight. The weight of any plant part after its water content has been removed by drying.

Drybed method. Raising seedlings on a dry seedbed. It is usually practiced in rainfed areas or where irrigation water is not adequate.

Drying rice. To remove or reduce the moisture content of rice to a safe level (below 14%) for storage, by placing the rice in the sun or using hot air machines.

Dry seeding or dry (DSR). A method of rice establishment where rice seeds are broadcast on dry soil (unpuddled). The technique is more popular in upland unbunded areas, where the soils are light-textured and easily drained. It can also be practiced in wetland areas, when the field is prepared at the time of early germinating rain.

Duplicate genes. Two independent genes that produce the same phenotypic effect when present individually or when present together.

Duplication. Occurrence of a segment of a chromosome twice in the haploid set.

Dust. A powder like formulation which may consist of (1) only toxic agents, such as sulfur, (2) toxic agents plus an active diluent which serves as a carrier, or (3) toxic agent plus an inert diluent which may be in the form of talc or clay.

Dusting. The process of applying insecticide or pesticide in dust formulation.

Dwarf disease. A disease caused by rice dwarf virus (RDV). The disease is transmitted by leafhoppers and is characterized by pronounced stunting, reduced tillering, and irregular chlorotic specks on leaves and leaf sheaths. These specks may fuse to form broken streaks that run parallel to leaf veins. Rice dwarf disease occurs in Japan, China, Nepal, and Korea.

Dwarf variety. A rice variety that is short in stature (with a plant height of 85±5 cm).

Dynamic model. A collection of regression models that together represent an entire dynamic system.

Dynamic system. A part of the real world that can be quantified with state and rate variables and parameters. The environment affects the system through rate variables. System behavior is the dynamic pattern of all state variables of a system.

E

Ear. A common name for the panicle of rice.

Early-maturing variety. A variety that matures early; variety that matures at least 10 days earlier than common types. In rice, varieties that mature in 100 days are called very early.

Ecology. The study of the interactions of organisms with their physical environment and with one another.

Economic injury or threshold level. The pest population density at which the loss caused by the pest is greater than the cost of controlling the pest. The pest density at which artificial control measures are economically justified.

Ecosystem. 1).A major interacting system that involves both living organisms and their physical environment. 2). A system of ecological relationships in a local environment, including relationships between organisms and the environment itself.

Ecotype. A population within a species that has developed distinct physiological characteristics (e.g., herbicide resistance) in response to a specific environment which persist even if individuals are moved to a different environment.

Edaphic factors. Soil conditions such as alkalinity, extreme acidity, iron toxicity, salinity, and zinc deficiency, which adversely affect plant growth.

Edaphology. The science that deals with the influence of soils on living things, particularly plants, including people's use of land for plant growth.

Effective rainfall. That portion of rainfall that can be stored within the rooting depth of the crop and on the paddy surface and can be subsequently utilized by the plants.

Electrical conductivity. A measure of the amount of salts in a solution expressed as deciSiemen per meter (dS/m).

ELISA. Enzyme-linked immunosorbent assay. A serological test in which one antibody carried within an enzyme releases a colored compound.

Elite lines. Breeding lines that possess most of the characteristics being sought for a particular environment or plant.

Elongation ability. (1). Deepwater rice: The ability of the internodes, leaves, and leaf sheath to elongate during the vegetative growth phase, thereby escaping submergence by rapidly rising floodwater. Floating rice can elongate as fast as 20 cm per day. (2). Grain: The ability of the rice kernel to stretch lengthwise upon cooking to sometimes twice its original length or more in some high quality Basmati rice varieties.

Elongation ratio. Ratio of grain length of 10 cooked rice grains to that of raw rice grains, preferably presoaked in water 30 min before cooking for 10 min.

Emasculation. The systemic removal of all of the male sex organ from an organism; in rice, the removal of the stamen of the floret.

Embryo. The generative part of a seed that develops from the fertilized (diploid) egg called zygote, which upon germination gives rise to a young seedling. It is located on the ventral side of the seed. It is easily detached and removed in the milling process as part of the bran.

Embryo rescue. An in vitro technique used to culture the hybrid embryos which otherwise abort in vivo.

Embryonic. Rudimentary, in an early stage. Embryogenic callus- An undifferentiated cell mass that produces somatic embryos. Embryonic shoot- The early leaves coming from the seed or the plumule.

Emergence. Coming out or rising of seedlings from the ground; coming out of panicles from the boot (panicle exsertion); coming out of adult insect from cocoon or pupa.

Emulsifiable concentrate. (EC). A concentrated solution and an emulsifying agent in an organic solvent which will form an emulsion when added to water.

Emulsifier. A substance that promotes the suspension of one liquid in another.

Emulsion. One liquid suspended as minute globules in another liquid, e.g., oil dispersed in water.

Enation. Tissue malformation or overgrowth induced by certain virus infections.

Endemic. Native to a country, geographical region, or area.

Endophyte. A plant growing within a plant. The association may be symbiotic or parasitic.

Endosperm. The nutritive tissues of the ripened ovary, consisting of the aleurone layer and the starchy endosperm. It serves as food for the germinating embryo; triploid in chromosome number.

Endotrophic. Nourished or receiving nourishment from within, e.g., fungi or their hyphae receiving nourishment from plant roots in a mycorrhizal association.

Entry. A test variety or a breeding line under evaluation.

Environment. The total external conditions and surrounding organisms, physical and biological factors which affect the growth and development of an organism.

Enzyme. Any of the numerous proteins or conjugated proteins produced by living organisms that function as biochemical catalysts within the organism's body. Some RNA enzymes were recently discovered (see ribozyme).

Epicotyl. The part of the stem of a seedling or embryo just above the cotyledons.

Epidemic. An extensive development of a pest or disease in a geographical area or community, in a given time and space.

Epidermis. The outer layer of cells. A layer of primary tissue in higher plants that is commonly one cell thick, often cutinized on its outer surface, and continuous in young plants except over the stomata. The epidermis provides protection to the underlying parts against mechanical injury and desiccation and is largely replaced in older plants except on leaves and herbaceous stems.

Epiphyte. A plant attached usually to another plant solely for support; not a parasite.

Equilibrium moisture content. The condition wherein the moisture-retaining tendency of the grain is the same as the moisture-withdrawing tendency of the air.

Erect growth habit. A tendency of the plant to grow upright without spreading. The culms are erect and closely grouped. Erect growth is recessive to the spreading or procumbent habit.

Ergonomics. An applied science which considers human characteristics in designing machinery and arranging things for effective interaction, comfort, and safety; also called human engineering.

Erosion. The loss of soil particles and nutrients when the surface soil is carried away by wind, water, or other agents.

Essential amino acids. Those amino acids that are not made by the human body but must be taken in as part of the diet. Examples are lysine, tryptophan, and valine.

Essential elements. Those elements such as N, P, and, K that are required for plant growth.

Estuary. The mouth of a river where the salty ocean tide meets the current of a stream.

Etiolation. Growth of plants from seeds in darkness. A yellowing condition of the plants due to light deficiency characterized by an elongated stem, small leaves, and lack of chlorophyll.

Etiology. The science dealing with the causes of diseases.

Eukaryote. Organism characterized by the presence of a nucleus. Also other organelles such as mitochondria and/or chloroplasts may be present in eukaryotes. Includes all plants, animals, green algae, and fungi.

Euploid. Having a chromosome number that is a whole multiple of the monoploid number.

Eutrophication. The process of becoming rich in nutrients.

Evaporation. In rice cultivation, the moisture lost in vapor form from the free water surface. It is one of the important factors that determine the effectivity of rainfall, particularly in arid or semiarid areas.

Evaporites. The class of materials including gypsum and all more soluble species, precipitated by evaporation.

Evapotranspiration. The loss of water from a given area and during a specified period of time by evaporation from the soil surface and by transpiration from plants.

Ex situ. Away from the natural or original position/place.

Exchange capacity. The total charge of the adsorption complex active in the adsorption of ions.

Exotic. An organism or a plant introduced from one country to another.

Expressivity. The degree of manifestation of a certain character in the offspring. The degree of expression of a given gene. Genes that always produce the same phenotype have 100 percent expressivity.

Exserted panicle. Panicle that emerges fully from the flag leaf sheath.

Exsertion. (1). When the spikelets of the rice plant become distinguishable and the panicle extends upward inside the flag leaf sheath. (2). Degree of emergence of the panicle from the flag leaf sheath; maybe positive or negative depending on whether the neck node is above or below the collar of the flag leaf. (3). Ability of the panicles to arise from the boot.

F

Fl. Abbreviation for the first filial generation, usually the hybrid between two homozygous types.

FCC. Fertility Capability Classification; a classification based on quantitative topsoil and subsoil parameters directly relevant to plant growth; various soil condition 'modifiers' define specific types of soil adversity or complexes of edaphic problems.

Fl hybrid. Denotes the first generation offspring from the mating of two parents.

F2. The second filial generation obtained by self-fertilization of F_1 individuals.

F2 population. Population of individuals present in the F2 generation of a cross.

F2 ratio. The segregation pattern of a character or characters in F2 generation.

F2 seed. The second generation progeny from the mating of two parents.

F3. Progeny obtained by self-fertilizing F2 individuals.

Facultative saprophyte. An organism living on another living organism during part of its life cycle or living on the dead tissues of its host.

Fahrenheit (°F) A temperature scale in which the freezing point of water is taken as 32°F and the boiling point of water as 212°F (or 100°C) under standard atmospheric pressure.

Fallow. Land that is ordinarily used for crops but allowed to lie idle between crops.

Farming system. (1). The manner in which a particular set of farm resources is assembled within its environment, by means of technology, for the production of primary agricultural products. (2). A collection of distinct functional units, where crops, livestock, and marketing activities interact because of the joint inputs received from the environment and management personnel. (3). A unique and reasonably stable arrangement of farming enterprises that a household manages according to well-defined practices in response to the physical, biological, and socioeconomic environment and resources. These factors combine to influence output and production methods.

Female parent. The parent which contributes its egg in a cross. The plant with the ovary, style, and stigma that may produce fruits or seeds.

Fermentation. Any of the chemical reactions induced by living or nonliving agents that split complex organic compounds into relatively simple substances, e.g., the anaerobic conversion of sugar to carbon dioxide and alcohol by yeast.

Ferrolysis. A soil-forming process resulting in acidification and clay destruction due to alternating reduction and oxidation in practically flooded soils.

Fertile. (1). Land that has enough nutrients to support abundant plant growth. (2). Plants that are reproductive and can produce fruits, seeds, spores or pollen grains.

Fertility (soil). The status of a soil with respect to its ability to supply nutrients essential to plant growth.

Fertilization (genetics). Fusion of the nuclei of male and female gametes. The union of the pollen (male) and the egg (female) initiating reproduction.

Fertilization (soil). Making soil fertile with the application of fertilizer.

Fertilizer. Any substance added to the soil to supplement elements required in the nutrition of plants.

Fertilizer, complete. Used formerly to denote a fertilizer containing appreciable amounts of N, P, and K; it now includes secondary and micronutrients essential to plant growth.

Fertilizer response function. A function that relates yield (output per hectare) to the amount of fertilizer used (input per hectare), holding all other inputs constant.

Fertilizer grade. The guaranteed minimum analysis of the major plant nutrient elements contained in a fertilizer material. It is expressed in percentage of N, P_2O_5, and K_2O.

Fertilizer requirement. The quantity of certain plant nutrient elements needed, in addition to the amount supplied by the soil, to increase plant growth to a designated level.

Fibrous root system. Adventitious roots from the base or stem of the rice plant that form the mat roots.

Field capacity. The amount of water remaining in a field soil that has been thoroughly wetted and drained until free drainage has practically ceased. The soil can no longer absorb additional moisture.

Field resistance. Resistance observed in the field as distinguished from resistance observed in the laboratory or greenhouse. It may involve seedling resistance as well as adult plant resistance and often involves resistance to all locally occurring biotypes.

Field sampling. (1). Fields may be chosen randomly for sampling from a map locating the coordinates and the cultivator interviewed. Samples are then taken according to the plan. (2). A field is divided into particular proportions and a grid is used for even sampling of different sections of the field.

Field screening. Evaluating varieties for resistance in the field in contrast to greenhouse and screenhouse evaluations.

Filament. In a flower, the slender stalk of the stamen.

Filled grain. A seed with a fully developed endosperm.

Filler. (1). A material that does not contain any fertilizer element but is added to obtain the desired weight of a mixed fertilizer. (2). Rice variety used to fill vacant areas of an experimental plot.

Fixation. The process by which certain nutrient elements in a soil are converted from available form to unavailable form essential for plant growth, being converted from a soluble or exchangeable form to a much less soluble or to a nonexchangeable form. Examples are potassium, ammonium, and phosphorus fixation.

Fixation, nitrogen. Conversion of available nitrogen to insoluble or unavailable nitrogen.

Fixative. A preservative or chemical used to keep plants for subsequent study.

Fixed ammonium. The ammonium in soil that cannot be replaced by a neutral potassium salt solution.

Fixed line. A breeding line that continues to breed true to type and does not segregate.

Fixed phosphorus. That phosphorus which has been changed to an unavailable form as a result of reaction with the soil.

Flag leaf. The uppermost leaf originating just below the panicle base.

Flash flood. A rapid surge of flooding that subsides after several days and lasts no longer than 10-12 days.

Floating rice. Rice that can maintain their canopies above the water in slowly rising flood as deep as several meters. It can take gradually increasing water level up to several meters at a maximum rate of 15 cm per day, provided there is adequate growth at the seedling stage approximately six weeks before the onset of flooding. Examples are floating rice varieties grown where maximum water depth ranges between 1 and 6 m for more than half of the growth duration. In densely populated areas, floating rice is grown as a subsistence crop because no other crop will grow.

Flooding. (1). The normal process that floods the plains where deepwater rice is grown. (2). To apply water to the field for the benefit of saturating the soil for land preparation. (3). Establishing and maintaining an irrigated rice crop.

Flower. The reproductive part of the angiosperms. In rice, it consists of two lodicules, six stamens, and the pistil. Flowering- The stage when the anthers of the terminal spikelets protrude and shed pollen. Flowering date - The date recorded when the panicles emerge from the boot.

Foliar. Referring to leaves of a plant; application of chemicals to the leaves.

Foliar diagnosis. An estimation of mineral nutrient deficiencies (excesses) of plants based on examination of the chemical composition of leaves and the color and growth characteristics of the foliage of the plants.

Formulation. Mixture of a pesticide with diluents, solvents, wetting agents, or other inert ingredients which can be used in pest control.

Foundation seed. Seed stock produced from breeder seed by or under the direct control of an agricultural experiment station. Foundation seed is the source of certified seed, either directly or through registered seed organizations.

Fumigant. Highly volatile chemical that kills mice, insect, weeds, and other living organisms.

Fumigation. The process used to destroy bacteria, insects, or pests by smoke or exposure to poisonous gas.

Fungicide. A chemical agent that kills or inhibits fungal spores or mycelium. A pesticide used to treat or prevent diseases caused by pathogenic fungi.

Fungus. (1). An organism with no chlorophyll, reproducing by sexual or asexual spores, usually having mycelia with well-marked nuclei. (2). Heterotrophic microorganisms chiefly saprophytic or parasitic, that constitute the kingdom fungi.

G

Gall. An abnormal plant growth, swelling or tumor induced by another organism such as an insect.

Gall midge. An insect pest (*Orseolia oryzae*) that causes the youngest leaf or the shoot in a tiller to appear as light-colored tubular gall resembling an onion leaf. Tillers with gall midge do not produce panicles.

Gamete. A mature reproductive male or female germ cell (sperm or egg) specialized for fertilization.

Gametic (tissue or generation). Having n number of chromosomes (haploid) in contrast to zygotic tissue with 2n (diploid).

Gametocide. Organic or inorganic chemicals used for killing the functional sexual parts (pollen, ovule) of the plant. These may be selective for male or female parts.

Gel consistency. Degree of rice gel firmness measured as the length of cold milled rice gelatinized in 2 ml of dilute potassium hydroxide in a 13 x 100-mm test tube, placed horizontally for 30 minutes or 1 hour. Gel consistency may be classified as hard (27 to 40 mm), medium (41 to 60 mm), or soft (61 to 100 mm). It is correlated with hardness of cooked rice.

Gene. (1). A functional hereditary unit that occupies a fixed location in a chromosome, has a specific influence on the phenotype (usually) through encoding a single polypeptide chain or molecule), and is capable of mutation to various allelic forms. (2). DNA segment that performs a specific function and is a unit of inheritance.

Gene mapping (genome mapping). The assigning/locating of a specific gene to particular region of a chromosome; the determination of the sequence of specific genes and their relative distances from each other on a given chromosome.

Gene pair. The identical or nonidentical alleles of a specific gene at a given locus on homologous chromosomes in a diploid cell.

Gene pool. The sum total of genetic variability present in a crop species.

Genealogy. The noted history of the genetic ancestry of a variety.

General combining ability. The average performance of a parent in a series of crosses.

Generation. Offspring having a common parent or parents and constituting a single stage of descent.

Genetic. Resulting from, or produced by, soil-forming processes; for example, a genetic soil profile or a genetic horizon.

Genetic diversity. The genetic variability present in a population or in a species.

Genetic engineering. Technologies used to isolate genes from an organism, manipulate them in the laboratory, and insert them stably into another organism. It refers to all of the techniques used to manipulate DNA, including gene isolation and cloning, gene splicing, plasmid construction, and transformation.

Genetic map. The linear arrangement of genes on a chromosome.

Genetic marker. Any gene whose presence can be readily detected by its phenotypic expression and which is used to locate other genes. Also detectable RFLP used to identify a specific linked gene or an individual carrying a gene of interest.

Genetic purity. Trueness to type; seeds or plants genetic potential. The likelihood that the genetic material (improved or primitive) is capable of producing many new genotypic combinations from stored variousness by hybridization and subsequent segregation, recombination, and selection under varying pressures conforming to the characteristics of the variety, line, or hybrids.

Genetic recombination. The production of progeny that derives some of its genes from one parent and some from another genetically different parent; as a result, the combination of genes in the progeny is different from that of either of the parents. In higher organisms, recombination occurs by way of independent assortment or crossing over; in lower organisms, it occurs by way of transformation, conjugation, or transduction.

Genetic resources. Germplasm that includes the entire array of cultivars in the crop species, related wild species in the genus, and hybrids between the wild and cultivated species.

Genetics. The science of heredity and variation.

Genetic shift. Change in the genetic makeup of the line, variety, or hybrid if grown over a long period particularly in areas outside their adaptation.

Genetic tester. (1). A cultivar/line used for testing the presence/absence of a gene in a given line/cultivar. (2). A pure line or hybrid known to carry one or more genes that distinctly express themselves in morphological or physiological features. The mode of inheritance of such traits is known by prior genetic experiments.

Genetic variability. The state of being genetically variable, i.e. having more than one genetic state or allele at each gene locus. Such genetically variable populations are referred to as polymorphic.

Genome. A complete single set of the genetic material of a cell or organism; the complete set of genes in a gamete, the single DNA/RNA molecule of bacteria, phages, and most animal and plant viruses. In plants, composed of the nuclear genome, the mitochondrial genome, and the chloroplast genome.

Genome analysis. Procedure which helps to identify the chromosomes corresponding to the haploid set of species which contributed to the evolution of an allopolyploid species.

Genotype. The genetic constitution of an organism.

Genus. A class, kind, or group marked by common characteristics or by one common characteristic; a category or biological classification ranking between the family and the species, comprising of structurally or phylogenetically related species or an isolated species exhibiting unusual differentiation, and being designated by a Latin or latinized capitalized

Geomorphology. Study of the origin of landforms; largely replaces the term 'physiography'.

Germ. The embryo of a seed; pathogenic or harmful microorganism.

Germination. The resumption of active growth of the embryo in the seed and the process by which a seed sprouts to development into a seedling when a favorable environment exists, culminating in the development of a plant from the seed.

Germination to emergence. The period which is signified by the emergence (coming out) of the radicle or coleoptile from the germinating embryo in the seed.

Germinator. A piece of equipment used for germinating seed which provides the necessary conditions needed for germination.

Germplasm. (1). The sum total of genetic material in a species. (2). The material basis of heredity. (3). The potential hereditary materials within a species, taken collectively.

Germplasm bank. A facility to store plant genetic resources for custody, conservation, exchange, and use by present and future generations.

Germplasm collection. A collection of genotypes of a particular species, from different sources and geographic sites, used as source materials in plant breeding.

Gibberellins. Plant growth regulators responsible for cell elongation.

Glabrous. Having a smooth even surface, hairless condition. Hairs are scarce or absent on the hull and/or the leaf blade. In a glabrous strain, few hairs may be found on the margins of the blade.

Glossy. Leaf blades have a glossy surface to which water easily adheres in large droplets.

Glume. The two chaffy structures in the inflorescence of grasses. It is made up of the lemma and palea which usually cover the floret before blooming and often remain attached to the ripened "fruit" or grain of some cereals, such as rice.

Glume discoloration. Synonymous with grain discoloration. Rice grains maybe infected with various organisms causing discoloration ranging from black dots to brown or blackish blotches which may cover the entire glume.

Glutinous rice. Refers to waxy rice with only 0-2% amylose or mainly amylopectin in its endosperm and pollen starch. Raw glutinous endosperm is opaque and becomes moist, sticky, and glossy when cooked.

Grade. To separate milled grain according to size and quality--whole grain, broken grain, short grain, or long grain.

Grain (syn. rough rice, paddy, padi, caryopsis, and seed). 1). A fruit in which the pericarp is fused with the seed. 2). The ripened ovary and its associated structures such as the lemma, palea, rachilla, sterile lemmas, and the awn if present.

Grain characteristics. The different character and dimensions of the rice grain, e.g. length and shape and cooking qualities.

Grain chemical quality. The chemical analysis of the grain's content or nutritional value.

Grain dormancy. Physiologically, the inactive growth phase of the grain.

Grain length. The distance from the base of the sterile lemma to the tip or spiculus of the lemma or palea.

Grain milling quality. Physical and chemical characteristics of the grain after milling.

Grain number. The number of grains per panicle. This varies with the different varieties.

Grain shape. The physical shape of the rice grain--round or slender, etc.

Grain size. The dimensions and weight of the grain.

Grain smut. A disease of the rice grain caused by *Ustilaginoidea virens*. The grain becomes yellow-greenish or greenish-black velvety spore balls. Also called false smut.

Grain weight. The weight of the grain often recorded as grams per 1,000 grains.

Grain width. The measurement of the grain across the broadest section.

Grain yield. Weight of harvested grain expressed as tons/ha with a moisture content of 14%.

Grain-straw ratio. Ratio of grain weight to the remaining aboveground dry weight or straw weight at harvest.

Granular. Small grains or pellets. An aggregate similar in size to a crumb but more dense.

Granules. Chemical formulations in small pellets, formed from various inert clays or sand impregnated with pesticide or fertilizer.

Green manure. Plant material incorporated into the soil while green or soon after maturity to improve the soil.

Green revolution. A term used to describe the success in increased crop production throughout Asia, commencing in the 1960s as a result of high-yielding rice varieties developed by IRRI and wheat varieties by CIMMYT.

Greenhouse effect. The process whereby short-wave radiation passes readily through the earth's atmosphere to surface areas, whereas the longer-wave outgoing radiation is absorbed and reradiated by water vapor, droplets and carbon dioxide, thus retaining heat in the atmosphere.

Groundwater. That portion of the water below the surface of the ground whose pressure is greater than atmospheric pressure. Groundwater run-off- Water from a catchment area that moves freely under the influence of gravity and enters the soil.

Growing point. Mass of meristematic tissue at the stem or root tip where growth in length of the stem or root occurs.

Growing season. The period of a year when the environment enables farmers to produce a crop of economic value.

Growth. (1). The change in the size of a plant, resulting in increased volume, increased dry weight, and protein content. (2). The increased population or colony of a culture of microorganisms.

Growth analysis. A method of taking samples at defined intervals and determining the development patterns.

Growth duration. Length of time the crop takes to mature (seeding to harvest).

Growth habit. The usual sequences of the vegetative and reproductive phases of the plant.

Growth inhibitors. Suppressants of plant growth such as hormones or chemicals that inhibit the plant's ability to grow or develop.

Growth regulators. Chemicals that are used to control the plant's growth.

Growth retardants. Chemicals that delay or slow down growth.

Growth stages. The defined periods or cycles of the plant's growth, such as the seedling, tillering, and reproductive stages.

Growth stimulators. Chemicals or hormones that are used to accelerate or promote the rate of growth in plants.

Guard cell. One of two crescent-shaped epidermal cells united at the ends whose changes in turgidity determine the opening and closing.

H

Habit. The characteristics of growth and general appearance of a plant.

Habitat. The physical location or type of environment in which a plant or an organism is found.

Hair. A uni- or multi-cellular epidermal outgrowth.

Hand weeding. A practical and efficient method used to manually remove weeds within rows and hills; used in areas where a cultivating implement cannot be used.

Haploid (n). A cell or an organism with the gametic (n) chromosome number.

Haploid plant. Plant regenerated from anther or microspore culture with half the number of chromosomes of the diploid (symbol : n).

Hard dough stage. Stage when the rice grain is almost fully ripe.

Hardpan. A physically compacted soil layer restricting root growth and water movement through it.

Hardy. A plant that can withstand and grow under severe adverse climatic conditions.

Harvest index (HI). The ratio of grain weight to total aboveground plant dry weight.

Harvest. To cut and collect by hand or machine the grain or product of the plant.

Harvester. A person or machine that cuts and collects the product of a plant.

Harvesting. The process of gathering in a crop; in the case of rice, the gathering in of mature rice panicles from the field.

Head. (1). A dense cluster of sessile or nearly sessile flowers. (2). The inflorescence or panicles of cereals such as rice and sorghum.

Head rice. The whole grains of milled rice that can be obtained from a given quantity of clean rough rice (paddy). It is usually expressed as a percentage of rough rice. Broken rice larger than 3/4 of a grain is also considered as head rice. Head rice may vary from as low as 25% to as high as 65%.

Head-to-row. A process of growing seeds of individual plants to determine or maintain seed purity of a variety.

Heading. Growth stage of the rice plant marked by emergence of the panicle from the boot.

Hectare. A measurement of land area in the metric system equivalent to 10,000 m^2.

Herbaceous. Non-woody plants.

Herbicide. A chemical used for killing or inhibiting the growth of certain plants or weeds.

Heredity. The genetic transmission of characteristics from parents to their offspring.

Heritable. The characteristics of parents transmitted to their progeny.

Hermaphrodite. Flower which possesses both male and female organs; bisexual.

Heterocyst. A thick-walled, usually translucent cell of some blue-green algae known to be the site of nitrogen fixation.

Heterogeneous. Having different genotypes in a population.

Heterosis (standard). Refers to the phenomenon in which the Fl hybrid obtained by the crossing of two genetically dissimilar parents shows superiority over the best standard check prevailing at that time in one or a combination of characters.

Heterosis breeding. A method of breeding to develop Fl hybrids which exploit the phenomenon of hybrid vigor to increase yield potential and yield stability.

Heterotrophic. Obtaining nourishment from outside sources, specifically requiring complex organic compounds of nitrogen and carbon for metabolic synthesis.

Heterotrophic nitrification. Biochemical oxidation of reduced forms of nitrogen to nitrate by heterotrophic microorganisms.

Hexaploid. A plant that has six sets of chromosomes.

Hibernation. The inactive or dormant state of insects during winter.

High density grain. Grains having specific gravity greater than 1.20. The specific gravity of a grain indicates the degree to which the cavity of the hull is filled and the density of carbohydrate deposition in the grain. Hence, high density grains are filled to their potential size and weight.

High-yielding varieties (HYVs). Rice varieties developed through a breeding program that possess desirable agronomic characteristics, resistance to insects and diseases, and tolerance for environmental stresses like soil problems, temperature, and drought; nearly all HYVs are short-statured or semi dwarf.

Hill. A group of rice plants directly adjacent to each other because the seeds or seedlings were planted together. A hill may also consist of only one plant.

Hill rice. Rice grown in the uplands on the steep slopes of hills usually with high rainfall.

Hilum. The scar at the point of attachment of the seed of the plant.

Hoeing. A technique used with a hoe to loosen the soil, to remove weeds between the rows of rice or other plants.

Homogeneous. Uniform in appearance and similar in genetic composition due to descent from a common ancestor.

Homozygosis. The union of gametes identical for one or more pairs of genes.

Homozygote. A zygote derived from the union of genetically identical gametes. An individual with both alleles alike for any given gene; these are therefore pure for any given trait.

Homozygous. Having one type of allele in both chromosomes for a given gene or genes. Having the same members of the allelic gene pair, either dominant or recessive on both chromosomes, and producing only one kind of gamete.

Hopper burn. The complete drying up of rice plants caused by the extensive feeding by large number of plant hoppers which suck plant sap (e.g., brown plant hoppers *Nilaparvata lugens*).

Hormone. Regulatory substance that controls metabolism and development, acts at low concentrations (e.g., less than one ppm, micromolar) and at a distance from its site of synthesis.

Host. The organism on which a parasite lives or the plant on which a pest feeds.

Host plant. A plant that provides nourishment and protection to a parasitic organism, living plant, or animal.

Host plant resistance. The relative genetic ability of a cultivar to produce a larger or higher quality crop compared with other cultivars exposed to the same infestation level.

Host resistance. A plant that is resistant to or tolerant of attacks of parasites or pathogens by evolving morphological or physiological modifications.

Hot air emasculation. The use of hot air to open and force the floret to remove or kill the anther.

Hot spot. Site where the natural field infestation of a particular insect or disease is high, providing sufficient pressure for reliable results in varietal resistance tests.

Hot start. PCR modification with the reaction tube kept at 95°C prior to addition of primer and polymerase.

Hot water emasculation. The use of hot water to open and force the floret to remove the anther. Emerging the panicle in hot water to kill the pollen grains.

Hot water treatment. A water treatment at high temperature applied to seed samples to make them aseptic, used particularly for international shipments.

Household. Composed of the farmer and his family. Considered both the production and consumption unit of the social organization. The household can be managed by one person or operated collectively, as family members live, sleep, eat, share the same place, and divide household duties and general farm management and work.

Hull. The outermost covering of the rice grain which provides protection to the rice caryopsis composed of lemma and palea. It ranges from 17 to 24% of the rough rice content.

Hulling. The removal of the husks or hulls from the rice grain; converting rough rice or paddy into milled rice.

Humidity. Dampness, the amount of water or water vapor in the air.

Hybrid. The product of a cross between genetically dissimilar parents. hybrid rice. The Fl seed used as commercial variety.

Hyaline. Translucent or transparent.

Hybrid variety. (1). A variety of rice developed from crossing and selection to gain higher yields. (2). Fl offspring of two genetically dissimilar parents.

Hybrid vigor. The increase in vigor of hybrids over their parental inbred types; also known as heterosis.

Hybridization. A breeding method in which two varieties are crossed to generate new variability and to produce desired recombinants. The hybrids are allowed to self-pollinate and the segregating populations are handled by an appropriate method.

Hybridization block. A specific piece of land set aside for research in making specific crosses in breeding particular varieties of rice.

Hydration. The process where a compound combines with water in a definite ratio.

Hydrograph. A flooding pattern shown as a continuous curve of water depth based on daily depth values.

Hydrophilic. Water-loving.

Hydro tiller. Engine-powered, rotary, walk-behind tiller or puddler of the floating type.

Hydrogen ion concentration. (1) A measure of acidity/alkalinity of a solution. (2) Mathematically is equal to the negative logarithm of the molar concentration of hydrogen ion; measurement of acidity/alkalinity of a solution; pH less than 7 = acidic; above 7 = basic.

Hydrology. The science dealing with the distribution and movement of water.

Hydrolysis. Decomposition or alteration of a chemical substance by water.

Hydromorphic. (1). Pertaining to a soil showing characteristics associated with permanent or periodic excess water, e.g. gley phenomena. (2). Derived from hydro (water) and morph (form). Descriptive of soil developed in the presence of permanent or periodic excess of moisture.

Hydromorphic rice. Rice that is grown where the water table is very close to the surface.

Hygroscopicity. The ability to readily take up and retain moisture.

Hypersensitivity. Extreme sensitivity.

Hypertrophy. An abnormal increase in size of cells.

Hypocotyl. The part of the stem of a seedling or embryo below the cotyledons.

Hypodermis. The tissue immediately beneath the epidermis of a plant especially when lignified, suberized, or otherwise modified to serve as a supporting and protecting layer.

I

I1, I2, I3, etc. Symbol used to designate the first, second, third inbred generations, etc.

Ideotype. An ideal plant type used as a model for selective breeding to develop specific plant characteristics desired or required in particular ecosystems.

Imbibition. The process of absorbing or taking in water as in drinking.

Immature grains. Grains with still greenish caryopsis or brown rice.

Immobilization. The conversion of an element from inorganic form to organic form in microbial tissues so that the element is not readily available to plants.

Immune. Completely resistant to pest or disease of the plant.

Impervious. Incapable of being penetrated. Resistant to penetration by fluids or by roots.

Improved cultivar. A rice variety that has been bred to increase production.

Improved plant type. Specific morphological and physiological features of the rice plant that make it more efficient and higher yielding.

In situ. Latin: meaning, in the natural or original location.

In vitro. Latin: meaning, outside the living cell. In a test tube or other artificial medium. Denotes the growth of explants in glass or other artificial environment.

In vitro selection. Selection for phenotypes (traits) expressed at the cellular or callus level which usually possess genetic changes that control the trait.

In vivo. Latin: meaning, in real life; in the living cell.

Inbred. An individual resulting from the mating of closely related parents or selfing.

Inbred line. A nearly homozygous breeding line produced by continued self-fertilization.

Inbreeding. The interbreeding of closely related individuals occurring naturally (as in a closed population), or as a deliberately chosen system of breeding and serving especially to preserve and fix desirable characters of and to eliminate unfavorable characters from a suitably selected stock but tending to effect an unwanted decline (as in size, vigor, or fertility) through the fixation of undesirable and often recessive characters when the initial stock is any way defective.

Inbreeding depression. Continuous breeding of genetically related individuals resulting in decrease in reproductive capacity.

Incompatible. A plant that is unable to set seed or units when selfed, crossed, or grafted because of structural, physiological, or ontogenic reasons.

Incomplete dominance. Two different alleles (e.g., dominant and recessive gene) producing an intermediate effect compared with effect of the dominant alleles.

Incomplete fertilizer. A classification of fertilizers that contains only 1 or 2 of the major fertilizer elements N, P, and K.

Incorporate. To mix foreign materials, such as pesticides, fertilizers, or plant residues into the soil.

Incubate. To provide (for seeds, embryos of animals, or bacteria) prescribed and usually controlled conditions (as temperature and moisture) favorable for development or hatching.

Incubation period. Time after inoculation required for the symptoms of a disease to appear.

Indehiscent. 1. Does not split open when ripe. 2. Does not open at maturity to discharge seeds.

Indeterminate. When no flower bud growth ends the axis of the flower cluster on the inflorescence.

***Indica* rice.** One of the two major ecogeographical races of *Oryza sativa* (see also japonica). The major type of rice grown in the tropics and subtropics. It has broad to narrow, light green leaves and tall to intermediate plant stature (except for the semidwarf). Indica plants tiller profusely. Grains are long to short, slender, somewhat flat, and awnless. Indica grains shatter easily and have 23-31% amylose content. They grow mostly in the Philippines, India, Pakistan, Java, Sri Lanka, Indonesia, central and southern China, African countries, and other tropical regions.

Indicator plants. Plants showing particular characteristics in response to specific soil or site conditions.

Indigenous. Native to a particular area.

Induced resistance. A type of pseudoresistance in which a plant temporarily acquires increased resistance from some conditions of plant or environment, such as change in soil fertility.

Infest. To attack externally, e.g. by insect pests; attack on a part of a plant such as grain of rice. Also to contain the parasite.

Inert ingredient. Any ingredient in a formulation which has no pesticidal activity.

Infect. To introduce or contaminate the plant with a pathogen or disease-producing substance. To invade and establish a parasitic relationship with the host plant.

Infection. A stage in a disease cycle when the pathogen penetrates and establishes a parasitic relationship with the host.

Infestation. The presence of insects on a plant or on grain.

Infiltration. The downward entry of water into the soil.

Inflorescence. The flowering part of a plant including the bracts, flower stalks, and flowers.

Inheritance. The process of genetic transmission of characters or characteristics.

Inhibitory gene. A gene that inhibits the expression of other genes either in dominant or recessive form.

Initial evaluation period. The first trial conducted in a program of testing to identify varieties or lines that could be retained for future trials.

Inoculate. To introduce microorganisms into another organism or medium.

Inoculation. The introduction of an explant (any plant part) into a suitable medium for callus formation and ultimately plant production.

Inoculum. Material (as spores, bacteria, or contaminated fluids) used in or suitable for use in inoculating or inoculation, e.g., portion of a pathogen (spores, mycelium, hypha in the case of fungi) or insect (eggs, young larvae) capable of being disseminated and causing infection/infestation.

Inorganic. Of mineral origin.

Inorganic fertilizer. Fertilizer materials derived from minerals, atmospheric gases, water, and inert materials. Can be natural or synthetic products of chemical reactions.

Insecticide. A chemical used for killing insects.

Integrated pest management (IPM). A strategy that utilizes various tactics or control methods (cultural, biological, and chemical) in a harmonious way in which control actions are based on frequent monitoring of the pest. IPM depends on multidisciplinary ecological strategies to weigh the effect of each tactic, as part of the agroecosystem, in producing the least disturbance and yield loss in the long run.

Interaction. The differential response of two or more factors which is more than can be attributed to chance.

Interbreeding. The use of plants capable of gene exchange by hybridization.

Intercropping. Growing two or more crops simultaneously in alternating rows or sets of rows in the same plot (See also Mixed intercropping)

Interflow. Shallow and often ephemeral and perched groundwater moving laterally.

Internode. The portion of a stem between two nodes.

Intermediate. (1). The characters or traits of the plant between two different levels. (2). A substance which is in a metabolic pathway between two other substances.

Intermittent irrigation. A method of applying irrigation water by which the field is alternately watered and drained. The soil surface is allowed to dry prior to the next application of water.

Internode. The portion of the stem between two nodes. Smooth and solid (when young) or hollow; (when mature) part of the culm.

Interspecific cross. A cross between two different species of a genus.

Introgression. Incorporation of genes of one species into the gene pool of another species by hybridization and backcrossing.

Inundate. To be completely submerged, e.g.. by floodwater. Inundation- The time of onset or arrival of the floods.

Invasion. When an infectious agent attacks the host plant causing a disease.

Inversion. A rearrangement of a chromosome segment so that its genes are in reversed linear order. Reverse order, e.g. warm air near the ground, cool air above.

Iron toxicity. Disorder that occurs when the iron concentration in the soil solution is high. The symptoms of iron toxicity are many tiny brown spots on the green leaves which start from the tips and develop into a general browning of the leaf blades followed by the death of lower leaves.

Irradiation (genetics and plant breeding). Exposure of plants or plant parts to radiation to increase mutation rates.

Irrigated rice. Rice which is supplementary or fully supported with water supplied by artificial means.

Irrigated rice land. Rice areas that have assured irrigation for one or more crops per year. Some areas are served only by supplemental irrigation in the wet season.

Irrigation. The artificial supply of water to areas by ditches, pipes, etc. by pumping or gravitation to the soil for the purpose of growing crops.

Irrigation efficiency. The ratio of the water actually consumed by crops on an irrigated area to the amount of water diverted from the source onto the area.

Irrigation methods. The manner in which water is controlled and applied to an area.

Isogenic. Having the same genic constitution.

Isogenic lines. Lines (families) which differ by one gene only.

Isolation (barrier). (1). The growing into pure culture in artificial/synthetic medium of pathogenic microorganisms. (2). The separation of one group of plants from another so that fertilization between or among groups is prevented. Separation between two hybrid seed production plots can be provided by topographic surface features or artificial/natural barriers, to the height of at least 2.5 m., in the case of rice.

Isolation (time): The separation can also be provided by growing two groups at different times of the crop season so that one group is already mature (stopped providing pollen) when the other group comes to flowering. Generally a 21-day difference in flowering is sufficient in the case of rice.

Isoplasmic. CMS or restorer lines differing in nuclear genetic constitution but having a common cytoplasm.

Isozyme. Also called allozyme. An enzyme which exists in multiple electrophoretic forms either because of allelic variation within one polypeptide or multimeric associations of variant forms. Similar or identical catalytic activities occur within the same organism.

J

Japonica rice. One of the two major ecogeographical races of *O. sativa* (see also indica).A group of rice varieties from northern and eastern China grown extensively in some areas of the world. It has narrow, dark green leaves, medium-height tillers and short to intermediate plant stature. It is found in the cooler zones of the subtropics and in the temperate zones. Grains are short, roundish, spikelets awnless to long-awned, panicles low-shattering, and have 0-20% amylose content.

***Javanica* rice**. Designation for the bulu and gundil varieties of Indonesia and many upland rices. They belong to the japonica race of *O. sativa*. Rice varieties with broad, stiff, light green leaves. It is low-tillering and has a tall plant stature.

Grains are long, broad, and thick, awned or awnless. Javanica grains are low-shattering and have 0-25% amylose content.

Juvenile stage. The immature or young growth period.

K

Keel. The middle nerve of the lemma. The two lower united petals of a legume flower.

Kernel. The grain or seed of cereal or grass enclosed in a hard husk.

Kilogram. A metric measurement of weight equal to 1000 grams or the mass of one liter of water at 4°C.

Kneeing. The ability of the upper stem of deepwater or floating rice plants to bend upward after water has receded.

L

Labile. A substance that readily undergoes transformation or is readily available to plants.

Labile genes. Genes that are always mutating.

Lag phase. First stage of the sigmoidal growth curve where growth declines within a brief span of time.

Lag phase (vegetative). That portion of the vegetative phase affected by the photoperiod to which the plant is exposed. The period from end of basic vegetative phase to panicle initiation; also called photoperiod-sensitive phase.

LAI. leaf area index; the sum of leaf area of all leaves divided by the ground area covered by the leaves.

Lamina. The leaf blade.

Lanceolate. (1). Lance-shaped, several times longer than wide; widest towards the base and tapering at both ends. (2). Lance shaped, much longer than broad, widoning above the base, and tapering to the apex.

Landform. The morphology and specific character of the land surface resulting from the interaction of physical processes and the earth's crustal movements with the geology of the surface layers.

Land equivalent ratio (LER). The area needed under sole cropping to produce the same amount as 1 ha of intercropping or mixed cropping.

Land preparation. The process of preparing the soil for planting, to provide a soil environment favorable for plant germination and/or growth.

Landscape. The overall surface form of an area including both natural and man-made features; it includes various landforms among its properties.

Land utilization index (LUI). The number of days during which crops occupy the land during a year, divided by 365.

Land water requirement. The net seepage and percolation requirement of the soil.

Landholdings. Land owned or occupied or used by farmers or tenant farmers.

Larva. The immature, wingless, and often vermiform (wormlike) feeding form that hatches from the egg of many insects, alters chiefly in size while passing through several molts, and is finally transformed into a pupa or chrysalis from which the adult emerges.

Latent infection. The state in which a host is infected with a pathogen but does not show any symptoms.

Latent period. (1). In epidemiology, latent period refers to the time elapsed from arrival of a dispersal unit at a susceptible plant surface until the formation of the next generation of dispersal units. (2). Incubation period of a virus in an insect. The time between acquisition of the virus and the time when it becomes infective.

Latent. Something, e.g., a gene or disease that is present in a plant but is unexpressed.

Lateral. On the side.

Lax panicle. A panicle that is loose or open with a few lateral branches and sparsely distributed spikelets.

Layout. Plan showing the allocation of treatments among experimental units. A detailed plan showing the arrangement of replications and treatments in an experiment.

LD50 (lethal dosage). The dose that is lethal to 50% of the test animals or micro organisms in a given time period. It is usually expressed in milligrams of insecticides per kilogram of body weight in mammals, and micrograms of insecticides per gram of body weight in insects. The lower the LD value, the higher the toxicity.

Leaching. The removal or loss of nutrient elements in soil solution due to the downward movement of water.

Leaf. An outgrowth from the stem of a vascular plant, usually green. It consists of a leaf-base, petiole and flattened lamina, which is usually conspicuously veined.

Leaf angle. The angle of the leaf blade from the stem. The descriptive terms often used are erect (angle of 0-30°), intermediate (31-60°), horizontal (61-90°), and descending (more than 90°). The angle of the flag leaf on the rice plant is important in photosynthesis.

Leaf area index. The sum of the leaf area of all green leaves divided by the ground area above which the leaves were growing.

Leaf blade. The thin, elongated, flat portion of the leaf.

Leaf curling or rolling. Rolling of the leaves to control water loss affected by lack of atmospheric or soil moisture. Also the bending or twisting of the leaves due to viral disease (as in rice ragged stunt) or to other harmful agents.

Leaf mesophyll cells. Cells within a leaf that produce carbohydrates in photosynthesis.

Leaf scraping. The removal of the epidermal portion of a leaf by the feeding of an insect.

Leaf senescence. Change in color from green to yellow or brown due to a reduction of metabolic activities in the plant preceding the death of the plant. Natural drying of the leaves due to crop maturity.

Leaf sheath. The lower part of the leaf enclosing the stem, originating from a node and wrapping around the culm above the node.

Leaf veins. The vascular bundles of the leaf. In monocotyledonous plants such as rice, leaf veins are seen as longitudinal ridges.

Leaflet. A segment of a compound leaf.

Legume. Plant belonging to the family Leguminosae. Most legumes form a symbiosis with nitrogen-fixing rhizobia bacteria in specialized nodule structures that fix nitrogen.

Lemma (syn. outer glume). One of the two bracts enclosing the flower or spikelet of rice. It becomes the covering of two-thirds of the rice seed surface and has five vertical ridges, formed along the surface, from the vascular bundles.

Lesion. A localized spot of diseased tissue on a plant.

Lethal gene. A gene whose effect is sufficiently drastic to kill the bearer of certain genotypes.

Levee (also called bund or dike). An embankment of soil for retaining water inside the rice field. A dike made of soil to retain water in rice fields.

Level. A flat surface of a rice field, a horizontal line, or an index of altitude.

Leveling. Land preparation involving moving soil from high to low spots in the field to achieve a flat horizontal surface so that irrigation water will be evenly distributed throughout the field.

Life cycle. The stages in the growth and development of an organism that occur between the appearance of the individual and its reproduction or death.

Light soil. A coarse-textured soil such as sandy soil.

Light transmission ratio. Light intensity at the ground level of a population divided by the light intensity at the top of the population.

Light trap. A device for collecting insects, consisting of a light source which attracts insects at night and a mechanism that traps the insects.

Ligule. A thin, upright, membranous appendage at the top of the leaf sheath. It is attached to the base on the inside of the leaf collar of the rice plant and some grasses.

Lime. A compound containing carbonates, oxides and/or hydroxides or calcium and/or magnesium used to neutralize soil acidity.

Line breeding. A system of breeding in which a number of genotypes which have been progeny-tested with respect to some character or group of characters are composited to form a variety.

Line sowing. To sow the seed in lines with a drill or by hand.

Linkage. The association between two or more genes located on the same chromosome that tend to be inherited together.

Linkage map. A scale representation of a chromosome that shows the relative positions of all its known genes or DNA sequences.

Lobed. Partially divided into usually rounded or obtuse segments but not deeply enough to form separate entities.

Local check or control. Farmer's variety or method. Used in experiments as a check in a specific place or community, in rice production.

Locule. A cell.

Locus. The position of a gene on a chromosome.

Lodge, lodging. The falling down of rice plants in the field due to wind, rain, flooding, pest damage, or because the stems are too tall or too weak to stand erect during the grain-filling stage, usually causing yield loss.

Lodicules. The two scale like structures adjoining to the base of the palea which control the opening of the lemma and palea during anthesis.

Long glume (g). The sterile lemmas ("outer glumes") exceed one-third the length of the **lemma and palea.** Another dominant gene, Gm, controls the sterile lemmas which are longer than the lemma and palea. The two sterile lemmas may have unequal length.

Long-duration varieties. Varieties that mature in 150 days or more.

Loose. (1). An uncompacted soil. (2). Spreading panicles.

Loosen, loosening. To break up, to not be compacted. Decreasing soil bulk density and increasing porosity due to the application of mechanical forces to the soil.

Lowland field. A land area surrounded with dikes or levees to impound rain or irrigated water. An irrigated field surrounded by levees.

Lowland rice. Rice grown on fields where water is held by bunds. About 30% of the world's rice is grown as rainfed lowland; about 45% as irrigated lowland. Some areas are flooded lowlands.

Luxury uptake. The absorption by plants of nutrients in excess of their need for growth.

M

M1, M2, M3. First, second, and third generations following exposure to a mutagenic agent.

Macronutrient. A nutrient element absorbed by the plant in large amounts, often more than 0.1 % of its dry weight. The following are considered macroelements: carbon, hydrogen, oxygen, nitrogen, phosphorus, potassium, calcium, magnesium, and sulfur.

Macrophytes. Large plants, as opposed to microphytes such as algae and fungi.

Maintainer line. A pollinator variety used to pollinate the male sterile line and produce progenies which still retain male sterility. If there is no maintainer line, the male sterile line cannot be maintained and multiplied generation after generation.

Maintenance application. Application of fertilizer materials in amounts and at intervals to ensure a certain minimum level of an available nutrient.

Major gene. A gene that determines major characteristics.

Major gene resistance. Resistance in which genes show clear-cut and discrete segregation in the F2 or later generations of crosses between resistant and susceptible parents. Effects are thus qualitative.

Male flower. A flower with functional stamens but no ovary.

Male parent. The plant which is the source of pollen used in hybridization.

Male sterile. A plant having no pollen grains or nonfunctional pollen grains and therefore is unable to fertilize the embryo.

Mangrove. Tropical or subtropical vegetation (mainly trees) typical of tidal swamps.

Mangrove rice cultivation. Use of the cleared mangrove forest land for rice production.

Mangrove swamps. Low-lying coastal swamps where mangrove trees grow.

Manure. Material used as fertilizer (e.g., organic material and excreta).

Marginal cost. The increase in variable cost which occurs in changing from one production alternative to another; it is often measured relative to adding a marginal unit of input.

Marginal net benefit. The increase in net benefit which can be obtained by changing from one production alternative to another; it is often measured relative to adding a marginal unit of input.

Marginate. With a distinct margin.

Marker gene. A gene with a clear-cut phenotype (and often a known location on a chromosome) that is used as a point of reference when mapping or selecting another gene at a nearby locus.

Mass screening. Screening a large number of varieties in preliminary studies. Varieties selected as resistant in the mass screening trial are re-tested to confirm their resistance.

Mass selection. A breeding method wherein a large number of plants having the desirable traits are harvested individually from a standing crop. The seeds from all selections are then bulked. From the bulk, a seed sample is taken and used to plant a population from which desirable plants are selected at maturity. The procedure is repeated for several cycles until the population becomes uniform and homogeneous. A variety developed by mass selection is fairly uniform and contains fewer genotypes than the original population.

Mature grain. The stage when the rice grains in the panicle are yellow, fully developed, hard, and capable of sprouting.

Mechanical weeding. Removing weeds with the use of a mechanized interrow cultivator or other machines.

Mechanisms of resistance. Processes involved in the resistance of a plant to an insect, including nonpreference, antibiosis, and tolerance.

Meiosis. Specialized cell division associated with gamete production. It consists of two divisions, one of which is reductional in terms of chromosome number and the other, equational. Meiosis produces genetically different gametes in a heterozygote having one-half of the genetic material of the original cell.

Membranous. Thin, soft and pliable, often partly transparent.

Mendel's laws. (1). Law of independent segregation: Hereditary traits are determined by discrete factors (now known as genes) that appear in pairs, one of each pair being inherited from each parent. During meiosis, the pairs of factors are separated or segregated. Hence, each gamete that is produced by an offspring at maturity contains only one member of each pair that the offspring produces. (2). Law of independent assortment: When two pairs of alleles are brought together in a hybrid combination, they will segregate independently from each other in the F2 itself.

Meristem. Cells at the growing tips of stems or roots that are dividing and are undifferentiated but later become different parts of the plant. Region of actively dividing cells in plants.

Mesocotyl. The internode between the coleoptile node and the point of union of the culm and the root in the young seedling. The mesocotyl pushes the coleoptile above the soil surface. Mesophyll tissues- The soft tissue of a leaf between the top and bottom epidermis (outermost layer of cells or protective covering of a plant or plant part).

Messenger RNA (mRNA). A chain of single-stranded ribonucleotides that codes for a specific protein. Product from DNA by transcription which serves as the information carrier for translation into proteins in the ribosomes..

Metabolism. The physical, chemical or biochemical processes in living organisms for maintaining growth, development, and reproduction. A collective term for all chemical changes in living cells.

Microbe. Synonymous with microorganism. Very small living organism.

Microclimate. The climatic condition of a small area, such as a rice field, resulting from the modification of the general climatic conditions by local differences in elevation or exposure. Also the environmental conditions below the crop canopy.

Microelement or micronutrient. A nutrient element needed by the plant in very small amounts. Considered as microelements are iron, manganese, boron, molybdenum, copper, zinc, and chlorine.

Microfauna. Protozoa and smaller nematodes.

Microflora. Bacteria, including actinomycetes; viruses, and fungi.

Microorganism. A living entity which is visible only with the aid of a microscope.

Microplot. Small plot 1-2 m^2 covered with a cage, used to simulate large field plot conditions.

Microspore. In seed plants, it gives rise to the pollen grain, the male gametophyte.

Midrib. The prominent ridge in the middle of the underside of the leaf blade. Central vein of the leaf.

Milk grain stage. Stage in the ripening phase of rice growth and development when the contents of the caryopsis (starch portion of the grain) is at first watery, but later turns milky in consistency.

Milled rice. Rice from which the hull and bran have been removed.

Milling. The process of separating the hull or husk and bran from the paddy or rough rice into milled rice and bran-and-chaff.

Milling yield / recovery. The estimate of the quantity of head rice (whole rice or nearly whole kernels) and of total milled rice that can be produced from a unit of rough rice. It is generally expressed as a percentage.

Milling recovery rate. Total milled rice yield as a percentage of rough rice yield.

Mineralization. The conversion of an element from an organic form to an inorganic state as a result of microbial decomposition.

Minerals. Solid particles other than organic matter which act as major components of soils and sources of nutrient elements.

Minikit trial. An extremely small trial with a small number of varieties tested in farmers' fields.

Mismatching. Error during annealing of primer to template or probe to target, caused by low stringency.

Missing hill. The place where transplanted seedlings were damaged, are non-existing, or died before maturing.

Mitosis. A process of somatic cell division which produces daughter nuclei which are identical genetically to one another and to the parent nucleus. They contain the same number of chromosomes. Mitosis produces genetically equivalent cells in the growing somatic region of organisms. It is a process by which the diploid stage of a cell is maintained, i.e., there is no reduction in chromosome number. The five main stages are prophase, prometaphase, metaphase, anaphase, and telophase

Mixed cropping. A cropping system that uses more than one type of crop, growing simultaneously, on the same piece of land at the same time

Mixed farming. Raising of crops, animals (livestock), and/or trees simultaneously in the same piece of land.

Mixed fertilizers. Fertilizers containing two or more of the major fertilizer elements.

Moderate resistance. Intermediate levels of resistance, between resistant and susceptible.

Modern varieties (cultivars). Dwarf, semidwarf, stiff-stemmed, high-tillering, nitrogen-responsive, photoperiod-insensitive, high-yielding varieties. Their characteristics are different from those of traditional varieties.

Moisture content. The amount of water measured in the soil or plant. Expressed as percentage moisture based on total fresh weight of the substance.

Moisture stress. Occurs when the plant suffers from a lack of water to support good growth and development.

Moisture tester. Experimental equipment and instruments used to test the moisture (water) content of the soil, plant or seeds.

Molecular marker. A molecular signpost used in eukaryotic gene isolation. Usually a RFLP probe or a primer site for DNA amplification.

Molecular weight. The sum of the atomic weights of a molecule's constituent atoms.

Molluscicide. A pesticide that kills molluscs (such as snails).

Monocotyledonous. An embryo having a single cotyledon.

Monocropping. When only one type of crop is grown on the land.

Monocot. A plant with one cotyledon or seed leaf, parallel leaf venation, and fibrous roots.

Monoculture. Growing the same crop on the same piece of land year after year.

Monocyclic. Having one cycle per season.

Monoecious. Having male and female flowers on the same plant of the same species.

Monoploid (haploid). A plant or organism having only one set of chromosomes.

Monsoon. The cyclic surface winds blowing in tropical and subtropical latitudes for several months from the southwest, switching for several months from the northeast, and the rains that accompany them.

Morphogenesis. The combined processes of cell division, differentiation, organ initiation, and organ development.

Morphology. A branch of biology that deals with the physical form and structure of plants and animals. A study of the forms, relations, metamorphoses, and phylogenetic development of organs apart from their functions.

Mortar and pestle. A device for grinding, consisting of a vessel (mortar) in which substances are ground with a club-shaped implement (pestle).

Mosaic. The mottled patchy pattern on leaves caused by a virus. The arrangement of leaves on the upper stem to avoid shading.

Muck. Highly decomposed organic material in which the original plant parts are not recognizable.

Muck soil - An organic soil in which the organic matter is well decomposed.

Multicropping. The growing of more than one crop together in one field or sequentially.

Multidisciplinary. A combination of many disciplines in an assignment, not necessarily working in an integrated or coordinated manner.

Multidisciplinary approach. Approach that uses many experts from different disciplines working together on a research program to solve a problem.

Multilocation trials. The same experiment is conducted in a number of different locations or sites across a geographic area.

Multiphase sampling. Collecting information from different categories and from different subsamples.

Multiple alleles. A series of more than two mutational forms at a given locus.

Multiple cropping. Growing more than one crop in the same plot in 1 year.

Multiple cropping index (MCI). The sum of the areas planted to different crops harvested during the year, divided by the total cultivated area.

Multiple cross. A cross involving more than two parents.

Multiple-range test. A sequential test used to make pair wise comparisons among several means which provide the least significant ranges.

Multiple resistance. Resistance to several stresses such as insects, nematodes, diseases, drought, and nutritional deficiencies.

Multiplication. The growing of seed for sale or wide distribution as a planting material in a given community or for specific research.

Mutagen. A physical or chemical agent capable of producing a heritable change in the DNA sequence of an organism.

Mutant. A variant from the normal or wild type that is inherited; the product of mutation.

Mutation. A sudden heritable change in an organism other than that due to the incorporation of genetic material from other sources. Most mutations are within the limit of a single gene.

Mycorrhiza. The symbiotic association of the mycelium of a fungus with the roots of a seed plant. This symbiosis helps to extend the effective root surface area and promotes uptake of phosphorus.

N

n, 2n. The notations for the genetic and zygotic (diploid) or somatic chromosome number of a plant, e.g., 2n = 24 in rice.

Narrow cross. A cross between parents that have many genes in common and only differ in a few.

Narrow row planting. A method of planting in which the seed is planted in uncommonly narrow rows, closer than normal, for the given crop.

Natural cross pollination. Pollination of the plant by insects, wind, or floral parts without human intervention.

Natural selection. The growing and survival of the best adapted plants in a given location, with selection occurring naturally without human intervention.

Neck blast. A form of blast disease where the rice panicle base becomes dark thereby called "rotten neck"; the panicles may fall over; the symptoms appear also on panicle branches.

Neck node (panicle base). The nearly solid node between the uppermost internode of the culm and the panicle axis.

Necrotic mosaic. A viral disease caused by the rice necrotic mosaic virus (RNMV). *Polymyxa graminis* is considered to be the vector of the RNMV. The disease occurs in central and southern Japan. Infected plants have elongate or spindle-shaped, yellow flecks and streaks on the basal portions of stems and sheaths. Plants may be moderately stunted with reduced tiller number and spreading growth habit. It can be transmitted by mechanical means and through the soil. There has been no evidence that insects or seeds transmit the disease.

Nematicide. A chemical compound or physical agent that kills or inhibits nematodes.

Nematode. Any of the class or phylum Nematoda of elongated cylindrical worms parasitic in animals or plants free-living in soil or water. Generally microscopic, worm like animals that live saprophytically in water or soil.

Net assimilation rate. The rate of plant weight increase per unit leaf area.

Neutral soil. Specifically, a soil in which the surface layer, at least down to normal plough depth, is neither acid nor alkaline in reaction. Practically, a soil in which the pH of the surface soil ranges between 6.6 and 7.4.

Niche. Role and function of an organism specific in its habitat.

Nitrification. The biochemical oxidation of ammonium (NH4+) to nitrate (NO3-).

Nitrogen assimilation. The incorporation of nitrogen into organic cell substances by living organisms.

Nitrogen cycle. The series of chemical and biochemical changes of nitrogen and nitrogenous compounds taking place in the soil - plant - atmosphere continuum.

Nitrogen deficiency symptoms. Characteristics of a plant at suboptimal N supply usually indicated by yellowing leaves, reduced tillering, and stunted growth.

Nitrogen fixation. This process occurs in some free-living microorganisms, e.g., blue-green algae and in bacterial-induced nodules of legumes. The conversion of elemental nitrogen into organic nitrogen or to forms readily utilized in biological processes.

Nitrogen response. Increase in grain yield (and other growth parameters) after N fertilizer application.

No-tillage system. A procedure whereby a crop is planted directly into a seedbed, not tilled since the harvest of the previous crop.

Nocturnal. A term applied to organisms that move or are active at night.

Nodal roots. Roots developing from the upper nodes when deepwater rice varieties are flooded.

Nodal tillers. Tillers arising from the upper nodes in elongated deepwater rice stems while the field is flooded.

Node. (1). The joint of a stem from which leaves or branches arise. (2). The solid portion of the culm, panicle axis, and panicle branches. Leaves, tillers, and adventitious roots arise from nodes on the culm.

Node blast. Form of blast disease where the node becomes infected. The node rots with dark discoloration and breaks apart, remaining connected only by the nodal septum.

Nodules. Swelling caused by bacteria *Rhizobium spp.* which live on the roots or stems of leguminous plants.

Nonglutinous (nonwaxy) rice. Common nonwaxy rice grain that contains >2% amylose in the starch, with translucent endosperm.

Noninfectious disease. A disease that is caused by an environmental factor, not by a pathogen.

Nonpersistent virus. A virus that has short transmission time, is not recoverable from the hemolymph, is not transmitted following a molt of the vector, and when purified and inoculated into the hemocoele, does not make the vector infective.

Nonselective herbicide. A herbicide that kills all vegetation with which it comes in contact.

Nonsymbiotic nitrogen fixation. Biological nitrogen fixation performed by a group of autotrophic bacteria living free in the soil either aerobically or anaerobically and not dependent on plants.

Nonwaxy endosperm (rice). The endosperm that contains both amylose and amylopectin as in most common nonglutinous rices. Amylose content ranges from very low to high.

Noodles. A product of rice flour. A food paste made from rice flour, with egg, and shaped into long strands (ribbon form).

Nozzle delivery. The rate of water discharged from a nozzle during operation, usually expressed in liters per minute.

Nucleus. The distinctive component of eukaryotic cells. It is surrounded by a double membrane, and in which the majority of the genetic material of the cell is organized as chromosomes.

Nucleus seed. The seed derived by a plant breeder from selected core plants to produce breeder seed.

Null hypothesis. The statistical hypothesis that states there is no measurable difference between the observed information and measured data, i.e., it assumes that the treatments have no effect.

Nursery. A piece of prepared land where crop seedlings are raised for a short time pending their planting in permanent sites. A place where a large number of introduced or breeding varieties or lines are first looked at or screened under field conditions.,

Nutrient. A chemical element essential for the growth and development of an organism.

Nutrient antagonism. A reciprocal relationship among concentrations of two or more nutrients in plants.

Nutrient balance. An as yet undefined ratio among concentrations of nutrients essential for plant growth which permits maximum growth rate and yield.

Nutrient interaction. A statistical term used to denote departure from additive responses to two or more nutrients applied separately, and together.

Nutrient medium. Mixture of substances on or in which cells, tissues, or organs can grow, with or without agar.

Nutrient stress. A condition of plant growth when inadequate nutrient supply restricts growth.

Nutritional disorder. Any abnormality of the rice plant caused by a deficiency of any essential element; or toxicity caused by a high level of any substance or ion in the soil; or any retarded growth due to a high osmotic pressure of the soil solution.

O

Obligate. An organism that can sustain life only when thriving in another living organism.

Obligate parasite. A parasite that in nature can grow and multiply only under specific conditions such as on or in living organisms.

Oblique. Sloping, having unequal sides (leaves).

Oblong. Longer than broad with almost-parallel sides.

Observation. Record of the facts and information in a sequence of events exactly, during the research program.

Observation yield trial. A preliminary yield trial with small plots, often unreplicated.

Obsolete types. Varieties of rice that are considered of no importance at present or no longer popular and used by the community.

Obtuse. Blunt or rounded at the end.

Off-season crop. A crop that is planted after the main planting season has been completed.

Off-shoot. A lateral branch of the plant that may not be propagated.

Off-types. Plant that differ in morpho-agronomic characters from the majority or representative plants of a variety; admixtures in a field; obvious contaminants such as tall plants in a semi-dwarf cultivar or vice versa.

Oligogene. A gene controlling the inheritance of a qualitative character or one showing typical Mendelian distribution.

Oligogenic resistance. Resistance governed by a few genes. Major gene resistance.

Oligonucleotide. A polymer of nucleotides usually 5-30 base pairs long.

Oligophagous. Eating only a few specific kinds of food - used especially for an insect subsisting on a few usually related plants.

On-farm trial/on-farm research. Experiment conducted in farmers' fields.

Once-over tillage. An operation whereby land is tilled and planted in the same operation.

Open hull. The lemma and hull cannot close after opening at blooming time.

Opaque. Impervious to the rays of visible light: not transparent or translucent, e.g., the glutinous endosperm of the rice grain.

Open pollination. The natural process of pollinating grains in a plant (not controlled).

Optimum tillage. The combination of tillage operations which maximizes growth of crop plants.

Optimum photoperiod. The day length at which growth duration is at a minimum.

Orbicular. Almost circular.

Organ. A functionally distinct part of an organism consisting of several kinds of tissues such as leaf, root, flower, seed, etc.

Organelle. Membrane-bound cellular compartment (e.g.,nucleus, chloroplast, mitochondrion, Golgi apparatus).

Organic fertilizers. Fertilizer materials derived from plant and animal parts or residues.

Organic matter. The fraction of the soil that includes plant and animal residues at various stages of decomposition and substances synthesized by the soil populations of various organisms.

Organic soil. A soil which contains a high percentage (>15 or 20%) of organic matter.

Organic solvent. A chemical compound (usually liquid) containing carbon to dissolve another substance (the solute).

Organogenesis. Development of individual plant organs such as leaves, tillers, floral organs, or roots.

Origin. In referring to accessions in the germplasm collection, the site where the accession was collected.

Osmosis. The passage of water or another solvent through a semipermeable membrane from a region of low solute concentration to a region of high solute concentration.

Outbreeding. A system of breeding that involves crossing unrelated parents.

Outcrossing. A cross of different genotypes usually under natural conditions. This may also refer to cross-pollination.

Outbreak. A sudden increase in pest population resulting in economic damage to the rice crop.

Oval. Broadly elliptical.

Ovary. The bulbous, basal portion of the pistil which contains one ovule.

Ovate. Egg-shaped, widest toward the base.

Overdominance. An effect in the heterozygote, A/a, which is greater than that in the homozygous dominant, A/A.

Ovule. A minute structure which, after fertilization, becomes a plant seed. The immature seeds before fertilization.

Ovum. Female reproductive cell.

P

P. The parents of a cross. Used also to designate phosphorus.

PAB. Photosynthetic aquatic biomass; the mass of plants and algae which photosynthesize in the water.

Palea. (1). The inner of the two bracts enclosing the floret in grasses. (2). The hardened 3-nerved bract of the floret, which fits closely to the lemma. It is narrower than the lemma. (3). The upper bract that, with the lemma, encloses the flower in grasses.

Panicle. (1). An inflorescence with a main axis and at least primary and secondary branching. (2). The terminal shoot of a rice plant that produces grain.

Panicle axis. The main axis of the panicle from the base to the apex.

Panicle base. The solid node between the uppermost internode of the culm and the panicle.

Panicle development. The growth stage of the rice plant in which the spikelets become distinguishable and the panicle extends upward inside the flag leaf sheath.

Panicle exsertion. (1). Degree of emergence of the panicle from the flag leaf sheath. (2). Growth stage of the rice plant marked by the emergence of the panicle from the boot.

Panicle initiation. The growth stage of the rice plant which starts when the primordium of the panicle has differentiated and becomes visible, changing from the vegetative primordium to a reproductive primordium. Used to predict the flowering date of the rice plant.

Panicle length. The length of the panicle from its base to the apex.

Panicle primordium. The panicle at its rudimentary (earliest) stage of development.

Panicle shattering. When the rice grains fall from the panicle before harvesting.

Panicle type. Term used to classify the angle of the primary branches and spikelet density of the rice panicle, as compact, intermediate, or lax.

Parameter. The true value of an attribute or characteristic of a population.

Parasite. An organism living in or on another living organism, obtaining from it part or all of its organic nutrients, and commonly exhibiting some degree of adaptive structural modification.

Parboiled rice. Rough rice soaked overnight or longer in water at ambient temperature, followed by boiling or steaming the steeped rice at 100 °C to gelatinize the starch. The rice is then cooled and dried before storage or milling. In this way part of the vitamins and minerals of the bran permeate the endosperm and are thus retained in the polished rice.

Parenchyma. A tissue of higher plants consisting of thin-walled living cells that remain capable of cell division even when mature. These cells are agents of photosynthesis and storage and make up much of the substance of leaves and roots and the pulp of fruits as well as parts of stems and supporting structures.

Parthenocarpy. The development of fruit without seeds or fertilization.

Partial dominance. Incomplete dominance.

Partial factor productivity (PFP). The average productivity of a single factor, measured by grain output divided by the quantity of the factor applied. This measure is sometimes referred to as input use efficiency.

Particle density. Dry weight per unit volume of soil solids, normally expressed in grams per cubic centimeter or megagram per cubic meter.

Pathogen. An organism that incites a disease. Examples are pathogenic viruses, fungi, bacteria, and nematodes.

Peat soil. An organic soil consisting largely of undecomposed or only slightly decomposed organic matter accumulated under conditions of excessive moisture.

Pedicel. The stalk supporting a spikelet on the panicle branch.

Pedigree. The ancestral history of an individual or family.

Pedigree method (breeding). A system of breeding in which individual plants are selected in the segregating generations from a cross on the basis of their desirability judged individually and on the basis of a pedigree record.

Pedigree record. Record of the relationships of the varieties or family lines grown and their distinguishing characteristics.

Peduncle. (1). The stalk bearing the entire inflorescence or a solitary flower. (2). Main stalk of the panicle

Penetrance. The frequency with which a gene produces a recognizable effect in individuals which carry it.

Peptide bond or peptide linkage. The C-N bond between consecutive amino acids in a protein. A covalent bond formed between the carboxyl group of one amino acid and the a-amino group of the next amino acid with the release of a molecule of water.

Perianth. (1). A single term for calyx and corolla, usually used when these are not clearly differentiated. (2). The outer envelope of a flower, including the calyx (grouping of sepals) and corolla (grouping of petals).

Pericarp. The wall of the ripened ovary.

Percolation. The downward movement of excess water through the soil.

Perennial. Said of plants or weeds that live for many years, usually flowering each year.

Pericarp layer. The wall of the ripened ovary and the outermost layer of cells covering the caryopsis (seed).

Perimeter. Around the edge, outside border.

Perisperm. Distinguished from endosperm as the nutritive tissue of a seed derived from nucellus and deposited external to the embryo sac.

Permanent hybrid. A hybrid that breeds true to type and does not segregate.

Permanent wilting. Wilting from which a plant will not recover even if soil water content increases.

Permanent wilting percentage. The water content of a soil at which crop plants wilt and fail to recover after soil water content is increased. It is often estimated by the amount of water at 1.5 mega Pascals soil matric potential.

Permeability, soil. The ease with which gases, liquids, or plant roots penetrate or pass through a bulk mass of soil or a layer of soil.

Perennation. The survival of plants from year to year by vegetative means.

Persistent pesticides. Pesticides that remain unchanged in the environment for long periods and are not readily degraded by microorganisms, enzymes, heat, or ultraviolet light.

Persistent virus. A virus that has a long transmission time, is recoverable from the hemolymph of a vector, is transmitted following the molt of a vector, and when purified and inoculated into hemocoele, makes the vector infective.

Pest. An organism which competes with other organisms for food and shelter, or threatens their health, comfort, or welfare.

Pest control. The control of pests in the rice field by selective cultivation methods influencing natural factors or predators or by using chemical/physical control methods to reduce pest damage to rice plants.

Pest outbreak. A sudden increase in a pest population resulting in economic damage to the rice crop; epidemic or epiphytotic.

Pesticide. Any substance used to control pests. It is a broad term that covers insecticide, herbicide, fungicide, bactericide, rodenticide, nematicide, etc.

PET. potential evapotranspiration; the loss of moisture by direct evaporation and transpiration from vegetation; a pressure measured in mm.

Petal. One of the divisions of the corolla, usually colored.

Petiole. Leaf stalk.

pH. A measure of the degree of acidity or alkalinity of a soil or solution. A pH of 7.0 is neutral, below 7.0 is acidic, and above 7.0 is alkaline or basic. It is the negative logarithm of the hydrogen ion concentration of a soil or solution.

Phenology. Periodic developmental processes and events which include processes stimulated by annual environmental fluctuations.

Phenotype. The appearance of an individual or population produced by the genotype in interaction with a given environment.

Physiography. In one sense, an outmoded term for the study of landforms now replaced by the term 'geomorphology'; in another sense it means the combined study of geomorphology, pedology (naturally occurring soils) and biogeography (plant and animal distributions and geographical relationships with their environments over time).

Phloem. A complex tissue in the vascular system of higher plants consisting mainly of sieve tubes and companion cells; chiefly for translocation, structural support, and storage.

Phosphorus-efficient plants. Plants that are able to absorb, translocate, and utilize P more efficiently than other genotypes.

Phosphorylation. Biological process by which proteins are 'decorated' with phosphate groups derived from ATP. The process alters the biological activity of the protein, thereby facilitating a form of physiological regulation.

Photoperiod. The duration of the light period during a given day, which fluctuates during the year and according to latitude.

Photoperiod sensitivity. The degree or extent to which the rice plant responds to a short photoperiod (e.g., day length for flowering.)

Photoperiod-sensitive cultivars. Cultivars that do not flower unless exposed to day length longer or shorter than a critical period.

Photoperiodism. The response (e.g., flowering, germination) of organisms to the relative length of daily periods of light and darkness.

Photosynthesis. A complex process by which the energy of sunlight is absorbed and set to drive chemical reactions. In plants, the major role of photosynthesis is to reduce CO_2 to the level of carbohydrate in the presence of chlorophyll and light and release oxygen as a by-product.

Photosynthesis production. The flag leaf produces the photosynthesis necessary for the rice plant and the growth of the panicles.

Photosynthetic unit. The group of pigment molecules necessary for photochemical action.

Phototropic. Moving toward (positively) or away from (negative) a light source.

Phreatic. Pertaining to groundwater. To indicate groundwater that periodically rises to the rooting zone of cultivated plants.

Physical environment. The climatic factors, the soil, and free water surrounding the rice plant.

Physiological diseases. Diseases that are a result of nutritional disorders caused by the imbalance of the essential elements.

Physiological maturity. When a seed reaches its maximum dry weight. This usually occurs before harvest.

Physiological races. Pathogens that are morphologically inseparable but differ in their ability to attack and develop on their hosts.

Physiology. The processes, activities, and phenomena incidental to and characteristic of life or of living organisms. A branch of biology dealing with the study of the functions and activities of living matter (as of organs, tissues, or cells) as such and of the physical and chemical phenomena involved.

Phytometer. A plant or plants used to measure the physical factors of the habitat by its response in terms of physiological activities.

Phytotoxicity. Plant injury caused by chemicals or some other agent. Common symptoms are spotting, wilting, stunting, tiller spreading, and twisting of leaves.

Phytotoxin. A substance toxic to plants.

Piezometer. An open-ended tube placed vertically in the soil for measurement of the hydrostatic pressure level of groundwater.

Pinnate. (1). A compound leaf having the leaflets or segments arranged on either side of a common axis. (2). Resembling a feather's construction. Leaflets are arranged on each side of a common axis.

Pistil. The female reproductive organ of a flower consisting of the ovary, style, and stigma.

Plankton. Mostly small, floating organisms (flora and fauna) drifting passively, whose movements largely depend on water currents.

Plant anatomy. The study of the structure of plants.

Plant density. Number of plants per unit area.

Plant growth regulator. Broad class of chemicals that control the growth of plants. Many are also natural compounds found within plants, where they may act as hormones.

Plant growth substances. Natural and synthetic compounds that elicit growth, developmental or metabolic responses. These substances are usually not metabolites in the sense that they are not intermediates or products of the pathways they control, and they are active at very low concentrations.

Plant height. The distance from the base of the plant at ground level to the top of the highest leaf or the panicle of a mature plant, measured in centimeters.

Plant nutrient. A substance which is absorbed by plants and is necessary for completion of the life cycle.

Plant physiology. The study of the plant's processes, functions, and responses to the environment.

Plant population. Number of plants in a particular area or the number of different plant types in a particular area.

Plant quarantine. A restrictive measure imposed by a government to prevent the entry of injurious pathogens and pests by regulating the movements of plants, fruits, seeds, or other plant parts.

Plant residue. The stubble or remains of a plant left after harvest.

Plant resistance. Resistance of plants to pests and diseases.

Plant tissue culture. The culture of protoplasts, cells, tissues, organs, embryos, or seed in vitro.

Plant type. A set of plant characters (e.g., tillering, leaf and panicle characteristics, plant height) that give a variety its peculiar architecture and geometry.

Planting ratio. The ratio in which the male and female parental lines are planted in hybrid seed production or maintenance of the cytoplasmic male sterile line.

Plasmid. Nonchromosomal, self-replicating (circular) DNA in bacteria. A principal tool for inserting new genetic information into microorganisms or plants.

Plasticity. The capacity of organisms with the same genotype to vary in developmental pattern, in phenotype, or in behavior according to varying environmental conditions.

Plastid. Generic name for all chloroplast like organelles found in different plant cells.

Plate count. A count of the number of colonies formed on a culture medium which has been inoculated with a small amount of organisms.

Plot. A measured piece of land within a given larger area. It is the unit on which experimental treatments are assigned at random in replicated trials.

Plough layer. The top 15-20 cm of soil normally ploughed or disturbed with tillage operations.

Plough pan (hard pan). A hard layer of soil at 15-20 cm depth developed through continuous ploughing at the same depth. It reduces downward water loss during flooding and prevents penetration of roots.

Ploughless farming. Tilling soil without ploughing so that the crop residue will be left on the surface.

Plumule. The embryonic leaves of the young plant in the embryo. It is enclosed by the coleoptile.

Pluvial rice. Upland rice that depends only on rainfall for water supply.

Pollen. A mature reproductive male germ cell, microscoporocyte, specialized for fertilization.

Pollen grain. The minute spores (spheroidal structures) in the anthers of a floret.

Pollination. The transfer of pollen from the anther to stigma (the female organs) of a flower.

Polycaryoptic (mp). Several pistils with multiple ovaries are present in the spikelet.

Polycross. Intercrossing of a group of cultivars or clones in isolation so as to allow random pollination.

Polycross nursery. Plants specifically arranged to facilitate random outcrossing to develop an improved line or variety.

Polygenes. Numerous genes contributing equally toward a given phenotypic effect as found in the continuous variation of many quantitative traits, each having small and equal effects; mainly additive and subject to environmental influence.

Polygenic resistance. Resistance governed by many genes. Minor gene resistance.

Polymorphism. The occurrence together in the same breeding population of two or more distinct forms at frequencies too great to be explained by recurrent mutation. Occurrence of two or more types of variants.

Polyphagous. Feeding on or utilizing many kinds of food or parasitizing a number of different hosts; specifically feeding on various plants or animals.

Polyploid. An organism with more than two sets of chromosomes. Polyploids can be autopolyploid (duplication of the same set) or allopolyploid (combinations of different sets). More than diploid by multiples of the haploid number.

Polysaccharide. Polymer molecule of sugars (e.g., starch, glycogen).

Popped. Said of rice kernels that have burst open due to exposure to heat.

Population. (1). In genetics, a community of individuals which share a common gene pool at a given site. In statistics, a hypothetical and infinitely large series of potential observations among which actual observations constitute a sample.

(2). A group of individuals (plants) within a species or a variety that are found at one site or field. Plants in the population may or may not be genetically alike.

Pore space. The space in soils not occupied by solid particles.

Post emergence. Factors that are relevant after the emergence of the seedling or weed from the soil. post emergence herbicide- An herbicide applied after the crop and or weeds have emerged from the soil.

Potassium. A reactive alkali metal element essential for plant growth and proper development.

Potassium fixation. The process of converting exchangeable or water-soluble potassium to moderately soluble potassium.

Potassium-supplying power of soils. The capacity of the soil to supply potassium to growing plants in both the exchangeable and the moderately available forms.

Potential, soil water. The potential energy of a unit quantity of water produced by the interaction of the water with such forces as capillary (matric), ion hydration (osmotic), and gravity, expressed relative to an arbitrarily selected reference potential. In practical application, potentials are used to predict the direction and rate of water flow through soils, or between the soil and some other system, such as plants or the outer atmosphere.

Potential yield. The maximum yield of a cultivar in the absence of any abiotic or biotic stress. e.g., yield obtained in the absence of yield-limiting factors such as water, fertilizer, proper cropping practices and in the absence of yield-reducing factors such as pests and natural calamities.

Pounded. Reduced to small particles by beating or striking heavily and repeatedly.

Pre-emergence. The period after sowing before the germination of the seed.

Pre-emergence herbicide. A herbicide applied after the crop has been planted but before the crop and or weeds emerge.

Precipitation. (1). Formation of a solid from a solution as a result of a physical or chemical **change.** (2). A deposit on the earth of hail, mist, rain, sleet or snow. 3). The quantity of water deposited.

Predator. An animal that attacks and feeds on the other animals, such as a bird or spider which feeds on many different species of insects and attacks a number of prey by quickly eating or sucking their body fluids.

Pregerminate. To germinate seed before sowing by exposing it to moist conditions.

Pregerminated seeds. Seeds soaked in water for 24 hours and incubated for 24-48 hours until sprouting is visible.

Pregermination. The process of making the seed sprout before sowing or direct seeding. Soaking and incubating the seed hasten sprouting.

Preliminary trial or experiment. A test or trial to gain initial information that can be used in more detailed and critical experiments later.

Preplant herbicide. A herbicide applied before the crop is planted.

Preventative measures. Actions taken in advance to prevent any undesirable effects from lack of nutrients or pests and diseases.

Primary. (1). Essential or related to first principles, the first objectives (2). The first plant parts such as the root, leaf and tillers (e.g. primary leaf, primary tillers).

Primary growth. Growth initiated by the stem and root meristems.

Primary infection. First infection by a pathogen after going through a resting or dormant period.

Primary leaf. The first seedling leaf, without a blade.

Primary panicle branches. The structures arising from the panicle axis. They bear the secondary branches and primary spikelets.

Primary root system. The first root and its branches.

Primary tillage. Initial cutting of soil where a crop has been grown or harvested. It is commonly referred to as ploughing.

Primary tillers. Tillers arising from the lowermost nodes of the main culm.

Primitive forms. Unimproved plants having features or traits that are akin to wild rice relatives. Plants having primitive features such as pigmented plant parts, long awns, lax panicles, extreme shattering, and perennial growth habit.

Probability. The proportion of times in which an event occurs in an infinitely large and hypothetical series of cases, each capable of producing an event.

Procumbent. Trailing or lying flat but not rooting.

Production function. A statistically estimated function that relates the output of a production system (e.g., rice) to the inputs used in its production (e.g., labor, capital, fertilizer, pesticides).

Productive tiller. A tiller that produces a panicle.

Productive soil. A soil in which the chemical, physical, and biological conditions are very favorable for the production of crops.

Progeny. Offspring; individuals resulting from mating.

Prokaryote. Organisms characterized by the absence of major organelles such as nucleus and plastids.

Propagate. Produce an identical organism from a given organism. To multiply by sexual or asexual reproduction through seeds or cuttings.

Propagule. Part of an organism that may be disseminated.

Prostate. Lying flat on the ground.

Protein. A biopolymer consisting of many different amino acids linked by peptide bonds. Proteins can be more than 100 amino acids long and are very complex in structure. Protein content- The amount of protein in a living organism based on the analysis for total N. The percent protein is computed by multiplying the percent N by 6.25.

Protoplasm. That portion of cellular tissue active in life processes.

Protoplast. Plant cell from which the cell wall has been removed, usually by enzymatic process.

Protoplast fusion. An in vitro technique of producing somatic hybrids between two cultivars, species, or genera. Protoplast fusion provides the unique opportunity to exploit cytoplasmic variability in two parental lines.

Prune. To cut off what is not needed, such as to prune tillers so that the number of tillers in each entry is uniform.

PSP. photoperiod-sensitive phase; the growth stage when a plant becomes sensitive to photoperiod.

Pubescent. Having a fuzzy surface; specifically, covered with fine soft short hairs.

Puddle. To stir up; make muddy or turbid. Puddled soil- A soil whose structure has been mechanically destroyed, allowing the soil to run together when saturated with water. A soil that has been puddled is usually in a massive nonstructural state. Puddling. Turning the soil into a muddy or watertight paste through secondary tillage.

Pulse (crop). The dried edible seeds of cultivated legumes; applied especially to leguminous crops grown in tropical and subtropical Asia.

Pure culture. The growth of a single species or strain of an organism in the absence of any other living species or strain.

Pure line. (1). A line that has been made almost completely homozygous by repeated self-pollination and selection of a specific type (or by the removal of off-types) over generations. (2). A group of identical individuals that always produce offsprings of the same phenotype when intercrossed.

Pure line selection. A method of selecting the best progeny for breeding from individual plants of a variety or cultivar.

Pure seed. Has a high level or purity with little or no contamination of other varieties or species. Seeds give rise to genetically similar plants.

R

Race. A genetically and often geographically distinct mating group within a species; or a group of pathogens that infect a given set of plant varieties.

Raceme. A more or less conical inflorescence within flowers arising alternately from a common axis, the youngest towards the tip.

Rachilla. The small axis below the lemma and palea, bearing the single floret.

Rachis. (1). The axis of a spike or raceme, or of a compound leaf. (2). The elongated axis of a panicle.

Radicle. The embryonic primary root unsheathed by the coleorhiza and the root cap, persisting only for a short time after germination.

Rainfed field. An unirrigated field surrounded by levees and depending on rainfall for moisture supply.

Rainfed lowland. Bunded fields where water depth does not exceed 50 cm for more than 10 consecutive days and the fields are inundated for at least part of the season. Such fields have no access to an irrigation system, but may have on-farm rainwater conservation facilities.

Rainfed rice. Rice grown on leveled bunded fields to allow an accumulation of flood water on the surface during heavy rains. Grown in areas that depend entirely on rain for moisture for its water supply.

Random. Arrived at by chance without the exercise of any choice.

Range. The variation of data from highest to the lowest value.

Rating. Classifying test entries based on degree of plant damage, number of insects, etc., by expressing on a numerical scale, usually 0-9. It is relative estimate.

Ratoon. New tillers that grow from the stubble of harvested plants and may bear grains. These new tillers constitute the ratoon crop. Ratoon cropping-- Cultivation of growth from stubble after a crop has been harvested to develop another crop. Conducted in level bunded fields where water is readily available for irrigation.

Raw rice. Dehulled and milled rice not subjected to steaming or parboiling.

Ray floret. One of the outer tubular flowers of a composite flower head when different from those of the center.

Rear. To raise or multiply organisms such as insects.

Receptive. When the plant's physiological or physical status allows the entrance or growth of a foreign body.

Receptor. Protein molecule, usually on the cell surface, that is able to receive and interpret an external signal.

Recession. Complete draining of flood water from the land.

Recessive gene (recessive allele). A gene the expression of which is either partially or entirely suppressed when a dominant allelic gene is present. Recessive gene is the alternate form of the dominant gene.

Reciprocal cross. A set of crosses between two parents where either is used as male or female, or where the source of male and female gametes are reversed e.g., A x B & B x A.

Reclamation. The act or process of restoring land for cultivation or use.

Recombinant. One of the progeny formed as a result of genetic recombination.

Recombinant DNA. DNA molecule composed of segments from different individuals or species and formed by a combination of in vitro and in vivo manipulations, typically in bacteria. Genetically engineered DNA brings together new combinations of genes not found in nature.

Recombinant DNA technology (syn. gene cloning, plasmid engineering, genetic engineering, biotechnology, gene therapy). The techniques by which genetic recombination is carried out in vitro and in vivo. It entails the breakage and rejoining of DNA molecules from different organisms and the production and isolation of the modified DNA or fragments thereof.

Recommended rate. The amount of fertilizer materials required for application to a field expressed in kilograms nitrogen (N), phosphoric acid (P_2O_5) and potash (K_2O) per hectare or number of bags of FM per hectare.

Recommended variety. The variety that has the desirable traits and attributes for a particular environment and can be expected to produce and grow well in that environment, recommended by competent authority/agency for cultivation in that environment.

Recovery. Plants that have the ability to grow after being exposed to unfavorable seasonal conditions or after being attacked by pest or disease.

Recurrent parent. The parent to which successive backcrosses are made in backcross breeding.

Recurrent selection. A method of breeding designed to concentrate favorable genes scattered among a number of individuals.

Recurved. Bent backward.

Red rice. A rice kernel that has a red seed coat frequently found in African rice *Oryza glaberrima* or some *Oryza sativa* cultivars.

Redistribution (water). The process of soil water movement to achieve an equilibrium energy state throughout the soil profile.

Reduced tillage. A tillage sequence in which the primary operation is performed in conjunction with planting procedures in order to reduce or eliminate secondary tillage operations.

Reduced zone. A portion of a flooded soil devoid of free oxygen.

Reduction division. A nuclear division where the chromosomes are reduced from the diploid to the haploid number.

Regeneration. The growth of whole plants from cultured cells or calli.

Registered seed. The progeny of foundation seeds normally grown to produce certified seed.

RGR. Relative growth rate; the increase in plant material per unit of material in a specified time.

Regression. The dependence of one variable on one or more other variables.

Regression, coefficient of. A numerical measure of the rate of change of the dependent on the independent variable.

Regression line. A straight or curved line that indicates the nature of a regression on a graph with the dependent variable on the y axis and the independent variable on the X axis.

Rejuvenate. To re-grow a variety or stock of seed when the supply has been exhausted, lost, or destroyed.

Relative humidity. The ratio of the amount of water vapor in the air at a specific temperature to the maximum moisture-holding capacity of the air at that temperature.

Relay cropping. Growing two or more crops in sequence, planting the succeeding one after the flowering but before the harvest of the former.

Remnant seed. Seed that is in excess and remains after sowing for a given test.

Repellants. Chemicals that prevent pest damage to living organisms or materials by rendering them unattractive, unpalatable, or offensive.

Replicate/replication. (1). The assignment of a treatment to more than one experimental unit. It provides a means by which to measure experimental error, which is necessary for detecting real differences between treatments. (2). The process of producing complete copies of the genetic material (DNA or RNA) of a cell or virus.

Reproduction. The sexual or asexual process or mechanism by which the species multiplies and is maintained.

Reproductive phase. The period of crop growth from panicle initiation to panicle maturation.

Reproductive system. All the organs and structures involved in the production and delivery of gametes or reproductive cells.

Residual. The remainder of a chemical or a quality after the original application, or in statistical analysis of variance, the remainder when all true values are subtracted from observed or calculated values or values due to known sources of variation.

Residual effect. The effect of chemical fertilizers, herbicides, insecticides, or fungicides that remains in the following crop season without fresh application of the chemical.

Residue. The plant part left after the economically useful part has been harvested.

Residue processing. Operations that cut, crush, or otherwise break (fracture) residues in a step preparatory to tillage, harvesting, or planting operation.

Resistance. The inherent ability of an animal or plant body to resist (oppose, counteract) untoward circumstances such as insect attack, diseases, toxic agents, or infection. Resistance to pesticides-- The ability of populations of pests to survive doses of a pesticide which are normally lethal. Resistant- Possessing the ability to resist or counteract.

Resistant variety. A variety which has genetic resistance to an adverse factor or pest. It is genetically able to suppress or retard the activity of a pathogen or insect. A variety which has genetic resistance to or tolerance for all adverse factors or pests.

Respiration. (1). The process by which living organisms obtain energy through oxidation of carbohydrates to produce carbon dioxide. (2). The metabolic process by which a plant (or animal) oxidizes its food and provides energy for assimilation of breakdown products. (3). The process by which cells or tissues obtain oxygen.

Response. The reaction of a plant or group of plants to a treatment, e.g., application of fertilizer, stimulus, or water loss.

Restorer gene. The gene that restores fertility to an otherwise sterile plant.

Restorer line. A pollinator variety used to pollinate a male sterile line to produce Fl progeny which are male fertile and thus produce seeds on selfing. An inbred line that permits restoration of fertility to the progeny of male sterile lines to which it is crossed.

Resurgence. The occurrence of significantly more damage or more insects in an insecticide-treated crop after insecticide application.

Reticulate. In the form of a network; net-veined.

Reverse mutation. The process whereby the original state is restored for an inherited or mutant gene or chromosome.

Reversible mutation. A mutation that can be restored to the original state.

Reversion. The changing of essential plant nutrient elements from soluble to less soluble forms as a result of interaction with, or reactions in, the soil.

Rhizobia. Bacteria capable of living symbiotically in roots of legumes from which they receive energy and often utilize molecular nitrogen.

Rhizome. An underground stem--distinguished from a root by the presence of buds and scales--which is capable of creeping and producing new shoots.

Rhizosphere. The zone of soil where the microbial population is altered both quantitatively and qualitatively by the presence of plant roots.

Rib. The ridge of a fruit.

Ribbed. Having prominent ribs.

Ribonucleic acid (RNA). (1). Occurs in three major forms -- as messenger RNA, transfer RNA, and ribosomal RNA. Some RNA molecules are the genetic molecules of viruses (e.g., tobacco mosaic virus and HIV) and viroids. Some RNA molecules may have enzymatic activity (see ribozymes). (2). Any of various nucleic acids that yield ribose as one product of hydrolysis, that are found in the cytoplasm and sometimes in the nucleus. (3). Required for the synthesis of proteins in all living systems It consists of a linear array of the nucleotides cytidine-, guanosine-, adenosine-, and uridine-s'-monophosphates.

Ribozyme. Enzyme made entirely of RNA.

Rice blast. A fungus disease of rice which is characterized by the presence of spindle-shaped lesions. The lesion develops a grayish center and a brownish margin. The disease can have different forms: leaf blast, node blast, or neck blast.

Rice bug. An insect that sucks the sap of the developing grains, causing serious rice crop losses. The species of rice bugs of major economic significance are *Leptocorisa acuta, L. oratorius,* and *L. chinensis.* The nymphs and adults are difficult to recognize in the rice field because of their color. The freshly hatched nymphs are tiny and green but become brownish as they grow. Both nymphs and adults feed on rice plants.

Rice gall midge. The Asian rice gall midge, *Orseolia oryzae,* is a serious pest of rice in South and Southeast Asia. The adult gall midge is about the size of a mosquito. Gall midge attack starts in the seedbed and continues until the booting stage, but most of the damage is limited to the vegetative growth stages. The damage changes rice tillers into tubular galls, which dry without bearing panicles. New tillers are initiated as the older ones are infested, causing profuse tillering and stunting of the plants. A related species, *O. oryzinova,* is prominent in Africa.

Rice leaffolder. This insect, *Cnaphalocrocis medinalis Guenee,* is known to occur in all rice-growing countries of Asia. The full grown larva is yellowish green with a dark brown head. The damage is caused by the larvae which fold the leaf blades into a tubular shape and feed on green leaf tissue within the tubular structures.

Rice processing. The milling, polishing, and grading of rough rice. The processing of rice by parboiling.

Rice stalk. The stem or culm portion of the plant.

Rice transplanter. (1). A person transplanting rice. (2). A machine used for transplanting rice.

Rice whorl maggot. Dull gray in color, Hydrellia sp. is widespread in the Philippines. The maggots feed on the unopened whorl of the leaf by remaining in the center of the whorl and nibbling the innermost margin of the leaf. The symptoms manifest as small chewed discolored areas on the innermost margin of the central whorl. Severe infestation stunts the plant and reduces tillers in a given hill. The infestation and damage are limited to vegetative stages of the crop.

Ridge. The relatively higher part of floodplain landscapes (in contrast to basins and channels). Raised strips of ploughed ground, rows of soil lying between furrows.

Ridge planting. A method of planting crops on ridges formed through tillage operations or banked by hand tools.

Ripening phase. (syn. maturity phase, grain-filling phase). The period from milky stage of the grain to maturity.

Rodents. Rats, mice, and guinea pigs.

Rogue. A plant that is different from the standard variation or strain.

Roguing. Judicious removal of infected and undesirable individuals from a pure population (variety) to prevent spread of disease and to purify the stock.

Root. The underground part of a seed plant body that originates usually from the hypocotyl, and functions as an organ of absorption, aeration, and food storage or as a means of anchorage and support. Types: (a) primary. the root developing from the radicle that first appear from the seed; (b) secondary. developing from the primary root; branch roots; (c) tap. a primary root that enlarges and grows downward.

Root gall. Abnormal enlargement of the roots caused by root-knot nematode.

Root hair. Small hairs that grow from a main root. The tubular or filamentous outgrowth of epidermal cell or root that has immediate contact with soil particles assisting in the absorption of water and minerals on the surface of the root. The root hair is short-lived and replaced.

Root nodules. The small enlargement or swelling on roots of legumes produced as a result of infection by nitrogen-fixing bacteria.

Rootlets. The lateral roots that grow indigenously from the mature roots.

Rosette. A cluster of radiating leaves, usually arising at or close to the ground.

Rotary hoeing. A tillage operation employing ground-driven rotary motion of the tillage tool to shatter and mix soil and control small weed seedlings.

Rotary tiller. (1). An implement that accomplishes primary and secondary tillage in one operation. (2). A tillage operation employing power-driven rotary motion of the tillage tool to loosen, shatter, and mix the soil.

Rotate. To move around, to grow crops in succession or in a regular sequence.

Rotation. A cropping system using the rotation of different plant species on the same plot of land.

Rough rice. (syn. paddy rice) Called paddy, threshed, unmilled or whole rice grain, or caryopsis.

Row. The lines along which a number of plants are planted or arranged.

Row marker. A marker or device used for seeding in rows or aerial seeding of large areas using satellite marking.

Row ratio. The proportion of seed parent and pollen parent rows planted to maintain cytoplasmic male sterile line or to produce Fl hybrid seed in a seed production plot.

Row seeding or sowing. To be sown in rows and not broadcast by hand.

Rudimentary. A development that is arrested at an early stage or imperfectly developed.

Rudimentary auricles. (au) Auricles reduced to protuberances with 1 or 2 cilia or without cilia.

Rudimentary glumes. The laterally enlarged cup like apex of the pedicel; they are beneath the sterile lemmas.

Rugose. Wrinkled.

Run-off. The overflow of water from a field carrying with it dissolved nutrient or nutrient-containing soil particles.

Runner. A slender, trailing stem rooting at the nodes.

Rupture. To burst open along irregular lines.

S

S1, S2, etc. Symbols designating first-selfed generation (progeny of S0 plant), second-selfed generation (progeny of S1 plant), etc.

Saline soil. A soil that contains sufficient salt in the root zone to impair plant growth. Its electrical conductivity is greater than 4 dS/m and exchangeable sodium percentage is less than 15. Salinity - The state of consisting of or containing salt.

Saline-sodic soil. A soil containing soluble salts with sufficient exchangeable sodium to interfere with the growth of most crop plants. Soil whose saturation extract has an electrical conductivity greater than 4 dS/m, sodium adsorption ratio greater than 15, and pH less than 8.5 (See saline soil and sodic or alkali soil).

Salinization. The process whereby soluble salts accumulate in the soil.

Salt balance. The relation between the quantity of dissolved salts carried to an area by irrigation water and the quantity of dissolved salts removed by drainage water.

Salt tolerance. Ability of a plant to maintain productivity as salinity increases beyond and EC of 4 dS/m.

Salt-affected soil. Soil that has been adversely modified for the growth of most crop plants by the presence of soluble salts, exchangeable sodium, or both.

Sample. A finite series of observations taken from a population; a portion of a population. Individuals taken from a population to represent it.

Sample size. The number of sampling units to be measured per plot, e.g., number of plants to be measured per unit area.

Sampling unit. The unit designated within the experimental plot for the measurement of a character to be made. A good sampling unit must be uniform, easy to identify, and easy to measure.

Sand bar. A bank or ridge of sand deposited in a river channel.

Saprophyte. An organism that subsists on dead organic matter and inorganic materials; nonpathogenic.

Saturate. (1). To be totally wet, to fill all the voids between soil particles with a liquid. (2). To fill completely or load to capacity.

Saturation extract. An increment of solution obtained from a saturated soil paste.

Scabrous. Rough.

Scarification. Treatment to rupture the seed coat sufficiently to permit entry of water to initiate germination.

Scarify. To loosen the topsoil aggregates by means of raking the soil surface with a set of sharp teeth harrows or other instrument.

Scarious. Applied to leaves or leaflike parts that are not green, but thin, dry, membranous, and often more or less translucent.

Scented kernel. Grains containing aromatic substances in the endosperm which give a strong aroma when the polished rice is cooked.

Screen. Process of accepting or rejecting rice cultivars or breeding lines for future use.

Screening. Evaluation of varieties or breeding lines for resistance where the resistant ones are selected for further studies and possible use as donors in the breeding program and the susceptible are eliminated.

Scurfy. Scaly or a scaly coating.

Scutellum. A shield-shaped organ of the embryo of grass. It is often viewed as a highly modified cotyledon in monocotyledons.

Second leaf (of a seedling). The first differentiated leaf, with a blade and sheath.

Secondary adventitious roots. Roots arising from the nodes at the base of the plant.

Secondary infection. An infection resulting from inoculum arising from a primary infection.

Secondary panicle branches. The branches that bear the spikelets arising from the primary branch of the panicle.

Secondary pests. Pests which because of natural control agents normally have low populations and do not require chemical applications for control.

Secondary tillage. The cultivation that follows primary tillage; it breaks clods, incorporates plant materials into the soil, and levels the soil. It is often called harrowing or hoeing.

Secondary tillers. Tillers that arise from the primary tillers.

Sedges. Members of the family Cyperaceae. They bear a close resemblance to the grasses and can be distinguished by a thin triangular stem, the absence of ligules, and the fusion of leaf sheaths forming a tube around the stem.

Sediment. Solid particles transported and deposited by water, glaciers and wind.

Seed. The fertilized and ripened ovule of a seed plant, comprising a miniature plant usually accompanied by a supply of food (endosperm) enclosed in a protective seed coat, often accompanied by auxiliary structures, and capable, under suitable conditions, of independent development into a plant similar to the one that produced it. In rice, the grain is the common form of seed; the caryopsis is the true seed.

Seed box. A rectangular box specially constructed with the open top filled with good soil for germinating seeds.

Seed coat. In rice, the seed coat or testa is firmly adhered to the maternal pericarp as one of the protective layers of the seed kernel.

Seed distribution. Dispatch of seed to other users. The way in which seed is distributed on the soil.

Seed dormancy. A physical or physiological condition of a viable seed that prevents germination even under favorable germination conditions.

Seed dressing. An application of chemicals or fungicides to cover and protect seeds from parasites, pathogens, or adverse environmental conditions that cause damage to the seed or seedling.

Seed health. Refers primarily to conditions of the seeds or seedlots as planting materials which may be affected by the presence of pathogens, insects and contaminants such as weeds.

Seed incubation. Process by which moisture, temperature, and/or light is provided to ensure optimum conditions for seed germination.

Seed multiplication. Growing the seed of a particular variety on a large scale for sale to farmers or for use in experiments.

Seed physiology. The study of processes involved with seed development, in germination and its control, and in the utilization of seed reserves during the early stages of seedling growth.

Seed purity. The seed being free of contamination from other genotypes, dirt, or foreign matter.

Seed quality. (1). The properties of the seed such as cleanliness, germination percentage, moisture content, shape, purity, and desirable characters for research. (2). The export qualities of the seed, the size, shape, moisture content, protein content, color, chalkiness, and other properties as desired by the millers or traders.

Seed setting. The production of seeds after pollination.

Seed treatment. The process of coating or impregnating seeds with a chemical.

Seed viability. The ability of seeds to germinate under favorable conditions and develop normal seedlings.

Seed vigor. Seed property that determines its potential for rapid uniform emergence and development under a wide range of field conditions.

Seed-borne. Carried within, by, or through the seed.

Seedbed. The bed on which rice seeds are sown, consisting of soil (wet bed method) or banana leaves, plastic sheets or concrete floor (dapog method).

Seeder. A machine that plants the seed in the field. Small seeders are used for small experiments; very large seeders are used on large farms.

Seeding. Planting of the seed with machine or by hand.

Seeding rate. The amount of seed planted per unit area of land.

Seedling. The stage of the plant after seed germination until the development of the fifth leaf.

Seedling blight. A rice disease caused by *Sclerotium rolfsii* or *Drechslera oryzae* and characterized by yellowing, wilting, and finally, drying of seedlings in seedbeds. White mycelia and/or sclerotia are sometimes observed at the base of infected seedlings or germinating seeds.

Seedling emergence. Coming up of the seedling from the surface of the soil or water.

Seedling height. The height of a seedling from the base of the shoot to the tip of the tallest leaf blade.

Seedling rate. The number of seedlings planted per unit area of land.

Seedling stage. The period when approximately the first five leaves and tillers begin to develop.

Seedling vigor. Seedling properties which determine the potential for rapid growth and development of uniform plants under a wide range of environmental conditions.

Seepage. The lateral flow of water into or from a soil, as from a body of water into neighboring soil, or the reverse.

Segregant. Any plant from a segregating population (F2 or F4 generations).

Segregation. The separation of alleles that occurs typically during meiosis.

Selective herbicide. An herbicide that kills or stunts some plant species but causes little or no injury to others.

Selective marker. A marker that permits the selection of recombinants over the parental types.

Selective pesticide. A pesticide that kills a certain pest or group of pests and causes little or no injury to others.

Selectivity. The quality of being able to select, to take what you want.

Self. Self-pollination by natural or artificial means.

Self-incompatibility. The genetic or physiological reason that fruit cannot be formed by selfing (fertilization).

Self-pollination. Pollination between the pollen and stigma within the same flower of the same plant.

Self-fertilization. Fusion of male and female gametes from the same plant.

Senescence. The phase of plant growth that extends from full maturity to actual death; characterized by an accumulation of metabolic products, increase in respiratory rate, and a loss in dry weight especially in leaves and fruits.

Sensitive period. The plant's growing period when it is most affected by the prevailing environmental factors such as day length and disease or insect attack.

Sensory test/evaluation. Method for the assessment of cooked rice 1 hour after cooking by judges or panel members. The characteristics assessed are aroma, flavor or taste, tenderness or hardness, cohesiveness or stickiness, appearance, and color using a six-point scoring scale.

Sepal. A segment of the calyx, usually green.

Septate. Having cross walls. Divided into cells or segments by partitions.

Septum. (1). A cross wall in a fungal hypha or spore. (2). The cross-sectional layer inside the node which separates adjoining internodes.

Sequence (biotechnology). Linear order of the 4 nucleotide bases along a DNA or RNA molecule.

Sequencing. The determination of the order of amino acids in a peptide, polypeptide chain, or protein, or the determination of bases (nucleotides) in a nucleotide, polynucleotide strand, or nucleic acid.

Sequential cropping. Growing two crops in rapid sequence, planting one immediately after the harvest of the first, on the same piece of land.

Sequential varietal release. A breeding technique that aims at incorporating different resistance gene(s) into similar improved varietal backgrounds and releasing them sequentially, enabling farmers to choose the variety on the basis of biotype/race identified. As the biotype/pathogen population shifts and the variety becomes susceptible, a new variety with a different gene is released.

Serrate. Saw-toothed.

Sessile. Resting on a main stem or branch without an intervening stalk.

Sex ratio. The number of males in relation to the number of females.

Sexual reproduction. Reproduction involving the union of two compatible gametes.

Shattering. Separation of grains from the panicle. Grain shattering or easy threshing are partly correlated with the degree of development of the abscission layer between the spikelet and the facet of the pedicel.

Sheath. The lower part of the leaf enclosing the stem, originating from the node and wrapping around the culm above the node.

Sheath blight. A disease caused by the fungus *Rhizoctonia solani / Thanatephorus cucumeris*. The disease is characterized by ellipsoidal, necrotic lesions occurring initially at the leaf sheaths near the water line. Lesions may enlarge and spread on leaves and ultimately on panicles.

Sheath blotch. A disease incited by *Pyrenochaeta oryzae*, symptoms are brown blotches on the leaf sheath, leaf blade, and glumes.

Sheath rot. A fungal disease caused by *Acrocylindrium oryzae.* Rotting occurs on the uppermost leaf sheaths enclosing the young panicles. The lesions start as oblong or somewhat irregular spots, 0.5-1.5 cm long, with brown margins and gray centers, or they may be grayish brown throughout. Powdery growths may be found inside affected sheaths.

Shed. (1). To drop or disperse pollen grains or the falling of unharvested rice grains from the panicles. (2). A structure used for storage.

Shoot. The vegetative parts (leaves, stem) of a plant above the ground level.

Short-term crops. Crops that grow for three months or less.

Short-day plant. A plant that needs shorter daylight to flower.

Short-duration varieties. Varieties that mature within 105 days or less.

Shrub. A small perennial tree with no main trunk. A woody plant with spreading branches arising from the base.

Side dressing. Fertilizer usually applied near, beside, or in between the plants.

Sieve tube. A tube consisting of an end to end series of thin-walled living cells characteristic of the phloem, having no nucleus when mature, and believed to function chiefly in translocation of organic solutes.

Sieve. A meshed device through which material is strained to separate particles of different sizes.

Significance, test of. A statistical procedure designed to distinguish real treatment difference from that due to chance (i.e., random variation). To distinguish differences due to sampling error from differences due to discrepancy between observations and hypothesis.

Significance. The clear difference between that observed and that expected.

Significant difference. A statistical reality used to determine the differences among plant comparisons; usually expressed to have a certain level of probability of being wrong due to chance.

Silt. A soil separate consisting of particles between 0.05 and 0.002 mm in diameter.

Silvershoot. A symptom of gall midge infestation, the growing points and tillers elongate and form like a white tube.

Simulation. The statistical process or act of resembling or having the characteristics of something.

Single cropping. The growing of only one crop in a given place in one year.

Single cross. A cross between two genotypes, usually two inbred lines.

Single fertilizers. Fertilizers that contain only one of the major fertilizer elements.

Single-copy sequence. A sequence of nucleotides that occurs only once in a genome.

Sinica (japonica) race. Rice grown in temperate zones and subtropics.

Slow-release fertilizer. Fertilizer that releases the wanted element at a slow rate. A term used interchangeably with delayed release, controlled release, controlled availability, slow acting, and metered release to designate a rate of dissolution (usually in water) much less than is obtained for completely water-soluble compounds.

Soak. To be completely covered by water for a determined length of time or until the object is saturated.

Sod planting. A method of planting in sod with little or no tillage.

Sodication. The process whereby the exchangeable sodium content of a soil is increased.

Sodicity. The quality or degree of having sodium as a component.

Sodium adsorption ratio (SAR). A relation between soluble sodium and soluble divalent cations which can be used to predict the exchangeable sodium percentage of soil equilibrated with a given solution.

Soil. The unconsolidated mineral matter on the surface of the earth that serves as a natural medium for growth of land plants. This has been subjected to and influenced by genetic environmental factors of parent material, climate (including moisture and temperature), macro- and microorganisms, and topography, all acting over a period of time and producing a product--soil--that differs from the material from which it is derived in many physical, chemical, biological, and morphological properties.

Soil auger. A tool for taking soil samples.

Soil fertility. The ability of the soil to supply nutrient elements in the amounts, forms, and proportions needed for maximum plant growth.

Soil fumigation. A method of soil treatment using chemicals wherein the pesticide, in gaseous form, is introduced into the soil by a soil injector; the soil or field position to be treated is first covered with a plastic sheet to prevent volatilization of the chemicals; the soil fumigant is highly toxic as it kills all forms of life--fungi, bacteria, nematodes, snails, and weeds.

Soil heterogeneity. The case where soil in a relatively small area varies greatly in texture, fertility, topography, moisture content, and drainage. Lowers the precision of results in rice field experiments.

Soil horizon. A soil layer with clear and relatively uniform character running roughly parallel to the soil surface.

Soil microbiology. A sub-specialization of soil science concerned with soil-inhabiting microorganisms and their relation to agriculture, including both plant and animal growth.

Soil moisture. Water contained in the soil, expressed as a percentage of weight of water per unit weight of dry soil or the percentage of volume of water per unit volume of bulk soil.

Soil pH. A measure of the degree of acidity or alkalinity of a soil. A pH of 7.0 is neutral; below 7.0 is acidic, and above 7.0 is alkaline.

Soil population. All organisms living in the soil, including plants and animals.

Soil productivity. The capacity of the soil, in its normal environment, to produce a specified plant or sequence of plants under a specified system of management. It emphasizes the capacity of the soil to produce crops and is expressed in terms of yield.

Soil profile. A vertical section of the soil cutting through all its horizons and extending into the parent material.

Soil saturation extract. The solution obtained through suction of soil saturated with water.

Soil science. The science dealing with soils as a natural source on the surface of the earth including soil formation, classification and mapping, and the physical, chemical, biological, and fertility properties of soils per se and these properties in relation to their management for crop production.

Soil structure. The combination or arrangement of individual soil particles into definable aggregates or peds which are characterized and classified according to size, shape, and degree of distinctness.

Soil test. A chemical, physical, or microbiological operation which estimates the properties of the soil pertinent to the suitability of the soil to support plant growth.

Soil texture. The relative proportions of sand, silt, and clay particles in a soil.

Soil tilth. The quality of soil, keeps it from packing together.

Solar radiation. The energy given off by the sun. The sun exerts an energy level of about 2 cal/cm^2 per minute on the earth's surface.

Sole cropping. Growing one crop alone or in pure stand, either as a single crop or as a sequence of single crops within the year.

Soluble ion. The electrically charged form of an atom or group of atoms that exists in solution.

Solution. A homogeneous mixture formed by mixing a solid, liquid, or gaseous substance with a liquid or sometimes with a gas or solid.

Somaclonal variation. Increase in genetic variability in plants regenerated from tissue culture.

Somatic. Relating to or affecting the cell body.

Somatic cell. (1). A cell that is not destined to become a gamete, a "body cell" whose genes will not be passed on to future generations. (2). The somatic cell divides to form tissues, etc. It is a body cell that has full chromosome content.

Somatic cell hybridization. (1). The fusion of non-germ cells in cell culture under certain treatments and the formation of viable hybrid cells. (2). A breeding method that uses protoplast fusion to produce somatic hybrids between otherwise sexually compatible species.

Source nursery. Breeding nursery where all genetic material including sources imparting cytoplasmic male sterility, genotypes with specific traits useful for hybrid breeding program, and elite rice lines showing high general and specific combining ability are maintained for use in a hybrid breeding program.

Spacing. Distance of planting between hills and between rows.

Spathe. A large bract partially or wholly enclosing a group of flowers.

Species. The unit of taxonomic classification into which genera are subdivided. A group of similar individuals different from other similar arrays of individuals. The main characteristic is reproductive isolation.

Specific combining ability. Deviation in the performance of a cross from the performance predicted on the basis of general combining ability of its parents.

Specific gravity. The ratio of the density of a substance to the density of another substance (usually water at 4°C) taken as a standard when both densities are obtained by weighing in air.

Specific leaf area. Leaf area per unit dry weight (cm^2/g).

Specific leaf weight. Weight per unit leaf area (g/cm^2).

Specimen. The sample and reference material.

Spectroradiometer. An instrument used for high-accuracy, automated spectral measurements over the UV to the infrared range.

Spike. An unbranched elongated inflorescence with sessile (without a stalk)or almost-sessile flowers.

Spike-tooth harrow. A tillage implement used to cover seeds that are broadcast or drilled in furrows.

Spikelet. (1). The basic unit of the rice inflorescence consisting of two sterile lemmas, the rachilla, and the floret. (2). The unit of the grass flower that includes the two basal glumes subtending one to several florets.

Spikelet abortion. Spikelet development checked due to some stress factor.

Spikelet sterility. A situation in which there is no grain within the glumes of the rice plant.

Split application. Application of chemical or fertilizer divided into two or more parts and applied at given intervals.

Spore (fungi). A single- to many-celled reproductive structure of a fungus.

Sporocarps. (1). The spore-containing structures. (2). The fruit cases of certain ferns containing sporangia or spores.

Sporophytic. A condition in which sterility/fertility is imparted to the pollen by the plant upon which the pollen is borne and the genotype of the pollen has no bearing per se. It may be controlled by more than one gene with multiple alleles.

Spreader. An agricultural implement used to disperse seed or fertilizer in the field.

Sprout. A germinated seed; to grow from the seed or to grow from another part of the plant.

Stable line. A breeding line that does not change its characteristics.

Stages of flowering. The four phases of a flower's development--floral initiation, floral organization, floral maturation, and anthesis.

Staggered planting. Planting different fields in a community or on a farm over a period of several weeks, in contrast with simultaneous planting wherein all fields are planted over a period of a week or less.

Stained grain. The discolored grain usually due to fungal disease or infection.

Stamen. The part of the flower bearing the male reproductive cells; composed of the anthers on the filament (stalk).

Stand. A group of plants or trees growing together in a given area.

Standard check. A rice variety or breeding line whose response to a treatment is known and which is repeatedly included in tests as basis for comparing the reactions of tests entries. A resistant and a susceptible check are usually included.

Staple crop. The crop grown as the basic diet of the population of a particular country or area.

Starch. (1). The major storage carbohydrate of plants; a polymer composed of D-glucose units occurring as amylopectin and amylose and found in chloroplasts, amyloplasts, and endosperm. (2). Major carbohydrate source for animals.

Statistics. (1). An analysis and interpretation of numerical data used to define scientific **experiments.** (2). An orderly arrangement of facts collected for study in general information--e.g., production of a crop or a country.

Stem. The main support structure of the aboveground part of a plant from which leaves, tillers or branches, flowers, and fruits develop.

Stem elongation. (1). Increase in length of the upper stem internodes which begins before panicle initiation. This process is associated with exsertion of the panicles from the sheath to allow flowering and pollination. (2). Increase in length of internodes of deepwater rice in response to flooding.

Stem lumen. The hollow, tubular section of the plant internode, also called the internodal lacuna, and pith or medullary cavity.

Sterile. Failing to produce or incapable of producing offspring.

Sterile glumes. Glumes at the base of the spikelet of the rice plant which do not produce a flower.

Sterile lemmas. The two flowerless bracts at the base of the spikelet.

Sticky. Grains that stick together when cooked. A characteristic of glutinous rice.

Stigma. The terminal part of the pistil to which pollen grains adhere.

Stigma receptivity. The characteristic of a stigma that facilitates germination and penetration of pollen upon deposition on its surface.

Stipule. Leaf-like appendage at the base of the leaf stalk.

Stock. Purified seed in storage for future multiplication; the available amount of seed.

Stock culture. An insect culture which serves as a source of insects for beginning a rearing program.

Stock seed. Seed stored for multiplication.

Stolon. A modified aboveground stem creeping and rooting at the nodes.

Stomata. Small openings in the epidermis of the leaf or other plant parts which control exchange of gases and water vapor; controlled by the guard cells.

Straggling. Spreading out irregularly.

Strain. A group of plants within a variety that share common, recognizable features in morpho-agronomic characters, physiological differences, or reactions to specific disease pathogens or insect pests. A strain, may include several lines with similar features.

Stratification. The process of dividing a population into relative homogeneous subgroups to increase sampling efficiency.

Straw. The dry stems and leaves of a rice plant that have been harvested and discarded from threshing.

Stress. The state or condition of injury caused by detrimental effects of environmental factors such as drought, excess water, temperature, and others.

Striate. Marked with fine longitudinal lines or ridges.

Stripper harvesting. A method of harvesting seed or leaf material where the seed or leaf is mechanically removed from plant in situ, usually by a comb like device.

Stubble. The lower portion of the stems remaining in the field after the rice has been harvested.

Stubble mulch. The stubble of crops or crop residues left on the soil as a surface cover before and during the preparation of the seedbed and at least partly during the growing of a succeeding crop.

Stunting. Dwarfing due to disease or physiological disorder .

Style. The elongated stalk of the pistil connecting the stigma and the ovary.

Submerge. To place or plunge under water or other liquid.

Submergence (of plant). A plant standing in water with at least part of the terminal above the water (partial submergence); or completely covered with water (complete submergence).

Submergence tolerance. Ability of a rice plant to survive after being completely under water from a flash flood.

Substrate. The material or substance on which a microorganism feeds and develops; also a substance acted upon by an enzyme.

Subsurface tillage. Tillage with a special sweep like plough or blade which is drawn beneath the surface at depths of several inches and cuts plant roots and loosens the soil without inverting it or without incorporating the surface cover.

Succulent. Soft, juicy, and fleshy.

Sucker. A shoot of the plant arising from below the ground; a parasite from a host plant.

Summer fallow. A special case of fallowing in which all vegetative growth is prevented by shallow cultivation during the summer months in order to conserve moisture.

Supplementary pollination. The increase in pollination of the CMS spikelets done artificially by shaking the male parent at anthesis time particularly when wind velocity is less than optimum (2-3 m/s).

Suppressive soils. Soils in which certain diseases are prevented because of the presence in the soil of microorganisms antagonistic to the pathogen.

Surface runoff. Water from a catchment area that is discharged or lost without entering the soil.

Survival. The act of remaining alive or in existence.

Susceptible. Having little or no resistance to a specific infectious disease, insect pest, or other biological and physical stresses. When the host plant is unable to suppress or retard an injurious insect.

Suspended load. Finer particles of silts and clays carried long distances by rivers; as opposed to bed load (stones and coarse sand) and dissolved load (compounds in solution).

Suspension. A relatively coarse, noncolloidal dispersion of solid particles in a liquid.

Sustained. Maintained regularly, ongoing over a period of time.

Swamp. An area saturated with water throughout much of the year but with the surface of the soil usually not deeply submerged.

Swamp rice. Rice that usually grows in low-lying swampy conditions without water control or with little water control.

Symbiosis. The living together in intimate association of two dissimilar organisms, the cohabitation being mutually beneficial.

Symptom. A visible or otherwise detectable change in the plant or associated parts arising from disease.

Systemic. Common to a plant system, generally distributed throughout an organism.

Systemic fungicide. A chemical substance absorbed by the plant; it remains in the developing tissues & protects the plant from fungal attack for a length of time.

Systemic herbicide. Chemical absorbed into the plant parts which inhibits growth. It is toxic to particular plants only and designed to affect unwanted plants or weeds.

Systemic infection. Occurs when a pathogen is translocated throughout the plant tissues, causing extreme damage to the plant.

Systemic insecticides. A chemical substance absorbed by the plant tissues that affords protection for a reasonable amount of time against insect pests without harming the plant.

T

T. Aman. This rice is used in Bangladesh and eastern India for transplanted lowland rice, grown in the wet season when water depth does not exceed 0.5 m, approximately from July to September.

Tag. A small marker usually tied to a plant, bag, or packet and used for identification and classification of samples.

Target crop. One that has the potential to be a major crop, increasing the farmer's production and income.

Tempering. This means drying the grain on a drying pad for a number of hours or in a grain dryer, then tempering the grain by allowing it to cool for a number of hours in a bin or bag. This process should be repeated at least twice until the moisture is consistently at 14%.

Tendril. A slender, clasping or climbing organ of a stem or leaf.

Tensiometer. An instrument for measuring the amount of energy needed to extract water from the soil (usually expressed in kPa).

Terrace. (1). Bench-like portions of a former flat river floodplain left at higher levels by the down cutting action of a river; alluvial terraces and river terraces are generally considered as synonymous. (2). An artificially leveled field or series of fields made on sloping land.

Tertiary tillers. Arise from secondary tillers and are the third group of tillers to grow during the plant's development.

Test entries. Varieties or breeding lines being evaluated.

Testa. Hard external coating or integument of a seed.

Test cross. A cross made between cytoplasmic male sterile line and a test variety to identify maintainers and restorers.

Tester. A cross between a selection, line, or clone and a common pollen parent which may be a variety, inbred line, or a single cross. The common pollen parent is called the top cross or tester parent.

Tetraploid. An organism with four basic (n) sets of chromosomes.

Thermosensitive genic male sterile line (TGMS). The genic male sterile plant which responds to temperature for expression of its pollen sterility/fertility. *Thermus aquaticus*. *Thermophylic bacterium* found in hot springs. Its DNA polymerase enzyme is thermostable and is used in PCR, RAPD, and DAF.

Thinning. Removal of some plants to attain the desired population density.

Three-line breeding. Breeding methodology where three lines--cytoplasmic male sterile, maintainer, and restorer--are used to produce Fl hybrids.

Three-way cross. When an F1 hybrid is crossed to a third variety or breeding line.

Threshing. Operation of detaching or separating the rice grains from the panicle by hand or mechanical means.

Threshold. (1). A phenomenon that can obscure the gene-character relationship. (2). A safe level of economic viability.

Tidal flats. Areas of nearly flat, barren mud periodically covered by tidal waters. Normally these places have an excess of soluble salt.

Tidal wetlands. Areas where water levels in the rice fields fluctuate under the influence of tides.

Till. To plough and prepare for seeding; to seed or cultivate the soil.

Tiller. A vegetative branch of the rice plant composed of roots, culm, and leaves which may or may not develop a panicle. Shoot arising from the main culm (stem).

Tillage. (1). The mechanical alteration of some physical properties of the soil to provide a condition appropriate for crop growth. (2). To cultivate the land, to gain moisture in the soil.

Tiller number. Number of tillers present in a rice plant.

Tillering ability/capacity. The genetic potential of a given variety to produce a certain number of tillers.

Tillering stage. Growth stage of the rice plant that extends from the appearance of the first tiller until the maximum tiller number is reached.

Tilth. The physical condition of the soil as related to its ease of tillage, fitness as a seedbed, and its impedance to seeding emergence and root penetration.

Tissue culture. A technique of growing large pieces of tissue (explant) from different parts of a plant in semisolid or liquid medium under aseptic conditions. After the culture, masses of undifferentiated tissue are produced or plants are regenerated.

Tolerance. Ability of a plant to withstand stress without yield reduction or injury.

Toposequence. A sequence of soils in the landscape, from the crest to the valley bottom.

Topdressing. The application of fertilizer materials after seeding or transplanting or after the crop has been established.

Topsoil (surface soil). The uppermost part of the soil ordinarily moved in tillage, or its equivalent in uncultivated soils. Topsoil ranges from 8-10 cm to 20-25 cm in depth. It is frequently designated as the plough layer.

Total factor productivity. The productivity of all inputs taken together.

Total milled rice. Quantity of whole plus broken grains of milled rice that can be obtained from a given quantity of clean rough rice.

Toxic. Of, relating to, or caused by a poison or toxin, poisonous, causes illness, reduces plant growth.

Toxicity. The quality of being poisonous.

Toxin. Any of the various poisonous substances produced by certain plants and animal cells including bacterial toxins, phytotoxins, and zootoxins.

Traditional cultivars. Tall, weak-stemmed, long-duration, low-yielding cultivars grown by farmers for many years.

Traffic pan (handpan). A 5-10cm thick compacted subsurface horizon between the 10- to 40-cm depths; common in paddies.

Trait. A distinguishing quality or character of a plant.

Transfer RNA (tRNA). The RNA that becomes attached to an amino acid and guides it to the correct position on the ribosome for protein synthesis; there is at least one tRNA molecule for each amino acid.

Transformation. The genetic modification induced by the incorporation of "foreign" DNA into a cell.

Transgenic. Descriptive of an organism that contains some genetic material that has been experimentally transferred into it from some other source.

Translocation. (1). Genetic. Change in position of a segment of a chromosome to another location in the same or a different chromosome. (2). Physiological. The movement of assimilates (carbohydrates or nutrients) from one plant organ to another in response to stress or ontogeny. (3). Breakage and reunion of chromatid at a different point of the chromosome.

Translucent. Transmitting light but causing sufficient diffusion to eliminate perception of distinct images; intermediate between clear and opaque.

Transmitted nutrients. Nutrients moved to growing points during seed germination.

Transpiration. The process by which plants release water vapor to the atmosphere through surface pores (stomatal openings) in the plant foliage in response to atmospheric demand. The water of transpiration usually reaches a maximum value in the afternoon and a minimum value just before sunrise. Soil moisture content and plant characters, such as location and distribution of stomata, reduction of transpiration surface (leaf rolling), and plant age, affect transpiration rate.

Transplant. To remove seedlings from the nursery (seedbed) and plant them in the field either by hand or machine.

Transverse. Across the long axis.

Trap crop. A crop that is planted specifically to attract pests, rodents, or birds away from the main crop.

Treatment. A general term denoting different effects which are to be compared or the effects or factors that are to be measured. (e.g., variety, fertilizer rates) or procedures (e.g., method of nitrogen application).

Triple crossing. Sequential crossing of three cultivars, (e.g., for A, B and C, A x B, AB x C, ABC x A, ABCA x B, and so on.

Trisomic. An organism diploid except for one kind of chromosome which is present in triplicate, hence, having 2n + 1 chromosomes.

Tuber. A short thickened underground stem.

Tubercle. Any small, rounded projection.

Tuberous. Bearing a tuber.

Tungro. A viral disease transmitted by green (*Nephotettix* spp.) and zigzag leafhoppers (*Recilia dorsalis*). The diseased plants are characterized by stunting and reduced tillering. The leaf color ranges from light yellow to orange-yellow. Yellowing usually starts from the tips of the leaves. The panicles are often not fully exserted. The grains are usually covered with dark brown blotches.

Two-line breeding. Hybrid rice breeding methodology where only two lines, a male sterile (either photosensitive, thermo-sensitive, or chemically induced) and a pollen parent are used for the production of Fl hybrids.

U

Udbatta disease. A rice disease caused by *Balansia oryzae-sativae* (*Ephelis pallida* Pat). The panicles emerge as small cylinders covered with white mycelium.

Umbel. An inflorescence in which a number of divergent flowers arises from the same point.

Undulate. With a wavy surface.

Unfilled spikelet. Grain that has not filled and matured due to adverse environmental conditions or other factors affecting its development.

Uniformity trial (blank test). A trial or experiment where all the operations and treatments are the same except that samples of different sizes are used to evaluate heterogeneity.

Unimproved plant type. A traditional plant type; tall, leafy, and susceptible to lodging.

Univalent. An unpaired chromosome in meiosis.

Upland field. A rainfed land area with no levees or dikes surrounding it to impound water; sowing is by direct seeding.

Upland rice (dryland rice). Rice grown on both level and sloping fields that are not bunded, that are prepared and seeded under dry conditions, and that depend on rainfall for moisture (without surface water accumulation).

Upland. In general, land lying above the flood plain.

Uproot. To pull up by the roots.

Utricle. A small, bladdery, one-seed fruit or any small bladder like structure.

V

Variables. Data or some characteristics that show variability. The characteristic may be numerical (quantitative) or non-numerical (qualitative).

Variance. A measure of dispersion which is the mean of the squares of deviations of the observations from the population mean. Estimated as the ratio of a sum of squares to the corresponding number of degrees of freedom.

Variant. A plant that is different from the majority or from normal plants.

Variate. A single observation or measurement.

Variation. The occurrence of differences among cultivars due to differences in their genetic composition and/or the environment in which they were cultivated.

Varietal diversity. The measurable differences among varieties grown in a particular area or the genetic diversity among varieties of the rice plant.

Varietal improvement. The process of improving varieties to meet the standards required.

Varietal release. The procedure for releasing varieties from breeding stations to farmers for commercial cultivation.

Variety. (1). A group of cultivated plants within a species which is distinguished from another variety (group) by any characters (morphologic, physiological, biochemical, or other) of significance to agriculture and which, when reproduced, retains its distinguishing characters. A variety may be derived from several pure lines which have many common features and are reasonably uniform in appearance (but not necessarily genetically pure). (2). A group of similar plants which, by structural features and performance, may be identified from other varieties (groups) in the same species. It differs from a breeding line in that it has been named and made commercially available to farmers. (3). A subdivision of a species; a group of individuals within a species which are distinct in form or function from other similar arrays of individuals in commercial production. Variety is synonymous with cultivar.

Variety group. The classification of accessions into groups based mainly on the morphological features of the adult rice plant and on grain appearance.

Variety trial. A trial in which treatments are different in the varieties that are grown and all other factors are treated equally and in a uniform system.

Vascular. Pertaining to, or having vessels that convey, fluids.

Vector. (1). An organism that carries pathogens from one host to another. (2). An insect that transmits a disease. (3). A self-replicating DNA molecule that serves to transfer a DNA segment into a host cell in recombinant DNA technology.

Vegetative. Referring to asexual (stem, leaf, root) development in plants in contrast to sexual (flower, seed) development.

Vegetative phase/stage. The period from seed germination to the panicle initiation stage.

Vernalization. Exposure of germinated seedlings to low temperatures to promote flowering.

Vertical resistance. Resistance controlled by one or a few major genes, in which varieties with this type of resistance are usually highly resistant to one or several pathogen races or several disease races or insect biotypes of a given species but are susceptible to others. A type of resistance which is expressed against only some biotypes of a pest species and is governed by one or more genes in the host plant, each of which corresponds to a matching gene for parasitic ability in the pest species, sometimes called gene-for-gene resistance (See also Horizontal resistance).

Vesicular arbuscular (VA) mycorrhiza. A common endomycorrhizal association produced by phycomycetous fungi of the genus Endogone. Host range includes many agricultural and horticultural crops.

Viability. The ability to grow and develop into an adult.

Viable. Said of spore or seed that is able to germinate or grow under favorable conditions.

Vigor. (1). Plants. To have active, healthy, well-balanced growth. (2). Seeds. The capacity for natural growth and survival.

Virescent. Young seedlings that are nearly white or slightly yellow as a result of delayed development of chloroplasts. Later the plant gradually turns green.

Virgin land. Land that has not been used for agricultural purposes or disturbed by humans at all.

Virgin soil. Soil that has never been cultivated or disturbed for crop production or any other purpose.

Virulence. Capacity of a pathogen or insect to incite a disease or injury to the host. Degree or measure of pathogenicity.

Virulent. Strongly pathogenic.

Virulent gene. A gene in pest which is able to break down the gene for pest resistance in the plant and thus allows the pest to use the plant as a host.

Virus. A submicroscopic infectious agent consisting of particles made up of DNA or RNA which are usually covered by protein and reproduce only in living cells; cannot be separated by filters.

Volunteer plant. A plant that is growing from an unintentionally included seed, a seed that is shed or dropped by a previous crop. Also called a rogue.

W

Water control. To control the amount and depth of water in the rice field during the required time necessary for crop growth.

Waterlogged field. A field which continuously has standing water and cannot be drained. The water level may not be deep.

Waterlogging. Impeded drainage; or soil saturated with water but not necessarily with standing water.

Watershed. The total area from which a single river collects surface water runoff; the catchment or drainage area of a river system.

Water table. The upper surface of ground water and the level below it where the soil is saturated with water.

Water use efficiency. (1). Expressed in terms of dry matter or harvested portion of the crop produced per unit of water consumed. (2). Ratio of the water used for evapotranspiration, seepage, and percolation to the sum of rainfall and irrigation supply.

Waxy (glutinous) rice. Milled rice with negligible or no amylose in the grain.

Waxy endosperm. Glutinous or waxy type of starchy endosperm, in which the starch fraction is composed of nearly 100% amylopectin, 0-2% amylose, stains reddish brown with weak potassium iodide-iodine solution.

Weed control. Prevention or eradication of weeds by physical removal (hand weeding) or limiting their growth by mechanical or chemical means.

Weedicide. A term used for chemicals that control weeds.

Weeding. Removing unwanted plants by hand from a crop, or tillage action which lightly cultivates the soil for the purpose of destroying the weeds.

Well puddled. Said of muddy, very wet rice field ready for sowing.

Western hybridization. Also called western blotting. Method to detect protein by use of an antibody directed against it.

Wetbed method. Raising seedlings on a seedbed where land is prepared and puddled.

Wetbed seeding. Planting of pregerminated seed or small seedlings into a wet seedbed.

Wetland. (1). Pertaining to soils flooded for at least several weeks each year, or to crops growing in such soils. (2). Land of which the rooting zone can be kept saturated for a substantial part of the growing season, where necessary, by encouraging accumulation of water on the soil through puddling and the use of bunds or levees.

Wettable powder (WP). A pesticide formulation consisting of talc or clay and a wetting agent mixed with the pesticide. When water is mixed with the formulation, a suspension is formed and is used to spray on the target pest.

Wetland tillage. Preparing the soil by ploughing of lowland fields and harrowing the soil in a saturated or flooded condition.

Wetting agent. A substance that renders a surface non-repellent to a wetting liquid.

Wheel track planting. A practice of planting in which the seed is planted in tracks formed by wheels rolling immediately ahead of the planter.

White center. The chalky area in the center of the endosperm of the milled nonglutinous rice.

White tip. A disease of rice caused by the nematode *Aphelencoides besseyi*, with symptoms consisting of chlorosis of the leaf tips which become brownish and tattered. The infected plants are stunted and produce a few, small spikelets/panicles.

Whitehead. White, empty panicles resulting from the attack of stem borers that cut the lower portion of the stem. Whitehead can also be caused by drought, dry wind, or desiccation.

Whole kernel (white rice). The unbroken kernel of rice that has been milled.

Whorl. (1). A ring of similar organs arising from a node. (2). The arrangement of leaves in a circle.

Wide cross. Cross between two distantly related species.

Wide hybridization. Hybridization between plants belonging to different species or genera. It is also referred to as interspecific, intergeneric, remote or distant hybridization.

Wild rice. Species of Oryza that are not cultivated.

Wilt. A characteristic of the plant indicated by drooping, folding, rolling, or collapsing leaves due to an unfavorable water balance.

Wilting. The loss of turgidity in plant tissue where the intake of water is insufficient to replace that lost by transpiration or other means, causing a deflation of the plant cells.

Wilting coefficient. The percentage of water necessary for a plant that has begun to wilt to recover, if that water is supplied.

Wilting point. Reached when the soil cannot supply enough water to balance the plant's losses by transpiration.

Woody. Hard in texture.

Y

Yellow leaf (y). Seedlings which have pale yellow leaves, but this condition is non-lethal.

Yellow mottle. A disease caused by rice yellow mottle virus. The virus disease can be transmitted by mechanical inoculation or vectored by the adult beetle *Sesselia pusilla.* It is characterized by stunting and reduced tillering of the infected rice plant; crinkling, mottling, and yellowish streaking of the leaves; malformation and partial emergence of the panicles and sterility.

Yield. The amount of a specified substance produced.

Yield components. The factors that contribute to grain yield--number of panicles per square meter, spikelets per panicle, percentage of fertile spikelets, and weight of each single grain.

Yield decline. A decrease in grain yields over a period of at least several years.

Yield growth rate decline. A slowdown in the (percentage) rate of increase of grain yield over time.

Yield potential. The maximum grain yield of a given variety in a given environment without constraints involving water, nutrients, competition, pests, diseases, or climatic conditions.

Yield, sustained. A continual annual or periodic yield of plants or plant material from a given area. This implies that the management practices are such, that they will maintain the productive capacity of the land.

Yield trials. Trials in which the main objective is to determine and compare the yields of a cultivar against a check.

Yielding ability. The expected capacity of the plant to produce a certain quantity of grain.

Z

Zero tillage. A practice where planting or seeding is directly done in untilled land.

Zinc deficiency. Insufficient zinc in the plant tissue causes the midribs of the young rice leaves to become chlorotic especially at the base, brown blotches and streaks occur in the lower leaves, growth is stunted, tillering is reduced, leaf blade size is reduced, and uneven growth occurs.

Zygote. Cell formed by the union of two gametes and the individual developing from this cell.

REFERENCES

Abrol, I.P. and SP. Palaniappan 1988. Green manure crops in irrigated and rainfed rice based cropping systems in south Asia. Sustainable agriculture – Green manure in rice farming. Int. Rice Res. Inst., LosBanos, Laguna, Philippines. pp.71-82.

Abrol, I.P and R. Prasad (Eds.) 1985. Cultural practices for growing rice in alkali soils. Better farming in salt affected soils. CSSRI, Karnal, Bull No.9, pp.4

Abrol, I.P., K.S. Dargan and D.R. Bhumbla 1973. Reclaiming alkali soils, ICAR Bull. No.2., pp.25 – 58.

Acharya, C.L. 1979. Agronomy in A Decade of Research, (S. Chandra, R. Prasad and R.K. Gupta Eds., CSSRI (ICAR) Karnal, pp. 67-79.

Acharya, K.T. 1998. Food processing and nutrition. In: Textbook of Human Nutrition. Oxford and IBH publishing co pvt. Ltd., New Delhi.

Agrawal, V.P., K.P. Shrestha, G.K. Agrawal, S. Shakya, A.K. Keshari, R.L. Singh, S.K. Agrawal and J.B. Baral 1993. Integration of conventional breeding, anther culture and molecular methods for improvement of cold tolerant rice in Nepal. VI th Ann. Meeting of on "Rice Biotechnology" Feb 1-5, 1993, Chiang Mal, Thailand, pp. 96-97.

Ahamed, P. and M. Sivaswamy 1994. Appropriate rice farm mechanization technology for remunerative rice production. Proceedings of the Sixth Kerala Science Congress, Thiruvananthapuram.pp. 183-184.

Ahmed, J. and S. Gupta 1991. Germination and growth of some salt resistant selections in high salt concentration solutions. IRRN 16: (5), 15.

Ahmed, M. and A.B. Rab 1992. Feasibility of adopting aquaculture to increase resource productivity in existing Bangladesh farming systems. NAGA, The ICLARM Quarterly, 15:21-22.

Ahn, S.N., C.N. Bollichi and A.M. McClung and S.D. Tanksley, 1993. RFLP analysis of genomic regions associated with cooked kernel elongation in rice. Theo. Appl. Genet., 37: 27-32.

Ahn, S.N., C.N. Bollichi and S.D. Tanksley, 1992. RFLP tagging of a gene for aroma in rice. Theo. Appl. Genet., 84, 82-828.

AICRPWC, 1989. Project Coordinator's report (1978-1989). All India Coordinated Research Programme on Weed Control (I.C.A.R).

Aiyer, R.S., S. Salajudeen and G.S. Venkataramanm, 1972. Long term algalisation field trial with high yielding variety of rice. Indian J. Agric. Sci. 42: 80-83.

Alagusundaram, K. 2004. Food processing Industries in India- Current trends and Future thrusts. An invited lecture given in the National Conference on "Food Technology", Organised by Ministry of Food Processing Industries, Government of India held at Thiruvannamalai, Tamil Nadu.

Alagusundaram, K. 2005. A radio talk on "Velan Kalivugal Moolam Tholil Vaipugal, in All India Radio, Trichy, Tamil Nadu.

Alice, D. and A. Venkatarao, 1981. Antifungal effects of plant extracts on *Drechslera oryzae* in rice. IRRN, 12: 28.

Alice, D., 1984. Studies on antifungal properties of some plant extracts. M.Sc., (Ag.) thesis, submitted TNAU, Coimbatore, India.

and rew, D., Hanson, E. Nell, Hoffman and Catalina Samper, 1986. Identifying and manipulating metabolic stress-resistance traits. Hort. Science, 21(b):1313-1317.

and riesse, J.P. 1974. Tropical lowland peats in South – East Asia. R. Trop. Inst. Comm. No. 63, Amsterdam. 63 p.

Angus J.F, C.F. St.-Groth and R.C Tasic 1990. Between-farm variability in yield responses to inputs of fertilizers and herbicide applied to rainfed lowland rice in the Philippines. Agri. Ecosyst. Environ., 30:219-234.

Annadurai K and SP. Palaniappan 1995. Phosphorus management for lowland transplanted rice: A test verification trial with 2% DAP spray. Madras Agric. J., 82 (6, 7 and 8) : 452-456.

Annadurai, K., SP. Palaniappan, C. Chinnusamy and C. Jayanthi. 1994a. Integrated farming system. Kisan world, 21(8) : 51

Annadurai, K., SP. Palaniappan, C. Chinnusamy and C. Jayanthi. 1994b. Integrated farming system: Need of the hour. Kisan world, 21(9) : 39

Annadurai, K. and SP. Palaniappan. 2000a. Phosphorus management with diammonium phosphate spray on productivity of lowland transplanted rice. Indian J. of Agronomy, 43(2): 278-283.

Annadurai, K. and SP. Palaniappan. 2000b. Foliar nutrition of nitrogen and phosphorus on total chlorophyll content and grain yield of lowland direct seeded rice cv. ADT. 36. Oryza, 37(3): 241-243.

Annadurai, K., R. Kavimani and P. Masilamani. 1999. Effect of distillery effluent and organic amendments on rice yield and soil fertility status. Madras Agric. J., 86 (10-12): 572-577.

Annadurai, K., SP. Palaniappan, P. Masilamani and R. Kavimani. 2000. Split application of Potassium in rice- A review. Agricultural Reviews, 21(1) : 36-44.

Annadurai. K and R. Kavimani. 2004. A report on SRI trails conducted at Lalgudi block, Trichirappalli Dt. (unpublished).

Anon, 1994-95. Annual report on hybrid rice research. TNAU, Coimbatore.

Anon. 1974. Grow pumper crops in alkali soils. CSSRI, Karnal Information Bull. Sept. 1974 pp.4

Anonymous, 1970. Rice Production Manual, International Rice Research Institute, Philippines.

Anonymous, 1997. Proceedings of brainstorming session on "Indian aromatic rice: Recent status and future needs: March 29-30, 1997. IRRI-Indian office, New Delhi.

Arora, S.P. 1980. Fodder and feed resources requirement, availability and development. The Indian Dairyman, 32: 821-828.

Arunachalam, G. 1984. Nel Sakupattiku Elia Vilakkam (Tamil), IRRN-TNAU, Publication. 221p.

Asoaka, M., K, Okuno and H. Fuwa, 1985. Effect of environmental temperature at milky stage on amylase content and fine structure of amylopectin of waxy and non waxy endosperm starches of rice. Agril. Bio. Chem., 49: 373.

Babong, B. and M.S. Swaminathan, 1995. Magnitude of hybrid vigour retained in double haploid lines of some heterotic rice hybrids. TAG, 90: 253-257.

Bainu, T.K., M. Sivaswamy and G. Mathew, 1994. Modification and performance evaluation of six row rice transplanter for conventional seedlings. Proceedings of the sixth Kerala science Congress, Thiruvananthapuram. pp. 214.

Baki, B.B. 1993. Spatial pattern analysis of weeds in selected rice fields of Samarahan, Sarawak. MARDI Research Journal, 21(2):121-128.

Balagopal, K. 1994. Role of Chicken, Pigeon, Japanese quails and Duck rearing in sustainable IFS. In: Proceedings of Summer Institute on "Integrated farming systems research and development for sustainable agriculture" organized by ICAR and Tamil Nadu Agri. Univ., Coimbatore, India during 6-15 June,1994. pp: 235-240.

Balakrishna, P., 1994. Salt tolerant rice cultivars. Rice Biotech. Quarterly, 21: 11-12.

Balasubramaniam, V., A.C. Morales, R.T. Cruz and S. Abdulrachman, 1999. On-farm adaptation of knowledge-intensive nitrogen management technologies for rice systems - Nutrient Cycling Agro-ecosystems, 53: 59-69.

Balasubramanian, A. 1994. Farming system research and development in western zone of Tamil Nadu. In-Proceedings of the Summer institute on "Integrated farming systems research and management for sustainable agriculture". Tamil Nadu Agric. Univ., Coimbatore. pp: 278-284.

Balasubramanian, P. 2003. Studies on the utilization of rice straw as a source of silicon and potassium for lowland rice grown in Udic Haplustalf. Ph.D Thesis submitted to TNAU, AC & RI, Madurai, India.

Balasubramanian, P., S. Subramaniam, S. Anthony raj and P. Saravana Pandiyan 2002, Silicon nutrition for rice. AC & RI, TNAU, Coimbatore, India. 74 p.

Balasubramanian, T.N., 1988. Farming systems research survey for new Moriones village in Philippines, p. 9. In. Abst. National seminar on "Farming systems for Semiarid tropics". Indian Council of Agricultural Research and Tamil Nadu Agric. Univ., Coimbatore.

Balusamy, M. and V.S. Shanmugasundaram. 1994. Integrated farming system studies at Agricultural Research Station, Bhavanisagar. In: Proceedings of Summer institute on "Integrated Farming system research and management for sustainable agriculture". Tamil Nadu Agric. Univ., Coimbatore. pp: 270-277.

Bandyopadhyay, A.K. 1986. Varieties screened for acid saline soils. Intl. Rice Res. Newsletter. 11 (6):14.

Bandyopadhyay, A.K. 1988. Performance of some acid tolerant rice varieties in two acid saline soils of Sunderbans. Intl. Rice Res. Newsletter. 13(3) : 19.

Barea, J.M. 1991. Vesicular arbuscular mycorrhizae as modifiers of soil fertility. Adv. Soil. Sci., 15: 1-140.

Bates, G.W. 1989. In: Biotechnology in agriculture and forestry – Plant Protoplasts and Genetic Engineering II (Bajaj, Y.P.S. Ed), Springs – Verlag, Berlin, p. 241.

Behra U.K., P.K. Jha and I.C. Mahapatra 1997. Low cost weed management technology for rainfed upland rice of Orissa. Indian Fmg. 47 : 13-15.

Berge H.F.M., T.M. Thiyagarajan Quinghua Shi, M.C.S Wopereis, H. Drenth and M.J.W. Jansen 1997 Numerical optimization of nitrogen application to rice. Part I. Description of MANAGE-N. Field Crops Res, 51:29-42.

Bertrand, R. 1973. Contribution a Petude hydrologique pedolofique, et agronomique des sols gris sableux hydromorphes de Casamance (Senegal). Agric. Trop, 27 (12) : 1145-1192.

Bhadona, P.B.S. 1986. Effect of compaction and tillage on the yield and percolation loss of rice in lateritic sandy loam soil. J. Agron. Crop Sci. 156: 45 – 49.

Bharadwaj, K.K.R. 1980. Effect of different incorporation of wheat straw and water hyacinth on the yield of wheat. Symp. Proc. Recycling residues of agriculture and industry, pp. 57-75.

Bhardwaj, K.K.R. and A.C. Gaur 1985. Recycling of organic wastes. ICAR Publication, New Delhi. 5p.

Bhol, B.B and K.N. Singh 1987a. Effect of methods of seeding and herbicide application in upland direct seed rice. I. Composition and distribution of weed flora. Indian J. Agron., 32 : 132-139.

Bhol, B.B and K.N. Singh 1987b. Effect of methods of seeding and herbicide application in upland direct seed rice. II. Growth, yield attributes and yield. India J. Agron. 32 : 140 – 143.

Bhumbla, D.R. 1971. Management of saline and alkali soils. Oryza, (suppl.) 8:283-296.

Binh Doquang, E.G. Heszky and A. Csillag 1992. Plant regeneration of NaCI-pretreated cells from long term suspension culture of rice (*Oryza sativa* L.) in high saline conditions. Plant Cell, Tissue and Organ Culture, 29: 75-82.

Binh Doquang, L.E. Heszky and A. Csillag 1993. Rice regenerated from salt stressed cultures. Rice Biotech. Quarterly, 13: 5-6.

Bini, S. 1992. Development and testing of a manually operated paddy dibbler. M.Tech. thesis, Kelappaji College of Agricultural Engineering & Technology, Tavanur, Malapuram Dt., Kerala.

Biswas, C.R. and A.K. Bandyopadhyay 1987. Agronomy of rice coastal saline soils- a review. J. Indian Soc. Coastal. Agri. Res., 5:15-23.

Blight, H.F.J., R.L. Till and C.A. Jones, 1995. A microsatelite sequence closely linked to the waxy gene of *Oryza sativa*. Euphytica, 86: 83-85.

Borgohain, M. and L.P. Upadhyay, 1980. Efficacy of butachlor, propanil and 2,4-D in controlling weeds in upland rice. Indian J. weed Sci.,12:145 – 50.

Bouharmont, J., A. Dekeyser, V. Sint Jan, Van and Y.S. Dogbe, 1991. Application of somoclonal variation and in vitro selection to rice improvement. In: Rice Genetics II-Proceedings of the 2nd Int. Rice Genet. Symp. 14-18 May, 1990, IRRI, Manila, Philippines, p. 271-277.

Bouman, B.A.M. 2001. Water efficient management strategies in rice production. IRRN, 26(2):17-11.

Brady, Nyle, C. 1984. The Nature and Properties of soil (9th Edition), Mac Millan Publishing Co. USA. 750 p.

Brammer H. and R. Brinkman 1977. Surface – water -Gley soils in Bangladesh. Environment, land forms and soil morphology. Geoderma 17 : 91 – 109.

Brammer, H. 1971a. Soil survey project, Bangladesh : Agricultural development possibilities. Tech. Rep. 2, FAO, Rome. 153p.

Brammer, H. 1971b. Soil Survey Project, Bangladesh: Soil Resources. Tech. Rep. 3, FAO, Rome. 211p.

Brar, L.S., J.S. Kolar and L.S. Brar 1997. Chemical control of *Caesulia axillaris* in transplanted rice. Indian J. Agron., 42 : 82 – 85.

Brinkman, R. 1977. Ferrolysis, a hydromorphic soil forming process. Geoderma 3 (3): 199 – 206.

Brinkman, R. and L.J. Pons. 1968. A pedomorphological classification and map of the Holocene sediments in the coastal plain of the three Guianas. Soil survey Paper 4, Wageningen, 44p.

Broadbent, F.E. 1978. Nitrogen, Pages xix – xx in International Rice Research Institute. Soils and rice. Los Banos, Philippines.

Brown, F.B. 1969. Upland rice in Latin America. Int. Rice Comm. Newsletter. 18:1-15

Buchanan, K. 1970. The transformation of the Chinese earth. Bell, London. 336 p.

Buol, S.W., P.A. Sanchez, R.B. Cate, JR. and M.A. Granger, 1975. Soil fertility capability classification : A technical soil classification for fertility management.

In: Soil management in tropical America, North Carolina State University, Raleigh. pp.126-141.

Buresh, R.J. and S. K. De Datta 1991. Nitrogen dynamics and management in rice legume cropping systems. Advances in Agronomy, 45, 1-52.

Byerlee, D., L. Harrington and D.L. Winkelmann 1982. Farming systems research: Issues in research strategy and technology design. Amer.J. Agric. Econ. 64: 897-904.

Byerlee, D. 1996. Knowledge intensive crop management technologies: Concepts, Impacts and Prospects in Asian agriculture. Paper presented at the International Rice Research Conference, Bangkok, 3-5 June, 1996. Bangkok.

Cassman K.G and P.L. Pingali 1995. Intensification of irrigated rice systems: Learning from the past to meet future challenges. Geo Journal., 35:299-305.

Cassman K.G, S. Peng and A. Dobermann 1997. Nutritional physiology of the rice plant and productivity decline of irrigated lowland rice systems in the tropics. In: Plant nutrition - for sustainable food production and environment. (Edited by T. and o, K. Fujita, T. Mae, H. Matsumoto, S. Mori and J. Sekiya). Kluswer Academic Publishers, Tokyo,pp. 783-788

Cassman K.G., A. Dobermann, P.C Sta. Cruz, G.C. Gines, MI. Samson, J.P. Descalsota, J.M. Alcantara, M.A. Dizon and D.C. Olk 1996. Soil organic matter and the indigenous nitrogen supply of intensive irrigated rice systems in the tropics. Plant Soil. 182:267-278.

Castro, R.U. 1977. Zinc deficiency in rice : A review of research at the International Rice Research Institute. Int. Rice. Res. Inst. Pap. Ser., No. 9. 19 p.

Castro, R.V. and S.R. Sabado 1977. Influence of varying level of salt applied at different stages on the growth and yield of rice. Grains J., 2(3): 43 – 45.

Central Soil Salinity Research Institute (CSSRI). 1974. Annual report, Karnal, India. 95p.

Chakravarthy, D.K. and S.N. Pariya 1977. Inhibition of germination of phytopathogenic fungi in some Indian medicinal plants. Z. pflazenkrankh pflanzenschulz. 84: 221-223.

Champoux, M.C., G. Wang, S. Sarkarung, D.J. Mackill, J.C. Otoole, N. Huang and S.R. McCouch 1995. Locating genes associated with root morphology and drought avoidance in rice vig linkage to molecular markers. Theo. Appl. Genet., 90 : 969-981.

Chand, P. 1995. Agricultural and forest pests and their management. Oxford & IBH Publishing Co. Pvt. Ltd., New Delhi. 374p.

Chander, S. and J. Pandey 1996. Effect of herbicides and nitrogen on yield of scented rice under different rice culture. Indian J. Agron., 41.209 – 214.

Chander, S. 1995. Effect of nitrogen and herbicides on growth, yield and quality of scented rice (Pusa Basmati-1) under varying rice cultures. Ph.D. Thesis, P.G. school, IARI, New Delhi.

Chandrakar, B.L., B.R. Chandravansi and G.K. Das 1987. Efficacy of weed control measures under different methods of direct seeds rice. Indian J. Weed Sci. 19: 107 – 110.

Chang, T. T. and E.A. Bardenas 1965. The morphology and varietal characteristics of the Rice Plant'. Tech. Bull. 4,1965. International Rice Research Institute, Los Banos, Phlippines.

Chang, T.T 1976. The origin, evolution, cultivation, dissemination and diversification of Asian and African rice. Euphytica, 25 (2): 425 – 441.

Chela, G.S. and H.S. Gill 1980. Chemical control of *Echinochloa crusgalli* in transplanted rice. India J. Weed Sci., 12:7 – 14.

Chellaiah, S. and B. Chandrasekaran 1989. Integrated farming system for Thanjavur district, Tamil Nadu. Thanjai Ulavar Mani (Dec. 88-Jan, 89). pp: 5-16.

Chen, T.P. and Y. Li. 1980. Integrated agriculture- aquaculture studies in Taiwan. In: Proceedings of ICLARM – SEARCA conference on "Integrated agriculture – aquaculture farming system". 4th Aug, 1979. Manila, Philippines., pp: 239-241.

Chen, T.T., K.H. Houng and C.T. Wei 1969. Water management of a paddy as related to physicochemical properties of the soil. Soils Fert. Taiwan 16: 30- 45.

Chinnusamy, C. and S. Purushothaman 1992. Integrated farming systems under garden land and their economics. In the Proceedings of National training on "Integrated farming systems". GOI and Tamil Nadu Agric. Univ., Coimbatore. 10-17 Dec, 1992. pp: 46-48.

Chopra, V.L. and R.P. Sharma 1991. Biotechnology in crop improvement. Curr. Sci., 60 (9&10): 543-547.

Chowdhry, J.B., N.R. Yadav, A Gaur, R.K. Jain and D.V.S. Panwar 1992. Plant regeneration, protoplast culture and somaclonal variation in rice. National Rice Biotech Network, 2nd Ann. Meeting 1992, Directorate of Rice Research, Hyderabad, pp. 10-12.

Cocheme, Y. 1971. Notes on the ecology of rice in West Africa. In: Notes on the ecology of rice and soil suitability for rice cultivation. AGS / WARDA / 71 / 5. FAO, Rome. pp. 1-6.

Cocking, E.C. 1988. In : Proceedings Int. Cong. Plant Physiol. (S.K. Sinha, P.V. Sane, S.C. Bhagava and P.K. Agarwal Eds.)., Soc. for Plant physiology and Biochem., New Delhi, India.

Collinson, M.P. 1988. The development of African farming systems – Some personal views. Agric. Admin. and Extension, 29:7-22.

Conklin, H.C. 1957. Hanunoo agriculture : A report on an integral system of shifting cultivation in the Philippines. Food and Agricultural Organization of the United Nations. Rome., 299 p.

Coughlan, S.J. and R.G. Wyn Jones 1980. Some responses of *Spinacea olerace* to salt stress. J. Expt. Bot., 31: 883-893.

Csonka, L.N. and A. Baich 1983. In: Amino acids – biosynthesis and genetic regulation (K.M. Herrmann and R.L. Sommerville Eds.). Addition – Wesley, Reading, Mass, p.35.

Curfs, H.P.F. 1976. Systems development in agricultural mechanization with special reference to soil tillage and weed control. Communications Agricultural University, Wageningen. The Netherlands, 76-5. 179 p.

Dahiya, I.S., R.S. Malik and M. Singh 1982. Reclaiming a saline–sodic sandy loam soil under rice production. Agric. Wat. Mgmt. 5 (1) : 61 – 72.

Daniel Sundarraj D. and G. Thulaisidas 1993. Botany of field Crops. (2nd Ed.) Mac Millan India Ltd., Delhi. 586p.

Dawn Berkelaar 2001. SRI, the system of rice intensification: Less can be more. EDN (Echo Development Notes, January, 2001 No. 70. pp.1-6.

Dayanand and M.M. Hosmani 1985. Weed control in drill sown upland paddy. Indian J. Weed Sci. 13: 52 – 53.

Dayanand 1984. Studies on weed control in direct sown upland rice (*Oryza sativa*). Abstr. Ann. Conf. Indian Soc. Weed Sci., Varanasi, p. 6.

De Datta S.K, H.K. Krupp, EI Alvarez and S.C Modgal 1973. Water management in flooded rice. In: Water management in Philippines irrigation systems: Research and operation. Los Banos (Philippines): International Rice Research Institute. pp. 1-18.

De Datta, S.K. 1975. Upland rice and around the world. In: International Rice Research Institute. Major research in upland rice. Los Banos, Philippines. pp. 2-11.

De Datta, S.K. 1978. Fertilizer management for efficient use in wetland rice soils. *In :* Soils and Rice. International Rice Research Institute, Los Banos, Philippines. pp. 671 – 701

De Datta, S.K. 1981. Principles and practices of rice production. John Wiley, New York, 618p.

De Datta, S.K. and H.M. Beachell 1972. Varietal response to some factors affecting production of upland rice. In: International Rice Research Institute. Rice breeding. Los Banos, Philippines.pp.685-700.

De Datta, S.K., K.A. Gomez, R.W. Herdt and R. Barker 1976. Farm yield constraints. Laguna, Nueva Ecija, Camarines Sur and Iloilo province sites, Philippines. 1975. Paper No.5, Int. Rice Agro – Econ. Network Workshop, International Rice Research Institute, Los Banos, Philippines, 49 p.

De Dutt, S.K. 1979. Weed problems and methods of control in tropical rice. In symposium on "Weed control in tropical crops". Weed Sci. Society, Philippines, INC and Philippine Council for Agriculture and Research. p.244.

De Ridder, N.A. 1975. Hydrogeological investigations in Haryana State, India. International Institute land Reclamation and Improvement. Annual Report for 1975. Wageningen.

De. G.C. and S.K. Mukhopadhyay 1983. Efficiency of granular herbicides in saving the major plant nutrients removal by weeds in transplanted rice. Abstr. Ann. Conf. Indian Soc. Weed Sci. Varanasi, p. 10.

Debey A.N., G.B. Manna and M.V. Rao 1977. Studies on weed competition, weed control and varietal interaction with propanol and parathion in direct seeded rice. Abstr. Indian Soc. weed Sci. conf. Hyderabad. p. 172.

Dei, Y. and K. Maeda 1973. On soil structure of plowed layer of paddy field. JARQ, 7: 86 – 92.

Delmando, M.N. 1980. A review of Integrated livestock, bowls and fish farming systems. In. Proceedings ICLARM – SEARCA Conf. on "Integrated agriculture – aquaculture farming systems". 4th Aug. 1979, Manila, Philippines.

Deshpande, R.S. 2002. Suicides by farmers in Karnataka - Agrarian distress and possible alleviatory steps, Economic and Political Weekly, XXXVII(26): 2601-2610.

Devanathan, M., R. Sree Ramkumar and P. Narayanasamy 1995. Efficacy of different plant products on brown spot of rice cased by *Drechslera oryzae* In Mariappan, V. (ed.). Neem for the management of crop Diseases, Associate Publishing Co., New Delhi, India.pp.19-26.

Dhaliwal, G.S., M.D. Padhak and C.R. Vega 1990. Inhibition of growth of crop pathogens by plant extracts of rice varieties. Proceedings of National Symposium, on "Botanical pesticides in integrated pest management". January 21-22, 1990, Central Tobacco Research Institute, Rajamundry, India. pp. 466 – 468.

Dix, P.J., V.A. Mc Lysaght and R.S. Pearce 1984. In : Efficiency in plant Breeding, Proceedings 10th Cong. Eucarpia (W. Lange, A.C. Zeven and N.G. Hogenboom, Eds.),Wageningen, p. 219.

Djajadivedja, R., Z. Jangkaru and Mohamed Junu 1980. Freshwater aquaculture in Indonesia with special reference to small-scale agriculture. In. Proceedings ICLARM – SEARCA Conf. on "Integrated agriculture – aquaculture farming systems", 4th Aug, 1979, Manila, Philippines. pp: 143-165.

Dobermann A., C. Witt, S. Abdulrachman, H.C Gines, R. Nagarajan, T.T. Son, P.S. Tan, G.H. Wang, N.V. Chien, V.T.K. Thoa, C.V. Phung, P. Stalin, P. Muthukrishnan, V. Ravi, M. Babu, GC. Simbahan, M.A.A. Adviento and V. Bartolome 2003b. Estimating indigenous nutrient supplies for site-specific nutrient management in irrigated rice. Agronomy Journal., 95: 924-935.

Dobermann A., C. Witt and D. Dawe 2004. Increasing productivity of intensive rice systems through site-specific nutrient management. (Science Publishers, Inc., (USA) and International Rice Research Institute, Enfield, NH Los Baños, Philippines.

Dobermann A., C. Witt, S. Abdulrachman, HC. Gines, R. Nagarajan, TT. Son, PS. Tan, GH Wang, NV Chien, VTK Thoa, CV Phung, P. Stalin, P. Muthukrishnan, V. Ravi, M. Babu, G.C. Simbahan and M.A.A. Adviento 2003a. Soil fertility and indigenous nutrient supply in irrigated rice domains of Asia. Agronomy Journal, 95: 913-923.

Dobermann A., M.F. Pampolino and M.A.A. Adviento 1997b. Resin capsules for on-site assessment of soil nutrient supply in lowland rice fields. Soil Sci.Soc.Am.J., 61:1202-1213.

Dobermann, A. and K.G. Cassman 2002. Plant nutrient management for enhanced productivity in intensive grain production systems of the United States and Asia. Plant and Soil 247:153-175.

Dobermann, A. and T. Oberthuer 1997. Fuzzy mapping of soil fertility - A case study on irrigated rice land in the Philippines. Geoderma. 77:317-339.

Dobermann, A., J.L. Gaunt, H.U. Neue, I.F. Grant, M.A.A. Adviento, M.F. Pampolino 1994. Spatial and temporal variability of ammonium in flooded rice fields. Soil Sci. Soc. Am. J. 58:1708-1717.

Dobermann, A., P. Goovaerts and H.U. Neue 1997a. Scale dependent correlations among soil properties in two tropical lowland rice fields. Soil Sci. Soc. Am. J.

Dobermann, A. K.G. Cassman, S. Peng, Pham Sy Tan, Cao Van Phung, Sta. Cruz P.C. Bajita, J.B Adviento M.A.A and D.C. Olk 1996 Precision nutrient management in intensive irrigated rice systems. In Proceedings of the International Symposium on "Maximizing sustainable rice yields through improved soil and environmental management, November 11-17, 1996, Khon Kaen, Thailand Dept. of Agriculture, Soil and fertilizer society of Thailand Dept. of land Development, ISSS, Bangkok. pp. 133-154.

Dobermann, A., C. Witt, D. Dawe, G.C. Gines, R. Nagarajan, S. Satawathananont, T.T. Son, P.S. Tan, G.H. Wang, N.V Chien, V.T.K. Thoa, CV. Phung, P. Stalin, P. Muthukrishnan, V. Ravi, M. Babu, S. Chatuporn, M. Kongchum, Q. Sun, R Fu, G.C. Simbahan and M.A.A. Adviento 2002. Site-specific nutrient management for intensive rice cropping systems in Asia. Field Crops Research, 74: 37-66.

Douglas, G.C., W.A. Keller and G. Setterfied 1981. Somatic hybridization between *N. rustica* and *N. tobacum.* II. Protoplast fusion and selection and regeneration of hybrid plants. Can. J. Bot., 59: 220-227.

Draz, A.E., F.J. Zapata and G.S. Khush 1991. Development of dihaploid rice lines through anther culture (II), IRRN, 16(5): 6-7.

Duangpatra, P., P. Prabhuddham and C. Suwannarat, 1993. Agronomic effectiveness of phosphal – 34 on flooded rice in acid soils. Fert. Res., 34: 107 – 110.

Dudal, R. and F.R. Moormann. 1964. Major soils of Southeast Asia. J. Trop. Geogr., 18 : 54 – 80.

Dudal, R. 1958. Paddy soils. Int. Rice. Conn. Newsl., 7 (2) 19 – 27.

Dutta, R and A.K. Gogoi 1994b. Evaluating weed control practices in direct seeded rice. Indian. J. Weed Sci., 26 : 109 – 111.

Dutta, R. and A.K. Gogoi 1994a. Integrated weed control in direct seeded rice. Indian I. Agron., 39: 639 – 641.

Dwivedi, V.D., R.C. Yadav and A. Kumar 1994. Effect of weed control method and nitrogen on rice under midland situation of northeast Bihar. Indian J. Agron., 38 : 298 – 299.

Dwivedi, V.D., R.P. Pandey and K.N. Namdeo, 1991. Weed Management in upland rice (*Oryza sativa*) sown with different method. Indian J. Agron., 36:241-243.

Eguchi, J., K. Tsutsui and T. Hatta, 1978. Antifungal substances found in leaves of *Eucalyptus* species. Agric. & Biol. Chem., 40: 1431-1433.

Epstein, E. 1994. Proceedings of Nat. Acad. Sci. USA, 91 : 11-17.

Eraivana Arutkani Ayyanathan, K. and P. Narayanasamy 1988. Effect of neem oil and antiviral principles on the feeding activities of green leafhopper. Paper presented in the final workshop of IRRI ADB-EWC project on "Botanical pest control in rice based cropping system, December, 1988. International Rice Research Institute, Philippines.

Ezaki, K. 1975. Drainage of paddy field in Japan. Symposium on "Water management in rice field". Tropical Agricultural Research Center. Ministry of Agriculture and Forestry, Japan. Tech. Rep., 16, 18 p.

FAO, 1990, Comparative analysis of organic manures. Project field document. Vol.24, FAO, Rome, Italy.

Farooq, S., M. Asghar, N. Lqbal and T.M. Shah 1992. Variability in salt tolerance of accessions of wild rice species *Oryza punctata* and *O. officinalis*. IRRN, 17(6): 16.

Ferguson, J. A., J.T. Gilmour and C.L. Griffis 1973. Distribution pattern of Ca and Mg precipitates in flooded rice fields. Arkansas Farm Rev. 22 (6) : 2.

Flach, K.W. and D.F. Slusher 1978. Soils used for rice culture in the United States. Pages xix – xx in International Rice Research Institute. Soils and rice. Los Banos, Philippines.

Food and Agricultural Organization, 1993. FAO Bull. Stat. 6(4): 22.

Food and Agriculture Organization of the United Nations. Soils Bull. 32. A framework for land evaluation. Rome 72 p.

Fowke, L.C. and O.L. Gamborg 1980. Application of protoplasts to the study of plant cells. Int. Rev. Cytol., 68 : 9-51.

Fragoso Carlos and Patrick Lavelle 1995. Are earthworms important in the decomposition of tropical litter?. In: Soil organisms and litter decomposition in the Tropics, (edited by M V Reddy), Oxford and IBH publication Pvt. Ltd., New Delhi, pp. 103-112.

Fukui, H. 1971. Environmental determinants affecting the potential dissemination of high yielding varieties of rice. Southeast Asian Stud. 9:348-374.

Furuya, S 1987. Growth diagnosis of rice plants by means of leaf color. Japan Agric. Res. Quarterly., 20, 147-153.

Galbreith, D.W. 1989. In : "Biotechnology in agriculture and forestry – Plant Protoplasts and Genetic Engineering – II "(Y.P.S. Bajaj edited.), Springer – Verlag, p. 304.

Galinski, E.A. and H.G. Truper, 1982. Betaine : a compatible solute in the extremely halophilic phototropic bacterium *Ectothiorhodospira halochloris*. FEMS Microbio. Lett., 13: 357-360.

Gamborg, O.L. 1981. Protoplasts in genetic modification of plants. Adv. Biochem. Engg., 19: 239-263.

Ganesan, A.T. 1985. Gene transfer systems in the improvement of cereals. In : "Biotechnology in Agriculture" (edited by S. Natesh, V.L. Chopra and S. Ramach and ran) p. 79-84.

Ganesan, S., K.N. Chinnasamy, B. Chandrasekaran, M.N. Budhar and N.J. Prince Jayaseelan. 1991. Duck-cum-fish culture in rice farming system in Cauvery delta region of Tamil Nadu. Indian Journal Agric. Economics., 46(2): 180-185.

Gao, T.F., 1994. Research and development of new paddy seeder. In : Proceedings of "International Agrl. Engineering Conference," Dec. 6-9, 1994, Bangkok, Thailand

Garrity, D.P., V.P. Singh and M. Hossain, 1996. Rice-ecosystems analysis for research prioritization. In: "Rice Research in Asia-Progress & Priorities". (R.E. Evenson, R.W. Herdt and M. Hossain eds). CAB International & IRRI, pp. 38-58.

Gaur, A.C. 1992. Bulky organic manure and crop residues. Fertilizers, Organic manures recyclable wastes and biofertilizers. H.L.S. Tandon (ed.) Fertilizer Development Corporation, New Delhi.147 p.

Gaur, A.C. and L.N. Verma, 1991. Integrated plant nutrition in prevailing crop mixes in India. FAO- Rapa Reports. 1991/7. Asian experiences in Integrated Plant Nutrition Report of the export consultation of the Asian network on bio and organic fertilizers. Sept. 18-2, 1990. Bangkok, Thailand pp. 60-79.

Gaur, A.C. and R. Singh, 1982. Integrated Nutrient supply system. Fert. News 27: 87-98.

Gaur, A.C. R.S. Mathura and T. S. Varshney. 1970. Decomposition of different types of added organic matter in soil. Agri. Chemic., 14, 524-532.

Gaur, A.C., 1987. Recycling of organic wastes by improved techniques of composting and other methods. Resources and Conservation, 13: 157-174.

Gaur, A.C., 1990. Phosphate solubilizing micro organisms as biofertilizers. Omega Scientific Publishers, New Delhi. 240 p.

Gaur, A.S., 1982. Organic manures and biofertilizers. Review of soil research in India. 12th International Congress of Soil Science, New Delhi. Feb. 8-16,1982. pp. 278-305.

George, T.U., P.J. Tomy and R.G. Pinhero, 1987. Vyttila-3 an new rice variety for acid saline areas. Intl. Rice Res. Newsletter., 12 (4):20.

Geriach, W.I., A.J. Pryor, E.S. Denms, R.J. Beri, M.M. Sachs and W.J. Peacock, 1982. cDNA cloning and induction of the alcohol dehydrogenase gene (Adhi) of maize. Proceedings of National Academic. Science. USA, 79: 2981-2985.

Ghose, R.L.M., M.B. Ghatge and V. Subramanyan, 1960. Rice in India. ICAR, New Delhi.

Ghosh, A. and A.R. Sharma, 1998. Agronomic practices for augmenting rice productivity under flood prone lowland conditions. Paper presented at First Int. Agron. Congress, New Delhi. November 23-27. Abst. pp 122-123.

Gilbert, F.H., D. Narman and F.E. Winsh. 1980. Farming systems research - A critical appraisal. Rural development paper No. 6, East Lansing, Michigan State Univ., Michigan, USA.

Gogoi A.K. and H. Kalita, 1989. Effect of new herbicides on direct seeded late sown flooded rice. Indian J. Weed Sci., 7: 42 – 48.

Gogoi, A.K. and K. Kalita, 1990. Integrated weed control in direct seed upland rice. Indian J. Agron., 35 : 433 – 434.

Gogoi, A.K. and H. Kalita, 1992. Response of transplanted rice to fertilizer doses, plant protection and butachlor. Indian J. Agron., 37 : 567 – 568.

Gogoi, A.K. and H. Kalita, 1996. Effect of butachlor on weeds and yield of transplanted rice (*Oryza sativa*). Indian J. Agron., 41:230 – 232.

Gogoi, A.K. and M. Sarma, 1993. Response of upland rice varieties to herbicides. Indian J. Agron. 38:475-477.

Gogoi, A.K. and M. Sarma, 1994. Effect of varieties and herbicides on growth in upland direct seeded rice. Indian J. Weed Sci., 26 : 89 – 92.

Gogoi, A.K. and P.K. Gogi, 1993. Weed control in transplanted rice. Indian J. Agron., 38 : 298 – 299.

Gogoi, A.K., 1995. Weed management in transplanted lowland rice. Indian J. Agron., 40: 415 – 419.

GOI, 2001. Economic survey, Ministry of Finance, New Delhi.

GOK, 1996 Report of the expert committee on stagnation of agricultural productivity in Karnataka during 1980s, Bangalore.

Gomez, A.A. 1988. A farming system research approach to identifying farmer's production problems. In: Farming Systems Research Approach. pp: 63-70.

Gomez, K.A. and A.A. Gomez. 1984. Statistical procedures for agricultural research. John Willey & Sons, Singapore. pp. 533- 561.

Gopalan C., B.V. Rama Sastri and S.C. Balasubramanian, 1991. Nutritive value of Indian foods, National Institute of Nutrition, ICMR, Hyderabad

Gosal, S.S., H.S. Dhaliwal, J.S. Sandhu, M.S. Klinda and P.K. Agrawal, 1993. Somatic cell, anther and protoplast culture in *indica* rices. VIth Ann. meeting of the Int. programme on rice biotech., Feb. 1-5, 1993, Chlang Mai, Thailand, p. 95.

Goswami, N.N. R. Prasad, M.C. Sarkar and S. Singh, 1988 Studies on the effect of green manuring in nitrogen economy in a ricewheat rotation using a ^{15}N – technique. J. Agric. Sci., Camb., 111:413-417.

Govina, L.D. 1990. Pig- Duck–Fish-Azolla integration in La union, Philippines. NAGA. The ICLARM Quarterly, Apr. 18-20.

Govindan, R. 1988. Role of Poultry- cum -Fish culture on the economics of farming systems in Thanjavur district of Cauvery delta. M.Sc. (Ag.) Thesis, Tamil Nadu Agric.Univ. Coimbatore.

Govindan, R., K.N. Chinnaswamy, B. Chandrasekaran, M.N. Budhar and J. Price. 1990. Poultry – Fish culture in rice farming systems in Cauvery delta regions of Tamil Nadu. Indian J. Agron., 35: 23-29.

Goxnzalez, M.C., L. Iglesias, N. Santana, C. Ismall and N. Perez, 1991. Analysis of the variability obtained from *in vitro* cultures of seed of the rice cultivar Amistad-82 under saline conditions. Cultivos Troplcales, 12: 83-85.

Grainge, M., L. Berger and S. Ahmed, 1985. Effect of extracts of *Artabotrytys uninatus* and *Allium sativum* on *Xanthomonas campestris pv. oryzae*. Curr. Sci., 54: 90.

Grant, C.Y. 1965. Soil characteristics associated with wet cultivation of rice. In: International Rice Research Institute. The mineral nutrition of the rice plant. Johns Hopkins Press, Baltimore, Maryl and pp. 15-28.

Grist, D.H. 1969. Rich production In : Technical monograph on rice published by CIBA agro chemicals division, Basle, Switzerl and pp: 4-11.

Grist, D.H., 1959. Rice, Longmans Green & Co., Ltd., London.

Gunjal, S.S. 1991. Organic and Natural Farming: Principles of Organic Farming. Kisan World. June 1991:10-12.

Guo, Q. and Z.C. Liang, 1985. Inhibition of bacterial growth by extracts from leaf, wood and its relation to bacterial wilt resistance. J. South China Agrl. Univ., 6: 49-57.

Gupta, R.K and I.P. Abrol, 1990. Salt affected soils - Their reclamation and management for crop production. Adv. Soils Sci., 11:223-288.

Gupta, S.K., R.D. Singh and R.N. Rai, 1992. Response of rice to input factors under farmers practice. Indian J. Agron., 37: 240 – 243.

Guy, R.D., P.G. Warne and D.M. Reld, 1984. Glycine betaine content of halophytes: Improved analysis by liquid chromatography and interpretation of results. Plant Physiol., 61: 195-202.

Haokip, D.D., A.K. Sinha and A.K. Singh, 1998. Effect of puddling and water regimes on soil physical properties, water use and yield of rice. Paper presented at First Int. Agron. Congress, New Delhi. November 23-27, Abst. pp 158.

Harada, T., H. Hashimoto and M. Hara. 1955. Studies on the organo – mineral colloidal complexes of paddy soil. Part 2. Humus contained in the GI of degraded paddy soils. Soil Plant Food, 1: 105 – 108.

Hauser, G.F. and R. Sadikin. 1956. The productivity of the soils of East Central Java, based on the yields of Jawa rice. Gen. Agric. Res. Stn., Bogor No. 144. 98p.

Hazell Peter, 2001 Agriculture and the Environment, Environment and Development Economics, 6 (Part 4) October, pp. 516-521.

Hector, J.H. 1937. Introduction to the Botany of field Crops, Central News Agency, Johannesburg.

Hegde, D.M., 1992. 29th Workshop of Project directorate for cropping system research, 1-4, June. TNAU, Coimbatore, pp. 40.

Heong, K.L., P.S. Teng and K. Moody,1995. Managing rice pests with less chemicals. Geo. Journal, 35:337-349.

Heuvelink, G.B.M. 1993. Error propagation in quantitative spatial modeling applications in geographical information systems. Ph.D. thesis. University of Utrecht.

Higgins, G.M. 1964. Soils of some rice growing areas in north Nigeria. Afr. Soils, 9 (2):247 – 265.

Hill, J.E., D.E, Bayer, S. Bocchi and W.S. Clampett. 1991. Direct seeded rice in the temperate climates of Australia, Italy and the United states. In : Direct seeded flooded rice in the tropics". International Rice Research Institute, publication, P.O. Box 933, Manila, Philippines. pp.91-102.

Hindu, 2003. Water problems persist – Alternative crop pumpkin. The Hindu News paper, 21.11.03, p.3.

Hoan, N.T., S.M. Balachandran, N.P. Sarma, A.K. Garg, K.L. Hakim, A. Ghosh and E.A. Siddiq. 1992. Biotechnology programme at Directorate of Rice Research, National Rice Biotech. Network, 2nd Ann. Meeting, 1992, Directorate of Rice Research, Hyderabad, India, p. 7-9.

Huizing, H.G.J. 1971. A reconnaissance study of the mineralogy of s and fractions from East Pakistan, Sediments and soils. Geoderma, 6:109 - 133.

Huke, R.E. 1976. Geography and climate of rice. In Climate and rice. International Rice Research Institute. Los Banos, Philippines. pp. 31-50

IARI. 1995. Basic research on rice -wheat cropping systems. NARP Reports for 1993-94 and 1994-95.

ICSIM (Institute for Comm and Area Studies and Irrigation Management), 1990. Organic farming in Karnataka and Tamil Nadu, Project report, Bangalore. (Unpublished).

Ikehashi, H. and F.N. Ponnamperuma 1978. Varietal tolerance of rice to adverse soils. Pages xx–xx in International Rice Research Institute. Soils and rice. Los Banos, Philippines.

ILACO – NEDECO. 1966. Report on the soil investigation in the Niger Delta special area. Niger Delta Development Board, Port Marcourt. 161p.

International Rice Research Institute (IRRI), 1970. Rice production manual, published by IRRI, Philippines.

International Rice Research Institute (IRRI), 1972. Annual report for 1971. Los Banos, Philippines. 239 p.

International Rice Research Institute (IRRI), 1975. Annual report for 1974. Los Banos, Philippines. 479 p.

International Rice Research Institute (IRRI), 1976. Annual report for 1975. Los Banos, Philippines. 418 p.

International Rice Research Institute (IRRI), 1977. Los Banos, Philippines. Ann. Report, 1976.

International Rice Research Institute (IRRI), 1978. Annual report for 1977. Los Banos, Philippines.

International Rice Research Institute (IRRI), 1985. Field problems of tropical rice (revised edition) published by IRRI, Philippines. 172p.

International Rice Research Institute (IRRI), 1989. IRRI - Towards 2000 and Beyond. International Rice Research Institute, Los Banos, Philippines.

International Rice Research Institute (IRRI), 1992. Program Report for 1991, Los Banos, Philippines.

IOBC, 1993. International organization for biological and integrated control. Technical bulletin, 16 (1), 1993.

IRRI-INSURF. 1991, Progress report of the different international Network on soil fertility and sustainable Rice Farming, phase 1. (1988-1990), p 32-168. IRRI, Philippines.

Ismunadji, M. 1978. Utilization of cereal crop residues and its agricultural significance in Indonesia. Center Res. Inst., Agric. Bagar. Indonesia, 37: 1-34.

Jackson, M.L. 1967. Soil chemical analysis. Prentice Hall of India (Pvt.) Ltd., New Delhi. pp: 162- 165.

Jaikumaran, U. and Beena Thomas 1996. Is mechanized transplanting possible in rice fields? Souvenir, Association of agricultural officers, Thrissur, Kerala. pp. 81-82.

Jain, S.C., V.K. Paradkar and R.K. Gupta 1984. Weed control in transplanted paddy under sodic soil conditions of Madhya Pradesh. Abstr. Ann. Conf. Indian Soc. Weed Sci. Varanasi.

James, P.S., P.K.S. Kumar and G.B. Pillai 1996. Effect of selective mechanization on the economics of rice production. Proceedings of the Sixth Kerala Science Congress, Thiruvananthapuam. pp.192-194

Janitzky, P. and L.D. Whittig 1964. Mechanisms of formation of Na_2CO_3 in soils. II. Laboratory study of biogenesis. J. Soil Sci., 15:145-157.

Jauhri, K.S. 1989. Evaluation of certain agroindustrial wastes as seed pelletants under adverse soil pH conditions. Biological Wastes 28: 233-236.

Jayakumar, R., K. Siddeswaran, M. Premsekar, N. Kempuchetty and S. Subramanian 1987. Nutrient uptake by certain crops and associated weed and its effect on yield. Pesticides, 21:37 – 395.

Jayanthi, C., A. Rangasamy, R. Venkidusamy and C. Chinnusamy 1994. Sustainable integrated farming systems research and management for sustainable agriculture. Summer Institute held at Tamil Nadu Agric. Univ., Coimbatore between 6-15th June 1994. pp: 97-109.

Jayarajan, R. 1996. Design fabrication and testing of a power operated dibbler. M. Tech. thesis. Kelappaji college of Agricultural Engineering & Technology, Tavanur, Malappuram dt., Kerala.

Jayaseelan, M.J.P., M.S. Venugopal and S. Chelliah 1988. Two innovative culture methods for increased integrated fish production. Farm forum, 11:23-26.

Jena, K.K. 1994. Production of intergeneric hybrid between *Oryza L.* and *Porteresia coarctata* T. Curr. Sci., 67: 744-746.

Jeyarajan, R. 1988. Biological control of plant diseases. Training of ecological agriculture for the participants of A.M.E., Course, Tamil Nadu Agricultural University, Coimbatore. p. 57.

Jhingran, U.G. and B.K. Sharma 1980. Integrated livestock fish farming in India. In. Proceedings of ICLARM – SEARCA Conf. on "Integrated agricultural – aquaculture farming systems", 9th Aug. 1979, Manila, Philippines. pp: 135-141.

John, K.C. 1991. Farming systems research in eastern India: A new approach. In : "New frontiers in rice research". (edited by) In: Siddique, E.A and K. Muralidharan), Directorate of Rice Research, Hyderabad.

John, M., 1991. Design development and evaluation of a low cost paddy thresher. M. Tech. Thesis, Kelappaji College of Agricultural Engineering & Technology, Tavanur, Malappuram Dt., Kerala.

John, M.J. and C.P. Mohammad 1989. Design development and testing of an animal drawn dibbler for paddy. Kelappaji College of Agricultural Engineering & Technology, Tavanur, Malappuram Dt. Kerala.

John, P.S., R.K. Pandey, R.J. Buresh and R. Prasad 1989. Lowland rice response to urea following three cowpea cropping systems. Agronomy Journal, 81: 853-857.

Jones , M.G.K. and A. Karp 1985. In : Advances in Biotechnological processes (A. Mizrahl Ed.). pp. 91-121.

Jones, M.P., S. Mande, A. Dalebas and H. Sehi, 1997. Using anther culture of fertile double haploid interspecific progeny, IRRN., 22:7-8.

Jose, C.M., 1990. Development and performance evaluation of a rotary tillage attachment to the KAU Garden tractor. M.Tech. thesis, Kelappaji College of Agricultural Engineering Technology, Tavanur, Malappuram Dt., Kerala.

Kakati, N.N. and V.S. Mani, 1977. Chemical weed control relation to fertilizer use. Abstr. Indian Weed Sci. Conf. Hyderabad, pp.170-171.

Kampen, J. 1970. Water losses and water balance studies in lowland rice irrigation. Ph.D. thesis, Cornell University.

Kanapathy, K. 1975. Growing crops on peat soils. Soils and analytical services Bull. 4. University of Agricultural Rural Development of Malaysia. 64p.

Kanauja, R.S., 1974. Studies on certain aspects of seed – borne fungi of some starchy seeds. Acto Bot. Indica,, 2: 129-135.

Kandasamy, O.S *et al.* 1998. In:Lecture notes for the short training course on "Recent advances in rice production technologies" held at Department of Agronomy, TNAU, Coimbatore during 3-28, August, 1998.

Kang Sek Hjon and S. Markert. 1959. Ein Beitrag zur Kenntnis nordostkoreanisher Reisboden. I. Albrecht Thaer Arch., 4 : 262 – 292.

Kannaiyan, S. 1993. Bio- Fertilizers for rice, Tamil Nadu Agricultural University, Coimbatore.

Kannaiyan, S. 1996a. Immobilization of cyanobacteria in solid matrix for ammonia production and development of immobilized substrate based cyanobacterial inoculants for rice crop. Paper presented in X Southern regional conference on Microbiol inoculants, A.V.V.M. Sri Pushpam College, Poondi, India.p.1.

Kannaiyan, S. 1996b. Summary report on EEC Project on immobilized cyanobacteria for continuous ammonia fertilization in rice production, Univ. of Seville, Seville, Spain. p. 29.

Kannaiyan, S. 1998. Mass production of polyurethane foam immobilized cyanobacteria as a biofertilizer for rice crops : Paper presented in the International Conference on Agricultural Biotechnology, University of Saskatchewan, Canada, p.4.

Kannaiyan, S., S. Sundaravarathan and SP. Palaniappan, 1998. N fixing Green manures for sustaining Agricultural Production.TNAU,Coimbatore. 289p.

Kanno, I. 1956. A scheme for soil classification of paddy fields with special reference to mineral soils. Bull. Kyushu Agric. Exp. Stn., 261 – 273.

Kanno, I. 1962. A new classification system for rice soils in Japan. Pedologist, 6:2 -10.

Kanno, I., Y. Honyo, S. Arimura and S. Tokudome. 1964. Genesis and characteristics of rice soils developed on polder land s of Shiroishi area, Kyushu. Soil Sci. Plant Nutr., 10 : 1 – 20.

Kanwar, J.S., S.M. Virmani and S.K. Das 1992. Farming systems research in India: A Historical perspective. Expl. Agrl., 28:1-17.

Karmanov, I.I. 1966. Alteration of tropical soils under cultivation. Sov. Soil Sci.,1:31-43.

Katherkhan, S. 1994. Role of dairy, sheep and goat, piggery and rabbit rearing in IFS development. In: Proceedings of Summer Institute on "Integrated farming systems research and development for sustainable agriculture" organized by ICAR and Tamil Nadu Agric. Univ., Coimbatore, India during 6-15 June,1994. pp: 228-234.

Kathiresan, G., G. Manickam and P. Ganamoorthy, 1997. Effect of time of seedling and method of herbicide application on pre – monsoon sown semi dry rice. Indian J. Agron., 42:618 – 621.

Kato, T. 1972. On the water soluble organic matter in paddy soils (Part I). J. Sci. Soil Manure, Jpn., 43 : 357 – 936.

Kaushik, S.K. and V.S. Mani, 1978. Weed control in direct seeded and transplanted rice. Indian J. weed Sci., 10:73 – 78.

Kawaguchi, K. and Y. Matsuo. 1957. Re-investigation on distribution of active and inactive oxides along soil profiles in time series of dry rice fields in polder land s of Kojima Basin. Soil Plant Food, 3: 29 – 34.

Kawaguchi, K. and K. Kyuma. 1969a. Lowland rice soils in Malaysia. Reports on research in Southeast Asia. Natl. Sci. Ser. No. 5, Kyoto. 154 p.

Kawaguchi, K. and K. Kyuma. 1969b. Lowland rice soils in Thailand Reports on research in Southeast Asia Natl. Sci. Ser. No. 4, Kyoto. 270p.

Kawaguchi, K. and K. Kyuma. 1974a. Paddy soils in tropical Asia. I. Description of fertility characteristics. Southeast Asian Stud., 12 (1) : 3 – 24.

Kawaguchi, K. and K. Kyuma. 1974b. Paddy soils in tropical Asia. II. Description of material characteristics. Southeast Asian Stud., 12 (2) : 177 – 192.

Kawaguchi, K. and K. Kyuma. 1977. Paddy soils in tropical Asia- Their material nature and fertility. The University Press of Hawaii, Honolulu.158 p.

Kenchaiah, 1992. Effect of organic manures and oil cakes on rice productivity. M.Sc. (Afg.) thesis, Tamil Nadu Agricultural University, Coimbatore,India (Unpublished).

Khanna, S.S and A.N. Pathak, 1982. Phosphorus. In : Review of Soil Research in India, Indian Society of Soil Sciences, New Delhi. pp. 323-330.

Kishore, P.B., Z. Kavi, G.H. Hong, Miao, Hu, A. Chein-An and S. Verma Desh Pal, 1995. Over expression of proline–5-carboxylate synthetase increases proline production and confers osmotolerance in transgenic plants. Plant Physiol., 108 : 1387-1394.

Koenigs, F.F.T. 1950. A "sawah" profile near Bogor (Java). Trans IV Int. Conger. Soil Sci., 1:297-300.

Kolhe, S.S. and B.N. Mittra, 1981. Weed control in direct seeded upland rice. Proc. 8th Asian Pacific Weed Sci. Soc. Conf., Bangalore.

Kostenov, H. 1975. Genetic and chemical characteristics of Meadow Gley Soils in the rice fields of Primoyre. Pochvovdeniya, 6 : 23 – 31.

Krishmamurthy, K.K. and G.V. Kothandaraman, 1982. Organic manuring and green manuring in tropical soils : Indian experience. Transactions of the 12th International Congress of Soil Science–Non symbiotic nitrogen fixation and organic matter in the tropics. Symposia papers I. New Delhi, Feb. 8-16, 1982. pp. 179-188.

Krishnaiah, K. 1998. Rice research in India. Indian J. Agric. Sci., 86:385-395.

Krishnamohan, G., P. Narayanasamy, K. Pillayarsamy and R. Jeyarajan, 1988. Efficacy of certain plant products on rice blast pathogen. National seminar on "Management of crop diseases with plant products/biological agents," held on Jan. 11-12, AC & RI, Madurai.

Krishnarajan, J. (Eds.) 1998. Lecture notes for the short training course on "Recent advances in rice production technologies" held at Department of Agronomy, TNAU, Coimbatore during 3-28, August, 1998.

Kropff, M.J., P.S. Teng and R. Rabbinge 1995. The challenge of linking pest and crop models. Agricultural systems., 49, (4):413-434.

Kuether, D.O. and F. Cabrales, 1979. Some effects of tillage systems on a wetland rice soil. Seminar paper, IRRI, Las Banos, Philippines.

Kulmi, G.S. 1989. Comparative performance of thiobencarb formulations in transplanted rice. Indian J. Weed Sci., 21:53 – 55.

Kulmi, G.S. 1991. Associated weed flora and their susceptibility to herbicides in transplanted rice. Indian. J. Agron., 36:113 – 116.

Kulmi, G.S. 1992. Analysis of growth and productivity of transplanted rice (*Oryza sativa*) in relation to methods of weeding. Indian J. Agron., 37 : 312 – 316.

Kumar, B., G.T. Menon, P. Jayasree and G. Mathew 1990. Design, development fabrication and testing of a power operated paddy weeder for dryland cultivation. B. Tech. Project work report, Kelappaji College of Agricultural Engineering & Technology, Tavanur, Malappuram Dt., Kerala.

Kumar, J. and R.C. Gautam 1986. Effect of various herbicides on yield and yield attributes of direct seeds rice on puddle soil. Indian J. Weed Sci. 18 : 54 – 56.

Kumar, K. and B.N. Singh 1984. Preliminary observation on utilization of palletized poultry litter as feed for common carp. Int. J. Acad. Ichtyol. 151-154.

Kumar, P. and H.S. Gill 1981. Herbicidal control of weeds in direct seeded rice (*Oryza sativa* L.) under non puddled conditions and their residual effect on wheat and linseed. Indian J. Weed Sci. 13: 50 – 55.

Kumar, P.K.S. 1996. Development and evaluation of a low cost power operated paddy thresher. M. Tech. thesis, Kelappaji College of Agricultural Engineering & Technology, Tavanur, Malappuram Dt., Kerala.

Kumar, V., G.C. Shrotriya and S.U. Kaore, 1990. Scope of urea super granules for rice. Fert. News, 35(7): 23-24.

Kuppusamy, M. 1989. Feasibility of introducing sericulture as a component in the farming system of Thanjavur district, Tamil Nadu. M.Sc. (Ag.) Thesis, Tamil Nadu Agric. Univ., Coimbatore, India.

Kyuma, K. and K. Kawaguchi. 1996. Major soils of Southeast Asia and the classification of soils under rice cultivation. Southeast Asian Stud., 4: 290 – 312.

Lampkin, N. 1990. Organic Farming. Farming Press Books. IPSWICH, UK.

Le Buanec, B. 1975. La riziculture pluviale en terrain draine. Situation, problemes et perspectives. Agric. Trop., 30(4): 358-381.

Leakey, C.L.A. and J.B. Wills 1977. Food crops of the lowland tropics, Oxford University Press, Oxford,UK.

Lentini, Z., C.P. Martinex and E. Nossa 1997. Towards Introgression of high response to anther culture into *Indica* rices. IRRN., 22 : 18-19.

Lentini, Z., P. Reyes, C.P. Martinex and W.M. Roca, 1995. and rogenesis of highly recalcitrant rice genotypes with maltose and silver nitrate. Plant Sci., 110 : 127-138.

Lentini, Z., C.P. Martinez, L.R. Sanint, A. Ramirez and P. Reyes, 1993. Anther culture of recalcitrant rice genotypes and the cost/benefit implications in varietal development. Ann. Meeting of the Int. programme on "Rice Biotech", Feb. 1-5, 1993, Chiang Mai, Thailand , p. 92.

Leung, K.W. and C.Y. Lai. 1973. The characteristics and genesis of paddy soils is northern part of Taiwan. Agric. Res., 22:77-97.

Lewin, J. 1969. Ann. Rev. Pl. Physiol., 20: 289 – 304.

LI, Wen-Bin, Wang Ge-Jlao, E. Stanmchina, S. Cartiglione, Sun Yong-Ru and F. Sala, 1995. Transgenic rice plants produced by PEG mediated plasmid uptake into protoplasts. Acta Botanica Sinica, 37 : 409-412.

LI, H.J., 1990. Study on selection of salt resistant variants of rice. Journal of Shenyang Agril. Univ., 21 : 53-59.

Li, Y.H., 2001. Research and practice of water- saving irrigation for rice in China. In: Proceedings of the International Workshop on "Water saving irrigation for paddy, 23-25 March, Wuhan University, Wuhan, (China) pp. 135-144.

Lianqun, E.M. and F. Skoog, 1965. Organic growth factor requirement of tobacco tissue cultures. Physiological Planta, 18: 100-127.

Lightfoot, C. and D.R. Minnick. 1990. Farmer – first qualitative methods: Farmers' diagrams for improving methods of experimental design in integrated farming systems. A FSRE Symposium on "The role of farmers in FSRE and sustainable Agriculture". 14-17 October 1990, Michigan State University, East Lansing, USA.

Lightfoot, C., M.N. Bimbao, J.P.T. Dalsgaard and R.S. Pulhin. 1992. Aquaculture and sustainability through integrated resource management. Quarterly Agric., 22: 143-150.

Loganathan, P. 1989. Mushroom cultivation as a component in mixed farming system in Thanjavur district, Tamil Nadu. M.Sc. (Ag.) Thesis, Tamil Nadu Agric. Univ., Coimbatore.

Lynn, W.C. and L.D. Whittig. 1966. Alternation and formation of clay minerals during cat clay development. In: Proc. 14th National Conference on Clays and Clay Mineral. pp.241-248.

Machhi, R.M., S. Raman and P.L. Naik 1987. Optimum dose and effective time of application of nitrogen for the crop of paddy variety Mahsuri under reclaimed saline – sodic soils of coastal region. J. Indian Soc. Coastal Agri. Res., 5(2): 415-418.

Mahadevaswamy, M. and N.V. Nanjappa, 1991. effect of herbicides on weed control in drilled sown rice (*Oryza sativa*). Indian J. Agron., 36 (Supp.) : 247 – 248.

Maherchandani, N., 1991. Callus culture and adult plant response to salt stress. Annals of Biology (Ludhiana), 7: 105-117.

Manaligod, H.T. and G.R. Quick, 1991. Engineered equipments for timeliness in intensified cropping systems. In : Proceedings 22 nd Asian rice farming systems working group meeting, Sep. 30-Oct 6, 1991, Beijing, China. pp. 386-398.

Mandal, A.B. and A.K. Bandyopadhyay, 1996. Anther culture in rice at Bay Islands. Rice Biotech. Quarterly, 25 : 13.

Mandal, A.B., 1990-91. Screening breeding and evaluation of crops for coastal saline soils. CSSRI, Karnal, Ann. Report, p. 126.

Mandal, A.B., N.D. Majumdar and A.K. Bandyopadhyay, 1993. Screening rice for tolerance to salt stress and submergence. IRRN., 18: 34.

Mariappan, V., E. Rajeswari and A. Kamalakannan, 1995. Management of rice blast *Pyricularia oryzae* by using *Azadiracta indica* and other plant products. Neem for the management of crop diseases. (edited by V. Maliappan) Associated Publishing company, New Delhi, pp. 3-10.

Mariappan, V., V. Narasimhan and G. Sriman Narayana, 1987. Effect of plant derivatives on the insect vector and rice tungro virus transmission. in Proceedings symposium, March 30-31, 1987, Madurai, India. pp.55-59.

Mariappan, V., V. Narasimhan and S. Jeyaraj, 1988. Urd bean powdery mildew control by neem oil. Final workshop on IRRI – ADB-EWC project on Botanical pest control – rice based cropping system" held at IRRI, Los Banos, 12-16 Dec. 1988.

Marschner, H., H. Oberle, Cakmar and V. Romheld. 1990. In: "Plant Nutrition-Physiology and Application" (M.L. Van Bensichem, ed.). [Kluwer Academic Dordr cent.], pp. 241-249

Maskina, M.S., Y. Singh and B. Singh 1988. Response of wetland rice to fertilizer N in a soil amended with cattle, poultry and pig manures. Biological Wastes, 26: 1-8.

Mass, E.V. and G.J. Hoffman 1977. Crop salt tolerance current assessment. J. Irrigation Drainage. 103 : 115 – 134.

Matsuo, H., A.J. Pectot and J. Riquier. 1978. Rice soils of Europe. International Rice Research Institute. Soils and rice. Los Banos, Philippines. pp.193-198.

Matsuzaka, Y. 1969. Study on the classification of paddy soils in Japan [in Japanese, English summary]. Bull. Natl. Inst. Agric. Sci. Ser. B., 20: 155 – 349.

Matthews, R.B., M.J. Kropff, T. Horie and D. Bachelet 1997. Simulating the impact of climate change on rice production in Asia and evaluating options for adoption. Agricultural systems, 54:399-425.

Maxwell, S. 1984. Farming systems research: Hitting a moving target. Discussion paper No.199. SUSSEX Institute of Development studies.

Mc Couch, S.R., G. Kochert, Z.H, Yu, Z.Y, Wang, G.S. Khush, W.R. Coffman and S.D. Tanksley. 1988. Molecular mapping of rice chromosomes. TAG, 76 : 815-829.

McCauley G.N. 1990. Sprinkler Vs flooded irrigation in traditional rice production?

Mencher J P., 1994. Agricultural labour and pesticides in rice regions of India: Some Health Considerations, Economic and Political Weekly, XXVI (39):2263-68.

Menon, D.K., G.K. Maji and J. Jacob, 1993. Design, fabrication and testing of a manually operated paddy dibbler. B.Tech. Project report, Kelappaji College of Agricultural Engineering & Technology, Tavanur, Malappuram Dt., Kerala.

Merwin, H.D. and M. Peech, 1950. Exchangeability of soil K in the s and , silt and clay fractions as influenced by the native of the complimentary exchangeable captions. Soil Sci., Soc. America Proceedings, 15:125- 128.

Miah, M.A.A and M.S. Pathan 1993. Development of salt tolerant lines using biotechnology. VIth. Ann. Meeting of the Int. Programme on Rice Biotechnology", Feb. 1-5. 1993, Chiang Mai, Thailand, p.98.

Miah, M.A.A., E.D. Earle and G.S. Khush, 1985. Inheritance of the callus formation ability in anther cultures of rice(*Oryza sativa* L). Theo.Appl.Genet., 70: 113-116.

Mikkelsen, D.S., S.K. DeDatta and W.N. Obcemea, 1978, Ammonia losses from flooded rice soils. Soils Sci. Soc. Am. J., 42:725-730.

Mishra, B.P. and D. Lanka, 1973. A note on minimum tillage in rice cultivation. Indian J. Weed Sci., 5 : 57-59.

Mishra, B. and R.K. Singh, 1990-91. Evaluation and development of plant genetic resources tolerant to salinity, sodicity and water logging stresses. Rice. Ann. Report, 1990-91, CSSRI, Karnal, India.

Mishra, A. and S. Pasupalan. 1994. Integrated farming systems for rain fed lowland The Hindu. 14th Sept., 1994.

Mishra, B.P. and V.K. Singh, 1989. Studies on weed control in transplanted rice. Indian J. Agron., 34 : 457 – 458.

Misra, M., V. Padma and N. Saluja 1989. Silkworm litter: use as nitrogen replacement for vegetable crop cultivation and substrate for mushroom cultivation. Biological Wastes, 27: 209-216.

Mitsuchi, M. 1960. Profile differentiation of surface water type paddy soils under different drainage conditions [in Japanese]. J. Soil Maure, Jpn. 39 : 233 – 276.

Mitsuchi, M. 1974a. Pedogenic characteristics of paddy soils and their significance in soil classification. Bull. Inst. Agric. Sci. B., 25:29-112.

Mitsuchi, M. 1974b. Chloritization in lowland paddy soils. Soil Sci. Plant Nutr., 20:108-116.

Mitsuchi, M. 1975. Permeability series of lowland paddy soils is Japan. JARQ., 9:20 - 23.

Modgal, S.C. and M.P. Singh, 1980. Weed control through herbicides in direct seeded rice. Indian J. Weed Sci., 12: 80 – 86.

Mohamed Ali, A and S. Sankaran, 1984. Crop weed competition in direct seede lowland and upland rice. Indian J. Weed Sci.16:90-96

Mohamed Ali, A and S. Sankaran, 1977. A review of varietal tolerance of rice to herbicides., Abstr. Indian Soc. Weed Sci. conf., Hyderabad. p.170.

Mohamed Ali, A., K. Siddeswaran, R. Jegannathan, S. Muthusamy and S. Subramanian. 1998. Efficiency of mixed farming in the tribal areas of Sherveray Hills. Madras Agric. J., 71: 294-298.

Mohan, S. and R. Jeyarajan. 1994. Mushroom cultivation. In: Proceedings of summer Institute on "Integrated farming systems research and development for sustainable agriculture" organized by ICAR and Tamil Nadu Agric. Univ., Coimbatore, India during 6-15 June,1994. pp. 220-227.

Mohapatra, D., L.K. Bose and R.N. Mishra, 1992. Wide hybridization for rice improvement. National. Rice Biotech. Network, 2nd Ann. Meeting, 1992, Director of Rice Research, Hyderabad, India. pp. 29-31.

Moody, K. 1989. Weeds reported to occur in India. In: Weeds reported in rice in South East Asia, IRRI, Los Banos, Philippines. pp. 115-162

Moormann, F.R. and B.T. Kang. 1978. Micro variability of soils in the tropics. In Proc. Symposium on soil heterogeneity in the tropics. (ASSS Journal).

Moormann, F.R. 1961. The soils of the Republic of Vietnam. Ministry of Agriculture, Saigon 66 p.

Moormann, F.R. and L.J. Pons. 1974. Characteristics of mangrove soils in relation to their agricultural land use and potential. In: Proceedings of the International symposium on biology and management of mangroves. Honolulu, Hawaii. pp. 529-547.

Moormann, F.R. and N. Van Breeman, 1978. Rice : Soil, Water, land In: Soil and Rice. Los Banos, Philippines

Moormann, F.R. and R. Dudal. 1968. Characteristics of soils on which paddy is grown in relation to their capability classification. 11th session IRC, K and y, Paper S/68/7B. 14 p.

Moormann, F.R., W.Y. Veldkamp and Y.C. Ballaux. 1977. The growth of rice on a toposequence – a methodology. Plant soils, 48 p.

Moormann, R.F and N.V. Breemen 1978. Rice soil, water and land IRRN, Philippines, 185p.

Moorthy, B.T.S. and A.N. Dubey, 1981. Relative efficiency of different herbicides for control of weeds in upland rain fed rice. Indian J. Weed Sci., 13: 56-62.

Moorthy, B.T.S. and G.B. Manna, 1984. Weed control in transplanted rice by herbicides in summer season. Indian J. Weed Sci., 16: 82 – 89.

Mosier, A.R., S.K. Mohanty, A.Bhadrachalam and S.P. Chakravorti, 1990. Evolution of dinitrogen and nitrous oxide for the soil to the atmosphere through rice plants. Biol. Fert. Soils., 9:61-67.

Motomura, S. 1969. Dynamic behavior of ferrous iron in paddy soils. JARQ., 4:12-17.

Motomura, S., F. M. Lapid and H. Yokoi 1970. Soil structure development in Asiatic polder soil in relation to iron forms. Soil Sci. Plant Nutr., 16: 47-54.

Moya, P.F., D. Dawe, D. Pabale, M. Tiongco, N.V. Chien, S. Devarajan, A. Djatiharti, N. X. Lai, L. Niyomvit, H. X. Ping, G. Redondo and P. Wardana 2004. The economics of intensively irrigated rice in Asia. p. 29–58. *In* : Increasing productivity of intensive rice systems through site-specific nutrient management, Dobermann *et al.* (ed.). Science Publishers, Inc., Enfield, NH and International Rice Research Institute (IRRI), Los Baños, Philippines.

Mukhopadhyay, S.K., A.B. Khara and B.C. Ghosh 1972. Nature and intensity of weed competition with direct seeded upland IR-8 rice crop. Int. Rice Com. News letter, 21:10-14.

Mukhopadhyay, S.K. 1974. Increasing fertilizer efficiency through weed control. Fert. News., 19(12) : 56-58.

Mukhopadhyay, S.K. 1978. Weed control in lowland rice under submergence. Proc. Nat. Symp. on increasing rice yield in kharif. CRRI., pp. 301 – 310.

Murashlage, T. and F. Skoog 1962. A revised medium for rapid growth and bio assay with tobacco tissue cultures. Physiologia Planta., 15: 473-497.

Murthy, R.S. 1978. Rice soils of India. International Rice Research Institute, Soil and rice. Los Banos, Philippines.pp.3-17.

Murthy, R.S., H.S. Shankaranarayana and L.R. Hirekerur 1976. Distribution, genesis and classification of acid soils of India. In : Acid Soils of India – Their genesis, Characteristics and management. J. Indian Soc. Soil Sci. Bulletin, 11, pp.1-17.

Nagaraja, V.N. 1964. Hydrometeorolgical and tidal problems of the deltaic areas in India. In: Proc. of the UNESCO symposium on scientific problems of the humid tropical zone deltas, Dacca.pp.115-121.

Nagaraju, A.P. 1994. Efficiency of anilofos 2,4 – DEE and other herbicides on transplanted rice and their economic. Indian J. Weed Sci., 26:87 – 88.

Nakagawa, S. 1975. Water requirements and their determination. In: Symposium on 'Water management in rice fields', Tropical Agricultural Research Center, Technical Report. (Mimeo).

Nandal, D.P. and C.M. Singh, 1994. Weed management in directed seeded puddle rice wheat cropping system. Indian J. Weed Sci., 36 : 21-26.

Nandal, D.P. and C.M. Singh, 1995. Effect of herbicides on weed control and nutrient uptake in direct seeded puddle rice. Ann. Agric. Res., 16 : 12-15.

Narasiman, V., A. Chandrasekaran and A. Abdulkareem, 1994. Efficacy of botanicals in the management of sheath rot disease of rice. P. 375-378. In : Crop Disease – Innovative Techniques and Management, edited by Sivaprakasam, K. and K. Seetharaman. Kalyani Publishers, New Delhi, India.

Narasiman, V., K. Rajappan, C. Ushamalini and A. Abdulkareem, 1997. Efficacy of new EC formulation of neem and pungam oil for the management of sheath rot of rice. Phytoparasitica

Narasiman, V., R. Selvam and V. Mariappan, 1998. Efficacy of plant products for the management of bacterial plant pathogens. Neem for the management of crop diseases, (edited by V. Mariappan), Associated publishing company, New Delhi, India.

Narayan Rao. P.V. and B.S. Sirohi, 1975. Comparative study of improved puddlers. J. Agric. Engg., 12 (3-4):1-3.

Narayanan, K.K. and S.R. Rangasamy, 1990. Genetic analysis for salt tolerance of in *vitro* regenerated progenies of rice. In : Rice Genetics II-Proceedings of the 2nd Int. Rice Genetics Symp., 14-18 May 1990, IRRI, Manila, Philippines, pp.289-294.

Natarajan, 1992. Latest technologies in Sericulture and its economics. Paper presented in National training on "Integrated Farming System" held at TNAU, Coimbatore during December 11-18, 1992.pp. 76-80.

Nayak, B.C. and D. Lanka, 1988. Studies on cultural practices for rain fed shallow drought prone medium land Indian J. Agron., 33:1-6.

Noble, R.P. and B. Rshide. 1990. Aquaculture technology transfer to small farmers in Malawi, Southern Africa. NAGA, The ICLARM Quartly. 16: 14-16.

Norman, D.W. and M.P. Collinson. 1985. Farming systems research : Theory and Practice. ACIAR Proceedings, No. 11: 16-30.

Ohja, T.P., 1988. Farm mechanization: Need for a long-term policy. In : Survey of Indian Agriculture (1988), The Hindu, Chennai, India.

Oldeman, L.R., H.T.A. Makkeling and W.G. Sombroek, 1990. World map on the status of human induced soil erosion: Explanatory notes. ISRTC, Netherlands, UNEP, Kenya.

Olk, D.C, K.G. Cassman, G. Simbahan, P.C. Sta. Cruz, S Abdulrachman, R. Nagarajan, Pham Sy Tan and S. Satawathananont. 1996. N fertilizer management by farmers and soil N supply in tropical irrigated lowland rice systems. In: Proceedings of the International Symposium on "Maximizing sustainable rice yields through improved soil and environmental management", November 11-17, 1996, Khon Kaen, Thailand Dept. of Agriculture, Soil and Fertilizer Society of Thailand , Dept. of land Development, ISSS, Bangkok. pp. 29-38.

Page, A.L. 1982. Method of soil analysis. Agron. Part 2: Chemical and microbiological properties. Amer. Soc. of Agron, Madison, Wisconsin, pp: 1044 – 46.

Palaniappan, SP and K. Annadurai, 1999. Organic farming : Theory and Practice. Scientific Publishers (India). Jodhpur, 257 p.

Palaniappan, SP and K. Sivaraman, 1995. Cropping systems in the tropics: Principles & Management, New Age International Publishers (Pvt.) Ltd., New Delhi.

Palaniappan, SP. and K. Siddeswaran. 1994. Organic sources of nitrogen to rice. Paper presented in the training course on "Systems approaches for nitrogen application to rice (SANAR)", TRRI, Aduthurai, India. May 23-27, 1994.

Palaniappan, SP., K. Siddeswaran and D. Srinivasalu Reddy. 1990. Evaluation of green manures for the sustainable rice farming systems. International symposium on "National Resources Management for a sustainable Agriculture", New Delhi. Abst., p. 220.

Palaniappan, SP., K. Siddeswaran, C. Ramaswami, N.S. Radha and M. Premsekar. 1991, Evaluation, management and utilisation of green manures for rice-based cropping systems. Paper presented at the planning meeting, 7-10 March, 1991, Bali, Indonesia.

Palaniappan, SP., R. Loganathan and K. Annadurai 2000. Effect of method of incorporation of green manure and nitrogen levels on lowland rice. Oryza, 37(3): 247-248.

Palaniappan, SP., S. Kannaiyan and K. Natarajan 1992. Progress Report on Integrated use of green manure, INSUFR-Joint meeting of the ARFSN at Vietnam, Oct. 12-17th, 1992.

Palanisamy, G. 1998. History and origin of rice. In: Leture notes of short training course on Recent advances in rice production technologies. Organised by TNAU and ICAR during 3-28, August, 1998.pp.13-15.

Palaniswamy, K.M. and K.A. Gomez. 1974. Length width method for estimating leaf area for rice. Agron. J., 66: 430-433.

Panabokke, C.R. 1978. Rice soils of Sri Lanka. International Rice Research Institute. Soils and rice Los Banos, Philippines. pp.19-33.

Panda, N and D. Sahoo, 1989. Long-term effect of manures and fertilizers in rice based cropping systems in sub humid lateritics soils. Fert. News., 34(4):39-44.

Panda, N and S.S.K. Nanda, 1985. Soils of Orissa and their Management. In: Soils of India and their Management. Fert. Assoc. India. pp.284 – 305.

Pande, H.K. and V.M. Bhan 1966. Effect of deep tillage on yield of upland rice and on associated weeds. Exp. Agric., 2: 225 – 232.

Pande, H.K., S.P. Bose and R.S. Tripathi 1974. Management of upland soils for the cultivation of rice in the lateritic tract of west Bengal, Oryza, 11:1-5.

Pandey, J. 1991. Effect of weed management, soil manipulation and micro-nutrients on weed and yield of rice. Indian J. Weed Sci., 23: 72 – 37.

Pandey, J. 1984. Control of aquatic weeds in deep-water paddy. Pesticides, 18:66-67.

Pandey, J. 1993. Annual Report, IARI, New Delhi pp.49.

Pandey, J. and B.P. Singh, 1981. Effect of pre-emergence of applied herbicides on micro nutrient removal by lowland rice and associated weeds in calcareous soil. Pesticides, 15 (11): 32 – 34.

Pandey, J. and K. Shukla 1990. Herbicides for weed control in transplanted and puddle seeded rice. Indian J. Weed Sci., 22: 34 – 37.

Pandey, J. and K.N. Thakur 1988. Effect of herbicides and nitrogen on nutrient removal by weeds in transplanted rice Indian J. Agron., 33: 259 – 363

Pandey, J. and K.N. Thakur 1989. Increasing rice yield through weed control Indian Fmg., 38: 278-81.

Pandey, J. and S.J. Singh 1984. Varietal reaction of rice to butachlor and pendimethalin. Abstr. Ann. Conf. Indian Soc. Weed Sci., Varanasi, p.80.

Pandey, J., R.P. Singh and K. Sukla 1991. Study on the chemical weed control in upland rice. Indian J. Weed Sci., 23:7-9.

Pandey, O.K.A., N.N. Asthana, Tripathi and S.N. Dixit 1983. Fungitoxicity in pollen grains. Grains, 22: 21-31.

Pandey, R.K., S.K. Sharma and D.M. Hegde 1992. Approach to farming system research in ICAR : Concepts and methodology. Project Directorate for Cropping systems research. Bulletin. No. 2. pp. 1-16.

Parashar, K.S. and A. Prasad 1985. Indian Sugar, 35: 121-125.

Patel, G.K. and G.C. Tosh 1973. Relative efficacy of propanil, MCPA and 2,4-D on control of weeds in paddy Indian J. Weed Sci., 5:140 – 147.

Patel, C.L and Z.G. Patel 1983. Efficacy of weed control methods in transplanted rice. Indian J. Weed Sci., 15: 203-206.

Patel, Z.G., C.L. Patel, R.M. Desai and R.B. Patel 1988. Integrated weed control in low and transplanted rice. Indian J. Weed Sci, 20:88-89.

Patil, S.K. D. Panda and S.K. Mohanty 1990. Leaching loss of N from urea based fertilizer applied to rice under different hydrological conditions. Oryza, 22 : 342 – 345.

Patil, V.D., Y.S. Nerker, M.B. Misal and S.R. Harkal 1997. Parag – 401, a semi dwarf rice variety developed through anther culture. IRRN, 22: 19.

Patnaik, S. and A. Bhadrachalam 1985. Amelioration and management of low – productive and problem soils for optimizing rice yields. In: Rice Research in India. ICAR, New Delhi, pp.280 – 308.

Patron, W.J. J.A. Morgan, J.M. Altennoter and L.A. Harper 1988. Ammonia volatilization from spring wheat plants. Agron. J. 80: 419 – 425.

Paulraj, S. 1984. Composite fish culture by pig-dung. Geobios., 11:265-267.

Pearson, G.A. and A.D. Ayers 1960. Rice as a crop for salt affected soils in process of reclamation. USDA Prod. Res. Rep., 43:13 p.

Peng, S, F.V Garcia, R.C Laza, A.L Sanico, R.M Visperas and K.G. Cassman 1996. Increased N-use efficiency using a chlorophyll meter on high-yielding irrigated rice. Field Crops Res., 47:243-252.

Phillis, E. 1963. Profiles and permeability in Malayan padi soils. Malay. Agric. J., 44:3-17.

Pillai, G.K. and M.V. Rao 1974. Current status of herbicide research on rice in India. Paper presented at the Int. Rice Res. Conf. IRRI. Philippines. pp. 23-27.

Pillai, K.G. V.K. Vamadevan and S.V. Subbaih, 1976. Weed problems in rice and possibilities of chemical weed control. Indian J. weed Sci., 8: 77-87.

Pinar, Y., A. Sessiz and A. Guzel, 1994. Possibility of paddy sowing in rows by seed sowing machine. In : Proceedings of "International Conference on mechanization with energy in agriculture. Oct. 12-14, (1993), Turkey.

Pingali P.L., M. Hossain and R.V. Gerpacio, 1997, Asian Rice Bowls: The Returning Crisis? Wallingford: CAB International.

Pingali P.L. 2001. Environmental consequences of agricultural commercialization in Asia. Environment and development economics, 6 (Part 4) October, pp. 483-502.

Plucknett, D.L., J.L. Dillon and G.J. Vallaeys. 1987. Review of concepts of farming systems research: What, Why and How?. In : Proceedings of workshop on "Farming Systems Research", ICRISAT, Hyderabad.pp. 2-9.

Polak, B. 1952, Veen en veenowtginning in Indonesia. M.I.A.I.5 and 6, Vorrink, Bandung, Indonesia.

Ponnamperuma, F.N. 1972. The chemistry of submerged soils. Adv. Agron., 24:29- 88.

Ponnamperuma, F.N. 1976. Specific soil chemical characteristics for rice production in Asia. Int. Rice Res. Inst. Res. Pap. Ser. No.2. 18 p.

Ponnamperuma, F.N. 1984. Straw as a source of nutrient for wetland rice. In: organic matter and rice. International Rice Research Institute, Los Banos, Laguna, pp. 117-136.

Ponnamperuma, F.N. T. Attananada and G. Beye. 1973. Amelioration of three acid sulfate soils for lowland rice. In: Proc. Acid Sulfate Soils Symp. ILRI Publ.18 Vol. II. Wageningen. pp.391-406.

Ponnuswamy, A.S. 1996. Identification of alternative to GA_3 in hybrid rice seed production. Paper presented in the National Seminar on Crop physiology held during Jan.29-31. 1996 at TNAU, Coimbatore.

Ponnuswamy, A.S., K. Thiyagarajan, P. Rangasamy and M. Rangaswamy 1996. Effect of GA_3, on enhancing outcrossing in hybrid rice seed production. Paper presented in National Seminar on Crop physiology held during Jan. 29-31, 1996 at TNAU, Coimbatore.

Pons, L.J. 1973. Outline of the genesis, characteristics, classification and improvement of acid sulfate soils. In: Proc. Acid sulfate Soil Symp, ILRI Publ. 18 Vol I. Wageningen. pp. 3-27.

Potrykus, I. and R.D. Shilito 1985. Protoplasts: Isolation, culture and plant regeneration. In : Methods In enzymology-Plant molecular biology (A. Weissbach and H. Weissbach Eds.). Academic Press, Orlando, Florida. USA, pp.118.

Potrykus, I., R.D. Shillito, M.W. Saul and J. Paszkowski 1985. Direct gene transfer : State of the art and future potential. Plant Mol. Biol. Reports., 3: 117-128.

Prabhakar, A.S., K. Krishnamurthy and K.S. Ramadas 1973. Studies on weed control in paddy with herbicides. Indian J. Weed Sci., 5 : 140 – 147.

Pradhan, A.C. and C.N. Choudhary 1989. Performance of herbicides and their economics for direct seeded upland rice production. Indian J. Agron., 34 : 268-269.

Prakasa, Rao, E.V.S. and P.R. Bhat 1984. Ammonia volatilization loss from prilled urea, super granules and sulphur coated urea when surface applied and deep placed. Fert. News, 29(7):31-32.

Prakash, K.V. and M. Sivasamy, 1995. Development of a rower Tiller operated paddy Transplanter for conventional seedlings. M. Tech. thesis, Kelappaji College of Agricultural Engineering & Technology, Tavanur, Malappuram Dt., Kerala.

Prasad, M.N., S.S. Virmani and A.D. Gamutan, 1988. Substituting urea and boric acid for gibberellic acid in hybrid rice seed production. Int. Rice Res. Newsletter, 13(6): 9-10.

Prasad, R and J.F. Power, 1995. Nitrification inhibitors for agriculture, health and environment. Adv. Agron., 54.233-281.

Prasad, R. 1996. Management of fertilizer N for higher efficiency. In: Nitrogen Research and Crop Production. (HLS Tandon ed). FDCO, New Delhi. pp.104 – 115.

Prasad, R. and SP. Palaniappan, 1987. Pulse crop residue as N source in rice-based cropping system. International Rice Research Newsletter 12:31.

Prasad, R., G.B. Rajale and B.A. Lakhdive 1971. Nitrification retarders and slow release n fertilizers Adv. Agron., 23:337-383.

Prasad, R. 1990. Nitrogen management in soil fertility and fertilizer use. IFFCO, New Delhi pp. 41-51.

Premsekar, M and A. Rangasamy 1994. Sustainable IFS research on lift irrigated garden lands. In: Proceedings of summer Institute on "Integrated farming systems research and development for sustainable agriculture" organized by ICAR and Tamil Nadu Agric. Univ., Coimbatore, India during 6-15 June, 1994. pp:106-114.

Purseglove, J.W., 1972. TropicalCrops. Monocotyledons. Longmans, London.

Purushothaman, S., P. Balasubramanian and B.J. Pandian 1996. Production technologies for wetseeded rice. TNAU, Madurai, India.

Pushpavalli, R., K. Natarajan and SP. Palaniappan. 1993. Use of non-conventional green manures. The Hindu, March 24, 1993.

Quimio, C.A. and F.J. Zapata, 1990. Diallel analysis of callus induction and green plant regeneration in rice anther culture. Crop Sci., 30: 188-192.

Raghupathy, B. 1993. Int. Rice Res. Notes, 18(3): 27 – 28.

Raina, S.K., 1992. Rice tissue culture and transformation : Progress at IARI. National. Rice Biotech Network, 2nd Ann. Meeting, 1992, Directorate of Rice Research, Hyderabad, India.

Rajan, K.M., 1980. Soil amendments plant disease control. IRRN., 5: 15.

Rajendra Prasad, 1999. A text book of Rice agronomy. Jain Brothers, New Delhi. 238

Rajendran, S. 1998. An exploratory study on organic farming system in India, In Living Resources for the Millennium. (edited by J. William) Loyola College, Chennai.

Rajendran, S. 1993, Of algae, worms and cash flow, Down To Earth, 2(15):22-24.

Rajendran, P. and A. Tajuddin, 2002. Paddy cum green manure seeder. Agro India, August, p. 36.

Rajendran, P., A. Tajuddin and J. Krishnarajan, 1998. Organic manuring of paddy. Paper presented at the 1st RDA/ ARNOA International Conference on “Asian

Organic Agriculture : Development of Basic Standard for Organic rice cultivation, held at Suwon and Chenonan/ Korea, 12-15, November, 2002.

Rajendran, P., A. Tajuddin, G. Duraisamy and T.V. Karivaradaraju 1999. Local tools for a good rice yield. The New Indian Express Newspaper dated November 2.

Rajendran, S. 1992. Water use is excessive in rice cultivation, Down To Earth, 1(14):41.

Rajendran, S. 2002. Pesticide spraying in Kerala - Human cost and environment loss, economic and political weekly, XXXVII (23): 2206-2207.

Rajendran, P.A. Tajuddin and C. Ramaswamy, 1998. Rice cum green manure culture with modified drum seeder under lowland condition. IRRN., 23 (3) : 31-32.

Rajkumar, S. 1988. Feasibility of introducing dairy as complementary enterprise for small farmers in Thanjavur delta, Tamil Nadu. M.Sc. (Ag.) Thesis. Tamil Nadu Agric. Univ., Coimbatore, India

Raju, R.A. and M.N. Reddy, 1995. Performance of herbicide mixture for weed control in transplanted rice. Indian J. weed Sci., 27 : 106 – 107.

Ramakrishnan, M.S., A. Velayudham and A. Mohamed Ali, 1992. Effect of seed rate and weed management on direct seeded lowland rice. Indian J. Agron., 37 : 808 – 809.

Ramalingam, J., N. Natarajan, C. Vanniarajan and P. Rangasamy, 1994. Natural outcrossing potential in cytoplasmic male sterile lines of rice. Int. Rice Res. Newsletter, 19(3): p.5.

Ramamoorthy, K. and A. Mohamed Ali, 1988. Competition and control of *Echinochloa colonum* (L.) Link in upland bunded rice and pulses system. Indian J. Weed. Sci., 20: 63 – 67.

Ramamoorthy, K., S. Kulandaisamy and S. Sankaran, 1974. Effect of propanil on weed growth and yield of IR20 rice under different seeding methods and rates. Madras Agric. J., 62 : 403 – 407.

Ramasamy, C., T.R. Shanmugam and D. Suresh, 1996. Constrains to higher rice yields in different rice production environments and prioritization of rice research in Southern India. In *Rice Research in Asia – Progress* and Priorities. (R.E. Evenson, R.W. Herdt and M. Hossain eds). CAB International & IRRI, pp.145 – 159.

Ramiah, K., 1953. Rice breeding and genetics, ICAR, New Delhi, Monograph No. 19, 1953.

Ramiah, S. and P. Muthurkrishnan, 1992. Effect of weed control on weed growth and grain yield of semi dry rice. India J. Agron., 37 : 317 – 319.

Rana, R.S., 1986. Breeding crop varieties for salt-affected soils. In : Approaches for incorporating drought and salinity resistance in crop plants (edited by V.L. Chopra and R.S. Paroda), pp. 24-55.

Rangarajan, S. and V. Veerabadran, 1993. State of art in rice production in Tambiraparani Command Area. Technical bulletin, TNAU, Killikulam, India.

Rangasamy, A., V.S. Shanmugasundaram, S. Sankaranand P. Subbaian, 1990. Integrated farming system management: A viable approach. Research bulletin. Tamil Nadu Agric. Univ. Coimbatore, India.

Rangasamy, A. 1992. Sustainable agriculture for rice based ecosystem. Proceedings of National training on "Integrated farming system" held during 10-17th December, 1992. GOI and Tamil Nadu Agric. Univ., Coimbatore, India. pp 5-12.

Rangasamy, A. and C. Jayanthi, 1994. Recycling of organic wastes in integrated farming systems. Proceedings of National training on "Organic farming". GOI and Tamil Nadu Agric. Univ. Coimbatore, India. pp: 114-118.

Rangasamy, A., M.N. Budhar, R. Venkidusamy and SP. Palaniappan 1988. Integrated farming system for wetlands. In: Absts. of National seminar on "Farming systems for Semi-arid tropics". Tamil Nadu Agric. Univ., Coimbatore, India. pp: 1-6.

Rangasamy, A., R. Venkidusamy, C. Jayanthi, S. Purushothaman and SP. Palaniappan. 1995. Integrated farming system for rice based ecosystem. Madras Agric. J., 83: 287-290.

Rangaswamy, M., K. Thiyagarajan, A.S. Ponnuswamy, P. Jayamani and S.R. Sree Rangasamy, 1993. Hybrid rice seed production manual.pp.1-18.

Rani Perumal, H. Francis, P. Duraisamy, P. Kandasamy and SP. Palaniappan. 1991. Integrated nutrient management in Tamil Nadu. Dept. of Soil Science & Agricultural Chemistry, Tamil Nadu Agricultural University, Coimbatore, 50p.

Rao, A.S and R.P. Singh 1997. Effect of herbicide mixtures and sequential application on weed control in transplanted rice. Indian J. Agron., 42 : 77 – 81.

Rao, A.S and R.P. Singh 1994. Effect of herbicide mixtures and sequential application on relative composition of weed flora in transplanted rice. Indian J. Weed Sci., 26: 115 – 119.

Rao, M.V. 1985. Cultural practices for rice. In: "Rice Research in India". ICAR, New Delhi. p.412

Rao, M.V. and A.N. Dubey, 1997. Probability of the existing of pre emergence herbicide moisture -variety interaction to direct seeded rice. Abstr. Indian Weed Sci. Soc. Hyderabad, p.177.

Rao, M.V. and I.C Mahapatra, 1978. Fertilizer Use. In: Rice Production Manual, ICAR, New Delhi. pp. 96-109.

Rao, P.Y. 1984. Mushroom farms in Himachal Pradesh. Intensive agric.,22:19-20.

Rao, D.L.N. and S.K. Ghai, 1987. Slow release urea fertilizers in sodic soils. Inlt. Rice Res. Newsletter., 12(1) : 32-33.

Ravichander, R., 1987. Studies on antifungal activity of some plant extracts. M.Sc. (Ag.,) thesis, TNAU, Coimbatore. India.

Ravichandran, V.K. 1993. Weed control in dry seeded rice. Indian J. Agron., 38:477 – 480.

Ray, B. and P. Mustafee, 1980. Weed control in rice with bifenox. Indian J. Weed Sci., 12:93-98.

Raymundo, M.E., E.B. Pantastico and V.T. Manaspac. 1962. Some physical and chemical properties of an upland soil and its associated paddy soil. Philipp. Agric., 46:560-569.

Reddy, K.R. and W.H. Patrick. 1975. Effect of alternate aerobic and anaerobic conditions on redox potential, organic matter decomposition and nitrogen loss in a flooded soil. Soil Biol. Biochem., 7:8 – 94.

Reddy, A.R., A.H.K. Rao, K. Joanny, V. Subba Reddy, U. Radharanl, C.H.R. Kumar, A.K. Garg, B. Karunasree, G. Madhuri, G. Sailaja and K.R.K. Reddy, 1992. Research programme on "Molecular biology and genetics of rice". Natl. Rice Biotech, Network, 1992, Directorate of Rice Research, Hyderabad, India, pp. 68-71.

Reddy, B.G.M., M.D. Kumar, K.S.C. Basappa and S.G. Patil 1995. Evaluation of anilofos plus in kharif transplanted rice. Indian. J. Weed Sci., 27 : 108 – 109.

Reddy, G.M. 1992a. *In Vitro* techniques in gene manipulation in improvement of rice and Triticale. Special Publication – Taichung District Agrl. Improv. Stn., 29: 187-193.

Reddy, G.M., 1992b. *In vitro* techniques in gene manipulation in improvement of rice and *Porteresia coarctata* and genetic instability in rice. Natl. Rice Biotech. Network, 1992, Directorate of Rice Research, Hyderabad, India, pp. 32-35.

Redona, E.D. and D.J. Mackill, 1996. Mapping of quantitative trait local for seedling vigour in rice using RFLPs. Theo.Appl.Genet., 92: 395-402.

Rice, O.W. 1969. Notes on soil temperature regimes in Thailand , land Dev. Dep., Soil Surv. Rep. No.79. Bangkok. 26p.

Robert, P.C., R.H. Robert, Rust and W.E. Larson (ed.), 1993 Soil specific crop management: A workshop on research and development issues. April 14-16, 1992, Minneapolis, Madison, WI, ASA, CSSA, SSSA,USA.

Robert, P.C., R.H. Rust and W.E. Larson, 1995. Preface. Page xiii-xiv *in* Site-specific management for agricultural systems. (edited by P.C. Robert, R.H. Rust and W.E. Larson), ASA, CSSA, SSSA, Madison, WI, USA.

Roy, B., D.N. Das and P.K. Mukahopadhayay. 1990. Rice-fish – vegetables integrated farming: Towards a sustainable ecosystem. NAGA, The ICLARM Quartly, 13: 17-18.

Sadaphal and D.E. Rajat, 1994. (Ed.) Resource management for sustained crop production. Proceedings of the XII National symposium held at Rajasthan Agrl. University. Bikaner during 25-28, 1992. 424p.

Sahrawat. K.L., V. Anjaiah and P.T.C. Nambiar, 1990. Effects of mineral nitrogen and Brady rhizobium inoculation on growth and iron nutrition of groundnut. Fertil. Res. 24: 47-52.

Sahu, B.N. and Jena, A.C. 1968. Weed control in lowland rice field – I. Effect of cultural and weedicidal treatments on control of weeds and yield of rice crop. Indian J. Agron., 8 : 4 – 12.

Sahu, S.K. 1990. Effects of silica and phosphorous application on yield and phosphorous nutrition of rice. International Rice Research Newsletter, 15: 25.

Sailaja, L., 1993. Evaluation and Modification of spike tooth and Rasp bar type paddy thresher. M. Tech. thesis, Kelappaji College of Agricultural Engineering & Technology, Tavanur, Malappuram Dt., Kerala.

Sanchez, P.A. 1976. Properties and management of soils in the tropics. John Wiley and Sons, New York. 618p.

Santhanagopalakrishanan, S., 1988. Feasibility of rearing goat under deep litter system in the farming system of Thanjavur delta, Tamil Nadu. M.Sc. (Ag.) Thesis, Tamil Nadu Agric. Univ., Coimbatore, India.

Saqlan Naqvl, S.M., C.V. Ozalp, H.A. Oktem and M. Yucel, 1992. Study of proteins synthesized in rice roots under salt stress conditions. IRRN, 17: 15-16.

Saraf, C.S., V.S. Shinde and R. Hegde. 1990. Agronomic research towards sustainable Agriculture. Indian Society of Agronomy, New Delhi. pp. 153-162.

Sarkar, R.H., M.A. Samad, Z.I. Seraj, M.I. Hoque and A.S. Islam, 1993. Pollen tube growth in crosses between *Porteresia coarctata* and *Oryza sativa*. Euphytica, 67: 744-746.

Satapathy, R.B. and P.K. Singh, 1980. Utilization of *Azolla pinnata* as an organic nitrogen source. National Symp. on "Biological nitrogen fixation in relation to crop productivity". Tamil Nadu Agrl. University, Coimbatore.

Satoto and B. Sutaryo,1981. Natural outcrossing of cytoplasmic male sterile line IR54752A in Indonesia. Int. Rice Res. Newsletter, 14(1):7.

Savant, N.K., G.H. Snyder and L.E. Datnoff, 1997. Adv. Agron., 58.

Saxena, R.C. and Boncodin, 1987. Effect of crude neem seed kernel extract (NSKE) and NSB through seedling root dip against GLH. In: Workshop on “Botanical pest control in rice based cropping system”. Int. Rice Res. Institute, Manila, Philippines.

Scheltema, W. 1974. Puddling against dry plowing for lowland rice culture in Surinam. Agric. Res. Rep. 828, Pudoc, Wageningen. 241 p.

Selvam, R., 1995. Studies on the management of bacterial leaf blight disease of rice caused by *Xanthomanas oryzae pv. Oryzae* (Ishiyama), M.Sc., (Ag.) thesis, TNAU, Coimbatore, India.

Selvaraj, C. and P. Narayanasamy, 1994. Antifungal effect of some plant extracts against *Helminthosporium oryzae* and *Sarocladium oryzae* in rice. In Sivaprakasam, K. and K. Seetharaman, (eds). Crop Disease innovative techniques and management. Kalyani publishers, New Delhi, India. pp. 325-330

Selven, P., 1995. Evaluation and modification of power tiller operated reaper, M. Tech. thesis, Kelappaji College of Agricultural Engineering & Technology, Tavanur, Malappuram Dt., Kerala.

Seraj, Z.I., M.O. Faruque, K.G. Hossain, R.H. Sarkar, T. Devi, Z. Islam and A.S. Islam, 1996. Attempted hybridization between *Oryza sativa* and *Porteresia coarctata* T. IRRN, 21: 35.

Shah, S.A., C.M. Singh, N.N. Angiras and S.K. Gautam, 1990. Weed management in different rice cultures. Indian J. weed Sci., 22:104-106.

Shaner, W.W., P.E. Phillips and W.R. Schmehl, 1982. Farming systems research and development: A guideline for developing countries. Boulder, Colorado: West view Press.

Shanmugasundaram, V.S, M. Balusamy and K.S. Subramaniam, 1992. Integrated farming system in wetland ecosystem. Proceedings of National training on "Integrated Farming System". GOI and Tamil Nadu Agric. Univ., Coimbatore, India. pp: 18-20.

Shanmugasundaram, V.S. and M. Balusamy, 1993. Rice-Fish–azolla integrated farming system in low-lying wetlands. Farming Systems, 9: 105-107.

Shanmugasundaram,V.S and Vijaykumar, 1989. Integrated use of organic and inorganic fertilizers on direct seeded rice (ADT-36). Unpublished-personal communication.

Sharma Navin, K., 1992. Automation in plant tissue culture: Problems and prospectives, Curr. Sci., 62: 507-511.

Sharma, A. and G. Mathew, 1990. Testing and evaluation of a vertical conveyor reaper – 1.0 m IRRI model. B.Tech. Project work report, Kelappaji College of Agricultural Engineering & Technology, Tavanur, Malappuram Dt., Kerala.

Sharma, A.R. and B.N. Mittra, 1988. Effect of green manuring and mineral fertilizer on growth and yield of crops in rice – based cropping on acid – lateritic soil. J. Agric. Sci. Camb.,110 : 605 – 608.

Sharma, A.R., B.T.S Moorthy, S.K. Rao and B. Venkateswarlu, 1995. Improvement of rice production technology in the conventional areas of Eastern India. In: Sustaining Crop and Animal Productivity -The challenges of the Decade (D.L. Deb Ed.). pp.157-178.

Sharma, B.K., M.K. Das and D.P. Chakraborty, 1985. Integrated fish- cum -poultry farming. Agriculture Extension Manual Newsletter Series, 5: 20-29.

Sharma, L.R., J.P. Bhati and R.Singh. 1991. Emerging farming systems in Himachal Pradesh: Key issues in sustainability. Indian J. Agrl. Econ., 46(3): 421-427.

Sharma, R.S., G.S. Rathi and J.P. Tiwari, 1986. Weed management studies in drilled paddy under Jabalpur condition. India J. Weed Sci.,18;111-118.

Sharma, S.K., K.S. Gangwar and J.P. Singh. 1991. Farming systems diagnosis and analysis. Project report for Directorate of cropping systems research on diagnostic team, Meerut, Uttar Pradesh, India. pp: 1-15.

Sharma, U.C., A.N. Choudhury and A.K. Khan, 1994. Effect of integrated weed management on yield and nutrient uptake by rainfed rice in acid soils of Nagal and Indian J. Agron., 39: 553-556.

Shelke, D.K., R.H. Bhosle and N.S. Jadhav, 1985. Studies on crop-weed competition in irrigated upland drilled rice (Var. Prabhaval). Abstr. Ann. Conf. Indian Soc. Weed Sci, p.116.

Shetty, S.V.K. and H.S. Gill, 1975. Efficacy of different herbicides for weed control in direct sown rice. India J. Weed Sci., 7:80-84.

Shiga, H. and W.V. Entura, 1976. Nitrogen supplying ability of paddy soils under field conditions in the Philippines. Soil Sci. Plant Nutr., 22:307-399.

Shinde, B.N. P.A Sarangamath and S. Patnaik, 1978. Efficiency of HCI and H_2SO_4 acidulated rock phosphates for rice on acid soils. Plant & Soil. 50: 575 – 584.

Shukla, V.K., R.S. Sharma and S.K. Shrivastava 1995. Effect of different planting geometry, weed management and nitrogen levels on weed control efficiency, yield and profit in transplanted rice. Indian J. Weed Sci., 27:1-4.

Siddeswaran, K and SP. Palaniappan. 1990. Role of integrated farming systems in minimising the purchased input usage in small and medium farm holdings. International symposium on Asian Farming system research and extension 1990- sustainable farming systems in 21 century Asia. Nov. 19-22, 1990. Asian Institute of Technology, Bangkok, Thailand

Siddeswaran, K. 1992. Integrated nitrogen management with green manure and grain legume in rice-based cropping systems. Ph.D. Thesis, TNAU, Coimbatore.

Siddiq, E.A., M. lyas Ahmed and B.C. Viraktamath, 1995. Status of hybrid rice research in India. Hybrid rice research and development.pp.139-158.

Sidhu, B.S. and V. Beri. 1989. Effect of crop residue management on the yields of different crops and on soil properties. Biological wastes, 27: 15-27.

Simmonds, N. 1985. The State of the art of Farming Systems Research. Washington, D.C. USA: World bank.

Simmonds, N.W. 1986. A short review of Farming systems research in the tropics. Exptl. Agric., 22: 1-13.

Simmonds, N.W. 1985. Farming Systems research – a review. Technical Paper No. 43. World Bank, Washington, D.C.,USA.

Sin. A.W.C. 1980. Integrated animal- fish husb and ry systems in Hong Kong with case studies on duck–fish and goat-fish systems. In: Proceedings of KLARM – SEARCA Conf. on "Integrated agriculture – aquaculture farming systems. 4th Aug, 1979, Manila, Philippines.

Sinanuwong, S. and Y. Takaya, 1974a. Saline soils in northeast Thailand Southeast Asian Stud., 12(1): 105-120.

Sinauwong, S. and Y. Takaya. 1974b. Distribution of saline soils in Khorat basin of Thailand Southeast Asian Stud., 12(3):365-382.

Singh B, Y. Singh, J.K. Ladha, K.F. Bronso, V. Balasubramanian, J. Singh and C.S. Khind, 2002. Chlorophyll meter and leaf color chart-based nitrogen management for rice and wheat in northwestern India. Agronomy Journal, 94 :821-829.

Singh, I.B. 1988. Effect of weed control methods on performance of rice varieties under upland conditions of Nagal and Indian J. Weed Sci.,20: 54-54.

Singh, K.K. and P.C. Kharwara, 1984. Chemical weed control in transplanted rice under mid-Hill conditions. Abstr. Ann. Conf. Indian Soc. Weed Sci, Varanasi. p.13.

Singh, M.P. and S.B. Biswas, 1981. Weed control through herbicides in different rice cultures. Proc.8th Asian-Pacific Weed Soc.Conf. Bangalore, pp.91-97.

Singh, O.B. and V.R. Ramach and ran, 1992. Testing and performance evaluation of hold on type power paddy thresher. B.Tech. Project Report, Kelappaji College of Agricultural Engineering & Technology, Tavanur, Malappuram Dt., Kerala.

Singh, O.P. and R.K. Bhandari, 1985. Relative efficiency of different herbicides in transplanted rice. Indian J. Weed Sci., 17:47-49.

Singh, P.K., Y. Singh and K. Jitendra. 1994. Effects of tillage and planting management of yield and economics of rice – wheat cropping system. Agric. Sci. Digest., 14(1); 41-43.

Singh, R P. 1997. Keynote address on People's Technology and its Relevance in the Changing Scenario of Indian Agriculture at the Indian Social Science Congress, held at Tamil University, Thanjavur from 1-6 December.

Singh, R. and C.S. Patel, 1989. Effective weed control measure for upland rice in NEH region. Indian J. Agron., 34:406-409.

Singh, R. and C.S. Patel, 1994. Efficiency of herbicides for weed control in transplanted rice under north eastern hill region. Indian J. Weed Sci., 26:93-96.

Singh, R.B. 1992. Integrated farming with magur fish is profitable to small farmers. Indian Fmg., (11): 5-12.

Singh, R.K. and R.P. Singh, 1988. Efficiency of herbicides under different water management practices in transplanted rice. Indian J. Weed Sci., 20:15-19.

Singh, R.K., A. Rafi, A. and G.P. Saha, 1997. Weed control in direct sown upland rice under rainfed condition. Abstr. Indian Soc. Weed Sci. Conf. Hyderabad, pp.176-177.

Singh, R.K., T.R. Paris and R.D. Singh, 1991. ICAR-IRRI collaborative research project on rainfed rice ecosystem summary report on gender concerns in FSR. Proceedings 22nd Asian Rice Farming systems working group meeting, Sep. 30-Oct 4, 1991. Beijing, China. pp. 219-230.

Singh, R.P. and G.L. Sharma, 1981. Studies on rice weed competition under different method of rice cultivation and weed control. Proc. 8th Asian-Pacific Weed Sci. Soc. Conf., II:109-111.

Singh, S.J., K.K. Singh and S.S. Mishra, 1989. Weed management studies in different rice cultivars. Indian J. Weed Sci., 21:27-30.

Singh, S.P. and K.G. Pillai, 1996. Performance of rice varieties under various weed control measure in rainfed upland condition. Ann. Agril. Res.,17:158-162.

Singh, T., J.S. Kolar and S and hu, 1990. Control of *Ischaemum rogosum salisb* (*Wrincle grass*) in transplanted rice. Indian J. Weed Sci., 22:46-50.

Singh, U, J.T. Ritchie and D.C. Godwin, 1991. A user's guide to CERES-RICE - V2.10. Muscle Shoals, AL, IFDC. Spectra Precision (1997) Precision land leveling; Concept and benefits. Spectra Precision, Dayton, Ohio. USA.

Singh, N.T and R. Chhabra, 1988. Soil problems with emphasis on salinity / alkalinity. Paper presented at the Rice Workshop held at HAU, Hissar during 9-22, April, 1989.

Singh. Y., B. Singh, O.P. Meelu and M.S. Maskina. 1990. Nitrogen equivalence of green manure for wetland rice on coarse textured soils. International Rice Research Newsletter. 15: 23.

Singnuwong, S. and Y. Takaya, 1974b. Distribution of saline soils is Khorat basin of Thailand Southeast Asian Stud., 12(3):365-382.

Sinha, B.P., S.N. Singh and J. Prasad, 1977. Efficacy of different soil amendments as reclaiming agents and their responses to paddy and wheat in calcareous saline – alkali soils. Fert. Tech., 14:348 – 351.

Sirohiet, A.S., B.M. Sharma, Ikbal Singh and Rameshwar Dayal. 1980. Role of dairy and poultry enterprises for increasing income and employment of farm in the Union Territory of Delhi. Indian J. Agrl. Econ., 35: 112-119.

Sivaraj, S. 1989. Economics of poultry-cum-fish culture as a component in the farming system under Thanjavur district, Tamil Nadu. M.Sc. (Ag.) Thesis. Tamil Nadu Agric. Univ., Coimbatore, India.

Sivasubramanian, K. and A. Dejonge. 1990. Sustainability analysis of ecological agriculture in South India. Agriculture Man and Ecology Programme, Pondicherry, India.

Soepraptohardjo, M. and P.M. Driessen. 1976. The lowland peats of Indonesia. A challenge for the future. Soil Res. Inst. Bogor 3:11-19.

Somasundaram, E., K. Annadurai, M.L. Manoharan, R. Kavimani and R. Vijayalakshmi. 2000. H and book on Rice production technology. Golden net printers, Trichy-2,122p

Soo Swee Weng. 1975. Semi detailed soil survey of the Kelantan plain. Soil Surv. Rep. No. 2. Ministry Agric Nat. Res., Kuala Lumpur.58 p.

Spencer, D.S.C. 1991. Institutionalizing the farming systems perspective in multi commodity research institutes. The role of systems based research groups. Exptl. Agric., 27: 1-9.

Sree Rangasamy, S.R., S. Sukumar and S. Manonmani, 1992. Rice Improvement through anther culture. Natl. Rice Biotech. Network, 2nds Ann. Meeting 1992, DRR, Hyderabad, India, pp. 44-45.

Sreekantan, L and SP. Palaniappan. 1990. Integrated phosphorous management in a rice-based cropping system. International Rice Research Newsletter, 15: 18-19.

Sreeramkumar, R., 1990. Efficacy of plant products on brown spot of rice caused by *Drechsiera oryzae.* M.Sc.(Ag.) thesis, Faculty of Agriculture, Annamalai University, Tamil Nadu. India.

Srilakshmi, B. 2002. Food science (2nd Edition), New age International (p) Ltd., New Delhi. 375p.

Srinivasalu Reddy, D. 1988. Integrated nitrogen management in a rice-based cropping system. Ph.D. Thesis, Tamil Nadu Agric. Univ., Coimbatore.

Srinivasan, G., 1997. Bank's break through in rural development. The Hindu. Dec. 97.

Srinivasan, G., P. Pothiraj and S. Sankaran. 1992. Effect of management practices on weed dynamics rice based cropping system. Indian J Agron., 37 (1) : 13-17.

Srinivasan, V.M. 1994. Role of Agro-forestry for sustainable agriculture. In: Proceedings of Summer Institute on "Integrated farming systems research and development for sustainable agriculture" organized by ICAR and Tamil Nadu Agric. Univ., Coimbatore, India during 6-15 June,1994. pp:168-173.

Srivastava, N.S.L., 1990. Mechanization for grain saving. In : Survey of Indian Agriculture 1990. The Hindu, Madras, India. pp. 171-175.

Subba Rao, N.S., A.N. Sen and K.R. Dadarwal, 1982. Rhizobium research in India. Review of Soil Research in India. Part I. 12th International Congress of Soil Science, New Delhi, Feb. 8-16, 1982. pp. 211-224.

Subbaian, P. and SP. Palaniappan. 1992. Effect of integrated management practices on the yield and economics of crops under high intensity multiple cropping systems. Indian J. Agron., 37: 1-5.

Subbaian, P., K. Annadurai and SP. Palaniappan. 2000. Agriculture: Facts and Figures. Kalyani Publishers, 359p.

Subbaih, B.V and H.L. Asija 1956. A rapid procedure for the estimation of the available N in soils. Current Sci., 25: 259-260.

Subhash Chander, 1999. Insect management. In: A text book of Rice Agronomy. (edited by Rajendra prasad), Jain brothers publication, New Delhi. pp.173-190.

Subramanian, S. and A. Mohamed Ali, 1987. *Echinochloa crusgalli (L) Beauv* competition and control in transplanted rice. Indian J. Weed Sci.,19:77-82.

Subramanian, S., A. Mohamed Ali and R. Jeyakumar 1991. All about weed control. Kalyani Publishers, Ludhiana. 315 p.

Sundaresan, D. 1979. The role of improved buffaloes in rural development. Indian Dairyman, 31: 73-78.

Sushilendra, 1996. Dynamics of Power Transmission in Tractor Mounted Paddy Reaper. M. Tech. thesis, Kelappaji College of Agricultural Engineering & Technology, Tavanur, Malappuram Dt., Kerala.

Suzuki David, T., J.F. Griffiths Anthony, H. Miller Jeffrey and C. Lewonth Richard (Eds.), 1989. An Introduction to Genetic Analysis. Fourth Ed., p. 512.

Swaminathan, M. 1990. Food Science–chemistry and experimental foods. Bappco. Publishers, Bangalore. 468p.

Swaminathan, M.S. 1975. Research advances in agriculture in relation to dairy development in India. Indian Dairyman,. 27: 25-61.

Swaminathan, M.S. 1987. Inaugural address at the International symposium on "Sustainable agriculture". The role of green manure crops in rice farming systems. 25-29 May, 1987. IRRI, Philippines.

Tabbal, D.F. and T.H. Wickham. 1976. The effects of location and water supply on water shortages in an irrigated area. Paper presented at a Saturday seminar, 26 June 1976. International Rice Research Institute, Los Banos, Philippines. 8 p. (mimeo.)

TAC/ICGIAR, 1988. Sustainable agricultural production: Implications for international agricultural research. FAO, Rome, Italy.

TAC. 1990. Towards a Review of CGIAR- Priorities and Strategies. Program Rept, CGIAR, Rome.

Tajuddin, A and M. Subramanian. 1992. Development of a battery operated low volume knapsack sprayer. Oil Crops Newsletter (Ethiopia), 9 : 63-64.

Tajuddin, A. 1991. Improved battery operated very low volume sprayer. Kisan World, June,1991.

Tajuddin, A. and P. Rajendran, 2002. Devices for intercropping green manure in wet seeded rice. Agricultural mechanization in Asia, Africa and Latin America, 33 (3) : 9-10 & 20.

Tajuddin, A., 1998, Use battery – operated low volume knapsack sprayer to save fuel. Indian Farming, 48 (9) : 22-23.

Tajuddin, A., G. Gajendran, A. Regupathy and M. Balasubramanian, 1993. Performance evaluation of a low volume knapsack spinning disc sprayer. Pestology, 27 (7): 13 –16.

Tajuddin, A., R. Karunanithi, D. Asokan and A. Shanmugam, 1994. Lowland direct paddy seeder. Journal of Agricultural Engineering, 31 (1-4). 29-35.

Tajudeen, A. and T.V. Job, 1997. A light weight paddy harvester. Kisan world, Dec. 97. p. 48.

Tajudeen, S., 1995. IRRI -6 row manual rice transplanter. Central Inst. Agrl. Engineering, Coimbatore, India.

Takahashi, E. 1995. Sci. Rice Plant, 2: 58-71.

Takahashi, E., J.F. Ma and Miyaka. 1990. Comments Agric. Food. Chem., 2: 99 – 122.

Takaya, Y. 1971. Physiography of rice lands in the Chao Phraya Basin. Southeast Asian Stud., 9:375-397.

Takijima, Y. 1963. Studies on the behaviour of the growth inhibiting substances in paddy soils with special reference to the occurrence of root damage in the peaty paddy field [in Japanese]. Bull. Natl. Inst. Agric. Sci. ser. B, 13:117-397.

Tal M., 1993. In vitro methodology fir increasing salt tolerance in crop plants. Acta Horticulture, 336 : 69-78.

Tan, E.S.P. and K.H. Khoo, 1980. The integration of fish farming with agriculture in Malaysia. In: Proceedings of ICLARM – SEARCA Conf. on "Integrated agriculture – aquaculture farming systems". 4th Aug, 1979., Manila, Philippines. pp: 175-187.

Tanaka, A. 1978. Role of organic matter. Pages xx – xx in International Rice Research Institute. Soils and rice. Los Banos, Philippines.

Tanaka, A. and S. Yoshida. 1970. Nutritional disorders of the rice plant in Asia. Int. Rice Res. Inst. Tech. Bull., 10, 51 p.

Tandon, H.L.S and J.E. Shinde, 1993. Fertility management for rice in problem soils of India. In : New Frontiers in Rice Research. (Muralidharan, K. and Siddqi, E.A. eds). Directorate of Rice Research Hyderbad, pp. 315 – 322.

Tandon, H.L.S. 1992. Fertilizers, organic manures, recyclable wastes and bio-fertilizers. Fertilizer Dev. Corporation. New Delhi. pp. 15.

Tandon, H.L.S., 1989. Nitrogen recommendations for higher fertilizers efficiency. Fert. News, 34 (12):63-77.

Tanksley, S.D., N.D. Young, A.H. Paterson and M.W. Bonierbale, 1989. RFLP mapping in plant breeding: new tools for an old science, Biotech., 7: 257-265.

Te Tzu Chang, 1964. Present knowledge of rice genetics and cytogenetics', IRRI, Philippines, Tech. Bull. 1, 1964.

Tempelaar, M.J. and M.G.K. Jones, 1985. Directed electro fusion between protoplasts with different responses in a mass fusion system. Plant Cell Reports, 4: 92-95.

Thakur, K.N. 1984. Effect of herbicides on the transformation of the nutrients and their uptake by upland transplanted rice in calcareous soil. M.Sc. (Ag.). Thesis, R.A.U. Pusa, Bihar.

Thakur, K.S. and K. Bassi, 1994. Efficacy of promising herbicides in direct sown rain fed rice. Indian J. Weed Sci., 26:75-78.

Thangamani Narayanasamy and P. Narayanasamy, 1988. Role of neem products in the control of fungal diseases of rice. P. 6-7. Nat. Sem. On management of crop disease with plant products/biological agents. Agricultural College and Research Institute, Madurai.

Thanh Hung, L.E. 1992. Integration of aquaculture with coastal rice farming in Vietnam. NAGA, The ICLARM Quartly., 15(2): 2-29.

Thiagarajan, T.M., R. Sivasamy and M N. Budhar. 1995. Procedure for collecting plant samples at different growth stages of transplanted rice crop. In: SARP Research proceedings – Nitrogen management studies in irrigated rice. IRRI Publication, Philippines.

Thiyagarajan, K., P. Jayamani, A.S. Ponnusamy and M. Rangaswamy, 1994. Effect of GA_3 application on seed yield in hybrid rice. Abstract presented in IX. All India Seminar on seed storage, packing and marketing. December 7-9, TNAU, Coimbatore. Seed Tech News: 24(4): 59.

Thomas, J. and R. Prasad, 1982. On the nature of mechanisms responsible for the higher efficiency of urea super granules for rice. Pl. Soil, 69 : 127 – 130.

Thomas, J. and R. Prasad, 1983. Mineralization of urea, coated urea and nitrification treated urea in different rice growing soils. Pflanzanemahrung and Bodenkunde, 146: 341 – 347.

Tiwari, A.N. and K.K. Singh, 1989. Effect of removal of types of weed flora in upland direct seeded rainfed rice. Indian J. Weed Sci., 21:72-74.

Tiwari, A.N., K.S. Rathi, P.N. Pandey and K.K. Pandey, 1989. Studies on weed control in rice nursery raised under puddled condition through herbicides. Indian J. Weed Sci., 20:90-91.

Tiwari, K.N., A.N. Pathak and S.P. Tiwari. 1980. Fertilizer News, 25, 3-20.

Tiwari, P.N. 1993. Integrated farming research for sustaining food production. J. Nuclear Agric. Biol., 20: 1-13.

Tiwari, S.C. 1993. Effect of organic manures and NPK fertilization on earthworm activity in an oxisol. Biol. Fertil. Soils., 16:293-295.

Tiwari, V.N., A.N. Pathak and L.K. Lekhri. 1987. Effect of plant waste incorporation by different methods under uninoculated and inoculated conditions of wheat crops. Biological Wastes, 21, 267-273.

Tiwari, A.N. and S. Ram, 1991. Crop weed competition in upland direct seeded rainfed rice. Indian J. Weed Sci., 23:51-52.

Tosh, G.C. 1977. The relative efficacy of herbicides on weeds and the rice crop grown under upland condition. Abstr.Indian Soc. Weed Sci., Hyderabad, p.178.

Tosh, G.C., B.C. Jena and R.K. Nayak, 1984. Studies on varietal tolerance of ice to bentazon under direct seeded lowland condition. Abstr. Ann. Conf. Indian Soc. Weed Sci., Varanasi, pp.6-7.

Tripathi, N.N., N.K. Dubey, A. Dikshi, R.D. Tripathi and S.N. Dixit, 1978. Fungitoxic properties of *Alpinia galanga* oil. Trop. Plant Sci. Res.,1: 49-52.

Tripathy, S.K. and J. Kar, 1995. Rice root cell plasma membrane – its role in saline tolerance. In : Natl. symp. On 'Role of Plant Biotech. In Improving Agril. – Challenges & Opportunities and physiological & Biochemical basis of Crop Yield', 1995, Jaipur, India, p. 46.

Tuong, T.P., R.J. Cabangon and M.C.S. Wopereis, 1996. Quantifying flow processes during land soaking of cracked rice soils. Soil Sci. Soc. Am. J. 60:872-879.

Tuteja, S.S., B.L. Chandrakar and R.S. Tripathi 1995. Economic feasibility of different herbicides on direct seeded lowland rain fed rice. Indian J. Agron., 40: 286-287.

Udayasoorian, C. 1999. Methane emissions from paddy fields. In : Agricultural inputs and Environment" edited by SP. Palaniaapan, Published by Scientific Publishers, Jodphur

U.S.D.A. 1970. Soil Taxonomy, U.S. Deptt. Of Agriculture, Washington, D.C.

UNESCO (United Nations Educational, Educational, Scientific and Cultural Organization). 1974. FAO – UNESCO Soil map of the world 1:500,000 Vol. 1, Legend. Paris.

Unnikrishnapillai, K.P. 1988. Studies on the effect of plant extracts/products and antagonistic bacterium against rice brown spot pathogen *Drechslerg oryzae*. M.Sc. (Ag.) thesis, TNAU, Coimbatore.

USDA (United States Department of Agriculture), Soil Conservation Service. 1960. Soil classification. A comprehensive system. U.S. Government Printing office, Washington, D.C. 260p.

Vaishya, R.D. and V.K. Singh, 1989. Economics of weed control methods in upland direct seeded rice. Indian J. Weed Sci., 21:69-71.

Vaishya, R.D., V.K. Singh and A. Saxena, 1992. Mechanical and chemical weed control in upland direct seeded rice. Indian J. Weed Sci., 24: 11-16.

Vajrabhaya, M. and T. Vajrabhaya, 1991. Biotech. In Agril. and Forestry In : Rice (Bajaj, Y.P.S. Ed.), Springer Verlag, Berlin, p. 368.

Van Breeman, N. and L.J. Pons, 1978. Acid sulfate soils and rice. In : Soil and Rice. IRRI, Philippines, pp. 739 – 761.

Van Breemen, N. 1973. Dissolved aluminum in acid sulfate soils and in acid mine waters. Proc. Soil., Sci. Am., 37(5) : 694 – 697.

Van Breemen, N. 1975. Acidification and deacidification of costal plain soils as a result of periodic flooding. Proc. Soil. Sci. Soc. America, 39: 1153-1157.

Van Breemen, N. 1976. Genesis and solution chemistry of acid sulfate soils in Thailand Center for Agricultural Publishing and Documentation, Wageningen. 263 p.

Van Breemen, N., L.R. Oldeman and W.G. Wielemarker 1970. The Ifugao rice terraces. In: Aspects of rice growing in Asia and the Americas. Misc. Pap. 7. land bonwhogeschool, Wageningen. pp.39-74

Van Der Kevi, W. and B. Yenmanas. 1972. Detailed reconnaissance survey of the Southern Central Plain Area. Soil Survey Report 89. Soil Survey Division, Bangkok, 187 p.

Van der werf, K. and A.D. Jager. 1991. Ecological agriculture in South India. An agro-economic comparison and study of transition. land bouw. Economisch Institute, the Hague and ETC foundation, Lersden. The Netherlands.

Van Wambeke, A. 1975. Management properties of oxisols in Savannah eco-systems. In: Soil Management in tropical America. Raleigh, North Carolina. pp. 364-371.

Vangriensven, L.G. 1987. The cultivation of mushroom: Its present status and future development. Outlook Agric., 6: 131-135.

Varma, S. and R.S. Mathur. 1990. The effects of microbial inoculation on the yield of wheat when grown in straw amended soil. Biological Wastes, 33:9-16.

Varshney, C.K. and K.P. Singh, 1976. A survey of aquatic weed problem in India. Aquatic Weed in South East Asia, Proc. Regional Seminar on noxious aquatic vegetation. New Delhi. (C.K. Varshney and J. Rozoka eds.).

Vasil, I.K. (Eds.) 1984. Cell culture and somatic cell genetics of plants. Academic Press, New York.

Veerabadran, V. 1994. Farming systems research in southern parts of Tamil Nadu. In: Proceedings of Summer institute on "Integrated farming systems research and management for sustainable agriculture". 6-15th June, 1994. Tamil Nadu Agric. Univ., Coimbatore. pp: 258-269.

Velayudham, K. and Jagdish Seth. 1987. Effect of plant residue management in Kharif on the yield and quality of wheat. Indian J. Agron. 32: 341-346.

Venkatachalapathy, V. and V. Veerabadran, 1998. A contingent plan for delayed release of water in Tambiraparani command area. Abstracts of papers. Proceedings National Seminar on "Issues in irrigation Management- Prospects and Retrospects", Feb 5-6, 1998, IMTI, Tiruchy, India.

Venkataraman, G.S. and K.V.B.R. Till, 1990. Biofertilizers in sustainable agriculture in soil fertility and fertilizer use. Vol. IV. pp. 138-148. IFFCO, New Delhi.

Venkataraman, G.S., 1982. Nitrogen fixation by blue-green algae and its economic importance. In: Symposia papers I. Non-symbiotic nitrogen fixation and organic matter in the tropics. Transactions of the 12th International Congress of Soil Science, New Delhi. Feb. 6-16, 1982. pp. 69-82.

Venkateswaralu, J., 1987. Efficient resource management systems for dry lands of India. Adv. Soil Sci., 7: 165-221.

Venugopal, M.S. and S. Doraisamy, 1994. Plant derivatives in the management of insect vectors of rice diseases. Pages 355-360. In Sivaprakasam, K. and K. Seetharaman, (eds). Crop Disease-Innovative Techniques and Management Kalyani Publishers, New Delhi.

Vergara, B.S. 1988. Dhan Di Pika (Hindi), IRRN-ICAR, Publication. 221p.

Verma, L.N. and P. Bhattacharya, 1990. Role of biotechnology in supplying plant nutrients in the nineties. Fertilizer News, 35: 87-97.

Viji, M.M., M. Thangaraj, S.R. Sree Rangasamy and A. Jayapragasam, 1996. Performance of the progenies of embryo culture-derived rice plants to salinity. Oryza, 33 : 6-10.

Vikraman Nair, R., N.G. Pillai and Shaskanaman, 1979. Mixed farming in coconut gardens. Indian Farming, 26: 24-25.

Virmani, S.S., G.S. Khush, E.H. Bacolangco and Ran Cui Yang, 1980. Natural Outcrossing on cytoplasmic male sterile lines of rice under tropical conditions. Int. Rice Res. Newsletter, 5(5): 5-6.

Vlek. P.L.G. and E.T. Craswell, 1979. Effect of nitrogen source and management of ammonia volatilization from flooded rice systems. Soil Sci. Soc. Am. J., 43:352-358.

Wada, H. and S. Matsumoto. 1973. Pedogenetic process in paddy soils. Pedologist, 17:2 –15.

Wada, H., M. Okazaki and Y. TAKAI, 1975. Dissolution of ferri hydroxide with the leachate from the water logged soil. I. Role of water soluble substances in dynamics of paddy soil. J. Sci. soil manure (Japan), 46:201-244.

Wan, Y. and P.G. Lemaux, 1994. Generation of large numbers of independently transformed fertile barley plants. Plant Physiol., 104 : 37-48.

Wang, H.Q. and S.X. Tang, 2000. Upland rice production in China: its past, today and Future. Paper presented at the Aerobic Rice Workshop. 7-8 September 2000, International Rice Research Institute, Los Banos, Philippines.

WARDA (West African Rice Development Association). 1975. Rice vocabulary Monrovia. 207 p.

Watson, D.J. 1948. The dependence of net assimilation rate on leaf index. Ann. Bot., 22:37-54.

Welntroub Harold, M., 1990. Antisense RNA and DNA. Scientific American, January 1990, 40-46.

Wells, B.R., S.E. Baser and R.H. Dilday, 1991. In : Arkansas Rice Res. Studies (B.R. Wells Ed.), Fayettevilie Ark. Agric. Exp. Stn., pp. 142-144.

Werf, E. Van Der and B. Narayan. 1989. A Socio-Economic study of ecological agriculture in south India, AME and ICSIM, Project report, Bangalore, Unpublished.

Westcott, M.P and K.W. Vines, 1986. A comparison of sprinkler and flooded irrigation for rice. Agron. J. 78: 637-640.

Westfall, D.G. 1975. Rice soils, properties and characteristics. In: six decades of rice research in Texas. Texas Agric. Exp. Stn., pp. 22-30.

Whistler, F.D., J.C. Lance and R.S. Linebarger. 1974. Redox potentials in soil columns intermittently flooded with sewage water. J. Environ. Qual., 3:60-74.

Wickham, T.H. 1973. Predicting yield benefits in lowland rice through a water balance model. In: Water management in Philippines- Irrigation system; research & operations. International Rice Research Institute, Los Banos, Philippines. pp. 155-181.

Widawsky, D.A. and J.C. O'Toole, 1996. Prioritizing the rice research agenda for Eastern India, *In Rice – Research in Asia – Progress and Priorities.* (R.E. Evenson, R.W. Herdt and M. Hossain eds). CAB International & IRRI, pp.109-129.

Wijewardene, R. 1975. Engineering's research at MARDI for Malaysia's farming future. Agric. Mcch., Asia 6(1):20-24.

Williams J.F., S.R. Roberts, J.E. Hill, S.C. Scardaci and Tibbi. 1990. Managing water for weed control in rice. California Agric., 44:7-10.

Williams, R.F. 1946. Physiology of plant growth with reference to the concept of net assimilation rate. Ann. Bot., 10: 41- 72.

Wolf, Y.N. 1975. Soil – water relationships in oxisols from Puerto Rice and Brazil. In: Soil management in tropical America. Raleigh, North Carolina. pp.145-154.

Woyanarovich, E. 1980. Raising ducks on fishponds. In: Proceedings of ICLARM – SEARCA Conf. on "Integrated agriculture - aquaculture farming system". 4th Aug. 1979. Manila, Philippines.

Wu, P., G. Zhang, J.K. Ladha, S.R. Mc Couch and N. Nuang. 1995. Molecular marker – facilitated investigation on the ability to stimulate N2 – fixation in the rhizosphere by irrigated rice plants. Theo. Appl. Genet., 91: 1177-1183.

Xhikang, L.I., S.R.M. Pinson, M.A. Marchetti, J.W. Stansel and W.D. Park, 1995. Characterization of quantitative trait local (QTLs) in cultivated rice contributing to field resistance to sheath blight (*Rhizoctonia solani*). Theo. Appl. Genet., 91: 382-388.

Xiao, J., J. LI, L.Yuan and S.D. Tanksley, 1996. Identification of QTLs affecting traits of agronomic importance in a recombinant inbred population derived from a subspecific rice cross. Theo. Appl. Genet., 92 : 230-244.

Xu, D., X, Duan, B. Wang, B. Hong, T.H. David Ho and R. Wu, 1996. Expression of a late embryogenesis abundant protein gene. HVA 1, from barley confers tolerance to water deficit and salt stress in transgenic rice. Plant physiolo., 110: 249-257.

Yadav, D.V., T. Singh and A.K. Srivastava 1987. Recycling of nutrients in trash with N for higher cane yield. Biological Wastes, 20: 133-141.

Yadav, R.L. 1982. Expl. Agric., 16: 37-42.

Yadav, R.L. and S.R. Prasad, 1986. Ind. J. Sugarcane Techno., 3: 24-28.

Yaduvanshi, N.P.S., D.V. Yadav and T. Singh. 1990. Economy in fertilizer nitrogen by its integrated application with sulphination filter cake on sugarcane. Biological Wastes, 32: 75-79.

Yamada, Y., M. Ogawa and S. Yano, 1983. Cell and tissue culture techniques for cereal crop improvement (Han Hu and M.R. Vega Eds.). Science Press, Beijing, p. 229.

Yan, X. and K. Tan, 1991. Screening rice varieties for salt tolerance in the green house. IRRN. 16:16

Yanagisawa, M. and J. Takahasi.1964. Studies on the factors related to the productivity of paddy soils in Japan with special reference to the nutrition of the rice plants. Bull. Nalt. Inst. Agric. Sci. Jpn., Ser. B, 14:41-171.

Yang, W.H, S. Peng, J. Huang, A.L. Sanico, R.J. Buresh and C. Witt, 2003. Using leaf color charts to estimate leaf nitrogen status of rice. Agronomy Journal, 95:212-217.

Yano, S., M. Ogawa and Y. Yamada, 1982. In: Proceedings 5th Int. Cong. Plant Tissue & Cell Culture (A. Fujiwara Ed.) Maruzen Co. Tokyo. pp.495.

Yong Hwa Shin, 1978. Rice soils of Korea. In: International Rice Research Institute. Soils and rice. Los Banos, Philippines. pp. 179-191.

Yoshida S., J.H. Cock and F.T. Parao, 1972. Physiological aspects of high yields in rice breeding. International Rice Research Institute, Los Banos, Philippines. pp. 455-459.

Yoshida, S. 1975. The Physiology of silicon in rice. Technical bulletin No. 25. Food Fertilizer Tech. Centre, Taipei, Taiwan.

Yoshida, S. 1978. The availability of silicon in paddy soils. In: "Paddy soil Science" (K. Kawaguchi Ed.) Kodansha, Tokyo, Japan. pp. 293-299.

Yoshida, S., M. Ogawa, K. Suenaga and H. Chonya, 1983. Cell and tissue culture techniques for cereal crop improvement (Han Hu and M.R. Vega Eds.) Science Press. Beijing, China, p. 237.

Yoshino, T. and Y. Die. 1977. Prediction of nitrogen release in paddy soils by means of the concept of effective temperature. Cent. Afr. Exp. Stn., (Konosu Saitama, Japan) 25: 1 – 62.

Zhenhua, Z., F. Hongliang, Z. Chengmel, Z. Minfu and W. Wuyl, 1993. Progress on rice breeding by anther culture and related genetic research VIth Ann. Meeting of the Int. Programme on Rice Biotech., Feb. 1-5, 1993, Chiang Mai, Thailand, pp. 93-94.

Zhu shen You and Meng All Shen, 1991. Improvements of rice farming machinery for small farmers of China. In: Proceedings of 22nd Asian Rice Farming Systems working group meeting, Sep. 29-Oct. 4. 1991, Beijing, China. pp. 399-406.